Drug Targeting
Organ-Specific Strategies

Edited by Grietje Molema and Dirk K. F. Meijer

WILEY-VCH

Methods and Principles in Medicinal Chemistry

Edited by
R. Mannhold
H. Kubinyi
H. Timmerman

Drug Targeting
Organ-Specific Strategies

Edited by
Grietje Molema and Dirk K. F. Meijer

WILEY-VCH

Weinheim · New York · Chichester · Brisbane · Singapore · Toronto

Series Editors:
Prof. Dr. Raimund Mannhold
Biomedical Research Center
Molecular Drug Research Group
Heinrich-Heine-Universität
Universitätsstraße 1
D-40225 Düsseldorf
Germany

Prof. Dr. Hugo Kubinyi
BASF AG Ludwigshafen
c/o Donnersbergstrasse 9
D-67256 Weisenheim am Sand
Germany

Prof. Dr. Hendrik Timmerman
Faculty of Chemistry
Dept. of Pharmacochemistry
Free University of Amsterdam
De Boelelaan 1083
NL-1081 HV Amsterdam
The Netherlands

Volume Editors:
Dr. Grietje Molema
University Centre for Pharmacy
Department of Pharmacokinetics
and Drug Delivery
Antonius Deusinglaan 1
NL-9713 AV Groningen
The Netherlands

Prof. Dr. Dirk K. F. Meijer
University Centre for Pharmacy
Department of Pharmacokinetics
and Drug Delivery
Antonius Deusinglaan 1
NL-9713 AV Groningen
The Netherlands

Library of Congress Card No. applied for.

British Library Cataloguing-in-Publication Data: A catalogue record for this book is available from the British Library.

Die Deutsche Bibliothek – CIP Cataloguing-in-Publication-Data
A catalogue record for this publication is available from Die Deutsche Bibliothek

ISBN 3-527-29989-0

Printed on acid-free paper.

Composition: Datascan GmbH, D-64295 Darmstadt
Printing: betz-druck GmbH, D-63291 Darmstadt
Bookbinding: Wilhelm Osswald & Co., D-67433 Neustadt (Weinstraße)

Printed in the Federal Republik of Germany.

Preface

It is our prime intention to cover the topics of this series as comprehensively as possible. Thus, we are very pleased to introduce this volume focussing on organ specific strategies of drug targeting.

About hundred years ago Paul Ehrlich put forward his theory of "the magic bullet" as an approach to tame disease. Scientists have ever since worked on the principle of drug targeting based on this idea of specifically delivering drugs to diseased cells. Especially nowadays that by high-throughput screening and molecular modelling techniques highly potent drugs are being developed that interfere with general (signal transduction) processes in cells in the body, the need for their application by a drug targeting approach has almost become inevitable.

Progress in the field of drug targeting has been slow till thirty years ago. With the advent of the monoclonal antibody technology in the mid seventies of the last century as well as the development of liposomal devices as carriers did the drug targeting field expand and did the clinical application become a feasible aim.

Monoclonal antibodies, liposomes, polymers, proteins, and many other entities have ever since seen the light as carrier molecules. And, as with most technological developments, they have all encountered a vast array of difficulties, ranging from problems in the synthesis of the carriers and drug conjugates to unfavorable pharmacokinetics and toxicity. Furthermore, lack of knowledge on the anatomical and physiological barriers in the body hampered application. However, many problems have been solved, not in the least due to the advent of recombinant DNA technology to construct better defined carriers that can be produced in large amounts, and advanced pharmaceutical formulation technology. Similarly, the rapid developments in molecular biology, cell biology and immunology led to a better understanding of the processes taking place *in vivo* upon administration of carriers and conjugates. Important conclusion is that drug targeting has become a multidisciplinary research area.

What has been achieved until now? In the year 2001, several liposome and antibody based strategies have been or will soon be approved for clinical application, some for the treatment of cancer, some for the treatment of bacterial infections, some for chronic inflammatory diseases. Furthermore many monoclonal antibodies without a drug or pharmacologically active molecule attached are in the clinic. Their intrinsic targeting and effector function is obviously sufficient for the pharmacological effect.

Only a few polymer or protein based drug targeting strategies have reached the clinic and an important question in the coming years will be whether these strategies eventually will reach it. All will depend on their effectiveness and improved toxicity profiles as compared to free drug only and the ease of their production at large scale.

The present volume is in several respects unique. It provides a map of the body from the viewpoint of drug targeting/drug delivery. Potentials and limitations of targeting strategies are discussed in the light of organ related diseases for each organ separately. Furthermore, novel technologies are described that may be useful in the future to allow an even better product to be developed that can be clinically exploited at a more rapid pace.

The series editors are grateful to the contributors to this volume, in particular Grietje Molema and Dirk K. F. Meijer, as well as Wiley-VCH publishers, for the fruitful collaboration and the straightforward realization of this project.

January 2001

Raimund Mannhold, Düsseldorf
Hugo Kubinyi, Ludwigshafen
Henk Timmerman, Amsterdam

Foreword

It was in the mid-1970s I think, just a few years after Brenda Ryman and I introduced liposomes as a drug delivery system, when a well meaning colleague af mine advised me not to put all my eggs in one basket. The eggs were liposomes and the basket my career. At the same time there were all sorts of prophecies and rumours from a variety of quarters about liposomal stability problems, expense, toxicity, difficulties with large scale manufacture, etc. Some went as far as to dismiss the system as a flash in the pan phenomenon. Indeed, the yellow brick road to the magic bullet is littered with systems that once made the headlines and then fell by the wayside. So, such comments on liposomes, and later on on antibodies, were not surprising. I believe that what made many of us persevere throughout the decades in developing drug carrier systems such as liposomes, and associated technologies was the realization that, for the foreseeable future at least, molecular modelling is not the answer to drug selectivity for most therapeutics. The vagaries of the biological milieu *in vivo* ensures that optimal drug action (seen in the test tube) is compromised by such factors as opsonins and proteolytic enzymes in the bloodstream, membrane barriers, loss through the kidneys, and premature interception of therapeutics by the reticuloendothelial system. In the case of liposomes, monoclonal antibodies and some polymers, carrier development was greatly facilitated by their structural versatility which enabled the design of advanced versions of unique sophistication.

The first generation of liposome-based systems approved for clinical use are believed to function on the basis of their passive uptake by the target tissues (e.g. the AmBisome and the virosome vaccine Hepaxal) or by avoiding certain tissues (e.g. heart, kidneys) that are prone to damage by the drug when given as such (e.g. Doxil, Daunoxome). The next challenge is to create or build on the systems that can be actively targeted to specific tissues or circulation cells for which systems such as liposomes have little or no affinity. They include a variety of molecules with genuine targeting properties, for instance (neo-) glycoproteins, monoclonal antibodies and fragments thereof, applied either as a means to deliver drugs attached to these biopolymers, or as homing devices when attached to the surface of other drug delivery systems, for instance liposomes and other particle-type carriers. Success to that end will greatly enlarge the spectrum of therapeutics that can be selectively delivered, and widen the range of applications.

In this respect, Grietje Molema, Dirk K. F. Meijer and a team of drug delivery experts have taken an important step with the present book. Unlike previous volumes, this one is not devoted exclusively to liposome or antibody technologies. Rather, the book deals with organ-specific drug targeting strategies developed for the treatment of a wide spectrum of diseases and includes a collection of novel techniques applied to drug targeting research. Thus, the book provides a blueprint for both the experienced and the semi-experienced reader interested in drug targeting and related optimization strategies.

London, 2001 — Gregory Gregoriadis

List of Contributors

Jan-Willem Arends

Maastricht University
Department of Pathology
PO Box 616
6200 MD Maastricht
the Netherlands

Sigridur A. Ásgeirsdóttir

Groningen University Institute for Drug Exploration (GUIDE)
Department of Pharmacokinetics and Drug Delivery
Ant. Deusinglaan 1
9713 AV Groningen
the Netherlands
s.asgeirsdottir@farm.rug.nl

Leonie Beljaars

Groningen University Institute for Drug Exploration (GUIDE)
Department of Pharmacokinetics and Drug Delivery
Ant. Deusinglaan 1
9713 AV Groningen
the Netherlands
l.beljaars@farm.rug.nl

Ulrich Bickel

Texas Tech University HSC
School of Pharmacy
Pharmaceutical Sciences
1300 S Coulter
Amarillo, Texas 79106
USA
ubickel@ama.ttuhsc.edu

Anne H. de Boer

Groningen University Institute for Drug Exploration (GUIDE)
Department of Pharmaceutical Technology and Biopharmacy
Ant. Deusinglaan 1
9713 AV Groningen
the Netherlands
a.h.de.boer@farm.rug.nl

Maaike Everts

Groningen University Institute for Drug Exploration (GUIDE)
Department of Pharmacokinetics and Drug Delivery
Ant. Deusinglaan 1
9713 AV Groningen
the Netherlands
m.everts@farm.rug.nl

Henderik W. Frijlink

Groningen University Institute for Drug Exploration (GUIDE)
Department of Pharmaceutical Technology and Biopharmacy
Ant. Deusinglaan 1
9713 AV Groningen
the Netherlands
h.w.frijlink@farm.rug.nl

Arjan W. Griffioen

Maastricht University/
University Hospital Maastricht
Department of Internal Medicine
Tumor Angiogenesis Laboratory
Peter Debyelaan 25
6202 AZ Maastricht
the Netherlands
a.griffioen@intmed.unimaas.nl

Geny M. M. Groothuis

Groningen University Institute for Drug Exploration (GUIDE)
Department of Pharmacokinetics and Drug Delivery
Ant. Deusinglaan 1
9713 AV Groningen
the Netherlands
G.M.M.Groothuis@farm.rug.nl

Marijke Haas

Groningen University Institute for Drug Exploration (GUIDE)
Department of Clinical Pharmacology
Ant. Deusinglaan 1
9713 AV Groningen
the Netherlands
m.haas@med.rug.nl

R. Folgert G. Haverdings

Groningen University Institute for Drug Exploration (GUIDE)
Department of Pharmacokinetics and Drug Delivery
Ant. Deusinglaan 1
9713 AV Groningen
the Netherlands
r.f.g.haverdings@farm.rug.nl

Wijnand Helfrich

Groningen University Institute for Drug Exploration (GUIDE)
Department of Pathology and Laboratory Medicine
Medical Biology Section
Tumor Immunology Laboratory
Hanzeplein 1
9713 GZ Groningen
the Netherlands
w.helfrich@med.rug.nl

Hennie R. Hoogenboom

Dyax bv
PO Box 5800
6202 AZ Maastricht
the Netherlands
hhoogenboom@dyax.com

Jörg Huwyler

F. Hoffmann-LaRoche Ltd.
CNS Research
PRBN, Bldg. 68/323a
CH-4070 Basel
Switzerland
joerg.huwyler@roche.com

Young-Sook Kang

Physiology and Pathophysiology Laboratory
College of Pharmacy
Sookmyung Women's University
Chungpa-dong 2 ga 53-12
Yongsan gu, 140-742
Seoul
Korea
yskang@sookmyung.ac.kr

Yukio Kato

University of Tokyo
Graduate School of Pharmaceutical Sciences
7-3-1 Hongo
Bunkyo-ku
Tokyo 113-0033
kato@seizai-nl.f.u-tokyo.ac.jp

Robbert J. Kok

Groningen University Institute for Drug Exploration (GUIDE)
Department of Pharmacokinetics and Drug Delivery
Ant. Deusinglaan 1
9713 AV Groningen
the Netherlands
r.j.kok@farm.rug.nl

Jos G. W. Kosterink

University Hospital Groningen
Department of Hospital and Clinical Pharmacy
Hanzeplein 1
9713 GZ Groningen
the Netherlands
j.g.w.kosterink@apoth.azg.nl

Lou F. M. H. de Leij

Groningen University Institute for Drug Exploration (GUIDE)
Department of Pathology and Laboratory Medicine
Medical Biology Section
Tumor Immunology Laboratory
Hanzeplein 1
9713 GZ Groningen
the Netherlands
l.f.m.h.de.leij@med.rug.nl

Claudia S. Leopold

University of Leipzig
Department of Pharmaceutical Technology
Schönauer Str. 160
04207 Leipzig
Germany
cleopold@rz.uni-leipzig.de

Dirk K. F. Meijer

Groningen University Institute for Drug Exploration (GUIDE)
Department of Pharmacokinetics and Drug Delivery
Ant. Deusinglaan 1
9713 AV Groningen
the Netherlands
d.k.f.meijer@farm.rug.nl

Barbro N. Melgert

Groningen University Institute for Drug Exploration (GUIDE)
Department of Pharmacokinetics and Drug Delivery
Ant. Deusinglaan 1
9713 AV Groningen
the Netherlands
b.melgert@farm.rug.nl

Grietje Molema

Groningen University Institute for Drug Exploration (GUIDE)
Department of Pharmacokinetics and Drug Delivery and
Department of Pathology and Laboratory Medicine
Medical Biology Section
Tumor Immunology Laboratory
Ant. Deusinglaan 1
9713 AV Groningen
the Netherlands
g.molema@farm.rug.nl

Frits Moolenaar

Groningen University Institute for Drug Exploration (GUIDE)
Department of Pharmacokinetics and Drug Delivery
Ant. Deusinglaan 1
9713 AV Groningen
the Netherlands
f.moolenaar@farm.rug.nl

Ricardo Mutuberria

Maastricht University
Department of Pathology
PO Box 616
6200 MD Maastricht
the Netherlands

Peter Olinga

Groningen University Institute for Drug Exploration (GUIDE)
Department of Pharmacokinetics and Drug Delivery
Ant. Deusinglaan 1
9713 AV Groningen
the Netherlands
p.olinga@farm.rug.nl

Klaas Poelstra

Groningen University Institute for Drug Exploration (GUIDE)
Department of Pharmacokinetics and Drug Delivery
Ant. Deusinglaan 1
9713 AV Groningen
the Netherlands
k.poelstra@farm.rug.nl

Johannes H. Proost

Groningen University Institute for Drug Exploration (GUIDE)
Department of Pharmacokinetics and Drug Delivery
Ant. Deusinglaan 1
9713 AV Groningen
the Netherlands
j.h.proost@farm.rug.nl

S. Ramakrishnan

University of Minnesota
Health Science Center
Department of Pharmacology
321 Church Street
Minneapolis
Minnesota 55455
USA
sunda001@maroon.tc.umn.edu

Daisy W. J. van der Schaft

Maastricht University/
University Hospital Maastricht
Department of Internal Medicine
Tumor Angiogenesis Laboratory
Peter Debyelaan 25
6202 AZ Maastricht
the Netherlands
D.Vanderschaft@intmed.unimaas.nl

Astrid J. Schraa

Groningen University Institute for Drug Exploration (GUIDE)
Department of Pathology and Laboratory Medicine
Medical Biology Section
Tumor Immunology Laboratory
Hanzeplein 1
9713 GZ Groningen
the Netherlands
a.j.schraa@med.rug.nl

Yuichi Sugiyama

University of Tokyo
Graduate School of Pharmaceutical Sciences
7-3-1 Hongo
Bunkyo-ku
Tokyo 113-0033
Japan
sugiyama@seizai-nl.f.u-tokyo.ac.jp

Kokichi Suzuki

Meiji Seika Kaisha Ltd
760 Moro-oka-cho
Kohoku-ku
Yokohama City 222-0002
Japan

Willem R. Verweij

Groningen University Institute for Drug Exploration (GUIDE)
Department of Pharmacokinetics and Drug Delivery
Ant. Deusinglaan 1
9713 AV Groningen
the Netherlands
w.r.verweij@farm.rug.nl

Dick de Zeeuw

Groningen University Institute for Drug Exploration (GUIDE)
Department of Clinical Pharmacology
Ant. Deusinglaan 1
9713 AV Groningen
the Netherlands
d.de.zeeuw@med.rug.nl

Contents

Abbreviations and Acronyms

Aco	aconitylated (Chapter 4); aconitic acid (Chapter 11)
cis-Aco	*cis*-aconitic acid
ACE	angiotensin-converting enzyme
AD	Alzheimer's disease
ADCC	antibody dependent cellular cytotoxicity
ADEPT	antibody-directed enzyme pro-drug therapy
ADP	adenosine diphosphate
AEA	polyvinylacetal diethylaminoacetate
AIA	antigen-induced arthritis
ALL	acute lymphoblastic leukaemia
AML	acute myeloid leukaemia
ALT	alanine transaminase
AMP	adenosine monophosphate
ANP	atrial natriuretic peptide
AOX	alcohol oxidase (promoter)
AP	alkaline phosphatase
AP-1	activator protein-1
APL	acylated poly lysine
APP	amyloid precursor protein
APC	antigen presenting cell
APS	aerodynamic particle sizer
AS-ODN	antisense oligodeoxynucleotide
AST	aspartate transaminase
ATP	adenosine triphosphate
AUC	area under the (plasma concentration-time) curve
AVP	arginine vasopressin
AZTMP	azidothymidine-monophosphate
BBB	blood-brain barrier
B-CSF-B	blood-cerebrospinal fluid barrier
BDL	bile duct ligation
BDNF	brain derived neurotrophic factor
BDO	bile duct occlusion
BMEC	bovine microvessel endothelial cell
BSA	bovine serum albumin

Bs(M)Ab	bispecific (monoclonal) antibody
BUI	brain uptake index
C	proportionality constant (Chapter 3); complement (Chapter 8); (drug) carrier, any part of a drug-carrier conjugate which is not the pharmacologically active moiety (Chapter 13)
C_P	plasma concentration
C_{ss}	drug concentration at steady state
C_T	tissue concentration
CAM	chick chorio-allantoic membrane assay
CAT	catalase; chloramphenicol acetyl transferase (Chapter 3)
CBF	cerebral blood flow
CCA	cell cycle arrest
CD	cluster of differentiation
CDR	complementary determining region
CEA	carcinoembryonic antigen
CFC	chlorofluorocarbon
cfu	colony-forming units
CHO	Chinese hamster ovary cells
CL	clearance
$CL_{uptake,app}$	apparent clearance uptake
CL_{uptake}	clearance uptake
CLL	chronic lymphoblastic leukaemia
CMV	cytomegalovirus
CNS	central nervous system
COER	controlled onset extended release
COPD	chronic obstructive pulmonary disease
COS	African green monkey kidney cells
COX	cyclooxygenase
CPG2	pseudomonas carboxypeptidase-2
CSF	cerebrospinal fluid
CTDC	colon-targeted delivery capsule
CTL	cytotoxic T lymphocyte
CTLA-4	cytotoxic T lymphocyte associated protein-4
CVO	circumventricular organ
D	(active, free) drug, active form of the drug, not bound to drug carrier
D_A	aerodynamic particle diameter
D_E	equivalent volume diameter
DAB	diphtheria toxin enzymatic A domain and binding B domain
DC	dendritic cell (Chapter 1); drug-carrier conjugate, the conjugate of a drug and a drug carrier (Chapter 13)
DDI	drug delivery index

Dexa	dexamethasone
DIVEMA	divinyl ether and maleic anhydride copolymer
DOC system	dynamic organ culture system
DPI	dry powder inhaler
DSS	dextran sodium sulphate
DT	diphtheria toxin
DTH	delayed-type hypersensitivity
DTI	drug targeting index
DTPA	diethylenetriaminepenta acid
EC	energy charge
ECM	extracellular matrix
EF	edema factor
EF-2	elongation factor-2
EGF	epidermal growth factor
EGP-2	epithelial glycoprotein-2
ELISA	enzyme-linked immunosorbent assay
EMSA	electric mobility shift assay
EPOR	erythropoietin receptor
f_p	plasma unbound fraction
Fab'	antibody fragment with antigen binding capacity
$F(ab')_2$	antibody fragment consisting of two Fab'
FACS	fluorescent activated cell sorting
FBP	folate-binding protein
FEV1	forced expiratory volume in 1 s
(a/b)FGF	(acidic/basic)fibroblast growth factor (is FGF-1/-2)
FIR	flow increase rate
Form	formaldehyde-treated
FPF	fine particle fraction
Fu(A)	function of the cross section of a flow constriction
Gal	galactose
GDNF	glial cell-line derived neurotrophic factor
GFP	green fluorescent protein
GFR	glomerular filtration rate
GGT	γ-glutamyl transpeptidase
GI	gastrointestinal
Glc	glucose
Gludopa	γ-glutamyl pro-drug of l-dopa
GOX	glucose oxidase
gp	glycoprotein
GR	glucocorticoid receptor
GRE	glucocorticoid responsive element
GRO	growth related protein

GSH	glutathione
γ-GTP	γ-glutamyl transpeptidase
HAMA	human anti-mouse antibody
HDL	high-density lipoprotein
HDMEC	human dermal microvascular endothelial cell
HFA	hydrofluoroalkane
HGF	hepatocyte growth factor
HIV	human immunodeficiency virus
HPMA	N(-2-hydroxypropyl)methacrylamide
HRP	horseradish peroxidase
HSA	human serum albumin
HSC	hepatic stellate cell
HUVEC	human umbilical vein endothelial cell
IBD	inflammatory bowel disease
ICAM	intercellular adhesion molecule
i.c.v.	intracerebroventricular
IFN	interferon
IGFII/M6P	insulin-like growth factor II/mannose-6-phosphate receptor
IgG	immunoglobulin
IgSF	immunoglobulin superfamily
IκB	inhibitory factor κB
IKK	IκB-kinase
IL	interleukin
IP-10	interferon γ-inducible protein 10
IPTG	isopropyl-β-D-thiogalactopyranoside
IT	immunotoxin
JAB	JAK binding protein
JAK	janus kinase
KC	Kupffer cell
k_m	Michaelis-Menten constant of transport
LACHSA	lactosylated HSA
LAK	lymphokine activated killer cells
LAT	large neutral amino acid transporter
LDH	lactate dehydrogenase
(ox)LDL	(oxidized) low-density lipoprotein
LF	lethal factor
LH-RH	luteinizing hormone releasing hormone
LMWP	low molecular weight protein
LPS	lipopolysaccharide
LU	lucigenin

LRP	lung resistance related protein
LT	leukotriene
LZM	lysozyme
mAb/MAb	monoclonal antibody
MACS	magnetic activated cell sorting
Mal	maleylated (Chapter 4); maleic acid (Chapter 11)
Man	mannosylated (Chapter 4); mannose (Chapter 5)
MARCO	macrophage receptor with collagenous structure
MBP	maltose binding protein
MCP(-1)	monocyte chemotactic protein(-1)
(p)MDI	(pressurized) metered dose inhaler
MDR	multi-drug resistance
MHC	major histocompatibility complex
MIP	maximal inspiratory pressure (Chapter 3); macrophage inflammatory protein (–1α/β)
MLV-MTP-PE	multilamellar vesicles-muramyl tripeptide-phosphatidylethanolamine
MMAD	mass median aerodynamic diameter
MMP	matrix metalloproteinase
MPEG	monomethoxypolyethyleneglycol
MPTP	1-methyl-4-phenylpyridinium
MRP	multi-drug resistance related protein
MSLI	multi stage liquid impinger
MTT	3[4,5-dimethyl-thiazole-2-yl]-2,5-diphenyltetrazolium bromide
MUC-1	mucin 1
Nap	naproxen
Na/Pi-2 co-transporter	sodium/phosphate co-transporter
NBD	4-nitrobenz-2-oxa-1,3-diazole
NCE	new chemical entity
NCS	neocarzinostatin
NFκB	nuclear factor κB
NGF	nerve growth factor
NHL	non-Hodgkin's lymphoma
NIK	NFκB-inducible kinase
NK	natural killer cell
NLA	neutral avidin
NLS	nuclear localization sequence
NO	nitric oxide
NO_x	nitrite and nitrate
iNOS	inducible NO synthase
NSAID	non steroidal anti-inflammatory drug
ODN	oligodeoxynucleotide

OROS-CT	oral osmotic system for colon targeting
OX26-NLA/SA	conjugate of anti-transferrin receptor antibody OX26 and neutral avidin/streptavidin
PA	protective antigen
PAF	platelet activating factor
PB-PK	physiologically-based pharmacokinetic (modelling/models)
PBC	primary biliary cirrhosis
PBMC	peripheral blood mononuclear cell
PC	parenchymal cell/hepatocyte
PCNA	proliferating cell nuclear antigen
PD	Parkinson's disease (Chapter 2); pharmacodynamics (Chapter 13)
PDGF	platelet-derived growth factor
PDTC	pyrrolidine dithiocarbamate
PE(40)	Pseudomonas exotoxin (amino acid 1–40)
PECAM	platelet endothelial cell adhesion molecule
PEF	peak expiratory flow rate
PEG	polyethylene glycol
PET	positron emission tomography
$PG(E_2)$	prostaglandin (E_2)
PGA	poly-glutamic acid
P-gp	P-glycoprotein
PIFR	peak inspiratory flow rate
PK	pharmacokinetics
PKC	protein kinase C
PK/PD	pharmacokinetic/pharmacodynamic
PMN	polymorphonuclear cell
pro-drug	inactive form of the drug, which is converted within the body to the active drug
PS	phosphatidylserine
PS-product	permeability surface area product
PSC	primary sclerosing cholangitis
$Q\mathrm{r}$	renal plasma flow rate
R_{E}	external resistance (to airflow)
R_{I}	internal resistance (to airflow)
R_{TOT}	total resistance (to airflow)
RA	rheumatoid arthritis
RANTES	regulated upon activation, normal T-cell expressed and secreted
RB	Rhodamine B
Re	Reynolds number
RES	reticuloendothelial system
RGD	Arg-Gly-Asp

ROS	reactive oxygen species
RSV	respiratory syncytial virus
RV	residual volume
SA	streptavidin
scFv	single chain antibody variable fragment
SEC	sinusoidal endothelial cell
SELEX	systemic evolution of ligands by exponential enrichment
SHR	spontaneously hypertensive rat
S(L)T	Shiga(-like) toxin
SMA	styrene-co-maleic acid/anhydride
αSMA	α-smooth muscle actin
SMANCS	styrene-co-maleic acid/anhydride–neocarzinostatin conjugate
SOCS	suppressors of cytokine signalling
SOD	superoxide dismutase
SPARC	secreted protein acidic and rich in cysteine
SPECT	single photon emission computed tomography
SSI	STAT induced STAT inhibitor
STAT	signal transduction and activator of transcription
Suc	succinylated (Chapter 4); succinic acid (Chapter 11)
SV40	Simian virus 40
TA	therapeutic availability
TAA	tumour associated antigen
TEER	transendothelial electrical resistance
TES	time-controlled explosion system
TfR	transferrin receptor
TGF	transforming growth factor
TI	targeting index
Tie-2	angiopoietin receptor
tTF	truncated tissue factor
TIMP	tissue inhibitor of metalloproteinases
TLC	total lung capacity
TNBS	trinitrobenzene sulfonic acid
TNFα	tumour necrosis factor α
TOF	time of flight
TSP	thrombospondin
TTC	Tetanus toxin (C-fragment)
TUNEL	terminal transferase-mediated UTP nick-end labelling
unbound drug	active form of the drug not bound to plasma proteins or other macromolecular tissue compounds
V	variable; volume of distribution
*V*c	vital capacity

V_{max}	maximum transport (between compartments)
V_T	tidal volume
VCAM	vascular cell adhesion molecule
VEGF(R)	vascular endothelial growth factor (receptor)
VH	heavy chain variable domain
VIP	vasoactive intestinal polypeptide
VL	light chain variable domain
VMAD	volume median aerodynamic diameter
χ	dynamic shape factor
Φ	flow rate (air)
Φ_V	volumetric flow rate (of air)
ρ_A	density of air
ρ_P	particle density
ΔP	pressure difference

1 Drug Targeting: Basic Concepts and Novel Advances

Grietje Molema

1.1 Introduction

Since the early 1960s, many scientists have dedicated their research to the development of drug targeting strategies for the treatment of disease. In general, the aim of targeted therapies is to increase the efficacy and reduce the toxicity of drugs. The behaviour of the carrier molecules largely determines the pharmacokinetics and cellular distribution of the drug. Furthermore, selective delivery into the target tissue may allow a higher drug concentration at or in the target cells or even in specific compartments of the target cells. As a result, drug efficacy can be enhanced.

Whereas the majority of strategies studied so far have incorporated cytotoxic drugs for the treatment of cancer, it is believed that novel pharmacologically active substances will become more and more subject to study in the coming years. With the advent of molecular biological techniques, molecular mechanisms of disease become unravelled at an almost uncontrollable pace. As a result, new chemical entities (NCE) are generated that in principle can exert potent effects on disease processes but have a deficient distribution to the areas of disease. In addition, they can be highly toxic upon gaining access to healthy tissue. The chemical characteristics of NCEs may be such, that access to the site of action, in particular intracellular target enzyme systems, is minimal. By attaching an NCE to a carrier molecule, its whole body and cellular disposition can be considerably manipulated. Similarly, therapeutic macromolecules, gene transcription/translation modulating agents such as antisense oligonucleotides and genes themselves will progressively gain territory in the field of drug development. For these treatment modalities to become major successes, the delivery and/or targeting of these compounds will be an essential component [1].

The aim of this chapter is to introduce the basic principles of drug targeting as they have evolved over previous decades. The most important chemical features and biological behavioural characteristics of the carrier molecules exploited for drug targeting purposes will be addressed. Novel advances in the understanding of cellular routing of naturally-occurring entities such as viruses have in recent years been applied for cellular delivery purposes. Furthermore, a selection of drug targeting preparations that are either in the stage of clinical testing or have been approved for application in the clinic is discussed. As the basis of drug development lies in the understanding of the molecular basis of diseases, selective interference with regulatory processes in health and disease by drug targeting will become a powerful technology. Drug targeting can, in this respect, serve both as a therapeutic approach and as a research tool in unravelling the functions of these processes in normal physiology and under pathophysiological conditions.

1.2 Carriers used in Drug Targeting

Drug delivery and drug targeting research is blooming in a quantitative sense, as exemplified by the increase in research publications in international pharmaceutical and biomedical literature [1]. In the last three decades many strategies to deliver drugs in a controlled fashion have been developed. The aim of this section is to give a brief introduction on drug carriers employed in targeting strategies. For in-depth reviews on these subjects, the reader is referred to various (special) issues of journals such as *Advanced Drug Delivery Reviews, the Journal of Controlled Release* and *the Journal of Drug Targeting.*

The choice of carrier system to be used in drug targeting strategies depends on which target cells should be reached and what drug needs to be delivered. Carriers can be divided into particle type, soluble and cellular carriers. Particle type carriers include liposomes, lipid particles (low and high density lipoproteins, LDL and HDL respectively), microspheres and nanoparticles, and polymeric micelles. Soluble carriers consist of monoclonal antibodies and fragments thereof, modified plasma proteins, peptides, polysaccharides, and biodegradable carriers consisting of polymers of various chemical composition. For the site selective expression of genes, vectors such as liposomes, whole cells and viruses are widely exploited nowadays. With the advancement of chemical and recombinant DNA technology, combinations of strategies (e.g. antibody-targeted liposomes, or bispecific antibody-mediated cross linking of viral vectors and target cells) are now also under investigation.

1.2.1 Liposomes

Liposomes are small vesicles composed of unilamellar or multilamellar phospholipid bilayers surrounding one or several aqueous compartments. Charge, lipid composition and size (ranging from 20 to 10 000 nm) of liposomes can be varied and these variations strongly affect their behaviour *in vivo*. Many liposome formulations are rapidly taken up by macrophages. They are exploited either for macrophage-specific delivery of drugs or for passive drug targeting, allowing slow release of the drug over time from these cells into the general circulation. Cationic liposomes and lipoplexes have been extensively investigated for their application in non-viral vector mediated gene therapy.

The use of molecules such as polyethylene glycol (PEG) to prevent liposome recognition by phagocytic cells led to the development of so called 'stealth' liposomes with longer circulation times and increased distribution to peripheral tissues in the body [2]. Furthermore, a targeting device or homing ligand can be included at the external surface of the liposome in order to obtain target cell specificity as shown schematically in Figure 1.1. Although liposomes do not easily extravasate from the systemic circulation into the tissues, enhanced vascular permeability during an inflammatory response or pro-angiogenic conditions in tumours can favour local accumulation. Another approach is the design of target sensitive liposomes or fusogenic liposomes that become destabilized after binding and/or internalization to/into the target cells [2,3]. After two decades of development, the *in vivo* and pharmaceutical behaviour of liposomes is now better understood and forms the basis for further development of liposome-mediated drug targeting strategies for clinical application [4].

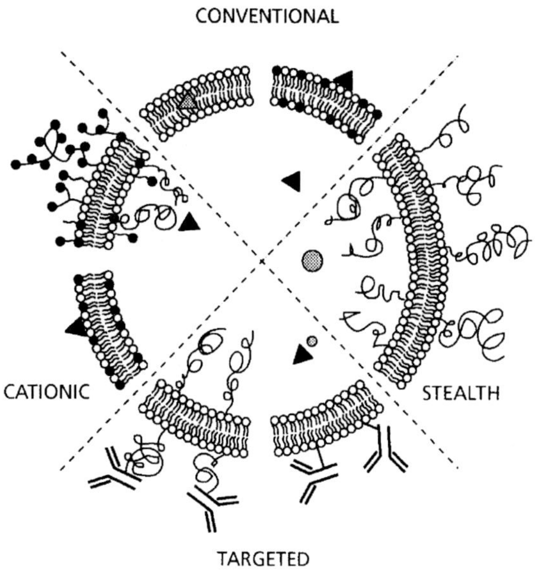

Figure 1.1. Schematic representation of four major liposome types. Conventional liposomes are either neutral or negatively charged. Stealth liposomes are sterically stabilized and carry a polymer coating to obtain a prolonged circulation time in the body. Immunoliposomes are antibody targeted liposomes and can consist of either conventional or sterically stabilized liposomes. Positive charge on cationic liposomes can be created in various ways. Reproduced from reference [112] with permission.

1.2.2 Monoclonal Antibodies and Fragments

The development of monoclonal antibodies by Köhler and Milstein in 1975 paved the way to antibody therapy for disease [5]. In the last 25 years, the number of pre-clinical and clinical studies with monoclonal antibodies and derivatives thereof have greatly increased. The majority of strategies based on antigen recognition by antibodies have been developed for cancer therapy. These strategies are mostly aimed at tumour associated antigens being present on normal cells but overexpressed by tumour cells [6]. More recently, antibodies against other molecules have been developed for clinical application. Examples are anti-TNFα antibodies for treatment of chronic inflammatory diseases and anti-VEGF (vascular endothelial growth factor) antibodies which inhibit new blood vessel formation or angiogenesis.

The advent of recombinant DNA technology led to the development of antibodies and fragments that are tailored for optimal behaviour *in vivo* [7,8]. Humanized and chimeric antibodies can be constructed to circumvent the human anti-mouse antibody response elicited by mouse antibody treatment of patients, which severely hampers the application of these powerful molecules. The treatment of rheumatoid arthritis patients with doses of as high as 10 mg kg^{-1} cA2 chimeric antibody specific for TNFα [9], emphasizes that at present the production and purification methods for these proteins have been optimized to such extent that clinical studies can be considerably intensified.

1.2.3 Modified (Plasma) Proteins

Modified plasma proteins are attractive carriers for drug targeting as they are soluble molecules with a relatively small molecular weight. They can easily be modified by covalent attachment of peptides [10] (see Figure 1.2), sugars [11,12], and other ligands, as well as drugs of interest. Particularly in the case of liver cell targeting, quite extensive modifications of protein backbones such as albumins have been carried out. The carriers and drug–carrier conjugates rapidly distribute to either the hepatocytes and/or the non-parenchymal cells, depending on the net protein charge and hydrophobicity. If the target cells are, however, for example tumour cells or vascular endothelial cells in tumours or inflammatory lesions, rapid distribution to the liver is an undesirable characteristic. As a consequence, only minor modifications are allowed in the protein backbone [13], which may pose a serious drawback in using these proteins for non-hepatic drug targeting.

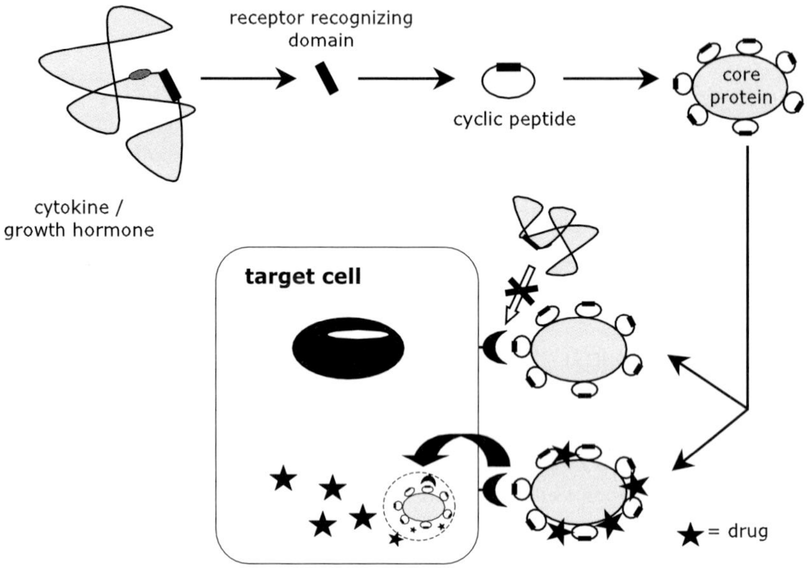

Figure 1.2. A novel strategy in the development of cell-specific carriers consists of the identification of a stretch of amino acids/peptides within a cytokine molecule that is specific for receptor binding. These peptides can serve as homing ligands for a macromolecular protein by covalent attachment to the protein backbone. The resulting carrier can subsequently be conjugated with drug molecules. Besides delivering the drug at or into the target cells, carrier or conjugate binding to the cytokine receptor may be able to inhibit or induce activation of signal transduction pathways. Adapted from Beljaars L, thesis Groningen University (1999) and reference [10].

1.2.4 Soluble Polymers

Soluble synthetic polymers have been widely employed as versatile drug carrier systems. Polymer chemistry allows the development of tailor made conjugates in which target moieties as well as drugs are introduced into the carrier molecule. In the case of enhanced permeability retention in e.g. tumour vasculature [14], the introduction of drugs into the polymer may suffice. As non-specific adherence to cells is an undesirable property, excessive charge or hydrophobicity should be avoided in the design of polymeric carriers.

For cancer therapy, the well established N(-2-hydroxypropyl)methacrylamide (HMPA) polymers have been extensively studied. PK1, a 28-kDa HPMA copolymer containing doxorubicin (Figure 1.3) is now in clinical testing [15]. Other drugs that have been incorporated

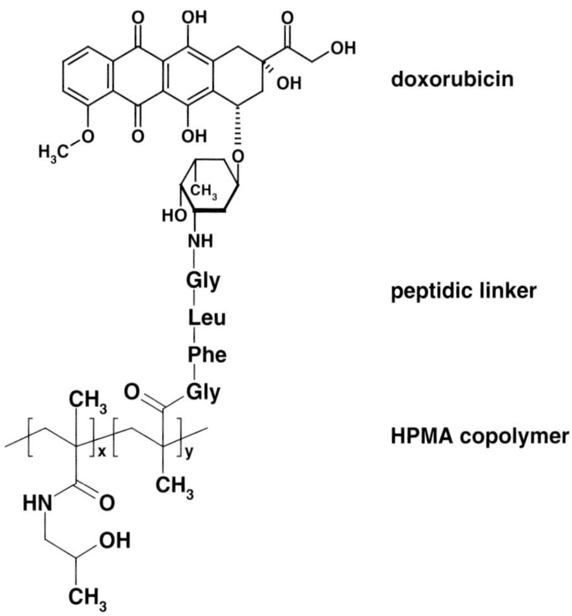

Figure 1.3. Structure of PK1 (HPMA copolymer doxorubicin), a 28-kDa polymeric carrier–drug conjugate investigated for its anti-tumour activity in a phase I clinical study. Adapted from reference [15].

in these polymers are platinates and xanthine oxidase, respectively [16,17]. Furthermore, conjugates (so called SMANCS) of the anticancer drug neocarzinostatin (NCS) and styrene-co-maleic acid/anhydride (SMA) have been developed for therapy of liver cancer (see Table 1.3). New polymers developed in the last few years include the cationic low molecular mass chitosan polymers for DNA delivery [18] and highly branched, low dispersity dendrimers consisting of various chemical origins [19]. Kopeček and colleagues furthermore reported on the application of Fab' antibody fragments that can copolymerize with HPMA and drug-containing monomers to yield a targetable HPMA copolymer–Mce_6 conjugate. *In vitro* studies showed that, as a result, the photosensitizer Mce_6 was more efficiently internalized by OVCAR-3 carcinoma cells than the non-targeted copolymer and hence had greater cytotoxicity [20].

1.2.5 Lipoproteins

Endogenous lipid particles such as LDL and HDL containing a lipid and apoprotein moiety can be seen as 'natural targeted liposomes'. The lipid core can be used to incorporate lipophilic drugs or lipophilic pro-drugs [21], covalent binding of the drug to the carrier is not necessary here. The apolipoprotein moiety of these particles can be glycosylated or modified with other (receptor) targeting ligands. Furthermore, modifications at the level of glycolipid incorporation can be employed to introduce targeting moieties. As with the liposomes, the size and charge of the particles determine their behaviour *in vivo*. Large particles will not easily pass the endothelial barrier of organs containing blood vessels with a continuous endothelial cell lining.

The majority of the research on the use of LDL and HDL particles has been devoted to the targeting of drugs to the liver [22]. With respect to hepatocyte targeting, antiviral agents and anti-malaria drugs are good candidates for being delivered to the site using lactosylated lipoproteins [23]. Kupffer cells and sinusoidal endothelial cells in the liver can more specifically be reached using oxidized and acetylated LDL. Uptake of both LDL derivatives takes place via scavenger receptors [24,25]. To overcome the difficulties in isolation and handling of the lipoproteins, various artificial supramolecular systems have been developed to mimic endogenous lipoproteins. Examples of these are lipoprotein-mimicking biovectorized systems [26] and lipid variants of the nanoparticles described below [27].

1.2.6 Microspheres and Nanoparticles

Microspheres and nanoparticles often consist of biocompatible polymers and belong either to the soluble or the particle type carriers. Besides the aforementioned HPMA polymeric backbone, carriers have also been prepared using dextrans, ficoll, sepharose or poly-L-lysine as the main carrier body. More recently alginate nanoparticles have been described for the targeting of antisense oligonucleotides [28]. As with other polymeric carrier systems, the backbone can be modified with e.g. sugar molecules or antibody fragments to introduce cellular specificity.

Nanoparticles are smaller (0.2–0.5 μm) than microspheres (30–200 μm) and may have a smaller drug loading capacity than the soluble polymers. Formulation of drugs into the nanoparticles can occur at the surface of the particles and at the inner core, depending on the physicochemical characteristics of the drug. The site of drug incorporation significantly affects its release rate from the particle [29]. After systemic administration they quickly distribute to and subsequently become internalized by the cells of the phagocytic system. Even coating of these carriers with PEG does not completely divert them from distribution to the phagocytes in liver and spleen. Consequently, intracellular infections in Kupffer cells and other macrophages are considered a useful target for these systems.

Besides parenteral application of microspheres and nanoparticles for cell selective delivery of drugs, they have more recently been studied for their application in oral delivery of peptides and peptidomimetics [30]. Immunological tolerance induction against beta-lactoglobulin could be achieved by application of this protein in a poly-lactic-glycolide microsphere formulation [31].

1.2.7 Polymeric Micelles

Polymeric micelles are characterized by a core-shell structure [32]. They have a di-block structure with a hydrophilic shell and a hydrophobic core. The hydrophobic core generally consists of a biodegradable polymer that serves as a reservoir for an insoluble drug. Non- or poorly biodegradable polymers can be used, as long as they are not toxic to cells and can be renally secreted. If a water-soluble polymeric core is used, it is rendered hydrophobic by chemical conjugation with a hydrophobic drug. The viscosity of the micellar core may influence the physical stability of the micelles as well as drug release. The biodistribution of the micelle is mainly dictated by the nature of the shell which is also responsible for micelle stabilization and interactions with plasma proteins and cell membranes. The micelles can contain functional groups at their surface for conjugation with a targeting moiety [32].

Polymeric micelles are mostly small (10–100 nm) in size and drugs can be incorporated by chemical conjugation or physical entrapment. For efficient delivery activity, they should maintain their integrity for a sufficient amount of time after injection into the body. Most of the experience with polymeric micelles has been obtained in the field of passive targeting of anticancer drugs to tumours [33]. Attachment of antibodies or sugars, or introduction of a polymer sensitive to variation in temperature or pH has also been studied [32].

1.2.8 Cellular Carriers

Cellular carriers may have the advantage of their natural biocompatibility. However, they will encounter endothelial barriers and can rather easily invoke an immunological response. Most of the approaches on cellular carriers have been applied to the field of cancer therapy. Antigen specific cytotoxic T lymphocytes have been studied as vehicles to deliver immunotoxins to cancer cells *in vivo*. To achieve this, a CD8 positive T-cell line that specifically recognized a murine leukaemia cell line was transfected with a retroviral vector encoding a truncated diphtheria-toxin molecule/IL-4 fusion protein. Intravenous injection of these transfected cells led to significant tumour growth inhibition without concomitant renal and hepatic toxicity common to this class of immunotoxins [34]. Whereas in this particular study the intrinsic capacity of T cells to home to tumour tissue was exploited, Wiedle *et al.* attached a homing device to a lymphoid cell line to improve homing ability [35]. By transfection of the lymphoid cells with a chimeric adhesion molecule consisting of the CD31 transmembrane domain and the disintegrin kristin, the lymphocytes specifically homed to $\alpha v\beta 3$ expressing pro-angiogenic tumour endothelium.

The identification of endothelial progenitor cells in peripheral blood [36], has led to the hypothesis that these cells may in the future be exploited as drug carriers. It has been shown that in diseases in which angiogenesis and/or vasculogenesis plays an important role, these progenitors represent a pool of cells that seed at the site of neovascularization [37]. In theory, one can isolate the progenitors from the peripheral blood and transfect them *ex vivo* with genes encoding e.g. anti-angiogenic proteins. Subsequent intravenous or local re-injection of the cells into the patient may lead to seeding of the transfected cells at the diseased site and hence, local delivery of the therapeutic protein of interest [38].

1.3 Intracellular Routing of Drug–Carrier Complex

Targeting of therapeutics, whether they are chemical entities, peptides, proteins or nucleic acid polymeric substances, relies on the release of the drug from the carrier and subsequent access to the molecular target. Advances in the understanding of membrane structure, functions and properties of the various cellular organelles is the basis for directing the pharmacologically active components to the correct cellular compartments [39].

1.3.1 Passive Versus Active Drug Targeting

In drug targeting, two types of strategies can be distinguished: passive targeting and active targeting. In the case of passive targeting, the carrier–drug complex is often delivered to macrophages and other cells of the monocyte-phagocytic system. This leads to gradual degradation of the carrier and (slow) release of the liberated drug from the cells either into the blood circulation or into the tissue environment. By size exclusion, extravasation of the carrier–drug complex can be limited, thereby preventing the drug from being distributed to non-target sites. As a consequence, toxicity can be reduced. Active targeting should lead to a higher therapeutic concentration of the drug at the site of action. This can be accomplished by cell specific delivery of the drug. In the ideal case, the dose of the drug can be reduced and side-effects can be diminished. The majority, if not all, active drug targeting strategies exploit receptor-based drug targeting principles, in which receptor-specific ligands attached to the carrier–drug complex or directly to the drug itself deliver the drug to the target cell of choice. Depending on the subsequent routing of the receptor complex, the drug will arrive in a specific compartment in the target cell.

1.3.2 Lysosomes as a Cellular Target Compartment

Ligands that are taken up via endocytosis or phagocytosis, are often transferred to lysosomes for degradation. Many drug targeting strategies exploit the decrease in pH and/or the presence of lysosomal enzymes for drug release from the carrier molecules (see Chapter 11 for a detailed discussion of acid and enzyme sensitive drug-carrier linkers). Only in the case of lysosomal infections and some metabolic disorders is the lysosome a relevant target compartment. Furthermore, lysosomal routing provides a pathway for presentation of peptides in major histocompatibility complexes (MHC) class I or II in macrophages or other antigen presenting cells.

In most other cases, the lysosomes are a transit compartment *en route* to the cytoplasm. In case the targeted agent is lysosomally unstable (e.g. DNA) this compartment should be avoided.

1.3.3 Cytoplasmic Delivery

The majority of drugs exert their action in the cytoplasm of the cell where their target enzymes are located. Consequently, they need to pass from one of the compartments of the endocytotic pathway into the cytoplasm. The endosomal and lysosomal membranes can be destabilized using fusogenic peptides derived from viruses, cyclodextrins and polyethyleneimine [39]. pH-sensitive liposomes or polymers become fusion competent at the acidic pH of the endosomes and subsequently release their contents into the cytoplasm. Particularly for the delivery of DNA into cells, this approach seems appropriate and quite successful [40]. Bacterial components such as listeriolysin O and alpha-haemolysin can form pores in phagosomal or plasma membranes [39]. It remains, however, to be established whether these components can be exploited for directing drug-targeting preparations *in vivo* to specific cellular compartments.

Schwarze *et al.* reported on the development of a recombinant fusion protein consisting of the protein transduction domain of HIV-derived TAT and the 120-kDa β-galactosidase. The TAT protein was able to deliver the large molecular weight protein to the interior of the cells *in vitro*. Interestingly, the enzymatic activity of intracellularly delivered β-galactosidase peaked about 2 h later than did the intracellular concentration. This likely reflects a slow posttransduction refolding of the protein by intracellular chaperones. Intraperitoneal injection of the fusion protein in mice resulted in delivery of the biologically active fusion protein to all tissues, including the brain [41]. Similarly, the *Herpes simplex* virus tegument protein VP22 is able to deliver proteins into the cytoplasm of cells [42]. Both approaches may prove useful to enhance the delivery of e.g. enzymes for pro-drug protocols.

1.3.4 Nuclear Targeting

The three major obstacles to DNA accessibility in the nucleus of the target cells are low uptake across the plasma membrane, inadequate release of DNA with limited stability, and lack of nuclear targeting. Delivery systems of the future need to fully accommodate all steps in the internalization and targeting routing in order to effectively guide the DNA into the nucleus. Due to space limitations, a complete overview of recent advances in this field will not be provided. For more in-depth reading, the reader is referred to a recent concise review by Luo and Saltzman on novel strategies to accomplish optimal gene delivery [43].

In short, increased targeting and uptake of DNA by the cell using better delivery systems is the basis for overcoming the plasma membrane hurdle. Increased stability of the DNA once inside the cell can be achieved by chloroquine or branched cationic polymers which facilitate the early release of the DNA from the endosomal pathway. Furthermore, bacterial subunits and adenoviral capsids are capable of bypassing the endosomes, although it should be realized that these approaches may be significantly hampered by inherent toxicity and/or immunogenicity. Stabilization of the DNA in the 'hostile' environment of the cytosol can be achieved using PEG, PEG-poly-l-lysine block copolymers and others [43]. Intermediate stability of DNA/delivery system interactions is most likely *the* prerequisite to achieve optimal liberation of DNA molecules once they have become available within the cytoplasm. To overcome the final obstacle, finding the nucleus, application of knowledge of viral infection processes have led to the application of viral nuclear localization signals. Furthermore, the aforementioned viral tegument protein VP22 also localizes in the nuclear compartment. In the early stages of cell mitosis, VP22 translocates into the nucleus, binds to the condensing chromatin and remains bound [44]. It will be interesting to see whether these principles of nuclear targeting can be exploited for the targeted delivery of DNA to the cell type of choice.

By assembling polypeptides that can code for all the necessary cellular transport tasks on a scaffold, Sheldon and colleagues developed so-called 'loligomers', branched squid-like peptides that can self-localize in the cytoplasm or nucleus [45]. *In vitro* application revealed good transfection properties of one of the nuclear localizing loligomers [46]. Its potential for application in drug targeting, i.e. the ability to combine cell specificity of DNA delivery with loligomer-orchestrated intracellular routing capacity, needs to be established.

1.3.5 Mitochondrial Targeting

Mitochondria are the ATP suppliers of the cells and have an important role in modulating intracellular calcium levels and cellular apoptosis. The mitochondrial respiratory chain is furthermore an important supplier of damaging free radicals. Evidence increases that mitochondria are heavily involved in numerous diseases and therefore they may become important targets for the development of new drugs and therapies [47].

Both the large membrane potential across the inner membrane and the protein import machinery of the mitochondria may be exploited for selectively delivering drugs to this cellular organelle. Lipophilic cations in particular, have been studied for mitochondrial targeting purposes based on their mitochondrial accumulation potential. Using triphenylphosphonium as a carrier, Smith *et al.* were able to selectively deliver antioxidant activity into the mitochondrial compartment of cells [48]. Similar to proteins that require nuclear localization sequences for homing to this compartment, cellular proteins that need to be targeted to mitochondria require mitochondrial localization sequences [49]. Fusion of these signal sequences with (model) proteins of interest redirects the proteins into the mitochondrial compartments. Whether these intracellular targeting entities can be combined with other targeting entities that specifically direct them into the desired cell type in the body, is however questionable. One option may be to package the mitochondrial targeting system in immunoliposomes that provide the cell specificity [47], but further research is awaited to provide insight into the potential and limitations of this approach.

1.4 Drug Targeting Strategies in the Clinic

Most of the drug delivery systems that have been studied for clinical application are capable of rate- and/or time-controlled drug release. The therapeutic advantages in these approaches lie in the *in vivo* predictability of release rate, minimized peak plasma levels, predictable and extended duration of action and reduced inconvenience of frequent re-dosing and hence, improved patient compliance [1].

With the development of more advanced drugs such as therapeutic proteins, antisense molecules and genes, not only controllable release, but also controllable delivery in the target cells becomes desirable. As a consequence, targeting modalities need to be incorporated into the vehicles. Although drugs and drug formulation based on proteins were first considered unfeasible, nowadays practice has proven this not to be the case. Since 1994, on average more than seven FDA approvals per year have been issued for protein-based therapies. It is furthermore believed that the increasing yield of protein molecules produced by recombinant techniques will boost application of protein drugs in the near future [50].

In recent years, various reviews have been published in which the current status of liposome [2] and antibody [6] based drug targeting strategies have been summarized. Similarly, the current status of gene therapy in both pre-clinical and clinical settings have been recently reviewed [51,52]. Without trying to be complete, the next sections will give an overview on recently published studies with drug targeting formulations in a clinical setting.

1.4.1 Liposome Based Therapies in the Clinic

Several liposomal formulations are either under investigation in phase I/II/III clinical trials or have been approved by the US Food and Drug Administration or cleared for empiric therapy (see also Chapter 8 for FDA approved formulations in cancer therapy). They are in use for the treatment of cancer or fungal infections and, consisting of non-targeted liposomes, aim at slow release of the encapsulated drugs over a prolonged period of time to circumvent dose-limiting (cardiac) toxicity. Recently reported clinical studies with liposome formulations are summarized in Table 1.1. The reader is also referred to references [4] and [53] for a more detailed description of these liposome-based therapeutic strategies.

The general outcome of these studies was that the liposomal drugs were well tolerated and exerted a clinical effect, although larger studies need to be carried out to demonstrate the extent of the effects. Furthermore, in the various studies of liposome-incorporated drugs, it was observed that cardiotoxicity was either absent or only limited, in contrast to studies with non-liposomally formulated drugs.

1.4.2 Monoclonal Antibody Therapies in the Clinic

Clinical studies on cancer therapy with antibodies have been elegantly summarized by Farah *et al.* [6] and more recently by Glennie and Johnson [54]. The majority of antibody-based therapies in the clinic have exploited the activity of the antibody *per se*. In the therapy of cancer, the highest response rates have been observed in patients with haematological malignancies that are easily accessible to antibodies. In recent years, numerous clinical studies with antibodies directed against the B cell non-Hodgkin lymphoma-associated epitope CD20 (Rituximab, Rituxan) and the breast carcinoma antigen Her-2 (also known as neu or erbB2) have been carried out. The mechanisms of action of these antibodies are believed to consist of complement-mediated lysis, antibody-dependent cellular cytotoxicity by macrophages and natural killer cells, and signal transduction leading to apoptosis or growth arrest [55]. Novel

Table 1.1. Overview of some recently reported clinical studies on the application of liposome-based drug formulations.

Liposome formulation	Drug incorporated	Disease	Clinical responsea	Remarks	Reference
DaunoXome[b]	Daunorubicin	Acute leukaemia	2/23 CR		[59]
		Heptocellular carcinoma	2/12 SD		[60]
		Recurrent progressive brain tumours in children	2/14 CR 3/14 PR 2/14 SD		[61]
Caelyx[c]	Doxorubicin	Hormone refractory prostate cancer	3/15 OR 2/15 SD	Severe mucocutaneous toxicities	[62]
		Advanced/metastatic soft tissue sarcoma	3/25 PR 2/25 MR 17/25 SD	Highest dose to treat not yet reached	[63]
		Non-small-cell lung cancer	3/15 CR	Combination with fractionated radio therapy	[64]
		Squamous cell head and neck cancer	11/15 CR	Tumour microvessel density correlated with degree of Caelyx accumulation	
TLC D-99[d]	Doxorubicin	Kaposi's sarcoma	6/40 PR 26/40 SD	Dose dependent response and toxicity	[65]
		Metastatic breast cancer	10/41 OR	Combination with cyclophosphamide and fluorouracil	[66]
AmBisome	Amphotericin B	Renal Tx recipients with viscer. Leishm.[e]	4/4 treated successfully		[67]
		Heart surgery with *C. tropicalis* endocardititis	1/1 treated successfully		[68]
		Bone marrow Tx or chemotherapy patients with fungal infections	Polyprophylactic treatment did not reduce fungal infections or requirement for systemic therapy		[69]
TLC C-35	Prostaglandin E1	Acute respiratory distress syndrome	348 patients evaluated: no difference in number of days to discontinuation of ventilation to placebo group; mortality not different to placebo		[70]

[a] Number of patients responding/total number of evaluable patients. CR, complete remission/response; PR, partial response; MR, minor response; OR, objective response; SD, stable disease.
[b] Daunorubicin citrate liposome injection formula.
[c] Also known as Doxil; PEG-coated liposome formulation with doxorubicin.
[d] Phosphatidylcholine/cholesterol liposomal formulation with doxorubicin.
[e] Tx, transplantation; visc. Leishm., visceral Leishmaniasis

advances in this field are the development of antibodies against angiogenesis-associated factors and receptors such as VEGF and integrin αvβ3 [56,57]. Furthermore, antibodies directed against the CD3 molecule on T lymphocytes, a treatment developed for prevention of organ rejection, are now under investigation for their immunomodulatory effects in cancer patients [58].

Whereas initially the focus on antibody-based therapies was on cancer, anti-TNFα antibodies in particular have recently proven powerful in the therapy of chronic inflammatory diseases such as inflammatory bowel disease and rheumatoid arthritis [71]. These antibodies complex serum TNFα, the clinical benefit to RA patients most likely being the reduction of pro-inflammatory IL-6 and acute phase protein levels [9]. Although they are directed against soluble proteins and as such will not serve as a drug carrier, they do show that targeted, i.e. selective, interference with a specific molecule or process can have a powerful effect without significant concomitant toxicity.

Although in the early 1990s several antibodies were developed that inhibited leukocyte–endothelial cell interaction to prevent e.g. allograft rejection or inflammatory processes [72], more effort is nowadays put into the development of small molecule antagonists and antisense oligonucleotides for this purpose [73,74]. A selection of more recently reported clinical studies with antibodies is summarized in Table 1.2.

1.4.3 Monoclonal Antibody Based Targeting Strategies in the Clinic

Not only can antibodies by themselves function as targeted effector molecules, they can also be used as carriers for the selective delivery of drugs, toxins, enzymes, radioisotopes, and adenoviral vectors. Most of the strategies have been applied in humans [6,75–77]. Besides these approaches, cellular cytotoxicity can be redirected towards the tumour cells using bispecific antibodies that consist of a recognition site for specific cells of the immune system and tumour-associated antigens [78,79]. This strategy circumvents the MHC restriction of antigen recognition and cellular cytotoxicity. It may therefore be exploited for the therapy of tumours which downregulate the expression of their MHC molecules and thereby avoid the normal immune response.

In the development of these antibody-based targeting strategies, modifications that have been applied to improve the efficacy of the therapy include the use of chimerized or humanized antibodies, more potent drugs and better linkages between drugs and carrier molecules. Furthermore, liposome encapsulation of drugs enabled a larger quantity of the drug to be delivered per antibody molecule [75]. The development of transgenic mice capable of making fully human antibodies now offers new opportunities for generating antibodies of therapeutic quality, and as a result has led to the revival of interest in antibody-based therapies [80].

1.4.4 Other Drug Targeting Strategies in the Clinic

Of the non-antibody, non-liposome based drug targeting strategies, most of the (limited) clinical experience has been obtained with polymer-based conjugates of anticancer drugs. The most widely employed drugs for this application are cytotoxic agents such as doxorubicin and

Table 1.2. Overview of recently reported clinical studies exploiting antibodies.

Antibody	Antigen	Disease treated[a]	Clinical response[b]	Remarks	Reference
Rituximab (rituxan)	CD20	NHL	21/39 OR 14/39 SD or MR	Similar results in follicular and small lymphocytic lymphoma	[81]
		PT-LPD after liver Tx	2/2 CR: polymorphic PT-LPD; 1/1 NR: large cell NHL		[82]
		PT-LPD after solid organ Tx	15/26 CR 2/26 PR		[83]
		PT-LPD after BMT	5/6 CR		[83]
		Mantle cell lymphoma (MCL), immunocytoma, small B cell lymphocytic lymphoma (SLL)	36/120 OR (varied with tumour type). CR only seen with MCL	Limited activity in SLL	[84]
Herceptin; trastuzumab	Her2/neu	Metastatic breast carcinoma	1/43 CR 4/43 PR 2/43 MR 14/43 SD	Responses seen in mediastinum, lymph nodes, liver, chest wall lesions	[85]
MKC-454	Her2/neu	Metastatic breast carcinoma	2/18 OR		[86]
rhuMAb HER2	Her2/neu	Advanced breast carcinoma	9/37 PR 9/37 MR – SD 19/37 DP	Combination therapy with cisplatin	[87]
rhuMab VEGF	VEGF	Metastatic renal cell carcinoma	Phase III study in progress		(clinical-trials.gov website)
hOKT3γ_4	CD3	Various malignancies	3/24 responses		[58]
Infliximab	TNFα	Rheumatoid arthritis	428 patients: 20–50% improvement in ~80% of patients	Concomitant methotrexate therapy	[71]
		Psoriasis lesions	Improved appearance of a patient 2 weeks after treatment		[88]
Synagis; palivizumab	RSV F - glycoprotein	RSV infection in infants	35 children < 2 years of age: significant reduction in tracheal RSV concentration, but not in nasal aspirate. No difference in disease severity compared to placebo		[89]
Abciximab	Platelet gp IIb/IIIa receptor	Ischaemic complications during balloon angioplasty or atherectomy	Platelet aggregation inhibition with abciximab provided long-term clinical benefits after coronary intervention compared to placebo	All patients received aspirin co-medication; patients with stent implantation received ticlopidine co-medication	[90]

[a] BMT, bone marrow transplantation; NHL, non-Hodgkin lymphoma; PT-LPD, post-transplantation lymphoproliferative disorders; RSV, respiratory syncytial virus; Tx, transplantation.

[b] Number of patients responding/total number of evaluable patients. CR, complete remission/response; PR, partial response; MR, minor response; NR, no response OR, objective response; SD, stable disease; DP, disease progression.

methotrexate, and the general aim of these conjugates is to lower the (dose-limiting cardio-) toxicity of these drugs. Enhanced permeability retention of the conjugates at the site of the tumour is most likely the basis of local accumulation. In the studies reported, some objective clinical responses were seen without concomitant toxicity of the drugs (Table 1.3). Although the clinical responses until now seem rather disappointing, one should realize that the patients who undergo clinical testing with these conjugates often carry large tumour burdens which have been non-responsive towards other therapies available.

Table 1.3. Overview of recently reported clinical studies with miscellaneous drug targeting strategies.

Carrier	Drug incorporated	Disease	Clinical response[a,b]	Remarks[c]	Reference
Antibody	^{90}Y	Metastatic colon cancer	2/25 OR 2/25 PR 4/25 SD	Pre-targeted strategy: mAb NR-LU-10 – streptavidin, followed by biotin-gal-HSA clearing and biotin-DOTA-^{90}Y	[91]
		Non-Hodgkin lymphoma	3/7 CR 1/7 PR 2/7 OR	Pre-targeted strategy: mAb Rituximab – streptavidin, followed by biotin-N-acetyl-galactosamine clearing and biotin-DOTA^{90}Y	[92]
HPMA copolymer (PK1)	Doxorubicin	Various malignancies	2/36 PR 2/36 MR	Clinical proof of principle of decrease of doxorubicin toxicity when polymer bound	[15]
Poly(styrene-co-maleic acid) polymers (SMANCS)	Neocarzinostatin	Renal cell carcinoma patients	Improved survival, depending on tumour size	SMANCS dissolved in lipiodol contrast medium, administered via the renal artery	[93]
Human serum albumin	Methotrexate	Cancer patients	17 patients: 1 PR and 1 MR in RCC patients; 1 MR in pleural mesothelioma patient		[94]

[a] Number of patients responding/total number of evaluable patients. CR, complete remission/response; PR, partial response; MR, minor response; OR, objective response; SD, stable disease.
[b] RCC, renal cell carcinoma.
[c] mAb, monoclonal antibody.

1.5 Vaccination Strategies for Enhanced Immunity

Vaccination can be considered as a therapeutic modality that actively engages the immune system of the patient. It encompasses numerous principles derived from drug targeting research. Modification of the responses of the immune system may be an effective approach to improve the disease status of patients with a variety of diseases. Either prevention of autoimmune diseases or allergic responses, or enhancement of immune responses against infectious agents and tumour growth can be induced by these strategies [95,96]. Vaccination strategies can be divided into gene-, peptide-, protein- and cell-based strategies.

Antigen presenting cells (APC), particularly dendritic cells (DC), play a central role in the induction of the desired immune response [97]. For successful (antitumour) vaccination therapy, either an *in vitro* or an *in vivo* approach can be followed. In the *in vitro* approach, DCs from an animal or a patient are isolated and manipulated by transfection with DNA or RNA encoding (tumour) antigens or pro-inflammatory factors, or by loading the cells with proteins or peptides. After transfer back into the animal or patient, the cells can evoke antigen-specific immune responses [98]. Similarly, *in vitro* transfected tumour cells encoding a combination of genes involved in regulation of immune responses (e.g. MHC class II, a co-stimulatory molecule and a superantigen [96]) may serve as a vaccine.

The major drawback in the use of isolated DCs is the time-consuming isolation and culture methods required to obtain the cell type of interest. Therefore, *in vivo* vaccination strategies employing DCs have been developed. For this purpose, fusion proteins of e.g. tumour antigens and molecules that are specifically recognized by DCs are under investigation. The DC targeting molecules consist of e.g. DC-specific chemokines, mannose, the Fc moiety of immunoglobulin, cytotoxic T lymphocyte-associated antigen (CTLA-4) molecule [99–103]. They enable efficient uptake of the fusion protein by the DCs, induction of DC migratory capacity into the lympoid organs and maturation into a dedicated APC. As an example, studies by Biragyn *et al.* showed that active targeting of a tumour antigen into a receptor-mediated uptake route in APCs by the fusion of the antigen to APC-specific chemokines, elicited superior protection against a large tumour challenge in mice [99].

For a more detailed discussion of vaccines and vaccination strategies, the reader is referred to Volume 170 of *Immunological Reviews* (1999), in which various aspects of this area of research are discussed in greater detail.

1.6 Drug Targeting as a Research Tool to Study Disease

From an historical point of view, 'magic bullets' were initially proposed as novel therapeutic tools for treatment of disease. Another, similarly attractive application of drug targeting is its use as a tool to study e.g. mechanisms of disease pathology. By selectively manipulating one specific cell type or one specific protein, the role of this cell type or protein in disease progression can be determined. The advent of knock-out and transgenic animal models has provided the researcher in the laboratory with a powerful tool to investigate the relative contribution of gene products of interest. Yet, counter-regulatory processes that compensate for the loss of a certain gene product may significantly hamper correct interpretation of the results obtained in these models. Manipulating a gene in an appropriately matured animal using drug targeting strategies can therefore be a valuable tool.

It is not always appreciated that the application of *in vivo* gene therapy actually represents a drug delivery protocol. By the incorporation of plasmids into e.g. adenoviral or liposomal coats they are protected from degradation. Subsequent interaction with the target cells leads to intracellular delivery and then to nuclear localization of the plasmid. Although transfection efficiencies in *in vivo* gene therapy protocols are still not without problems, the transcription of genes of interest by transduced liver cells and muscle cells are quite satisfactory when using a strong promoter. Liver cells can be targeted by i.v. injection of the encapsulated plasmids, whereas skeletal muscle cells can be locally transfected by intramuscular injections. As an example, i.v. administration of recombinant adenovirus AdCMV-PDX-1 into mice resulted in specific transfection of liver cells. Expression of the PDX-1 gene, encoding a protein that regulates insulin gene expression, led to increased plasma levels of insulin and ameliorated hyperglycaemia in diabetic mice treated with streptozotocin [104]. Transduction of skeletal muscle with liposome formulated human hepatocyte growth factor (HGF) gene induced high plasma levels of protein, phosphorylation of the HGF target receptor, subsequent suppression of transforming growth factor β production and improved survival rates of animals suffering from severe liver cirrhosis [105].

Proteins have also been selectively targeted to study a disease process. Selective induction of oxidative vascular injury in the lungs was accomplished by targeting glucose oxidase (GOX) to the vasculature using anti-PECAM-1 (anti-CD31)-antibody–GOX conjugates. As the pulmonary vasculature represents approximately 30% of endothelial cells in the body and receives all the cardiac output, the majority of anti-PECAM-GOX conjugate accumulated in the endothelium in the lung. By generating H_2O_2 locally from glucose, severe endothelial cell damage was induced. The injury was associated with the production of the oxidative marker iPF$_{2a}$-III isoprostane. This model of targeted induction of oxidative stress can now be applied to study the effects of pharmacological agents [106].

Inhibition of tumour blood flow or the outgrowth of solid tumour vasculature by angiogenesis inhibition has been proposed as a powerful strategy to eradicate cancer ever since the recognition of the importance of blood supply for tumour growth. Subsequent development of compounds which inhibited the formation of new tumour blood vessels, demonstrated the dependence of solid tumour growth on blood vessel formation. It was however the selective induction of blood coagulation in the vasculature of a well-developed tumour by a drug targeting strategy that really demonstrated the potency of instantaneous inhibition of blood flow as a therapy for solid tumours [107]. Based on these and other 'proof of principle' studies, enormous research effort is now put into the identification of target epitopes on human tumour vasculature to further develop this strategy for clinical application (see Chapter 9 and reference [57] for a more detailed discussion on tumour vasculature targeting).

The inhibition of TNFα by neutralizing antibodies in rheumatoid arthritis patients is one of the few examples of targeted interference with disease activity in humans that can provide us with new insights into the pathophysiology of the disease [108]. The concomitant reduction in systemic levels of acute phase proteins, endothelial cell adhesion molecule expression in synovial biopsies and inflamed joint-associated blood vessel density all suggest a central role for TNFα as a driving force in RA. Furthermore, it provides evidence for the existence of a relationship between pro-inflammatory activity and the occurrence of angiogenesis, although it is at present unfeasible to conclude on cause and effect relations due to limited knowledge on this subject.

In summary it can be said that besides being important in the development of therapeutic strategies to combat disease with minimal toxicity and maximal effects, drug targeting may be of interest for more basic studies on the mechanistic background of diseases and the identification of new molecules as targets for therapeutic intervention.

1.7 Challenges in Drug Targeting Research

In 1984, Poznansky and Juliano published a critical review on the biological approaches to the controlled delivery of drugs [109]. Already at that time, they had envisioned the use of molecular biology and molecular immunology in the delineation of structure and function of important proteins and nucleic acids as being the true frontier in the drug delivery field. The elucidation of signal transduction routes in cells responding to a variety of environmental factors of disease has indeed led to the development of numerous NCEs to be used in future therapies. Furthermore, the understanding of protein folding and minimal amino acid sequences required for receptor recognition provided significant added value in the development of better carriers and ligands for drug targeting.

Numerous problems in the construction of clinically applicable drug targeting moieties still need to be solved. Of these issues, immunogenicity after repeated administration, counterproductive liver clearance, and production yields are the most important. Although the problem of immunogenicity is believed to have been solved for monoclonal antibody therapy by the development of humanized and fully human antibodies [110], for other carrier systems such as modified plasma proteins and peptide modified polymers, this remains an important issue.

For many drug targeting approaches, extensive uptake by the liver is undesirable when the target cells are of non-hepatic origin. In this respect, defining the optimal physicochemical characteristics that would circumvent extensive hepatic clearance of the drug targeting preparation would be of great value. This may, however, be wishful thinking more than a realistic aim, as hepatic elimination of (slightly) modified proteins is one method of protecting the body from non-functional or aged proteins. Furthermore, drug targeting preparations have to be eliminated from the body, as do regular drugs, so that their pharmacological activity can be controlled.

In general, validation of the targeting concept (therapeutic effectiveness, lack of toxicity even after prolonged periods of time) for each drug targeting conjugate that is under development should be performed *in vivo* in animal models of disease. Not only is this the only unambiguous approach for showing proof of the targeting principle, but it also takes into account the pharmacokinetic characteristics of the conjugate which determine the added value of the targeting strategy compared to non-targeted therapy (see Chapter 13 for pharmacokinetic considerations in drug targeting).

Most of the drug targeting strategies explored so far have been aimed at tumour cell killing or targeting of drugs and genes to the liver (i.e. Kupffer cell or hepatocyte). More recently the potential of targeting other cells of interest has been put forward, e.g. endothelial cells lining tumour blood vessels (Chapter 9) or the vasculature in chronic inflamed tissue (Chapter 7), and hepatic stellate cells in the fibrotic liver (Chapter 4) [10,57]. As the number of target cells in these cases is often relatively low and at present little knowledge exists regarding basic cell regulatory processes such as responses to drugs or endocytotic activity, all aspects of drug targeting (pharmacological activity of the drug in the target cell, cellular handling and kinetics of release of the drug) need to be addressed in full.

One of the most frequently heard criticisms is that drug targeting preparations cannot be administered orally and are therefore either of secondary or no interest for development into therapeutics. Although significant research effort has been put into the development of orally available protein therapeutics [111], there is a good chance that these preparations will not be feasible because of the inherent limitations in manipulating the physiological barriers in humans. Possibly, the understanding of the structural requirements for drug targeting preparations may lead to the development of minimized proteins that may be orally administered or become systemically available via e.g. pulmonary delivery (see Chapter 3). Furthermore, major efforts are being invested in the development of transdermal administration and programmable infusion systems. These systems may lower the barriers for long-term parenteral administration of drugs and may become common practice in pharmacotherapy. It is noteworthy that if a single dose of a targeted protein is potent enough to silence a disease for a prolonged period of time, the requirement of parenteral administration will not hamper its development. This is exemplified by the development of the TNFα neutralizing antibodies and receptors in the therapy of chronic inflammatory diseases such as rheumatoid arthritis and inflammatory bowel disease.

Finally, drug targeting technology should be integrated in the earliest phases of drug development. Many drugs that are pharmacologically active at low concentrations are withdrawn from the R & D pipeline due to toxicity in non-target cells. Would it be possible during an early phase in the development of a drug to construct cell-specific targeting conjugates a large number of drugs could still be candidates for development into clinical therapeutic strategies.

1.8 Conclusions

At present, many of the problems encountered during the development of drug targeting strategies for clinical application, especially for cancer therapy, have been identified, analysed and solved. Several drug targeting preparations have entered the phases of clinical testing and/or have now been marketed. However, these strategies should be subjected to continuous evaluation in the light of advances in the understanding of the numerous processes occuring in response to administration of the carriers and/or the drugs. New strategies under investigation will need to be optimized and extensively evaluated, taking advantage of the 'bench to bed-side' experience available today. Furthermore, in the coming years, combining expertise in the drug targeting field with the technological developments in molecular biology and molecular medicine will facilitate the elucidation of the cellular and molecular processes underlying disease.

Acknowledgements

Drs E. S. Vitetta, D. K. F. Meijer and M. Everts are acknowledged for critically reading the manuscript. Drs G. Storm, R. J. Kok and L. Beljaars are thanked for their help with the preparation of the figures.

References

[1] Breimer DD, *Adv. Drug Deliv. Rev.* **1998,** *33*, 265–268.
[2] Mastrobattista E, Koning GA, Storm G, *Adv. Drug Deliv. Rev.* **1999,** *40*, 103–127.
[3] Duzgunes N, Nir S, *Adv. Drug Deliv. Rev.* **1999,** *40*, 3–18.
[4] Storm G, Crommelin DJA, *PSTT* **1998,** *1*, 19–31.
[5] Köhler G, Milstein C, *Nature* **1975,** *256*, 495–497.
[6] Farah RA, Clinchy B, Herrera L, Vitetta ES, *Crit. Rev. Eukaryot. Gene Expr.* **1998,** *8*, 321–356.
[7] Willuda J, Honegger A, Waibel R, Schubiger PA, Stahel R, Zangemeister-Wittke U, Pluckthun A, *Cancer Res.* **1999,** *59*, 5758–5767.
[8] Worn A, Pluckthun A, *Biochemistry* **1999,** *38*, 8739–8750.
[9] Charles P, Elliott MJ, Davis D, Potter A, Kalden JR, Antoni C, Breedveld FC, Smolen JS, Eberl G, deWoody K, Feldmann M, Maini RN, *J. Immunol.* **1999,** *163*, 1521–1528.
[10] Beljaars L, Molema G, Schuppan D, Geerts A, de Bleser PJ, Weert B, Meijer DKF, Poelstra K, *J. Biol. Chem.* **2000,** *275*, 12743–12751.
[11] Molema G, Meijer DKF, *Adv. Drug Deliv. Rev.* **1994,** *14*, 25–50.
[12] Ogawara K, Hasegawa S, Nishikawa M, Takakura Y, Hashida M, *J. Drug Target.* **1999,** *6*, 349–360.
[13] Stehle G, Sinn H, Wunder A, Schrenk HH, Schutt S, Maier Borst W, Heene DL, *Anticancer Drugs* **1997,** *8*, 677–685.
[14] Maeda H, Wu J, Sawa T, Matsumura Y, Hori K, *J. Control. Rel.* **2000,** *65*, 271–284.
[15] Vasey PA, Kaye SB, Morrison R, Twelves C, Wilson P, Duncan R, Thomson AH, Murray LS, Hilditch TE, Murray T, Burtles S, Fraier D, Frigerio E, Cassidy J, *Clin. Cancer Res.* **1999,** *5*, 83–94.
[16] Gianasi E, Wasil M, Evagorou EG, Keddle A, Wilson G, Duncan R, *Eur. J. Cancer* **1999,** *35*, 994–1002.
[17] Sawa T, Wu J, Akaike T, Maeda H, *Cancer Res.* **2000,** *60*, 666–671.
[18] Richardson SC, Kolbe HV, Duncan R, *Int. J. Pharm.* **1999,** *178*, 231–243.
[19] Malik N, Wiwattanapatapee R, Klopsch R, Lorenz K, Frey H, Weener JW, Meijer EW, Paulus W, Duncan R, *J. Control. Rel.* **2000,** *65*, 133–148.
[20] Lu ZR, Kopeckova P, Kopeček J, *Nature Biotechnol.* **1999,** *1*, 1101–1104.
[21] Kader A, Davis PJ, Kara M, Liu H, *J. Control. Rel.* **1998,** *55*, 231–243.
[22] Meijer DKF, Molema G, *Sem. Liver Dis.* **1995,** *15*, 202–256.
[23] de Vrueh RL, Rump ET, van De BE, van Veghel R, Balzarini J, Biessen EA, van Berkel TJ, Bijsterbosch MK, *Antimicrob. Agents Chemother.* **2000,** *44*, 477–483.
[24] Van Velzen AG, Da Silva RP, Gordon S, van Berkel TJ, *Biochem. J.* **1997,** *322 (Pt 2)*, 411–415.
[25] Ling W, Lougheed M, Suzuki H, Buchan A, Kodama T, Steinbrecher UP, *J. Clin. Invest.* **1997,** *100*, 244–252.
[26] Utreja S, Khopade AJ, Jain NK, *Pharm. Acta Helv.* **1999,** *73*, 275–279.
[27] Muller RH, Mader K, Gohla S, Eur. *J. Pharm. Biopharm.* **2000,** *50*, 161–177.
[28] Aynie I, Vauthier C, Chacun H, Fattal E, Couvreur P, *Antisense Nucl. Acid Drug Dev.* **1999,** *9*, 301–312.
[29] Soma CE, Dubernet C, Bentolila D, Benita S, Couvreur P, *Biomaterials* **2000,** *21*, 1–7.
[30] Allemann E, Leroux J, Gurny R, *Adv. Drug Deliv. Rev.* **1998,** *34*, 171–189.
[31] Pecquet S, Leo E, Fritsche R, Pfeifer A, Couvreur P, Fattal E, *Vaccine* **2000,** *18*, 1196–1202.
[32] Jones M, Leroux J, *Eur. J. Pharm. Biopharm.* **1999,** *48*, 101–111.
[33] Yokoyama M, Okano T, Sakurai Y, Fukushima S, Okamoto K, Kataoka K, *J. Drug Target.* **1999,** *7*, 171–186.
[34] Vallera DA, Jin N, Baldrica JM, Panoskaltsis-Mortari A, Chen SY, Blazar BR, *Cancer Res.* **2000,** *60*, 976–984.
[35] Wiedle G, Johnson-Leger C, Imhof BA, *Cancer Res.* **1999,** *59*, 5255–5263.
[36] Asahara T, Murohara T, Sullivan A, Silver M, van der Zee R, Li T, Witzenbichler B, Schatteman G, Isner JM, *Science* **1997,** *275*, 964–967.
[37] Asahara T, Takahashi T, Masuda H, Kalka C, Chen D, Iwaguro H, Inai Y, Silver M, Isner JM, *EMBO J.* **1999,** *18*, 3964–3972.
[38] Isner JM, Asahara T, *J. Clin. Invest* **1999,** *103*, 1231–1236.
[39] Moghimi SM, Rajabi-Siahboomi AR, *Adv. Drug Deliv. Rev.* **2000,** *41*, 129–133.
[40] Stayton PS, Hoffman AS, Murthy N, Lackey C, Cheung C, Tan P, Klumb LA, Chilkoti A, Wilbur FS, Press OW, *J. Control. Rel.* **2000,** *65*, 203–220.
[41] Schwarze SR, Ho A, Vocero-Akbani A, Dowdy SF, *Science* **1999,** *285*, 1569–1572.

[42] Dilber MS, Phelan A, Aints A, Mohamed AJ, Elliott G, Smith CI, O'Hare P, *Gene Ther.* **1999,** *6,* 12–21.
[43] Luo D, Saltzman WM, *Nature Biotechnol.* **2000,** *18,* 33–37.
[44] Elliott G, O'Hare P, *J. Virol.* **2000,** *74,* 2131–2141.
[45] Sheldon K, Liu D, Ferguson J, Gariepy J, *Proc. Natl. Acad. Sci. USA* **1995,** *92,* 2056–2060.
[46] Singh D, Bisland SK, Kawamura K, Gariepy J, *Bioconjug. Chem.* **1999,** *10,* 745–754.
[47] Murphy MP, Smith RAJ, *Adv. Drug Deliv. Rev.* **2000,** *41,* 235–250.
[48] Smith RA, Porteous CM, Coulter CV, Murphy MP, *Eur. J Biochem.* **1999,** *263,* 709–716.
[49] Renold A, Koehler CM, Murphy MP, *FEBS Lett.* **2000,** *465,* 135–140.
[50] Grindley J, Ogden J, **1999,** *SCRIP* 59–646.
[51] Vile RG, Russell SJ, Lemoine NR, *Gene Ther.* **2000,** *7,* 2–8.
[52] Li S, Huang L, *Gene Ther.* **2000,** *7,* 31–34.
[53] Lasic DD, *Trends Biotechnol.* **1998,** *16,* 307–321.
[54] Glennie MJ, Johnson PW, *Immunol. Today* **2000,** *21,* 403–410.
[55] Cragg MS, French RR, Glennie MJ, *Curr. Opin. Immunol.* **1999,** *11,* 541–547.
[56] Schlaeppi JM, Wood JM, *Cancer Metast. Rev* **1999,** *18,* 473–481.
[57] Griffioen AW, Molema G, *Pharmacol. Rev* **2000,** *52,* 237–268.
[58] Richards J, Auger J, Peace D, Gale D, Michel J, Koons A, Haverty T, Zivin R, Jolliffe L, Bluestone JA, *Cancer Res.* **1999,** *59,* 2096–2101.
[59] Cortes J, O'Brien S, Estey E, Giles F, Keating M, Kantarjian H, *Invest. New Drugs* **1999,** *17,* 81–87.
[60] Yeo W, Chan KK, Mukwaya G, Ross M, Leung WT, Ho S, Chan AT, Johnson PJ, *Cancer Chemother. Pharmacol.* **1999,** *44,* 124–130.
[61] Lippens RJ, *Pediatr. Hematol. Oncol.* **1999,** *16,* 131–139.
[62] Hubert A, Lyass O, Pode D, Gabizon A, *Anticancer Drugs* **2000,** *11,* 123–127.
[63] Toma S, Tucci A, Villani G, Carteni G, Spadini N, Palumbo R, *Anticancer Res.* **2000,** *20,* 485–491.
[64] Koukourakis MI, Koukouraki S, Giatromanolaki A, Archimandritis SC, Skarlatos J, Beroukas K, Bizakis JG, Retalis G, Karkavitsas N, Helidonis ES, *J. Clin. Oncol.* **1999,** *17,* 3512–3521.
[65] Cheung TW, Remick SC, Azarnia N, Proper JA, Barrueco JR, Dezube BJ, *Clin. Cancer Res.* **1999,** *5,* 3432–3437.
[66] Valero V, Buzdar AU, Theriault RL, Azarnia N, Fonseca GA, Willey J, Ewer M, Walters RS, Mackay B, Podoloff D, Booser D, Lee LW, and Hortobagyi GN, *J. Clin. Oncol.* **1999,** *17,* 1425–1434.
[67] Boletis JN, Pefanis A, Stathakis C, Helioti H, Kostakis A, Giamarellou H, *Clin. Infect. Dis.* **1999,** *28,* 1308–1309.
[68] Melamed R, Leibovitz E, Abramson O, Levitas A, Zucker N, Gorodisher R, *Scand. J. Infect. Dis.* **2000,** *32,* 86–89.
[69] Kelsey SM, Goldman JM, McCann S, Newland AC, Scarffe JH, Oppenheim BA, Mufti GJ, *Bone Marrow Transplant.* **1999,** *23,* 163–168.
[70] Abraham E, Baughman R, Fletcher E, Heard S, Lamberti J, Levy H, Nelson L, Rumbak M, Steingrub J, Taylor J, Park YC, Hynds JM, Freitag J, *Crit. Care Med.* **1999,** *27,* 1478–1485.
[71] Maini R, St Clair EW, Breedveld F, Furst D, Kalden J, Weisman M, Smolen J, Emery P, Harriman G, Feldmann M, Lipsky P, *Lancet* **1999,** *354,* 1932–1939.
[72] Haug CE, Colvin RB, Delmonico FL, Auchincloss Jr H, Tolkoff-Rubin N, Preffer FI, Rothlein R, Norris S, Scharschmidt L, Cosimi AB, *Transplantation* **1993,** *55,* 766–772.
[73] Kelly TA, Jeanfavre DD, McNeil DW, Woska Jr JR, Reilly PL, Mainolfi EA, Kishimoto KM, Nabozny GH, Zinter R, Bormann BJ, Rothlein R, *J. Immunol.* **1999,** *163,* 5173–5177.
[74] Glover JM, Leeds JM, Mant TG, Amin D, Kisner DL, Zuckerman JE, Geary RS, Levin AA, Shanahan Jr WR, *J. Pharmacol. Exp. Ther.* **1997,** *282,* 1173–1180.
[75] Trail PA, Bianchi AB, *Curr. Opin. Immunol.* **1999,** *11,* 584–588.
[76] Frankel AE, Kreitman RJ, Sausville EA, *Clin. Cancer Res.* **2000,** *6,* 326–334.
[77] DeNardo SJ, Kroger LA, DeNardo GL, *Curr. Opin. Immunol.* **1999,** *11,* 563–569.
[78] Molema G, Kroesen BJ, Helfrich W, Meijer DKF, de Leij LFMH, *J. Control. Rel.* **2000,** *64,* 229–239.
[79] van Spriel AB, van Ojik HH, van de Winkel JG, *Immunol. Today* **2000,** *21,* 391–397.
[80] Davis CG, Gallo ML, Corvalan JR, *Cancer Metast. Rev* **1999,** *18,* 421–425.
[81] Hainsworth JD, Burris HA, Morrissey LH, Litchy S, Scullin DC, Bearden JD, Richards P, Greco FA, *Blood* **2000,** *95,* 3052–3056.
[82] Zompi S, Tulliez M, Conti F, Leblond V, Gaulard P, Blanche P, Durand F, Ghandi D, Dreyfus F, Louvel A, Calmus Y, Bouscary D, *J. Hepatol.* **2000,** *32,* 521–527.

[83] Milpied N, Vasseur B, Parquet N, Garnier JL, Antoine C, Quartier P, Carret AS, Bouscary D, Faye A, Bourbigot B, Reguerre Y, Stoppa AM, Bourquard P, Hurault dL, Dubief F, Mathieu-Boue A, Leblond V, *Ann. Oncol.* **2000,** *11 (Suppl. 1)*, 113–116.
[84] Foran JM, Rohatiner AZ, Cunningham D, Popescu RA, Solal-Celigny P, Ghielmini M, Coiffier B, Johnson PW, Gisselbrecht C, Reyes F, Radford JA, Bessell EM, Souleau B, Benzohra A, Lister TA, *J. Clin. Oncol.* **2000,** *18*, 317–324.
[85] Baselga J, Tripathy D, Mendelsohn J, Baughman S, Benz CC, Dantis L, Sklarin NT, Seidman AD, Hudis CA, Moore J, Rosen PP, Twaddell T, Henderson IC, Norton L, *Sem. Oncol.* **1999,** *26*, 78–83.
[86] Tokuda Y, Watanabe T, Omuro Y, Ando M, Katsumata N, Okumura A, Ohta M, Fujii H, Sasaki Y, Niwa T, Tajima T, *Br. J. Cancer* **1999,** *81*, 1419–1425.
[87] Pegram MD, Lipton A, Hayes DF, Weber BL, Baselga JM, Tripathy D, Baly D, Baughman SA, Twaddell T, Glaspy JA, Slamon DJ, *J. Clin. Oncol.* **1998,** *16*, 2659–2671.
[88] Oh CJ, Das KM, Gottlieb AB, *J. Am. Acad. Dermatol.* **2000,** *42*, 829–830.
[89] Malley R, DeVincenzo J, Ramilo O, Dennehy PH, Meissner HC, Gruber WC, Sanchez PJ, Jafri H, Balsley J, Carlin D, Buckingham S, Vernacchio L, Ambrosino DM, *J. Infect. Dis.* **1998,** *178*, 1555–1561.
[90] Lincoff AM, Califf RM, Moliterno DJ, Ellis SG, Ducas J, Kramer JH, Kleiman NS, Cohen EA, Booth JE, Sapp SK, Cabot CF, Topol EJ, *N. Engl. J. Med.* **1999,** *341*, 319–327.
[91] Knox SJ, Goris ML, Tempero M, Weiden PL, Gentner L, Breitz H, Adams GP, Axworthy D, Gaffigan S, Bryan K, Fisher DR, Colcher D, Horak ID, Weiner LM, *Clin. Cancer Res.* **2000,** *6*, 406–414.
[92] Weiden PL, Breitz HB, Press O, Appelbaum JW, Bryan JK, Gaffigan S, Stone D, Axworthy D, Fisher D, Reno J, *Cancer Biother. Radiopharm.* **2000,** *15*, 15–29.
[93] Tsuchiya K, Uchida T, Kobayashi M, Maeda H, Konno T, Yamanaka H, *Urology* **1999,** *55*, 495–500.
[94] Hartung G, Stehle G, Sinn H, Wunder A, Schrenk HH, Heeger S, Kranzle M, Edler L, Frei E, Fiebig HH, Heene DL, Maier-Borst W, Queisser W, *Clin. Cancer Res.* **1999,** *5*, 753–759.
[95] Ramsay AJ, Kent SJ, Strugnell RA, Suhrbier A, Thomson SA, Ramshaw IA, *Immunol. Rev* **1999,** *171*, 27–44.
[96] Ostrand-Rosenberg S, Pulaski BA, Clements VK, Qi L, Pipeling MR, Hanyok LA, *Immunol. Rev* **1999,** *170*, 101–114.
[97] Banchereau J, Briere F, Caux C, Davoust J, Lebecque S, Liu YJ, Pulendran B, Palucka K, *Annu. Rev. Immunol.* **2000,** *18*, 767–811.
[98] Nair SK, Heiser A, Boczkowski D, Majumdar A, Naoe M, Lebkowski JS, Vieweg J, Gilboa E, *Nature Med.* **2000,** *6*, 1011–1017.
[99] Biragyn A, Tani K, Grimm MC, Weeks S, Kwak LW, *Nature Biotechnol.* **1999,** *17*, 253–258.
[100] Diebold SS, Kursa M, Wagner E, Cotten M, Zenke M, *J. Biol. Chem.* **1999,** *274*, 19087–19094.
[101] Machy P, Serre K, Leserman L, *Eur. J. Immunol.* **2000,** *30*, 848–857.
[102] van Bergen J, Ossendorp F, Jordens R, Mommaas AM, Drijfhout JW, Koning F, *Immunol. Rev.* **1999,** *172*, 87–96.
[103] Biragyn A, Kwak LW, *Nature Med.* **2000,** *6*, 966–968.
[104] Ferber S, Halkin A, Cohen H, Ber I, Einav Y, Goldberg I, Barshack I, Seijffers R, Kopolovic J, Kaiser N, Karasik A, *Nature Med.* **2000,** *6*, 568–572.
[105] Ueki T, Kaneda Y, Tsutsui H, Nakanishi K, Sawa Y, Morishita R, Matsumoto K, Nakamura T, Takahashi H, Okamoto E, Fujimoto J, *Nature Med.* **1999,** *5*, 226–230.
[106] Christofidou-Solomidou M, Pietra GG, Solomides CC, Arguiris E, Harshaw D, Fitzgerald GA, Albelda SM, Muzykantov VR, *Am. J. Physiol. Lung Cell. Mol. Physiol.* **2000,** *278*, L794–L805.
[107] Huang X, Molema G, King S, Watkins L, Edgington TS, Thorpe PE, *Science* **1997,** *275*, 547–550.
[108] Maini RN, Taylor PC, Paleolog E, Charles P, Ballara S, Brennan FM, Feldmann M, *Ann. Rheum. Dis.* **1999,** *58 (Suppl. 1)*, I56–I60.
[109] Poznansky MJ, Juliano RL, *Pharmacol. Rev.* **1984,** *36*, 277–336.
[110] Clark M, *Immunol. Today* **2000,** *21*, 397–402.
[111] Stoll BR, Leipold HR, Milstein S, Edwards DA, *J. Control. Rel.* **2000,** *64,* 217–228.
[112] Lasic DD, *Sci. Med.* **1996,** *3*, 34–43.

2 Brain-Specific Drug Targeting Strategies

Ulrich Bickel, Young-Sook Kang, Jörg Huwyler

2.1 Introduction

The brain is unique as a target organ for drug delivery: while it ranks amongst organs with the greatest blood supply and receives about 20% of the cardiac output in humans, access to the tissue is highly restricted by a tight vascular barrier, the blood–brain barrier (BBB). Due to the existence of the BBB, the transport of potentially neuroactive drugs from blood into brain is rarely blood-flow limited (for example for highly diffusible drugs like diazepam), but is in many cases extraction-limited. Therefore, drug delivery/targeting to the brain is primarily a permeability problem.

The major neurological diseases affecting the brain may be categorized as neurodegenerative, cerebrovascular, inflammatory (infectious or autoimmune) and brain tumours. Progress is being made for most of these disorders in terms of identification of epidemiological risk factors and of the aetiology and pathophysiological mechanisms, as will be briefly discussed. This information will facilitate the development of more efficient pharmacotherapy [1]. Solving the BBB delivery problem must be an integral part of these efforts.

The present chapter discusses approaches to brain drug delivery for small molecular weight drugs and the macromolecular drugs which become available as a result of advances in molecular biology and biotechnology, including peptides, monoclonal antibodies (mAb), and DNA- or RNA-based therapeutics (oligonucleotides, genes).

2.2 Overview of Central Nervous System Diseases

2.2.1 Neurodegenerative Diseases

2.2.1.1 Alzheimer Disease (AD)

AD affects an increasing number of elderly people. Its prevalence is currently estimated at about 4 million in the USA alone [2]. The lack of an early diagnosis is by itself impeding progress in the therapy of the disease. Although the aetiology is still not fully elucidated, overwhelming evidence has accumulated for the crucial involvement of the Aβ-peptide [3] in the pathophysiological process. Aβ deposits occur as cerebrovascular amyloid or as parenchymal plaques (senile plaques). The 39–43-amino acid peptide is derived from the amyloid precursor protein (APP) [4], an ubiquitous cellular transmembrane protein of unknown function which is present in several splice variants. APP mutations have been identified as rare causes of familial AD, but mutations in other proteins, the presenilins PS-1 and

PS-2, are found as the most frequent causes of early onset familial AD cases [5]. The presenilins are involved in APP processing by enzymes called secretases [6].

The massive amyloid deposition in the form of parenchymal plaques and/or in cerebrovascular amyloid (cerebral amyloid angiopathy) is associated with neuronal loss and dysfunction. In particular the cholinergic neurons of the basal forebrain, which are involved in the memory processes, are affected and neuron loss in these nuclei accounts for some of the AD symptoms.

According to clinico-pathological studies, the diagnosis of probable AD can be made with an accuracy of about 85% while the diagnosis of definite AD still relies on post mortem neuropathological proof, apart from the rare case where a brain biopsy can be performed. A more sensitive and accurate diagnostic method, which should also be non-invasive, would be highly desirable [2].

Currently the only specific pharmacological therapeutic option available for AD patients is treatment with cholinesterase inhibitors, which provide moderate benefits in a subset of patients for a limited period [7]. More efficient future therapeutic strategies may be directed at the metabolic events resulting in Aβ accumulation, for example by inhibition of β- or γ-secretase [7], or at the prevention of neuronal loss by neurotrophin therapy [6]. The availability of transgenic mouse models of the disease, such as mice overexpressing APP mutants [8], and the utilization of primate models of cerebral amyloid [9] permits preclinical testing of novel diagnostic and therapeutic approaches.

2.2.1.2 Parkinson's Disease

The etiology of the clinical syndrome is heterogeneous and ranges from physical insults (head trauma) and specific chemical toxicity (1-methyl-4-phenylpyridinium, MPTP) to as yet unknown causes in the majority of patients (idiopathic Parkinson's disease, PD). At the cellular level, PD manifests in a progressive loss of midbrain dopaminergic neurons of the substantia nigra over several years and a concomitant development of a dopaminergic deficit in the projection area, the striatum. Among the factors suspected of contributing to the preferential vulnerability of dopaminergic neurons, is the oxidative stress that is associated with dopamine metabolism. Several mutations in a single gene located on chromosome 4 were identified in rare cases of familial PD. The gene codes for a protein of unknown function, α-synuclein, which is deposited in neurons of brains afflicted with PD as the main constituent of intracellular deposits known as Lewy bodies [10].

Although current drug therapy of PD is more successful compared to AD, it does not stop the degenerative process, and pharmacotherapy unfortunately loses effectiveness with progression of the disease. Drugs aimed at the reduction of the dopaminergic deficit (L-DOPA, monoamine oxidase B inhibitors, dopamine agonists) remain the mainstays of symptomatic drug treatment. In addition to a loss of effect over time, the therapy is also accompanied by an increase in frequency and severity of side-effects [11]. Besides neurosurgical and electrophysiological based approaches, treatment options for the future include neuroprotective therapies. Analogous to efforts to save the cholinergic system affected in AD, neurotrophic factors with the ability to promote the survival of dopaminergic central nervous system (CNS) neurons have been identified *in vitro* and in animal models. Here again, the drug delivery dilemma of macromolecular drugs is an obstacle to the progression from experimental treatment to clinical testing.

Suitable animal models of the disease are available in both rodents and primates. A progressive dopaminergic degeneration of substantia nigra neurons is the result of intrastriatal 6-OH-dopamine injection in rats, and MPTP causes a Parkinson-like syndrome in monkeys [12]. The availability of transgenic mice [13] based on α-synuclein further expands the experimental options.

2.2.2 Cerebrovascular Disease

Ischaemic stroke is the third leading cause of death in industrialized countries. The debilitating or lethal consequences of transient or temporary reductions in cerebral blood flow are not only caused by necrosis in the infarct zone itself, but also by pathophysiological events in the peri-infarct zone [14]. Apparently, the release of inflammatory mediators such as cytokines and NO contributes to tissue inflammatory injury. There is also evidence for apoptosis in the peri-infarct zone. These processes offer novel targets for therapeutic strategies. In this respect, the potential of neurotrophic factor treatment is described in Section 2.4.2.6.

2.2.3 Brain Tumours

Gliomas are the most frequent primary malignant brain tumours and range among the second to fourth most frequent causes of cancer deaths in younger age groups of men and women (from ages under 15 to 54 years, [15]). They present a particular diagnostic and therapeutic problem due the existence of a tight vascular barrier and the poor response to chemotherapy. Although the blood–tumour barrier is compromised compared to the normal BBB, it is nevertheless tighter than vascular barriers in peripheral organs. Methods of opening the blood–tumour barrier are therefore required [16]. This point will be addressed in Section 2.4.2.4. Strategies for targeting to tumour cells and targeting tumour vasculature are discussed in detail in Chapters 8 and 9. A specific opportunity for targeting glioma cells is their tendency to overexpress functional epidermal growth factor (EGF) receptors, as described in Section 2.4.2.7.

2.2.4 HIV Infection

Although progress in antiretroviral therapy has reduced viral load in patients and also the frequency of HIV-1-related disease of the CNS (AIDS–dementia complex) [17] a principle problem remains. Once the virus has entered the brain via infected macrophages or as a free virus particle, it can infect microglial cells and astrocytes [18]. Current therapeutics penetrate the BBB poorly, and the brain is therefore a sanctuary from where endogenous reinfection may occur, even if systemic therapy was able to eliminate the virus in the periphery. Hence, if eventual virus eradication is a therapeutic goal in HIV-infected patients, the development of brain drug delivery strategies is crucial.

2.3 BBB Biology and Pharmacology

Two major barrier systems separate the central nervous system from the circulation: the BBB and the blood–cerebrospinal fluid barrier (B-CSF-B). These barriers have distinct morphological and physiological characteristics, according to their different tasks. Figure 2.1 highlights the salient features of both barrier systems.

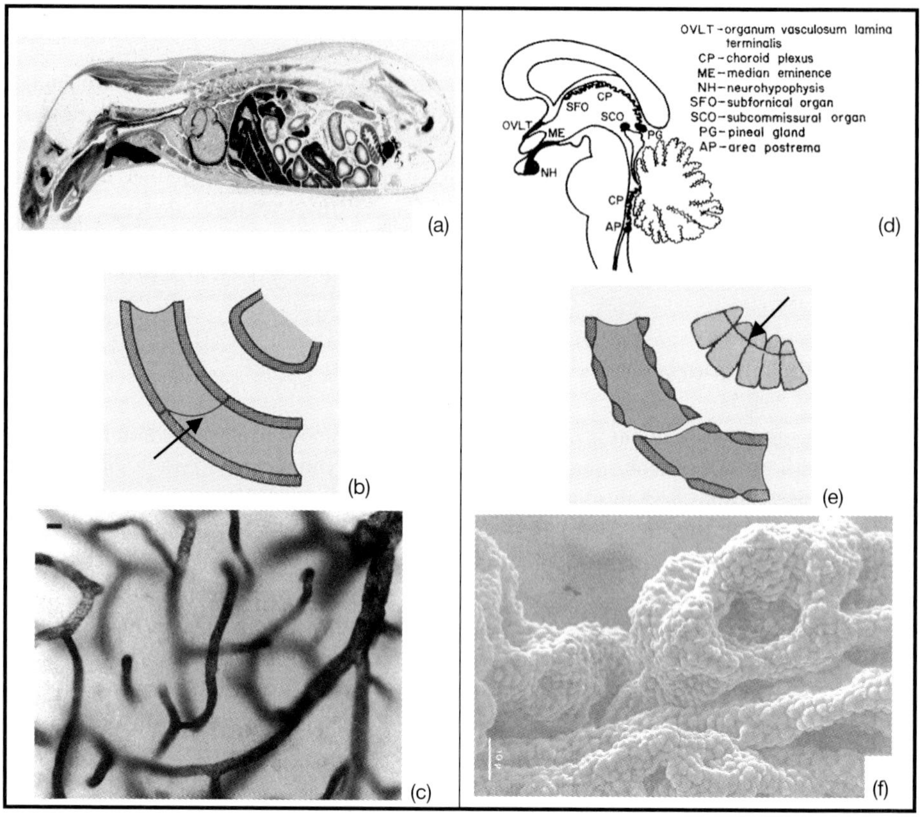

Figure 2.1. The two main barrier systems in the mammalian brain are the blood–brain barrier (BBB) (a–c) and the blood–cerebrospinal fluid barrier (d–f). (a) Autoradiograph of a sagittal section through a rat, which received an intravenous injection of [^{14}C]-histamine 15 min earlier. The tracer distributes to all organs except for the central nervous system, where passage is prevented by the tight BBB. (b) The site of the permeability barrier at the level of the tight junctions between microvascular endothelial cells (arrow), shown schematically. (c) Silver enhanced immunogold staining of the dense brain capillary tree. Endothelial cells were labelled by perfusion *in vivo* of a rat brain with a gold-conjugated antibody to the transferrin receptor followed by perfusion fixation. Scale bar = 10 μm. (d) Localization of the circumventricular organs on a schematic median sagittal section of the human brain. (e) Schematic diagram showing the difference between the barrier site and the BBB. The choroid plexus has leaky endothelium in its capillaries, yet a tight epithelial layer (arrow). (f) Scanning electron micrograph showing the epithelial cells of the ventricular surface of the choroid plexus. Reproduced with permission from references [37,118-120].

The dual purpose of the BBB is to ensure a constant internal milieu within the CNS and to provide the essential nutrient supply. The anatomical site of the BBB is the endothelial lining of the brain microvasculature [19]. Their endothelial cells are connected by complex strands of tight junctions and form an epithelial-like, high resistance barrier. *In vivo* estimates of the transendothelial resistance range up to values of 8000 ohm cm^{-2} [20]. Luminal and abluminal plasma membranes of the endothelial cells therefore represent the diffusion barrier for solutes between the circulation and brain interstitial fluid. The endothelium rests on a basement membrane of approximately 20 nm in thickness, and that same basement membrane encloses another cell type, the pericyte, whose numbers are about 1/3 of that of endothelial cells. The abluminal surface of the basement membrane of intraparenchymal microvessels is more than 99% invested by astrocyte foot processes. Glial sheathing and pericytes are essential for the induction and integrity of the BBB, although they do not directly control permeability [21,22].

The BBB may be imagined as a very thin membranous structure covering a large surface area that amounts, in the adult human brain with a weight of about 1200 g, to an area of the BBB of about 12 m^2 (approximately 100 cm^2 g^{-1} tissue on average). At the same time, the capillary volume and the endothelial cell volume itself constitute only approximately 1% and 0.1% of the tissue volume, respectively [23].

In contrast to other organs, the endothelial cells of the brain microvessels have no fenestrations or pores, and there is only very little pinocytosis [19]. An important angio-architectural feature is the mean intercapillary distance, which is in the order of only 40 µm in human brain [24], as is illustrated in Figure 2.1c. This allows for almost instantaneous solute equilibration throughout the brain interstitial space for small molecules such as nutrients (glucose, oxygen). Once the endothelial barrier has been passed, a diffusion distance in that range is no obstacle for macromolecules either.

The other barrier, the B-CSF-B, is found within the circumventricular organs (CVO), shown in Figure 2.1d–f. As the name implies, these are specialized parts of tissue in or around the brain ventricles. The largest CVO is the choroid plexus, which is found in the lateral ventricles, on the roof of the third ventricle, and in the fourth ventricle. The choroid plexus has the properties of a secreting epithelium, which actively forms the cerebrospinal fluid (CSF). Its leaky capillaries lack tight endothelial junctions. While the capillaries are similar to highly porous capillaries perfusing peripheral organs, the diffusion barrier is found at the level of the plexus epithelial cells which are connected by tight junctions. Though the composition of the CSF resembles a plasma ultrafiltrate, not all concentrations of electrolytes, nutrients, and proteins, show a simple linear relation to plasma concentrations [25]. The surface area of the B-CSF-B including the choroid plexus is estimated at 0.021 m^2 [26]. Other CVOs include the neurohypophysis, the median eminence, the organum vasculosum laminae terminalis, the subfornical organ, the subcommissural organ, the area postrema, and the pineal gland. The capillaries in these tiny areas are leaky and the diffusion barrier with tight intercellular junctions is found at the level of the covering ependymal cells, which locally seal the underlying tissue from the ventricular space. The ependyma in the rest of the ventricular surface is not connected by tight junctions.

2.3.1 Physiological Transport Systems

2.3.1.1 Nutrient Carriers Versus Diffusion-mediated Uptake

As many of the essential nutrients of the brain (glucose, amino acids, nucleotides and others) are highly hydrophilic and would not cross the BBB by diffusion in sufficient amounts, the endothelial cells are endowed with membrane-bound specific transport proteins for the facilitated uptake of these substances from the blood. Figure 2.2 gives examples of the brain uptake of typical transport substrates in comparison to (drug-) compounds that depend on passive diffusion-mediated uptake. Diffusional permeability is related to lipophilicity and size or molecular weight for compounds below 400–600 Da [27]. Transporters bind their substrate molecules and change their conformation or temporarily open up a pore, allowing passage across the plasma membrane. One of the transport proteins at the BBB is the hexose transporter, GLUT1 [28]. It is a member of the Na^+-independent family of glucose transporters and is highly expressed at the BBB and at the B-CSF-B. The K_M for glucose (2–5 mM) coincides with the physiological range of blood glucose and is lower than that for

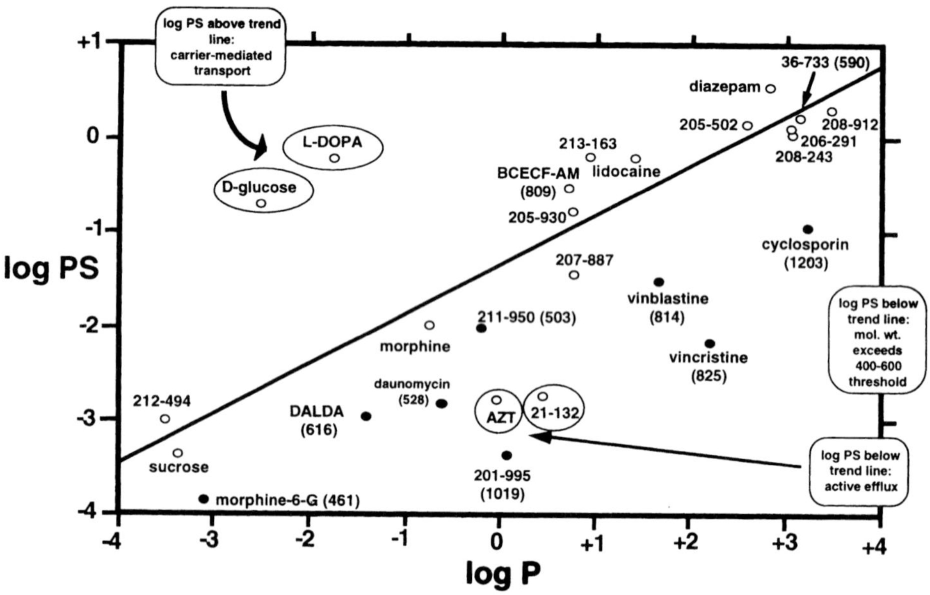

Figure 2.2. Specific transporters provide brain uptake of substrates or drugs, which exceeds the uptake by lipid-mediated diffusion through the blood–brain barrier (BBB). This is obvious from the position of these substrates 3–4 log orders above the regression line between lipophilicity (expressed as the log of octanol–water partition coefficient, log P) and BBB permeability (expressed as log of the permeability surface area product, log PS). The linear relationship between log PS and log P holds for substances with molecular weights below 400–600 Da. Drugs falling 1–3 log orders below the regression line are substrates of efflux mechanisms and/or have high molecular weights (given in parentheses). The numbered compounds are a series of somatostatin analogues. AZT = azidothymidine. Reproduced with permission from reference [121].

other monosaccharides. Accordingly, GLUT1 mediates facilitated glucose uptake, glucose being the main energy source for the brain.

The transport of amino acids at the BBB differs depending on their chemical class and the dual function of some amino acids as nutrients and neurotransmitters. Essential large neutral amino acids are shuttled into the brain by facilitated transport via the large neutral amino acid transporter (LAT) system [29] and display rapid equilibration between plasma and brain concentrations on a minute time scale. The LAT-system at the BBB shows a much lower K_M for its substrates compared to the analogous L-system of peripheral tissues and its mRNA is highly expressed in brain endothelial cells (100-fold abundance compared to other tissues). Cationic amino acids are taken up into the brain by a different facilitative transporter, designated as the y^+ system, which is present on the luminal and abluminal endothelial membrane. In contrast, active Na^+-dependent transporters for small neutral amino acids (A-system; ASC-system) and cationic amino acids (B^{0+} system), appear to be confined to the abluminal surface and may be involved in removal of amino acids from brain extracellular fluid [30]. Carrier-mediated BBB transport includes monocarboxylic acids (pyruvate), amines (choline), nucleosides (adenosine), purine bases (adenine), panthotenate, thiamine, and thyroid hormones (T3), with a representative substrate given in parentheses [31].

2.3.1.2 Efflux Systems

The sodium-dependent X^- system is localized to the abluminal membrane and apparently functions as an efflux mechanism for the excitatory amino acids aspartate and glutamate from brain interstitial fluid. In accordance with such a direction of transport, acidic amino acids do not undergo rapid uptake into the brain under physiological conditions. Other efflux pumps have recently been shown to exist at the BBB for substances belonging to diverse chemical classes. The most studied example to date is the multi-drug resistance protein (MDR), also known as P-glycoprotein or P-gp, with its broad substrate specificity [32]. P-gp and structurally related transporters are members of the large ATP-binding cassette family of membrane pumps. Several organic anion transport polypeptides have recently been cloned from animal and human tissue. The substrate specificity and direction of transport of these systems *in vivo*, which are expressed in liver, kidney, intestines as well as in the brain at the BBB and B-CSF-B [33], has not been fully elucidated. In any case, the importance of active transport systems at the BBB is rapidly emerging. Evidence supports the view that many more small molecular weight substances (including drugs) than previously thought, apparently do not cross the BBB by simple diffusion only, but are subject to active transport. The inhibition of BBB efflux systems as a strategy for brain drug delivery has therefore been proposed [34].

2.3.1.3 Receptor- and Absorptive-mediated Uptake

Macromolecules such as polypeptides and proteins are excluded from uptake by diffusion or through pores at the BBB. However, the presence of receptor-mediated uptake and transport into the brain has been found for a number of substances. Specific receptors for insulin and transferrin were initially described. These receptors have been identified at the molecular level as being identical to their counterparts in peripheral tissues [35,36]. However, at the BBB the receptors mediate transcytosis of their respective ligands through the endothelial cell as opposed to endocytosis in other cells.

The process of transcytosis is illustrated in Figure 2.3 for the transferrin receptor (TfR) [37]. The receptor is heavily expressed at the BBB compared to other vascular beds [38]. Transferrin or a monoclonal antibody to the extracellular domain of the receptor protein will bind from the luminal side of the BBB. This triggers cellular uptake by the mechanism of receptor-mediated endocytosis, i.e. the invagination and budding off of parts of the cell membrane as a result of the formation of small vesicles (endosomes). The transcellular passage of ligand (transcytosis) is completed by exocytosis at the abluminal membrane, and the whole process is completed within minutes *in vivo*.

Receptor-mediated uptake mechanisms have also been shown for insulin, insulin-like growth factors, and leptin. The fact that macromolecular complexes as large as LDL can un-

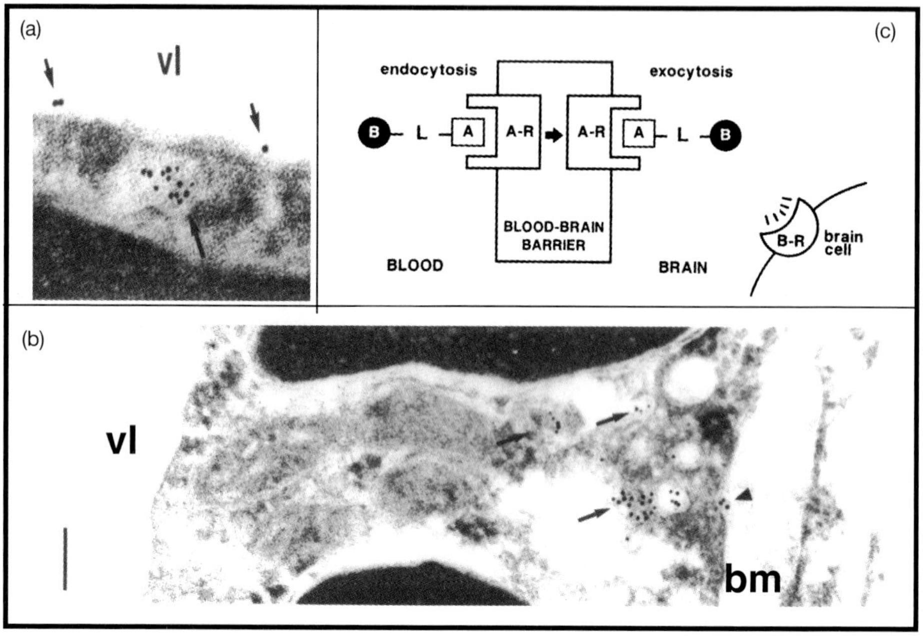

Figure 2.3. An example of receptor-mediated transcytosis through the blood–brain barrier. OX26, a monoclonal antibody to the rat transferrin receptor, was conjugated with 5-nm colloidal gold and perfused through the internal carotid artery in rats *in vivo*. The brain was then perfusion-fixed. (a) Arrows indicate binding to the luminal plasma membrane of capillary endothelial cells and internalized antibody in an endosome. (b) The endocytosed antibodies appear in multivesicular bodies (arrows) and are seen undergoing exocytosis at the abluminal cell membrane (arrowhead). vl, vascular lumen; bm, basement membrane. Scale bar = 100 nm. (c) A strategy for brain delivery can be based on receptor-mediated transcytosis. The non-permeable drug moiety 'B' is coupled to A by a linker, L. 'A' is a ligand or an anti-receptor antibody (e.g. OX26) which binds to its receptor on the luminal side and mediates endocytosis. The chimeric peptide enters brain interstitial space by exocytosis from the endothelial cell and is cleaved by local enzymes to release the drug. Reproduced with permission from references [37] (a,b) and [81] (c).

dergo transcytosis [39] underlines the difference between vesicular transport and carrier-mediated uptake.

Absorptive-mediated endocytosis and transcytosis of macromolecules through the BBB is related to receptor-mediated uptake, although not as specific. The process applies to certain lectins, for example wheat germ agglutinin [40], and to cationic proteins. Transport is triggered by glycoprotein binding (lectins) or by ionic interactions between negative charges on the endothelial plasma membrane and positive surface charges on the proteins. Both native proteins (histones) and chemically modified proteins (cationized albumin and IgG) can undergo absorptive-mediated transcytosis. While the physiological function of that process at the BBB remains to be identified, it offers a potential strategy for drug targeting/delivery [41].

2.3.2 Techniques for Measurement of Brain Uptake

2.3.2.1 *In vivo* Methods

For the correct interpretation of brain uptake studies in general and for the pharmacokinetic validation of a given delivery strategy, it is necessary to be familiar with characteristics and limitations of the applied technique [41]. *In vivo* methods remain the gold standard, as there are still no cell culture models available that fully represent the barrier characteristics.

Quantitative measurement of diffusional uptake and carrier-mediated transport of nutrients and drugs in experimental animals was greatly facilitated with the introduction of Oldendorf's brain uptake index (BUI) [42]. Test and reference tracers are injected as an intraarterial bolus into the carotid artery of the anaesthetized animal. After 5 s the animal is killed and the brain is removed for radioactivity counting. This method measures the ratio of the unidirectional brain extraction, E, of the test substance and of the reference ([^{3}H]-water, [^{14}C]-butanol), which are labelled with different isotopes, during a single passage through the brain capillary bed:

$$\mathrm{BUI} = E_{\mathrm{test}}/E_{\mathrm{reference}} \tag{2.1}$$

Advantages of the method include technical simplicity, and control over the composition of the injection fluid, making the technique suitable for competition and saturation experiments. Provided that the absolute value of $E_{\mathrm{reference}}$ is known, the absolute E_{test} may be calculated and, with the independently determined value of cerebral blood flow, F, E_{test} may be converted into a permeability surface area (PS) product. The latter conversion follows from the application of the Kety–Renkin–Crone equation of capillary physiology:

$$E = 1 - \mathrm{e}^{\mathrm{PS/F}} \tag{2.2}$$

The major drawback of the BUI is its limited sensitivity for measurement of compounds with low extraction.

Sensitivity was improved by at least two orders of magnitude with the internal carotid artery perfusion technique [43]. An outline of the method is given in Figure 2.4. Here, the extraction can be measured over a time frame of 15 s to 10 min or more, while maintaining the

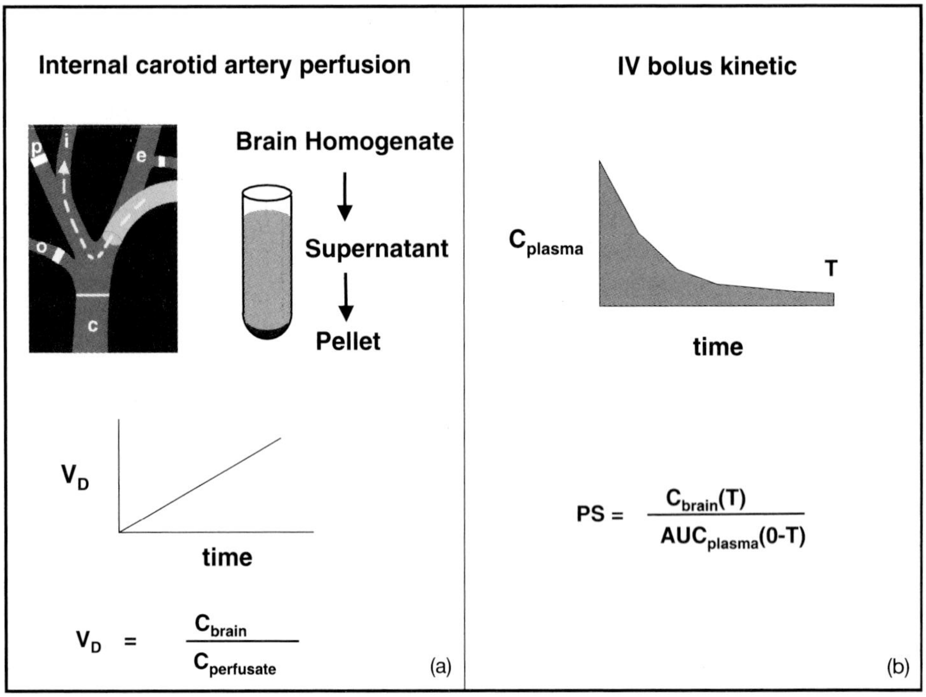

Figure 2.4. *In vivo* measurement of blood–brain barrier (BBB) permeability. (a) Internal carotid artery perfusion technique (i) in the rat. Other branches of the carotid artery are ligated or electrically coagulated (o, occipital artery; p, pterygopalatine artery). The external carotid artery (e) is cannulated and the common carotid artery (c) ligated. Perfusion time may range from 15 s to 10 min, depending on the test substance. It is necessary to subtract the intravascular volume, V_0, from V_D (apparent volume of distribution), to obtain true uptake values and this may be achieved by inclusion of a vascular marker in the perfusate, for example labelled albumin. Time-dependent analysis of V_D results in estimates of the unidirectional brain influx constant K_{in} ($\mu l\ min^{-1}\ g^{-1}$) which is equivalent within certain constraints to the PS product. BBB permeability surface area product PS can be calculated from the increase in the apparent volume of distribution V_D over time. Capillary depletion, i.e. separation of the vascular elements from the homogenate by density centrifugation, can discriminate capillary uptake from transcytosis. (b) i.v. bolus kinetics. The PS product is calculated from the brain concentration at the sampling time, T, and the area under the plasma concentration–time curve, AUC.

advantage of the experimental control of tracer concentrations and composition of the perfusion fluid. Gentle homogenization of brain tissue followed by density centrifugation, resulting in the separation into a vascular pellet and a 'postvascular' supernatant, differentiate the fractions of the test substance which have merely associated with the vasculature from those which have fully penetrated the BBB [44]. This 'capillary depletion' is applicable to substances, which bind to capillaries with high affinity and/or which are internalized by specific mechanisms such as receptor-mediated endocytosis. It could be shown that for example the acetylated form of LDL is only endocytosed and then sequestered by brain capillary endothelium. Erroneous data may be observed for substances without specific uptake but low affinity non-specific adsorption to the endothelium because of redistribution into the postvascular supernatant during processing and centrifugation [45].

The most sensitive technique for measuring brain uptake is the intravenous bolus administration or infusion and subsequent measurement of brain concentrations (Figure 2.4). Depending on the pharmacokinetics of the test compound in plasma, brain sampling may be performed after suitable circulation times ranging from minutes to hours or days.

The pharmacokinetic calculation of the unidirectional brain uptake rate, K_{in}, after intravenous injection, uses the relation of the brain concentration and the area under the curve of plasma concentration, AUC:

$$K_{in} = C_{brain}/\mathrm{AUC} \tag{2.3}$$

Here C_{brain} is the brain concentration after correction for intravascular content, and AUC is determined between time 0 and the final sampling time. Two assumptions must hold when interpreting the evaluation in the simple form described above: (1) the brain uptake of the compound is linear, meaning K_{in} is dose independent, and (2) the analysis is performed within a time-frame where the efflux from tissue is negligible (tissue concentrations are sufficiently low compared to plasma concentrations). Violation of these assumptions requires adjustments in experimental design and evaluation. For example, nonlinear kinetics may be accounted for by incorporation of a Michaelis–Menten term, while efflux can be treated by compartmental analysis [46].

The i.v. approach has the distinct advantage of measurements being carried out under the most physiological conditions. On the other hand, the caveats include confounding effects of peripheral metabolism, which may give rise to artifactual brain uptake of degradation products. To exclude such a possibility the application of suitable analytical techniques to tissue and plasma samples is required.

Unless a test compound is available in tracer form suitable for noninvasive quantification such as positron emission tomography (PET) or single photon emission computed tomography (SPECT), cerebrospinal fluid sampling is often used in human studies. It needs to be remembered that such measurements pertain to permeability at the B-CSF-B, not at the BBB, otherwise erroneous conclusions may be derived, in particular when specific transport processes are involved. For example the rapid penetration of the anti-HIV drug azidothymidine into CSF [47] is due to a carrier for pyrimidine nucleotides in the choroid plexus. In contrast, azidothymidine is subject to efflux at the BBB, which prevents it from reaching significant concentrations in the brain [48].

A caveat should be mentioned here concerning data relating to drug transport which has been obtained using intracerebral microdialysis. While the method is appealing for reasons such as its ability to investigate extracellular fluid concentration–time courses and concentration-dependent uptake rates in each single animal [49], it has limitations, in particular in the measurement of substances with low permeability. This is a consequence of the invasive nature of the placement of the microdialysis probe, which inevitably causes tissue damage, an inflammatory response and gliosis surrounding the dialysis probe. Therefore, the BBB may be locally compromised, obviously jeopardizing the valid interpretation of experimental data [50].

2.3.2.2 *In Vitro* Models

Freshly isolated capillaries from different species are a valuable model system of the BBB [51], because they retain the native repertoire of receptors and enzymes. In particular, receptor binding and carrier-mediated uptake or receptor-mediated endocytosis can be conveniently studied. On which side binding or uptake takes place cannot be easily determined, as both the luminal and abluminal side are exposed in such a preparation.

Numerous modifications of *in vitro* culture systems have been developed for the estimation of BBB transfer [52]. Culture systems in use are either primary cultures of brain microvessel endothelial cells (BMEC) or immortalized endothelial cell lines. BMEC may be grown in co-culture with astrocytes or in astrocyte-conditioned medium. Astrocyte-derived factors increase the tightness of the barrier as measured by transendothelial electrical resistance (TEER) and by the permeability of hydrophilic markers such as sucrose. They also up-regulate the expression of BBB-enriched enzymes such as γ-glutamyl transpeptidase (γ-GTP) and alkaline phosphatase. A setup of the *in vitro* technique in a transwell system for transport studies is depicted in Figure 2.5.

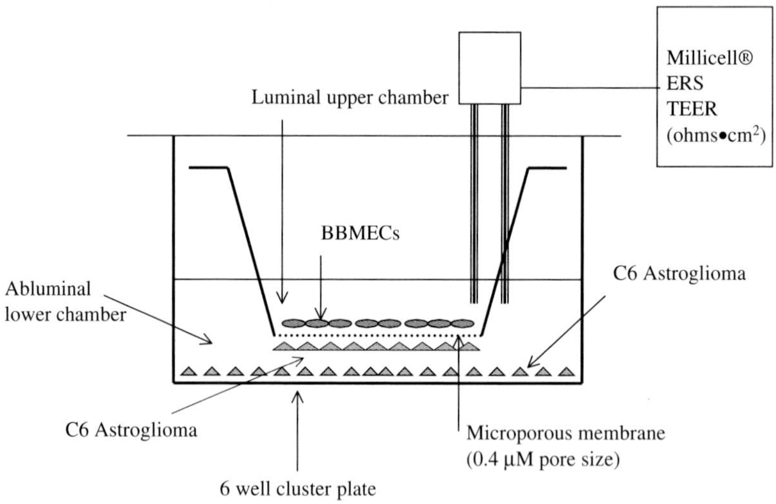

Figure 2.5. Setup for *in vitro* measurement of blood–brain barrier permeability with a co-culture of bovine brain microvascular endothelial cells (BBMEC) and an astroglioma cell line, C6. The BBMEC are grown on top of a filter insert. The C6 cells are either grown on the opposite side of the filter or on the bottom of the wells. Transport across the BBMEC monolayer is measured by adding the test substance to the upper chamber and sampling from the lower chamber. The tightness of the monolayer is also characterized by the transendothelial electrical resistance (TEER). Courtesy of T. Abbruscato.

2.4 Drug Delivery Strategies

The approaches for developing strategies for the delivery of drugs to the brain are shown in groups in the scheme in Figure 2.6. Pharmacological-based and physiological-based methods utilize noninvasive, systemic administration and a range of methods to enhance delivery through the BBB. Physical strategies in contrast rely on invasive (neuro-)surgical methods of drug administration.

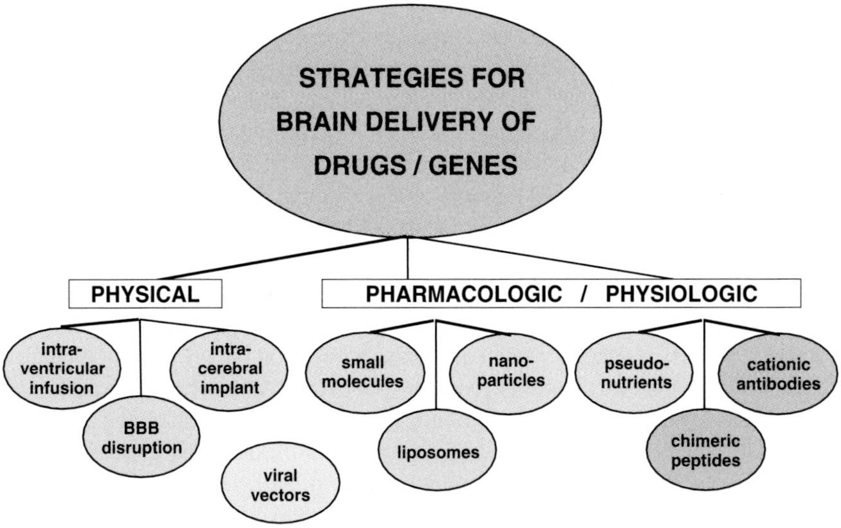

Figure 2.6. Strategies for drug delivery across the blood–brain barrier (BBB). The physical, pharmacological and physiological approaches are discussed in the text. Present experimental attempts at viral gene delivery would also be classified as invasive because of the intracerebral administration of the vector.

2.4.1 Small Molecule Drug Delivery

In accordance with the structure of the BBB as a double lipid bilayer, classical neuroactive drugs such as benzodiazepines, neuroleptics and tricyclic antidepressive agents, are all small lipophilic molecules. These small molecular weight neuropharmaceuticals were selected by a trial and error approach because their structural characteristics allow for diffusion-mediated, passive penetration through the BBB.

With the development of combinatorial chemistry and (ultra)high throughput screening methods, drug discovery is currently undergoing a rapid evolution. However the drug delivery aspect, in particular with regard to neuropharmaceuticals, is an area, which is lagging behind in efforts to integrate the overall drug development process [53].

The use of prodrugs with higher lipophilicity compared to the parent molecule is realized in the classical example of heroin and morphine. Heroin, the di-acetyl derivative of morphine, penetrates the BBB by one log order better than morphine and is cleaved by tissue esterases to release the active parent drug. As follows from the pharmacokinetic principles shown in Section 2.3.2.1 (Eq. 2.3), brain concentration is a function of both BBB permeability, reflected by K_{in}, and plasma area under the curve:

$$C_{brain} = K_{in} \times \mathrm{AUC} \tag{2.4}$$

While K_{in} increases with lipophilicity, AUC decreases due to higher uptake across all cell membranes including those of peripheral tissues. Therefore limits are imposed on the gain in brain delivery by the 'lipidization' strategy. In fact, for azidothymidine lipidization with the lipophilic adamantane moiety, CSF concentrations decreased by a factor of 10 as a result of a decrease in AUC [54].

Lipidization is also an integral part of the chemical delivery approach that is based on the concept of 'retro-metabolism' [55]. The underlying principle is the simultaneous derivatization of a drug with a 'redox targeter', e.g. dihydrotrigonellinate, and with a lipophilic moiety, e.g. cholesterol. This strategy has been applied to deliver a small enkephalin analogue to the brain [55]. Once inside the target tissue, the redox targeter will be enzymatically oxidized to a positively charged derivative and serve to lock-in the prodrug. Further enzymatic steps release the active drug by cleavage of the lipid modifier and the targeter. This strategy offers the potential of tissue-selective delivery, yet it requires simultaneous optimization of a multitude of rates, including the influx and efflux of the prodrug, biotransformation of the targeter and eventual release of the active drug.

An alternative delivery strategy for small molecules is based on the presence of the nutrient transporters. Drugs that are structurally similar to substrates of a carrier system can undergo facilitated brain uptake as pseudoneutrients. The best example of this is the therapeutic use of L-DOPA in Parkinson's disease. Unlike the neurotransmitter dopamine itself, which cannot cross the BBB in significant amounts, its precursor L-DOPA is a substrate for LAT, the transporter of large neutral amino acids [56]. Its uptake by the brain is saturable, and subject to competition by the other substrates of the carrier present in plasma.

L-DOPA therapy is an example of rational drug design based on knowledge of BBB transport biology. A number of other small molecular weight drugs are known to undergo carrier-mediated transport at the BBB. These include the substrates of LAT, such as the anticancer drugs melphalan and acivicin, or the $GABA_B$-agonist baclofen, and beta lactam antibiotics (e.g. benzylpenicilline, ceftriaxone, cefodizime) which are substrates for the organic acid carrier [31].

2.4.2 Macromolecular Drug Delivery

2.4.2.1 Intraventricular Route

At first glance, intracerebroventricular (i.c.v.) or intrathecal administration via catheters, typically connected to a subcutaneous drug reservoir, appears as a logical mode of drug delivery, because the percentage of the dose reaching the target organ would be 100%. The i.c.v. route is successfully used in disorders where pathogenetic events take place close to the brain surface. Clinical examples of intrathecal small drug delivery are the administration of glycopeptide and aminoglycoside antibiotics in meningitis [57], the i.c.v. treatment of meningeal metastasis [58], intrathecal baclofen injection for treatment of spasticity [59] and the infusion of opioids for severe chronic pain [60]. However, in addition to the invasive character of the procedure, it has to be taken into account that drug distribution into brain tissue is severely diffusion-limited and that the continuous production and reabsorption of the CSF results in rapid clearance from that compartment.

Although there is no cellular barrier preventing diffusion from the ventricular surface into brain tissue (there are no tight junctions between the ependymal cells lining the ventricular surface), the low speed of diffusion severely restricts tissue uptake of even small molecular weight drugs and practically prevents the penetration of large molecules such as peptides and proteins into deep tissue layers. Possible enzymatic inactivation and binding or sequestration by brain cells along the diffusion path may even lower the actual drug concentrations in brain interstitial spaces to levels lower than predicted from the molecular size and diffusion coefficients [61]. Figure 2.7a shows an example of a large molecule, the 26-kDa nerve growth factor (NGF), that could not penetrate into rat brain deeper than 1–2 mm from the infused ventricle. This might be expected, as even small drugs show very steep concentration gradients over a distance of only 2–3 mm from the ventricular surface (Figure 2.7b). Exceptions are found in areas to which retrograde transport occurs, e.g. into the neurons of the basal cholinergic nuclei in the case of NGF (Figure 2.7a).

After i.c.v. injection, the rate of elimination from the CNS compartment is dominated by cerebrospinal fluid dynamics. The CSF, which is secreted by the choroid plexus epithelium across the apical membrane, circulates along the surface and convexities of the brain in a rostral to caudal direction. It is reabsorbed by bulk flow into the peripheral bloodstream at the arachnoid villi within both cranial and spinal arachnoid spaces [62]. Of note is that the turnover rate of total CSF volume is species dependent and varies between approximately 1 h in rats and 5 h in humans. In adult human brain, the total CSF volume amounts to 100–140 ml and the production rate is 21 ml h^{-1} [63]. Accordingly, the entire cerebrospinal volume is exchanged regularly 4–5 times per day. This rapid drainage of CSF into peripheral blood leads to relatively high drug concentrations in the peripheral circulation. For instance, the concentration of methotrexate in peripheral blood reaches 1% of the ventricular CSF concentration following intrathecal administration of the drug [64]. In Rhesus monkeys, parenchymal concentrations of methotrexate of 1% of the intraventricular concentration have been measured at a distance of 2 mm from the ependymal surface [65]. Therefore, the concentration of methotrexate in the blood is actually higher than at tissue regions beyond 2 mm from the ependymal surface following intrathecal application.

2.4.2.2 Intraparenchymal Route

Restricted diffusion also limits tissue distribution after intraparenchymal drug administration, as shown in Figure 2.7c and d. Distribution has been measured in the rat brain after implantation of polymer discs containing NGF [66]. Drug concentrations decreased to less than 10% of the values measured on the disc surface within a distance of 2–3 mm, even after prolonged periods of several days. Therefore, applying this approach in the larger human brain would require the stereotaxic placement of multiple intraparenchymal depots, as has been evaluated in rat brain [67], on a repetitive schedule.

The same pharmacokinetic limitation is true in principle for the implantation of encapsulated genetically engineered cells, which synthesize and release neurotrophic factors [68].

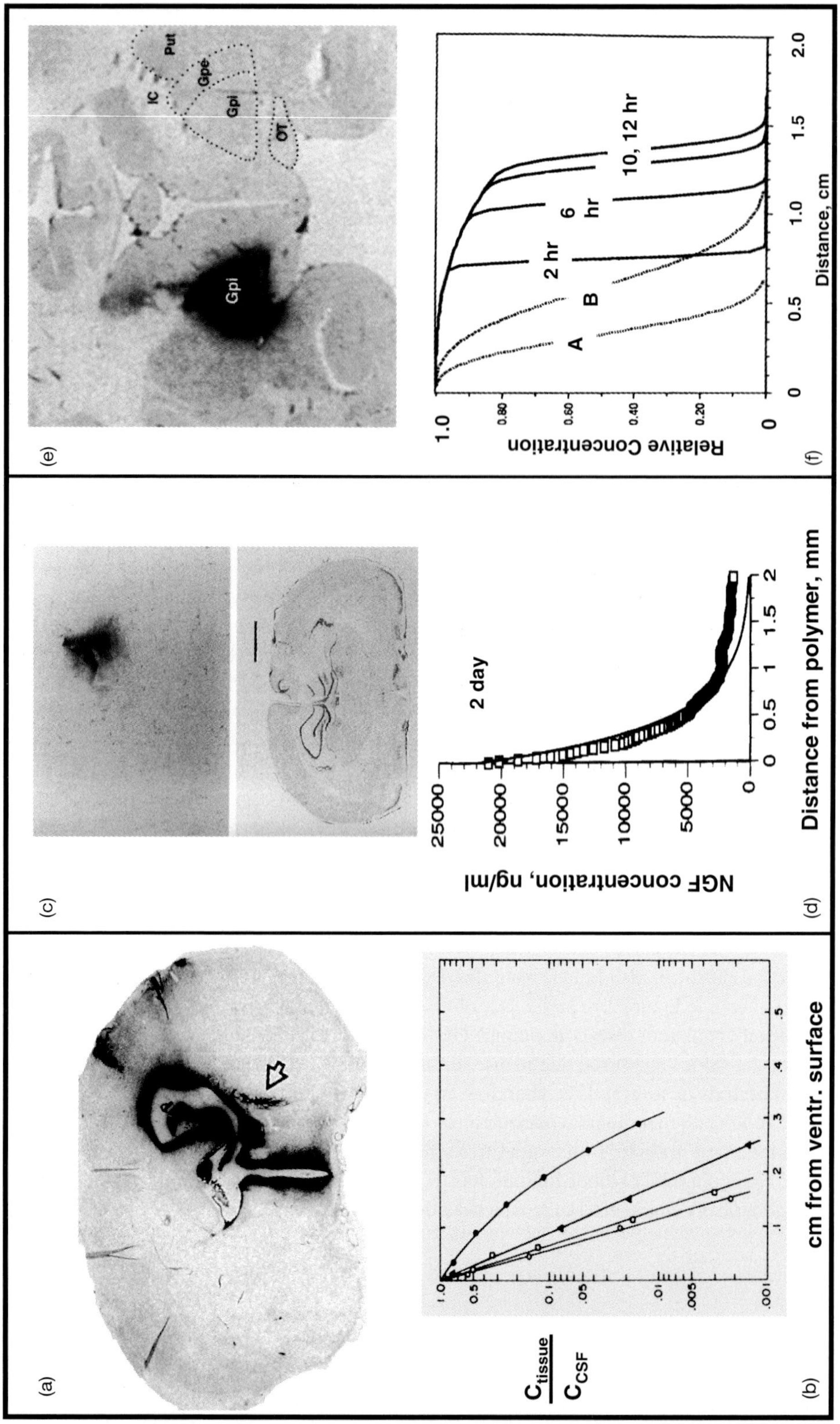
(a)
(b)
C tissue / C CSF
cm from ventr. surface
(c)
(d)
NGF concentration, ng/ml
Distance from polymer, mm
2 day
(e)
Put
Gpe
IC
Gpi
OT
(f)
Relative Concentration
Distance, cm
A
B
2 hr
6 hr
10, 12 hr

2.4.2.3 Convective Flow

Intraparenchymal high flow microinfusions with flow rates up to 4 μl min^{-1} result in almost homogenous tissue concentrations of macromolecules (transferrin, mw 80 kDa) over a large volume and over a distance of > 10 mm from the catheter tip within an infusion period of 2 h [69]. Figure 2.7 demonstrates the principle in an experimental study (Figure 2.7e) and the expected tissue concentration profiles (Figure 2.7f). The plateau-like concentration profile at the front of the convective flow is in contrast to the concentration gradients associated with diffusion-mediated distribution (Figure 2.7b, d). The method has been applied in clinical trials for treatment of gliomas and metastatic brain tumours with a transferrin-targeted mutant diphtheria toxin, Tf-CRM107 [70]. Current efforts are directed at overcoming the dose-limiting toxicities found in these studies in the CNS such as petechial haemorrhage and small vessel thrombosis [71].

2.4.2.4 Delivery by Barrier Disruption

The temporary physico-chemical disruption of brain endothelial barrier integrity is also considered an invasive strategy for drug delivery, because it typically involves the intracarotid infusion of a barrier-opening agent. Barrier opening for low molecular weight tracers and macromolecules (e.g. Evans blue-albumin) was experimentally demonstrated with intracarotid infusions of membrane active agents such as bile salts, oleic acid, cytostatic drugs etoposide and melphalan, and cytochalasin B. Intracarotid infusion of a low pH buffer also opens the BBB.

Most studies were performed with hyperosmolar solutions. Hypertonic disruption is under clinical evaluation for enhanced delivery of small molecular weight cytostatic agents to brain tumours. Technically, the procedure is performed as a high-flow short-term infusion of 25% mannitol or arabinose under general anaesthesia. The underlying mechanism is a sequelae of endothelial cell shrinkage, disruption of tight junctions and vasodilatation by osmotic shift [72].

◄

Figure 2.7. (a) Autoradiograph of a brain section showing the limited distribution of [^{125}I]-NGF (nerve growth factor) in rat brain 18 h after injection into the lateral ventricle. Exception: cholinergic neurons with retrograde transport to the cell body (arrow). (b) Brain–cerebrospinal fluid (CSF) concentration gradient of four drugs relative to the distance from the ependymal surface. Curves from left to right: BCNU, thiotepa, cytosine arabinoside, hydroxyurea. Drugs were infused by the intracerebroventricular route in a Rhesus monkey for a 1-h period. The gradients are steepest for highly diffusible substances (thiotepa, BCNU). (c) Autoradiograph (top) and unstained photograph (bottom) of coronal sections of rat brains following implantation of [^{125}I]-NGF-loaded polymers. Sections were obtained 2 days post-implanatation. Bar = 2.5 mm. (d) Concentration profile in the vicinity of the polymer implant in (c). (e) The convective-enhanced method allows homogenous delivery to precisely defined volumes, as demonstrated here in a coronal section of monkey brain stained for biotin (black region). The animal received an infusion of biotinylated albumin into the globus pallidus internus (Gpi). Gpe, globus pallidus externus; IC internal capsule; OT, optical tract; Put, Putamen. (f) Calculated concentration profiles for macromolecules in brain that can be achieved with high flow microinfusion. Dashed lines: steady-state profiles of molecules undergoing rapid metabolism with half-lives of 0.167 (A) or 1 h (B). Solid lines: profiles after 2, 6, 10 and 12 h for a molecule with long a half-life (33.5 h). The latter profiles show plateau-like tissue concentrations that extend with duration of infusion up to 1.5 cm from the tip of the infusion catheter. Reproduced with permission from references [61] (a), [65] (b), [66](c,d); [122] (e) and [123] (f).

Morphological studies in rats, where the induction of neuropathological changes by osmotic opening was examined, provided evidence of uptake of macromolecules by the brain. The extravasation of plasma proteins such as fibrinogen and albumin was shown immunohistochemically at the light microscopic level. Electron microscopy revealed ultrastructural changes such as swelling of astrocytic processes and severe mitochondrial damage in neurons [73]. There was also evidence of prolonged (24 h) cellular stress or injury in neurons and glia as expressed by the induction of heat shock protein (HSP-70). While the nonspecific opening of the BBB to plasma proteins harbours a risk of eliciting neuropathological changes, osmotic disruption has been tested for its potential as a delivery method for macromolecular drugs such as monoclonal antibodies against various tumour antigens or their Fab fragments. In other studies, uptake after intracarotid administration of nanoparticles (20-nm iron oxide particles) by normal brain, and uptake of recombinant adenovirus or herpes virus by normal brain tissue and brain tumour xenografts in nude rats was postulated [74,75].

Compared to small molecules, barrier opening for high molecular weight compounds is of shorter duration [72]. Furthermore, a characteristic difference exists in the degree of barrier opening in tumour versus normal brain tissue. Barrier disruption was consistently found to be more pronounced for the normal BBB, which may limit the clinical applicability of hyperosmolar barrier opening, at least for cytotoxic drugs [76].

BBB opening may also be achieved by receptor-mediated mechanisms. The vasoactive compounds prostaglandins, histamine, serotonin, leukotriene C_4 (LTC_4), and bradykinin have all been shown to increase BBB permeability [16]. The effects of LTC_4 and bradykinin are more pronounced on the blood–tumour barrier than on the normal BBB. In the case of LTC_4 this effect is ascribed to the presence of an enzymatic barrier in normal brain tissue due to the endothelial expression of γ-GTP. The enzyme metabolizes and inactivates LTC_4 to LTD_4. In contrast, tumour vessels are unable to express equivalent activities of γ-GTP, a fact that may be exploited for selective opening of the tumour barrier by intracarotid administration of LTC_4. However, the effect is restricted to small molecules, as there was no increase in the tumour accumulation of a dextran tracer of molecular weight 70 kDa. On the other hand, bradykinin also opens the barrier for the high molecular weight range. It acts on endothelial cells through B_2-receptors located on the abluminal side. Normal brain tissue is protected from barrier opening by bradykinin in the vascular lumen because the peptide cannot access these receptors. In tumour vessels the barrier integrity is sufficiently compromised to allow for additional bradykinin-mediated opening at low peptide concentrations. While bradykinin itself requires intracarotid administration, an analogue with prolonged half-life (RMP-7) is effective after intracarotid or intravenous application. A 4–5-fold increase in the delivery of the cytokines interferon-γ, tumour necrosis factor α and interleukin-2 to experimental RG2 glioma in rats was demonstrated after intracarotid infusion of RMP-7 [77]. The drug is currently being evaluated in the therapy of human malignant gliomas to enhance delivery of carboplatin to the tumour [78].

2.4.2.5 Vector-mediated Delivery

In this approach, 'chimeric peptides' [79] are generated as transportable drug derivatives targeting the receptor-mediated mechanism. Chimeric peptides are formed by linking a drug that is unable to cross the BBB to a vector (see Figure 2.8). Binding of the vector at the luminal membrane of brain capillary endothelial cells initiates receptor-mediated or adsorptive-mediated transcytosis (Figure 2.3c). Size and structure of the cargo may vary as long as binding and cellular uptake of the vector is not inhibited by the drug moiety, and may only be limited by the size of the endocytotic vesicles.

Initial studies of brain delivery based on the chimeric peptide strategy used the absorptive-mediated uptake of cationized albumin which was chemically coupled to the opioid peptide β-endorphin [80] or its metabolically stabilized analogue [D-Ala^2]β-endorphin. Tracer experiments in which the chimeric peptide was labelled in the endorphin moiety provided evidence of internalization by isolated brain capillaries and transport into brain tissue *in vivo* [81].

Endogenous ligands for receptor-mediated systems may be unsuitable as vectors due to competition for transport or undesirable pharmacological effects. For example, plasma concentrations of transferrin are in the range of 25 μM. Insulin as a vector would cause hypoglycaemia. A logical alternative as vectors are monoclonal antibodies specific to the extracellular domain of a peptide or protein receptor at the BBB. These antibodies can be designed as non-competitive, i.e. they bind to the receptors at a site distinct from the ligand binding site and do not interfere with the endocytosis process.

Brain uptake data for some vectors are compared in Table 2.1. Quantitative comparisons within the same species are possible for the rat with vectors derived from the anti-TfR monoclonal antibody OX26 and from cationized human serum albumin. To put the efficiency of brain delivery into perspective, the comparison to a classical neuroactive drug may be informative. In the rat, brain concentrations of morphine following systemic administration never exceed 0.08% of injected dose per gram [% ID g^{-1}] [82]. In contrast, OX26 easily reaches concentrations in rat brain that are three to four times higher. Vectors based on cationized hu-

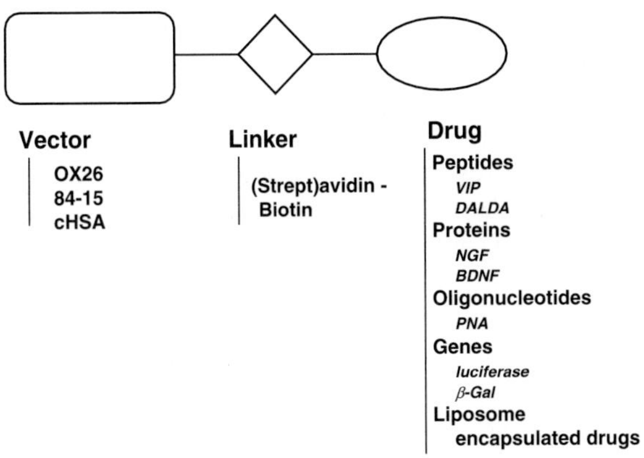

Figure 2.8. Scheme of a chimeric peptide with examples for each of the distinct domains. OX26, anti-rat transferrin receptor monoclonal antibody (mAb); 84-15, anti-human insulin receptor mAb; cHSA, cationized human serum albumin; VIP, vasoactive intestinal polypeptide; DALDA, dermorphin analogue; NGF, nerve growth factor; BDNF, brain-derived neurotrophic factor; PNA, peptide nucleic acid; β-gal, β-galactosidase.

Table 2.1. Brain concentration, blood–brain barrier PS product, and plasma AUC (0–60 min) of brain delivery vectors after i.v. bolus injection.

Vector (species)	%ID g^{-1} [a]	PS (µl min^{-1} g^{-1})	AUC 0–60 (%ID min ml^{-1})
^{3}H-OX26 (rat)	0.27 ± 0.04	1.92 ± 0.06	132 ± 19
OX26-Av[b] (rat)	0.041 ± 0.004	0.85 ± 0.02	49 ± 4
Anti-TfR IgG3-C_H3-Av[b,c] (rat, recombinant)	0.25 ± 0.09	2.25 ± 0.65	134 ± 29
OX26-NLA[b] (rat)	0.17 ± 0.04	0.70 ± 0.10	232 ± 25
OX26-SA[b] (rat)	0.20 ± 0.03	0.92 ± 0.10	216 ± 28
cHSA-Av[b] (rat)	0.015 ± 0.006	0.26 ± 0.13	64 ± 7
cHSA-NLA[b] (rat)	0.061 ± 0.012	0.20 ± 0.04	300 ± 14
^{125}I-8D3 (mouse)	3.1 ± 0.2	3.3 ± 0.1	932 ± 57
^{125}I-HIR MAb (rhesus monkey)	3.8 ± 0.4 (100 g brain)[d]	5.4 ± 0.6	5.9 ± 1.2 (0–180 min)

[a] Percentage injected dose per g brain 60 min after i.v. bolus injection.
[b] These proteins were labeled at the avidin moiety with [^{3}H]biotin.
[c] IgG3-C_H3 fusion protein with variable region of OX26 and avidin.
[d] %ID in total brain tissue after 3 h.
AUC, area under curve of plasma concentration; Av, avidin; NLA, neutral avidin; SA, streptavidin; cHSA, cationized human serum albumin; HIR MAb, human insulin receptor mAb; PS product, permeability surface area product.

man albumin reach lower brain concentrations compared to OX26 vectors. The difference is mainly caused by a corresponding difference in the PS products, i.e. the rate of uptake by absorptive-mediated transcytosis of cationized albumin at the BBB is lower than the rate of receptor-mediated uptake of OX26 (Table 2.1).

With regard to transport capacity, the introduction of the anti-human insulin receptor antibody (HIR MAb) 83-14 as a vector indicates the potential for future improvements in brain-specific delivery vectors. Compared to anti-TfR monoclonal antibodies, the brain delivery in primates is over 7-fold higher due to the high PS product of the HIR MAb.

A thorough characterization of vectors for drug delivery needs to take into account the saturable character of receptor-mediated processes and therefore requires investigation of dose dependence of uptake. While non-competing antibodies avoid the problem of competition by endogenous ligands, the saturability of the antibody binding site remains. The values in Table 2.1 were obtained in uptake experiments with doses of vectors in the low µg kg^{-1} range, corresponding to plasma concentrations in the low nM range. Linear pharmacokinetics cannot be expected at higher doses and hence changes in both plasma AUC and apparent PS product are likely to occur at higher doses. Experimental evidence of saturability *in vivo* was seen for the OX26 antibody in rats and the 8D3 antibody in mice. In the case of 8D3, for example, the brain uptake declined to virtually 0% ID g^{-1} brain at a dose of 4 mg kg^{-1} i.v. [83]. Proof of saturability in vivo is direct evidence for a specific uptake mechanism.

The coupling step between vector and drug moiety may be performed by either chemical or molecular biological approaches (see Chapter 11 for a more detailed discussion on conjugation strategies in drug targeting research). While the options offered by chemical methods provide rapid synthesis of conjugates, which is particularly suitable for animal experiments and 'proof of concept' studies, fusion proteins have the potential for bulk production of a defined molecular entity for future clinical development. With regard to chemical conjugation, the avidin–biotin technology as a linker strategy was introduced [84] as a highly versatile alternative to direct linkage of vector and drug moiety [85]. It exploits the broad availability of biotinylating reagents for a range of compounds and functional groups, and a single vector to be used for the delivery of different drugs. Moreover, the avidin–biotin bond is extremely stable. It proved advantageous for pharmacokinetic reasons to substitute the basic avidin with a chemically neutralized form of avidin, designated NLA [86] or with streptavidin (SA) [87].

Recently, a fusion protein between OX26 and avidin was engineered, in which the variable regions of an IgG3-antibody were substituted with those of OX26, and the C_H3 region was fused to the avidin monomer (anti-TfR IgG3-C_H3-Av). The fusion protein thus forms an avidin dimer, which displays high affinity biotin binding. The pharmacokinetics and brain uptake of the fusion protein were favourable compared to the chemical conjugate of OX26–avidin, as is evident from the values in Table 2.1 [88].

The biotin–avidin linker strategy is particularly suitable for synthetic peptide drugs. These can be designed to facilitate monobiotinylation at a site that does not interfere with bioactivity [89]. Monobiotinylation is recommended due to the multivalency of avidin. A 1 : 1 molar conjugate of vector and (strept)avidin can bind up to four biotin residues, and the genetically engineered fusion protein still has two biotin binding sites. Therefore, higher degrees of biotinylation of the drug moiety would result in the formation of high molecular weight aggregates, which are cleared rapidly from the circulation [90].

Chimeric peptides need to be stable in the circulation before brain uptake occurs, and either amide bonds, thioether or disulfide linkers fulfil that requirement in the plasma compartment. In addition, they must be stable during transcytosis through BBB endothelial cells. Finally they also need to retain binding affinity in their drug moiety. If binding of a peptide drug to the vector reduces binding affinity to the drug receptor on brain cells, then the release of free drug in the brain would be required. Disulfide reducing enzymes such as protein disulfide-isomerase are present in tissues intracellularly and on the plasma membrane [91]. Cleavage of (-S-S-) linked chimeric peptides in brain *in vivo* is possible. A biotinylated opioid peptide analogue ([Lys7]dermorphin-amide) with a disulfide biotin linker, N-hydroxysuccinimide dithiopropionate, was cleaved from the vector OX26-SA in brain but not in plasma [92].

An alternative coupling strategy that avoids potential steric hindrance of drug action and eliminates the need for cleavability utilizes long, flexible spacer arms, e.g. biotin-derivatized polyethylene glycol (PEG) linkers with molecular weights of 2000 or 5000 Da [93,94].

2.4.2.6 Pharmacological Effects of Chimeric Peptides

The cargo that is suitable for transport by chimeric peptides encompasses a wide array of substances. Table 2.2 gives examples of studies, which measured CNS effects after peptide drug delivery.

Among the peptide-based payloads that have been delivered is an analogue of the 28-amino acid peptide Vasoactive Intestinal Polypeptide (VIP) [89,95]. VIP is suitable for the

Table 2.2. Pharmacologic effects obtained with chimeric peptides in animal models.

Chimeric peptide	Dose	Mode of administration	Animal model	Effect
Biotinylated VIP analogue linked to OX26-Av	12 μg kg^{-1}	Intracarotid infusion	Rat; artificial ventilation under nitrous oxide anesthesia	Increase in CBF
Biotinylated VIP analogue linked to OX26-SA	20 μg kg^{-1} or 100 μg kg^{-1}	Single i.v. injection	Rat; conscious	Dose-dependent increase in CBF
NGF chemically conjugated to OX26	6.2 μg/ injection	i.v. injection 4× every 2 weeks	Rat; intraocular forebrain transplant	Survival of cholinergic neurons
NGF chemically conjugated to OX26	50 μg/ injection	i.v. injection, twice weekly for 6 weeks	Aged rat (24 months)	Improvement of spatial memory in impaired rats
NGF chemically conjugated to OX26	20 μg/ injection	i.v. injection daily 3 days + every 2 days 6×	Rat; quinolinic acid lesion	Rescue of striatal cholinergic neurons
NGF chemically conjugated to anti primate TfR mAb AK30		i.v. injection	Non-human primate	Upregulation of p75 NGF-receptor in striatum
GDNF chemically conjugated to OX26	5μg/ injection	i.v. injection 3× every 2 weeks	Rat; intraocular spinal cord transplant	Survival of motorneurons
Biotinylated PEG-BDNF linked to OX26-SA	250 μg kg^{-1}	i.v. injection daily for 7 days	Rat; transient forebrain ischaemia	Rescue of CA1 hippocampal neurons

BDNF, brain derived neurotrophic factor; CBF, cerebral blood flow; GDNF, glial cell line derived neurotrophic factor; NGF, nerve growth factor; TfR, transferrin receptor; VIP, vasoactive intestinal polypeptide.

demonstration of a pharmacological effect with a vector-mediated drug delivery strategy, because VIP-containing nerve fibres are abundant around intracerebral small arteries and arterioles. This peptide acts as a potent vasodilator when applied topically to intracranial vessels and plays an important role in the modulation of cerebral blood flow (CBF). However, as its receptors are expressed on the vascular smooth muscle cells, which are beyond the blood–brain barrier, no effects on CBF are usually seen after systemic administration of VIP.

A metabolically stabilized analogue of VIP was constructed which could be biotinylated at a single site. Brain delivery of the biotinylated VIP analogue by the OX26–avidin vector resulted in the desired pharmacological effect. A significant increase in CBF of 65% could be demonstrated after systemic administration of the chimeric peptide. The effect was seen both in anaesthetized rats under controlled respiration after intracarotid infusion as well as in conscious animals after i.v. bolus injection. When an equal dose of the peptide alone without a vector was injected (12 μg kg^{-1} for the intracarotid infusion or 20 μg kg^{-1} in the i.v. study) there was no measurable effect on CBF compared to control animals. In contrast, the well established peripheral effects of VIP on glandular blood flow in the thyroid gland or the sali-

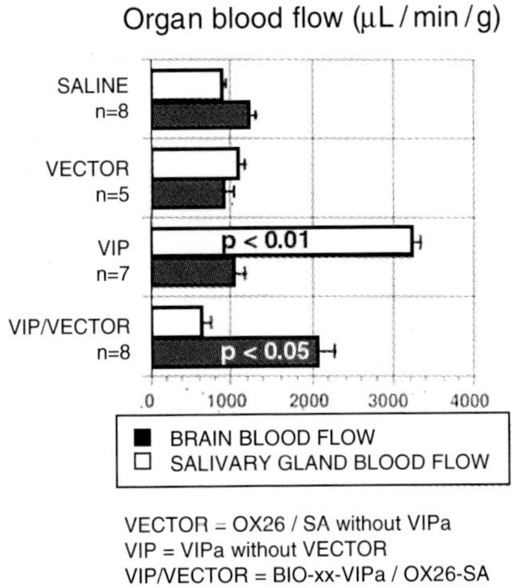

Figure 2.9. Differential pharmacological effect elicited by vector-mediated delivery of a VIP analogue. The organ blood flow in brain and salivary gland was measured in conscious rats after i.v. administration of vehicle (saline), the brain delivery vector OX26-SA, the VIP peptide alone, or the chimeric peptide. While cerebral blood flow increased in the chimeric peptide group by 60% compared to the saline control, the increase in salivary gland blood flow seen with the peptide alone was abolished by coupling to the vector. The VIP analogue was biotinylated with a non-cleavable 14-atom spacer (biotin-XX) for coupling to the vector. Data from reference [95].

vary gland were readily detectable [89,95], as shown in Figure 2.9. Notably, the effect on salivary gland blood flow was attenuated in animals treated with the chimeric peptide delivery system. Taking salivary gland blood flow in that respect as a potential adverse drug effect, the delivery strategy of the VIP analogue to the brain not only resulted in the desired pharmacological response at the target site, but it also diminished the effect at non-target sites and therefore increased the therapeutic index [95].

Demonstrations of pharmacological effects of chimeric peptides have been achieved with different neurotrophic factors in models of neurodegenerative diseases and ischaemia. The initial report by Friden *et al.* utilized an ocular graft model of fetal midbrain placed into the anterior eye chamber of adult rats [85]. The vasculature of the grafted tissue retained its BBB properties. Nerve growth factor was chemically coupled to the vector OX26 via a disulfide linker. Repeated i.v. administration (four times bi-weekly) of the chimeric peptide promoted survival of the cholinergic neurons within the graft. Further proof of pharmacological effect of the same conjugate was obtained in aged rats with spatial learning deficits. They responded to a 6-week treatment with twice weekly i.v. injections with improved performance in the so-called Morris water maze learning task and immunohistochemistry showed increased cell size of cholinergic neurons in the medial septal area of these rats [96]. The NGF-OX26 chimeric peptide was also effective in a quinolinic acid lesioning model of Huntington's disease [97]. Treatment for 2 weeks significantly reduced the loss of intrastriatal cholinergic neurons induced by stereotaxic injection of quinolinic acid.

Animal models of Parkinson's disease suggest that Glial Cell-Line Derived Neurotrophic Factor (GDNF) may be a suitable treatment modality for degenerative processes involving dopaminergic midbrain neurons, and traumatic injury of spinal motor neurons. Therefore, the effect of a GDNF-OX26 chimeric peptide was studied in another neural graft model [98]. The vector-mediated delivery of small i.v. doses equivalent to 5 μg of GDNF significantly promoted the survival of ocular implants of fetal spinal cord motor neurons in rats.

The potential therapeutic benefit of brain-derived neurotrophic factor BDNF for rescuing neurons after stroke was demonstrated in a forebrain ischaemia model [93]. In that study, transient forebrain ischaemia was induced in rats by bilateral clamping of the carotid arteries. In order to achieve BDNF delivery with the chimeric peptide approach it was necessary to modify the peptide by 'pegylation', i.e. the coupling of multiple PEG residues. Native BDNF is a basic peptide with rapid clearance from plasma. The poor pharmacokinetic properties persisted after coupling to OX26–streptavidin but could be overcome by pegylation. The PEG-BDNF could be delivered through the BBB by vector-mediated transport as efficiently as the OX26 antibody itself [90]. Animals treated for 1 week after the ischaemic insult with chimeric peptide (biotinylated PEG-BDNF coupled to OX26-SA) at a daily dose of 250 $\mu g\ kg^{-1}$ were fully protected from neuronal loss in the hippocampal CA1 region.

Oligodeoxynucleotides (ODN) represent another class of hydrophilic macromolecular drug candidates, which require transcellular as well as intracellular delivery where brain cell targeting is concerned. Due to their highly charged, anionic character they also have the potential to impair the pharmacokinetics of the delivery system when used as the drug constituent of chimeric peptides. Coupling of a biotinylated phosphodiester ODN to OX26-NLA increases hepatic clearance of the complex and limits brain uptake by lowering the AUC [99]. On the other hand, phosphorothioate-modified ODNs show high plasma protein binding which may contribute to the low BBB transport measured for a PS-ODN/OX26-SA chimeric peptide. In contrast, the neutral peptide backbone of peptide nucleic acids makes these compounds good drug candidates for chimeric peptides and allows for a substantial vector-mediated effect on brain targeting (28-fold increase, [100]). A potential therapeutic application of the ODN approach is the delivery of an antisense oligonucleotide to the *rev* gene of HIV-1 through the BBB. The feasibility of such an approach was recently demonstrated using the OX26–avidin fusion protein [88].

2.4.2.7 Chimeric Peptide Radiopharmaceuticals

The potential of chimeric peptides for delivery of radiopharmaceuticals across the BBB, either for diagnostic or therapeutic purposes, has been explored in studies with radiolabelled synthetic amyloid peptide and with EGF. Aß peptide in solution deposits specifically on preexisting amyloid plaques and vascular amyloid. A pharmacokinetic study in Rhesus monkeys with the insulin receptor antibody 83-14 as a vector showed brain accumulation of radiolabelled [^{125}I]-Aβ only after vector-mediated delivery. The peptide alone was unable to cross the BBB. In the monkeys, analysis of brain sections by phosphorimager quantitation of radioactivity resulted in images comparable to scans obtained with the non-metabolized glucose analogue 2-deoxyglucose [101]. Labelling with a suitable radioisotope should enable quantitative detection by a neuroimaging method such as SPECT.

EGF receptors are abundantly expressed by gliomas and present a target both for diagnostic imaging and radio-immunotherapy. A cerebral implant model in rats bearing human U87 gliomas was utilized to test the brain delivery of [^{111}In]-labelled EGF by vector mediated transport with OX26 following i.v. injection. Brains were sampled after 2 h and cryosectioned for subsequent autoradiography. The tumours were clearly visualized on these autoradiographs, but only when the labelled EGF was given as a chimeric peptide, not when injected without the vector [102].

2.4.3 Liposomes as Drug Carriers

2.4.3.1 Conventional Liposomes and Small Molecules

Liposomes, in addition to oligonucleotides [104], are often used as carriers for low molecular weight drugs and peptides [103]. It has been demonstrated that encapsulation within liposomes can dramatically alter the fate of the encapsulated drug *in vivo* [105]. Liposomal formulations may protect against metabolic degradation and can influence plasma clearance and tissue distribution of a variety of drugs. Loading efficiency, contents retention, plasma stability and pharmacokinetic properties can often be adjusted by appropriate formulation of conventional liposomal drug carriers [105,106]. However, conventional liposomes do not undergo significant blood–brain barrier transport [107]. This is also true for small unilamellar vesicles as demonstrated in a study where 60-nm liposomes radiollabeled with ^{111}Indium did not penetrate the blood–brain barrier of a normal brain [108]. In this study brain penetration was only observed following non-specific pharmacological disruption of the blood–brain barrier by infusion of high doses (25 mg kg^{-1}) of etoposide or at sites of brain tumours where the vasculature is porous.

2.4.3.2 Brain Targeting Using Immunoliposomes

Conventional liposomes are rapidly removed from the circulation by cells of the reticuloendothelial system [109]. This rapid accumulation of conventional liposomes in the liver and the spleen and the resulting high plasma clearance can be slowed down by coating the liposome surface with inert and hydrophilic polymers such as PEG [110]. The half-life of liposomes containing PEG-derivatized lipids increases up to 100-fold [106]. Such liposomes are often referred to as sterically-stabilized liposomes. The PEG polymers can also be used for covalent conjugation of an antibody or an antibody fragment to the liposome. In this case a chemically reactive linker lipid can be used (Figure 2.10) that consists of a bi-functional PEG molecule covalently bound at one side to a phospholipid headgroup and at the other side to a thiol-reactive maleimide group. Thus modified antibodies bearing a thiol group can be coupled under mild conditions to sterically-stabilized liposomes [111]. Such immunoliposomes retain both their prolonged circulation properties and their target specificity *in vivo*. Similar results can be obtained using alternative coupling techniques such as biotin–avidin conjugation [112].

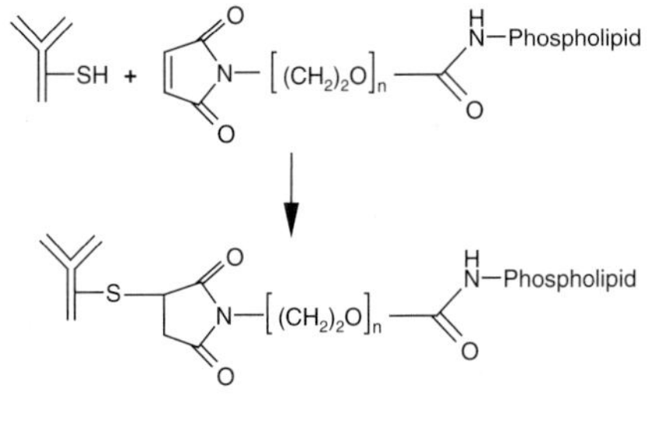

Figure 2.10. Schematic diagram of coupling of a thiolated antibody to a linker lipid (maleimide–PEG–phospholipid) which is part of a preformed liposome. The resulting thioether bond is metabolically stable. The strategy shown here was used to synthesize OX26-immunoliposomes [111].

An antibody used for brain targeting of immunoliposomes has to meet several requirements. First, the antibody should recognize a structure which is present exclusively at the blood–brain barrier. Second, the antibody should be able to cross the blood–brain barrier by an active transport mechanism such as receptor-mediated transcytosis. Third, the epitope against which the antibody is targeted should preferably not be species specific. Fourth, high quantities of the antibody should be available. The OX26 mAb [38] meets several (but not all) of the above requirements. *In vitro* experiments have demonstrated that OX26-immunoliposomes can be taken up specifically by living RG2 rat glioma cells overexpressing the rat transferrin receptor despite their particulate size of approximately 90 nm [113]. The fluorescent-labelled OX26-immunoliposomes accumulated within an intracellular (endosomal) compartment [114]. Similar results were obtained by incubation of fluorescent OX26-immunoliposomes with freshly isolated rat brain capillaries [115] which revealed binding to the luminal and basolateral membranes of the brain endothelium.

2.4.3.3 Drugs of Interest for Targeting to the Brain

Brain delivery of the anticancer drug daunomycin provides an example of the *in vivo* application of OX26-immunoliposomes [111]. Different formulations of [^{3}H]-daunomycin were i.v. administered to rats either as the free drug or encapsulated in conventional liposomes, sterically-stabilized liposomes, or PEG-conjugated immunoliposomes (Table 2.3). Plasma samples were taken at defined time points and after 1 h the animal was killed and drug concentrations in brain tissue were determined.

Free daunomycin and not PEG-conjugated liposomes containing daunomycin, disappear rapidly from the circulation. Plasma clearance of the liposome was reduced 66-fold by PEG-conjugation. Coupling 29 OX26 monoclonal antibodies per PEG-liposome partially reversed the effect of PEG-conjugation on plasma clearance.

Analysis of the blood–brain barrier permeability surface area (PS) product indicated that daunomycin, and to a lesser degree conventional liposomes, have the potential to penetrate the blood–brain barrier. However, brain tissue accumulation of free daunomycin or conventional liposomes was poor, being the result of their high systemic plasma clearance. The use of PEG-conjugated liposomes reduced the blood–brain barrier PS product value to zero. No brain uptake of the PEG-liposomes was observed, despite their marked increase in plasma circulation time. Conversely, the use of PEG-conjugated OX26 immunoliposomes increased the blood–brain barrier PS product, relative to PEG-liposomes, resulting in increased brain uptake. Thus, optimal brain delivery of daunomycin was achieved using OX26 immunoliposomes (see Table 2.3). Titration of the amount of OX26 conjugated per liposome (*n* between 3 and 197) revealed an increase in plasma clearance and a decrease in the systemic volume of distribution of immunoliposomes at higher OX26 concentrations. Highest PS product values and brain tissue accumulation was observed for immunoliposomes with 29 OX26 mAb. At higher OX26 densities on the liposome, a saturation effect was observed resulting in a reduction in volume of distribution, PS product and brain tissue accumulation of OX26 immunoliposomes.

Recently the OX26 immunoliposomes were used in a gene delivery approach to transport expression vectors for luciferase or β-galactosidase through the BBB [116]. The plasmids

Table 2.3. Pharmacokinetics of different formulations of [^{3}H]-daunomycin after i.v. administration to rats.

	Cl (ml min^{-1} kg^{-1})	**PS (µl min^{-1} g^{-1} tissue)**	**%ID g^{-1} tissue**
Daunomycin	44.7 ± 6.8	1.63 ± 0.20	0.009 ± 0.001
Liposomes	12.6 ± 6.3	0.21 ± 0.06	0.009 ± 0.004
PEG–liposomes	0.19 ± 0.01	0.001 ± 0.005	0.001 ± 0.003
29 OX26	0.91 ± 0.11	0.144 ± 0.038	0.029 ± 0.011
IgG_{2a}	0.37 ± 0.04	0.001 ± 0.006	0.001 ± 0.001

Plasma clearance (Cl), blood–brain barrier permeability surface area product (PS) and accumulation as % injected dose detected in brain tissue (%ID g^{-1} tissue) at 1 h after administration. Results show free [^{3}H]-daunomycin (Daunomycin), [^{3}H]-daunomycin encapsulated in conventional liposomes (Liposomes), sterically stabilized liposomes (PEG–liposomes), immunoliposomes (29 OX26, where 29 designates the number of OX26 mAb conjugated per liposome) and control immunoliposomes where the OX26 mAb was replaced by a non-specific isotype control antibody (IgG_{2a}). Values are means ± SEM of $n = 3$ experiments.

were physically entrapped inside the neutral liposomes rather than being complexed on the surface of cationic liposomes. Gene expression was demonstrated in brain cells beyond the BBB, indicating both penetration of the liposomes through the BBB *in vivo* and escape from the endosomal compartment by an as yet unidentified mechanism.

In conclusion, the use of an immunoliposome-based drug delivery system allows for targeted delivery of a small molecule such as daunomycin or plasmids to the rat brain *in vivo*. Further experiments will be needed to clarify the subcellular routes and compartments involved in the transcytosis mechanism, as well as the eventual release mechanism in the target cell.

2.5 Conclusions

Various strategies to circumvent or to overcome the BBB for brain-directed drug therapies are under evaluation. It can be predicted that for broad clinical application noninvasive methods will be required, in particular for chronic diseases where long-term treatment is necessary. The utilization of physiological transport mechanisms at the BBB in experimental models generated evidence that pharmacological effects can be achieved with this approach. In order to be useful as drug delivery systems in humans, several steps are necessary. The transport capacity must be increased, which is possible through improved vectors and optimized coupling strategies. In order to avoid potential immunogenicity of antibody-based vectors from murine sources, humanization techniques are now being applied [117]. Further developments may include specific targeting to neuronal or non-neuronal cells, and efficient intracellular release mechanisms.

References

[1] Shoulson I, *Science* **1998,** *282*, 1072–1074.
[2] Small GW, *Am. J. Geriatr. Psychiatry* **1998,** *6(2 Suppl. 1)*, S26–S33.
[3] Glenner GG, Wong CW, *Biochem. Biophys. Res. Commun.* **1984,** *120*, 885–890.
[4] Kang J, Lemaire HG, Unterbeck A, Salbaum JM, Masters CL, Grzeschik KH, Multhaup G, Beyreuther K, Muller-Hill B, *Nature* **1987,** *325,* 733–736.
[5] St George-Hyslop PH, *Biol. Psychiatry* **2000,** *47*, 183–199.
[6] Selkoe DJ, Wolfe MS, *Proc. Natl Acad. Sci. USA* **2000,** *97*, 5690–5692.
[7] Emilien G, Beyreuther K, Masters CL, Maloteaux JM, *Arch. Neurol.* **2000,** *57*, 454–459.
[8] Chen KS, Masliah E, Grajeda H, Guido T, Huang J, Khan K, Motter R, Soriano F, Games D, *Prog. Brain Res.* **1998,** *117*, 327–334.
[9] Walker LC, *Brain Res. Brain Res. Rev.* **1997,** *25*, 70–84.
[10] Polymeropoulos MH, Lavedan C, Leroy E, Ide SE, Dehejia A, Dutra A, Pike B, Root H, Rubenstein J, Boyer R, Stenroos ES, Chandrasekharappa S, Athanassiadou A, Papapetropoulos T, Johnson WG, Lazzarini AM, Duvoisin RC, Di Iorio G, Golbe LI, Nussbaum RL, *Science* **1997,** *276*, 2045–2047.
[11] Dunnett SB, Bjorklund A, *Nature* **1999,** *399 (Suppl.)*, A32–A39.
[12] DeLong MR, *Trends Neurosci.* **1990,** *13*, 281–285.
[13] Masliah E, Rockenstein E, Veinbergs I, Mallory M, Hashimoto M, Takeda A, Sagara Y, Sisk A, Mucke L, *Science* **2000,** *287*, 1265–1269.
[14] Dirnagl U, Iadecola C, Moskowitz MA, *Trends Neurosci.* **1999,** *22*, 391–397.
[15] Mikkelsen T, *Cancer Control* **1998,** *5*, 150–162.
[16] Black KL, *Adv. Drug Delivery Rev.* **1995,** *15*, 37–52.
[17] Tardieu M, *Curr. Opin. Neurol.* **1999,** *12*, 377–381.
[18] Nath A, *Semin. Neurol.* **1999,** *19*, 113–127.
[19] Brightman MW, *Exp. Eye Res.* **1977,** *25 (Suppl.)*, 1–25.
[20] Smith QR, Rapoport SI, *J. Neurochem.* **1986,** *46*, 1732–1742.
[21] Cancilla PA, Bready J, and Berliner J, In: *The Blood–Brain Barrier – Cellular and Molecular Biology* (Ed. Pardridge WM), pp. 29–46. Raven Press, New York, **1993.**
[22] Balabanov R, Dore-Duffy P, *J. Neurosci. Res.* **1998,** *53*, 637–644.
[23] Pardridge WM, Triguero D, Yang J, Cancilla PA, *J. Pharmacol. Exp. Ther.* **1990,** *253*, 884–891.
[24] Duvernoy H, Delon S, Vannson JL, *Brain Res. Bull.* **1983,** *11*, 419–480.
[25] Segal M, In: Introduction to the Blood–Brain Barrier: Methodology, Biology and Pathology (Ed. Pardridge WM), pp. 251–258. Cambridge University Press, Cambridge, 1998.
[26] Dohrmann GJ, Brain Res. 1970, 18, 197–218.
[27] Levin VA, J. Med. Chem. 1980, 23, 682–684.
[28] Drewes LR, In: *Introduction to the Blood–Brain Barrier: Methodology, Biology and Pathology* (Ed. Pardridge WM), pp. 165–174. Cambridge University Press, Cambridge, **1998.**
[29] Boado RJ, Li JY, Nagaya M, Zhang C, Pardridge WM, *Proc. Natl Acad. Sci. USA* **1999,** *96*, 12079–12084.
[30] Smith QR, Stoll J, In: *Introduction to the Blood–Brain Barrier: Methodology, Biology and Pathology* (Ed. Pardridge WM), pp. 188–197. Cambridge University Press, Cambridge, **1998.**
[31] Smith QR, In: *Frontiers in Cerebral Vascular Biology: Transport and its Regulation* (Eds Drewes LR, Betz AL), pp. 83–93. Plenum Press, New York, **1993.**
[32] Borst P, Schinkel AH, In: *Introduction to the Blood–Brain Barrier:Methodology, Biology and Pathology* (Ed. Pardridge WM), pp. 198–206. Cambridge University Press, Cambridge, **1998.**
[33] Gao B, Hagenbuch B, Kullak-Ublick GA, Benke D, Aguzzi A, Meier PJ, *J. Pharmacol. Exp. Ther.* **2000,** *294*, 73–79.
[34] Sugiyama Y, Kusuhara H, Suzuki H, *J. Control. Rel.* **1999,** *62*, 179–186.
[35] Pardridge WM, Eisenberg J, Yang J, *J. Neurochem.* **1985,** *44*, 1771–1778.
[36] Pardridge WM, Eisenberg J, Yang J, *Metabolism* **1987,** *36*, 892–895.
[37] Bickel U, Kang YS, Yoshikawa T, Pardridge WM, *J. Histochem. Cytochem.* **1994,** *42*, 1493–1497.
[38] Jefferies WA, Brandon MR, Hunt SV, Williams AF, Gatter KC, Mason DY, *Nature* **1984,** *312*, 162–163.
[39] Dehouck B, Fenart L, Dehouck MP, Pierce A, Torpier G, Cecchelli R, *J. Cell Biol.* **1997,** *138*, 877–889.
[40] Broadwell RD, Balin BJ, Salcman M, *Proc. Natl Acad. Sci. USA* **1988,** *85*, 632–636.
[41] Bickel U, Yoshikawa T,Pardridge WM, *Adv. Drug Deliv. Rev.* **1993,** *10*, 205–245.

[42] Oldendorf WH, *Brain Res* **1970,** *24*, 372–376.
[43] Takasato Y, Rapoport SI, Smith QR, *Am. J. Physiol.* **1984,** *247(3 Pt 2)*, H484–H493.
[44] Triguero D, Buciak J, Pardridge WM, *J. Neurochem.* **1990,** *54*, 1882–1888.
[45] Samii A, Bickel U, Stroth U, Pardridge WM, *Am. J. Physiol.* **1994,** *267(1 Pt 1)*, E124–E131.
[46] Rapoport SI, Fitzhugh R, Pettigrew KD, Sundaram U, Ohno K, *Am. J. Physiol.* **1982,** *242*, R339–R348.
[47] Klecker Jr RW, Collins JM, Yarchoan R, Thomas R, Jenkins JF, Broder S, Myers CE, *Clin. Pharmacol. Ther.* **1987,** *41*, 407–412.
[48] Takasawa K, Suzuki H, Sugiyama Y, *Biopharm. Drug Dispos.* **1997,** *18*, 611–622.
[49] de Lange EC, de Boer BA, Breimer DD, *Adv. Drug Deliv. Rev.* **1999,** *36*, 211–227.
[50] Morgan ME, Singhal D, Anderson BD, *J. Pharmacol. Exp. Ther.* **1996,** *277*, 1167–1176.
[51] Pardridge WM, In: *Introduction to the Blood-Brain Barrier: Methodology, Biology and Pathology* (Ed. Pardridge WM), pp. 49–61. Cambridge University Press, Cambridge, **1998.**
[52] Audus KL, Rose JM, Wang W, Borchardt RT, In: *Introduction To The Blood-Brain Barrier: Methodology, Biology And Pathology* (Ed. Pardridge WM), pp. 86–93. Cambridge University Press, Cambridge, **1998.**
[53] Pagliara A, Reist M, Geinoz S, Carrupt PA, Testa B, *J. Pharm. Pharmacol.* **1999,** *51*, 1339–1357.
[54] Tsuzuki N, Hama T, Kawada M, Hasui A, Konishi R, Shiwa S, Ochi Y, Futaki S, Kitagawa K, *J. Pharm. Sci.* **1994,** *83*, 481–484.
[55] Bodor N, *J. Control. Rel.* **1999,** *62*, 209–222.
[56] Wade LA, Katzman R, *Am. J. Physiol.* **1975,** *228*, 352–359.
[57] Nau R, Sorgel F, Prange HW, *Clin. Pharmacokinet.* **1998,** *35*, 223–246.
[58] DeAngelis LM, *J. Neurooncol.* **1998,** *38*, 245–252.
[59] Penn RD, Savoy SM, Corcos D, Latash M, Gottlieb G, Parke B, Kroin JS, *N. Engl. J. Med.* **1989,** *320*, 1517–1521.
[60] Lazorthes YR, Sallerin BA, Verdie JC, *Neurosurgery* **1995,** *37*, 422–428; discussion 428–429.
[61] Ferguson IA, Schweitzer JB, Bartlett PF and Johnson Jr EM, *J. Comp. Neurol.* **1991,** *313*, 680–692.
[62] Fishman RA, *Cerebrospinal Fluid in Diseases of the Nervous System*, W.B. Saunders Company, Philadelphia, **1980.**
[63] Davson H, Segal MB, *Physiology of the CSF and Blood–Brain Barriers*, CRC Press, Boca Raton, **1996.**
[64] Shapiro WR, Young DF, Mehta BM, *N. Engl. J. Med.* **1975,** *293*, 161–166.
[65] Blasberg RG, Patlak C, Fenstermacher JD, *J. Pharmacol. Exp. Ther.* **1975,** *195*, 73–83.
[66] Krewson CE, Saltzman WM, *Brain Res.* **1996,** *727*, 169–181.
[67] Mahoney MJ, Saltzman WM, *Proc. Natl Acad. Sc.i USA* **1999,** *96*, 4536–4539.
[68] Kordower JH, Winn SR, Liu YT, Mufson EJ, Sladek Jr JR, Hammang JP, Baetge EE, Emerich DF, *Proc. Natl Acad. Sci. USA* **1994,** *91*, 10898–10902.
[69] Bobo RH, Laske DW, Akbasak A, Morrison PF, Dedrick RL, Oldfield EH, *Proc. Natl Acad. Sci. USA* **1994,** *91*, 2076–2080.
[70] Laske DW, Youle RJ, Oldfield EH, *Nature Med.* **1997,** *3*, 1362–1368.
[71] Hagihara N, Walbridge S, Olson AW, Oldfield EH, Youle RJ, *Cancer Res.* **2000,** *60*, 230–234.
[72] Robinson PJ, Rapoport SI, *Am. J. Physiol.* **1987,** *253(3 Pt 2)*, R459–R466.
[73] Salahuddin TS, Johansson BB, Kalimo H, Olsson Y, Acta Neuropathol. (Berl.) 1988, 77, 5–13.
[74] Muldoon LL, Nilaver G, Kroll RA, Pagel MA, Breakefield XO, Chiocca EA, Davidson BL, Weissleder R, Neuwelt EA, *Am. J. Pathol.* **1995,** *147*, 1840–1851.
[75] Doran SE, Ren XD, Betz AL, Pagel MA, Neuwelt EA, Roessler BJ, Davidson BL, *Neurosurgery* **1995,** *36,* 965–970.
[76] Zunkeler B, Carson RE, Olson J, Blasberg RG, DeVroom H, Lutz RJ, Saris SC, Wright DC, Kammerer W, Patronas NJ, Dedrick RL, Herscovitch P, Oldfield EH, *J. Neurosurg.* **1996,** *85*, 1056–1065.
[77] Nakano S, Matsukado K, Black KL, *Neurol. Res.* **1997,** *19*, 501–508.
[78] Gregor A, Lind M, Newman H, Grant R, Hadley DM, Barton T, Osborn C, *J. Neurooncol.* **1999,** *44*, 137–145.
[79] Pardridge WM, *Endocrinol. Rev.* **1986,** *7*, 314–330.
[80] Kumagai AK, Eisenberg JB, Pardridge WM, *J. Biol. Chem.* **1987,** *262,* 15214–15219.
[81] Pardridge WM, Triguero D, Buciak JL, *Endocrinology* **1990,** *126*, 977–984.
[82] Wu D, Kang YS, Bickel U, Pardridge WM, *Drug Metab. Dispos.* **1997,** *25*, 768–771.
[83] Lee HJ, Engelhardt B, Lesley J, Bickel U, Pardridge WM, *J. Pharmacol. Exp. Ther.* **2000,** *292*, 1048–1052.

[84] Yoshikawa T, Pardridge WM, *J. Pharmacol. Exp. Ther.* **1992,** *263*, 897–903.
[85] Friden PM, Walus LR, Watson P, Doctrow SR, Kozarich JW, Backman C, Bergman H, Hoffer B, Bloom F, Granholm AC, *Science* **1993,** *259*, 373–377.
[86] Kang YS, Pardridge WM, *J. Pharmacol. Exp. Ther.* **1994,** *269*, 344–350.
[87] Saito Y, Buciak J, Yang J, Pardridge WM, *Proc. Natl Acad. Sci. USA* **1995,** *92*, 10227–10231.
[88] Penichet ML, Kang YS, Pardridge WM, Morrison SL, Shin SU, *J. Immunol.* **1999,** *163*, 4421–4426.
[89] Bickel U, Yoshikawa T, Landaw EM, Faull KF, Pardridge WM, *Proc. Natl Acad. Sci. USA* **1993,** *90*, 2618–2622.
[90] Pardridge WM, Wu D, Sakane T, *Pharm. Res.* **1998,** *15*, 576–582.
[91] Mandel R, Ryser HJ, Ghani F, Wu M, Peak D, *Proc. Natl Acad. Sci. USA* **1993,** *90*, 4112–4116.
[92] Bickel U, Kang YS, Pardridge WM, *Bioconjug. Chem.* **1995,** *6*, 211–218.
[93] Wu D, Pardridge WM, *Proc. Natl Acad. Sci. USA* **1999,** *96*, 254–259.
[94] Deguchi Y, Kurihara A, Pardridge WM, *Bioconjug. Chem.* **1999,** *10*, 32–37.
[95] Wu D, Pardridge WM, *J. Pharmacol. Exp. Ther.* **1996,** *279*, 77–83.
[96] Backman C, Rose GM, Hoffer BJ, Henry MA, Bartus RT, Friden P, Granholm AC, *J. Neurosci.* **1996,** *16*, 5437–5442.
[97] Kordower JH, Charles V, Bayer R, Bartus RT, Putney S, Walus LR, Friden PM, *Proc. Natl Acad. Sci. USA* **1994,** *91*, 9077–9080.
[98] Albeck DS, Hoffer BJ, Quissell D, Sanders LA, Zerbe G, Granholm AC, *Neuroreport* **1997,** *8*, 2293–2298.
[99] Boado RJ, Tsukamoto H, Pardridge WM, *J. Pharm. Sci.* **1998,** *87*, 1308–1315.
[100] Pardridge WM, Boado RJ, Kang YS, *Proc. Natl Acad. Sci. USA* **1995,** *92*, 5592–5596.
[101] Wu D, Yang J, Pardridge WM, *J. Clin. Invest.* **1997,** *100*, 1804–1812.
[102] Kurihara A, Pardridge WM, *Cancer Res.* **1999,** *59*, 6159–6163.
[103] Chonn A, Cullis PR, *Curr. Opin. Biotechnol.* **1995,** *6*, 698–708.
[104] Zelphati O, Szoka Jr FC, *Proc. Natl Acad. Sci. USA* **1996,** *93*, 11493–11498.
[105] Allen TM, *Drugs* **1998,** *56*, 747–756.
[106] Lasic DD, Papahadjopoulos D, *Science* **1995,** *267*, 1275–1276.
[107] Micklus MJ, Greig NH, Tung J, Rapoport SI, *Biochim. Biophys. Acta* **1992,** *1124*, 7–12.
[108] Gennuso R, Spigelman MK, Chinol M, Zappulla RA, Nieves J, Vallabhajosula S, Alberto Paciucci P, Goldsmith SJ, Holland JF, *Cancer Invest.* **1993,** *11*, 118–128.
[109] Derksen JT, Morselt HW, Scherphof GL, *Biochim. Biophys. Acta* **1988,** *971*, 127–136.
[110] Papahadjopoulos D, Allen TM, Gabizon A, Mayhew E, Matthay K, Huang SK, Lee KD, Woodle MC, Lasic DD, Redemann C, Martin FJ, *Proc. Natl Acad. Sci. USA* **1991,** *88*, 11460–11464.
[111] Huwyler J, Wu D, Pardridge WM, *Proc. Natl Acad. Sci. USA* **1996,** *93*, 14164–14169.
[112] Hansen CB, Kao GY, Moase EH, Zalipsky S, Allen TM, *Biochim. Biophys. Acta* **1995,** *1239*, 133–144.
[113] Huwyler J, In: *Brain Barrier Systems*, Vol. 45 (Eds Paulson OB, Moos Knudsen G, Moos T), Munksgaard, Copenhagen, **1999.**
[114] Huwyler J, Yang J, Pardridge WM, *J. Pharmacol. Exp. Ther.* **1997,** *282*, 1541–1546.
[115] Huwyler J, Pardridge WM, *J. Neurochem.* **1998,** *70*, 883–886.
[116] Shi N, Pardridge WM, *Proc. Natl Acad. Sci. USA* **2000,** *97*, 7567–7572.
[117] Coloma MJ, Lee HJ, Kurihara A, Landaw EM, Boado RJ, Morrison SL, Pardridge WM, *Pharm. Res.* **2000,** *17*, 266–274.
[118] Pardridge WM, Oldendorf WH, Cancilla P, Frank HJ, *Ann. Intern. Med.* **1986,** *105*, 82–95.
[119] Weindl A, In: *Frontiers in Neuroendocrinology* (Eds Ganong WF, Martini L), pp. 3–32. Oxford University Press, New York, **1973.**
[120] Clementi F, Marini D, *Z. Zellforsch. Mikroskop. Anat.* **1972,** *123*, 82–95.
[121] Pardridge WM, *J. Neurochem.* **1998,** *70*, 1781–1792.
[122] Lonser RR, Corthesy ME, Morrison PF, Gogate N, Oldfield EH, *J. Neurosurg.* **1999,** *91*, 294–302.
[123] Morrison PF, Laske DW, Bobo H, Oldfield EH, Dedrick RL, *Am. J. Physiol.* **1994,** *266(1 Pt 2)*, R292–R305.

3 Pulmonary Drug Delivery: Delivery To and Through the Lung

Anne H. de Boer, Grietje Molema, Henderik W. Frijlink

3.1 Introduction

The respiratory tract is one of the oldest routes used for the administration of drugs. Anaesthetics, aerosolized drugs, smoke or steam have been inhaled for medical purposes for centuries. Over the past decades inhalation therapy has established itself as a valuable tool in the local therapy of pulmonary diseases such as asthma or COPD (Chronic Obstructive Pulmonary Disease) [1]. This type of drug application in the therapy of these diseases is a clear form of targeted drug delivery: the major advantages are a rapid onset of the therapeutic effect, a lowering of the required dose (as compared to systemic administration) and a reduction in unwanted side-effects (increased therapeutic index). Currently, over 25 drug substances are marketed as inhalation aerosol products for local pulmonary effects and about the same number of drugs are in different stages of clinical development. Furthermore, there are some drugs that are not marketed as inhalation aerosols *per se* but are formulated as such by pharmacists.

The majority of the marketed products are used for asthma and COPD. Typical agents that are used for these indications are β_2-agonists such as salbutamol (albuterol), Terbutalin or formoterol, corticosteroids such as budesonide, Flixotide or beclomethasone and mast-cell stabilizers such as sodium cromoglycate or nedocromil.

Patients suffering from cystic fibrosis often use various aerosolized drugs. To reduce the viscosity of the mucus in the airways, recombinant human deoxyribonuclease is used. This enzyme is the first recombinant protein that has been developed for specific delivery to the lungs via the airways. It has a local action on the mucus in the airways and its absorption is minimal. Another drug that decreases the viscosity of the mucus is acetylcysteine. Aerosolized antibiotics are a further group of therapeutics that is widely used by cystic fibrosis patients. Solutions of antibiotics like tobramycin or colistin are used in nebulizers to prevent exacerbation of the disease. Pentamidine has been used for the prophylaxis of Pneumocystis pneumonia in patients infected with HIV virus, while chronic rejection of lung transplants provided a reason to develop an aerosol formulation of cyclosporine A.

The latest and probably one of the most promising applications of pulmonary drug administration is its use to achieve systemic absorption of the administered drug substances. Particularly for those drug substances that exhibit a poor bioavailability when administered by the oral route, as for example peptides or proteins, the respiratory tract might be a convenient port of entry [2]. For this application a more or less contradictory situation occurs: 'delivery (into the lung) is required to obtain systemic absorption followed by a non-targeted distribution of the drug substance'. In this case, the biopharmaceutical objective of improved bioavailability and not the improved therapeutic index for the drug is the rationale for organ delivery.

Systemic absorption of pulmonary-delivered peptides and proteins has been the objective of many investigations [2]. The most successful work in this field is the development of insulin formulations for inhalation. These dosage forms might, in the near future, become a suitable alternative for the current subcutaneous injection of insulin that is used to obtain meal-time glucose control [3]. In spite of the strict requirements regarding dose variability for insulin, the pulmonary products under development seem to be as safe as the subcutaneous injections.

Numerous other peptides and proteins have been, or are still in development as inhalation products with the objective of systemic absorption. Examples of these are: calcitonin, LH-RH antagonists, recombinant human granulocyte colony-stimulating factor and growth hormone.

Protein inhalation products that have been developed for local use are interferon, alpha-1-antitrypsin and secretory leukoprotease inhibitor. Other therapeutic products that have been investigated with regard to delivery to the lungs are genetic material (plasmid DNA) and vaccines. For example, the delivery of the gene encoding the cystic fibrosis transmembrane conductance regulator was extensively investigated [4]. Delivery of genes requires specific vector systems which enable the cellular transfection of the gene. Vectors that have been investigated are retroviral and adenovirus-associated vectors, recombinant adenoviruses, cationic liposomes and DNA–ligand complexes. However, none of these approaches was found to be successful in clinical studies up to now. Other diseases for which gene therapy via the lung was investigated are lung carcinoma malignant mesothelioma and alpha$_1$-antitrypsin deficiency [5].

Pulmonary drug delivery for local or systemic therapy comprises several aspects. Formulation of the drug, the generation of the aerosol, the lung deposition of inhaled particles and the passage of the drug substance over the epithelial membranes of the respiratory tract, all represent crucial aspects of pulmonary drug administration. During the last decade many of these aspects have been studied extensively [1,2,6–18]. This chapter summarizes the aspects that are of relevance in the development of new drug products for inhalation. Focus will be placed on the physical characteristics of the inspiratory flow curve since this is the driving force that finally brings the particles into the lung. This process is also relevant to understanding the generation of the aerosol cloud as for example in dry powder inhalers. Further aspects that will be discussed are the apparatus to generate the aerosol, lung deposition mechanisms, typical formulation types for the pulmonary route as well as the mechanism of transport over the alveolar membrane.

Principles of fluid and particle dynamics in the respiratory tract (physical and anatomical parameters) are also discussed, as they are the starting point for the development of drug products for inhalation. In fact, they set the conditions used for *in vitro* and *in vivo* testing of inhalation systems and define the specifications for new inhalation systems.

Finally, it should be noted that apart from the use of the pulmonary route for protein absorption (targeting *through* the lung), the lung can also be an object for organ selective delivery (targeting *to* the lung). The latter can in theory be achieved by two entirely different routes: deposition via the airways from the luminal side and targeting via the blood circulation. Whereas the main part of this chapter is dedicated to the former approach, the latter approach will also be briefly discussed (Section 3.12).

3.2 The Respiratory Tract

The human respiratory tract is a branching system of air channels with approximately 23 bifurcations from the mouth to the alveoli [8,18,19]. In Figure 3.1 a schematic representation of the human airways as described by Weibel [20] is shown. Furthermore, this figure shows some typical geometric features of the lung. The major task of the lungs is gas exchange, by adding oxygen to, and removing carbon dioxide from the blood passing the pulmonary capillary bed [21]. This task is facilitated by inhaling certain quantities of fresh air into the lungs at regular intervals and exhaling similar volumes of used air in between. The muscles that are responsible for this task can be divided into inspiratory and expiratory muscles. During inhalation, the chest is expanded mainly longitudinally by contraction of the main inspiratory muscle: the dome shaped diaphragm in the lower part of the chest. This enlargement of the chest volume creates an underpressure in the lungs which is the driving force for an airflow, entering through the mouth or nose. Expiration during quiet breathing occurs passively as a result of recoil of the lung. Only during heavy breathing are expiratory muscles, which depress the ribs, activated.

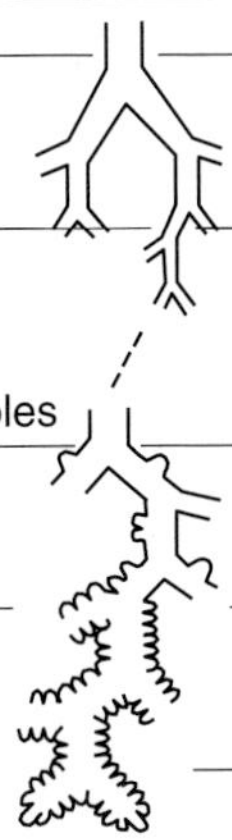

	Generation		diameter (cm)	length (cm)	Number	total cross sectional area (cm^2)
conducting zone	Trachea	0	1.80	12.0	1	2.54
	Bronchi	1	1.22	4.8	2	2.33
	↓	2	0.83	1.9	4	2.13
		3	0.56	0.8	8	2.00
	Bronchioles	4	0.45	1.3	16	2.48
	↓	5	0.35	1.07	32	3.11
		↓	↓	↓	↓	↓
	Terminal bronchioles	16	0.06	0.17	$6 \cdot 10^4$	180.0
transitional and respiratory zones	Respiratory Bronchioles	17	↓	↓	↓	↓
		18				
		19	0.05	0.10	$5 \cdot 10^5$	10^3
	Alveolar ducts	20	↓	↓	↓	↓
		21				
		22				
	Alveolar sacs	23	0.04	0.05	$8 \cdot 10^6$	10^4

Figure 3.1. Schematic representation of the lung according to the model described by Weibel [20].

3.2.1 Lung Capacities and Pulmonary Ventilation

The inhaled air volume (V in L) depends on the extent of chest enlargement. During normal breathing, the inhaled and exhaled volumes (tidal volume) are only part of the total lung volume [8,21]. The different parameters that describe pulmonary ventilation are shown in Figure 3.2. Table 3.1 presents a definition of the different parameters. Normal adults have a tidal

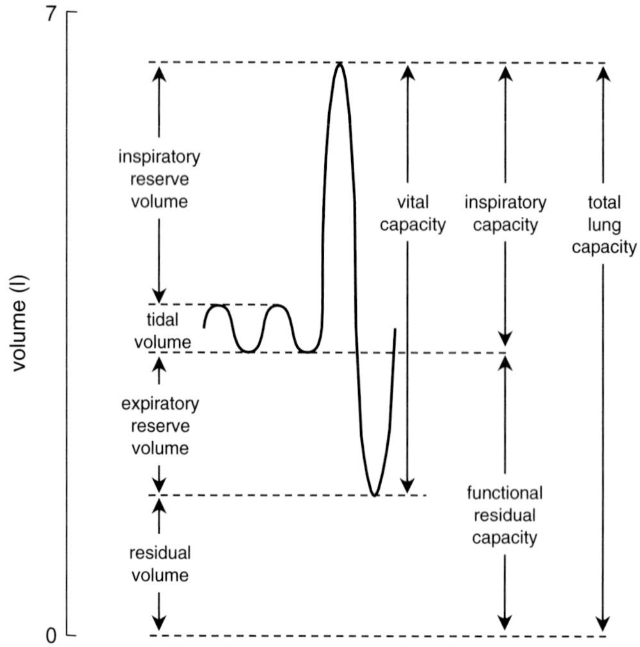

Figure 3.2. Schematic diagram of the different volumes describing pulmonary ventilation.

volume (V_T) of approximately 0.7 l, and inhale with a frequency of about 12 times a minute at rest [8]. The amount of air processed under these conditions is 12 m^3 per day (with a range of 10–20 m^3 per day). During heavy work, the tidal volume may be increased by a factor 3. A residual volume (*RV*) of 1 to 1.5 l is not exhaled during normal breathing: this volume is increased when a patient suffers from an obstruction e.g. in the case of asthma. Total lung capacities (*TLC*s) of adults are estimated to be 5 to 7 l [1], maximal inspired volumes (vital capacities: *VC*s) were found to be dependent on the external resistance and vary from less than

Table 3.1. Definitions of the different parameters describing pulmonary ventilation.

Parameter	Definition
Total lung capacity	The volume of air in the lung after a maximal inspiratory effort
Inspiratory capacity	The volume of air maximally inspired after a normal tidal expiration
Functional residual capacity	The volume of air remaining in the lung at the end of normal tidal expiration
Vital capacity	The maximum volume of air expired after a maximal forced inspiration
Inspiratory reserve volume	The maximum volume of air inspired after a normal tidal inspiration
Tidal volume	The volume of air entering or leaving the lung at each normal breath
Expiratory reserve volume	The maximum volume of air expired after normal tidal expiration
Residual volume	The volume of air left in the lung after a maximal forced expiratory effort

1 l (for high airflow resistances) to more than 2.5 l (for low resistances) for healthy adults [22].

Healthy subjects have hardly any alveolar dead space. However, disease may increase the dead space in the alveoli [15]. This might be of importance when alveolar deposition is desired, for example to obtain systemic absorption.

Attainable underpressures and inspiratory flow rates, which are especially relevant to the performance of dry powder inhalers, are discussed more in detail in Section 3.9 as function of the patient's effort, age and clinical condition and the (external) inhaler resistance. On the basis of Weibel's lung model (Figure 3.1), showing a strongly increasing total cross sectional area for airflow with increasing generation number, starting from the lobar bronchi, it can be calculated that the air velocity decreases with increasing penetration depth. At a common inspiratory flow rate of 60 l min^{-1} through a dry powder inhaler, air velocity first increases from approximately 4 m s^{-1} in the trachea to a maximum of 4.6 m s^{-1} in the lobar bronchi. But starting at the segmental bronchi, a steep decrease in velocity occurs to 0.5 m s^{-1} in the terminal bronchi and not more than 0.05 m s^{-1} in the terminal bronchioles. So, in the periphery of the lungs, the air is practically still. A similar falling off can be calculated for the Reynolds number, being 4800 in the trachea, 40 in the terminal bronchi and only 2 in the terminal bronchioles. This means that the flow is turbulent (at 60 l min^{-1}) in the upper respiratory tract and laminar in the central and deep lung. The decreasing air velocity is important for particle deposition in the lungs, as will be discussed in Section 3.3.

3.3 Lung Deposition and Particle Size

Airborne particles travelling through the respiratory tract are subjected to constantly changing forces as a result of bends and the decreasing air velocity (Section 3.2.1). In the absence of tribocharge, electrostatic forces play no role and particle behaviour is governed mainly by inertial forces, the drag force (Stokes' law) and the force of gravity [8]. In the relatively wide upper airways, where the air velocity is highest (Section 3.2.1), inertial forces are dominant. Particles enter the airway system with (near) air velocity, unless they have been accelerated to much higher speed, as by discharge from a metered dose aerosol [23]. They will have to follow the streamlines of the air in bends and bifurcations in order to penetrate deeper, but are unable to do so when their inertia is too high (either from a high mass or a high velocity, or both). Therefore, the largest particles are deposited by the mechanism of *inertial impaction* in the throat and first bifurcations. As the remaining small particles move on to the central lung, the air velocity gradually decreases to much lower values and the force of gravity becomes important. Settling by *sedimentation* is the dominant deposition mechanism in this part of the respiratory tract. However, settling velocity is too low and residence time too short to remove the smallest particles in the aerosol cloud from the air by this mechanism. So, the finest fractions are able to enter the periphery of the lung where they can make contact with the walls of the airways as the result of *Brownian motion* (particle diffusion). In rare cases, particle interception contributes to drug deposition, especially near obstructions in the smaller airways.

Particles entering the respiratory tract may not only vary in size and velocity, but also in shape and density, depending upon the type of drug and the inhalation system used for aerosol generation (see Section 3.5). In order to be able to compare the behaviour of different types of aerosol particles with each other, the aerodynamic particle diameter (D_A) has been introduced. By definition, the aerodynamic diameter of a particle is the diameter of a unit density sphere ($\rho_P = 1$ g cm^{-3}) having the same terminal settling velocity (in still air) as the particle under consideration. Irregular particles can also be expressed in terms of equivalent volume diameter (D_E) and dynamic shape factor (χ). The equivalent volume diameter is the diameter of a sphere having the same volume as the irregular particle, whereas the dynamic shape factor is the ratio of the actual resistance force on a non-spherical particle to the resistance force on a sphere having the same volume and velocity. The aerodynamic diameter can be calculated from the equivalent volume diameter, which is an expression of the geometric particle size, when particle density and dynamic shape factor are known, using Eq. 3.1:

$$D_A = D_E \cdot (\rho_P/\chi)^{0.5} \tag{3.1}$$

Deposition efficiencies for particles in the respiratory tract are generally presented as a function of their aerodynamic diameter (e.g. [8,12]). Large particles (> 10 µm) are removed from the airstream with nearly 100% efficiency by inertial impaction, mainly in the oropharynx. But as sedimentation becomes more dominant, the deposition efficiency decreases to a minimum of approximately 20% for particles with an aerodynamic diameter of 0.5 µm. When particles are smaller than 0.1 µm, the deposition efficiency increases again as a result of diffusional displacement. It is believed that 100% deposition due to Brownian motion might be possible for particles in the nanometer range.

Both from deposition studies and force balances it can be derived that the optimum (aerodynamic) particle size lies between 0.5 and 7.5 µm. Within this approximate range many different subranges have been presented as most favourable, e.g. 0.1 to 5 µm [24], 0.5 to 8.0 µm [25], 2 to 7 µm [26] and 1–5 µm [27–29]. Particles of 7.5 µm and larger mainly deposit in the oropharynx [30] whereas most particles smaller than 0.5 µm are exhaled again [31]. All inhalation systems for drug delivery to the respiratory tract produce polydisperse aerosols which can be characterized by their mass median aerodynamic diameter (MMAD) and geometric standard deviation (σ_G). The MMAD is the particle diameter at 50% of the cumulative mass curve.

3.4 Drug Absorption *via* the Lung

During the past decade the lung has been (re)discovered as a suitable port of entry to the systemic circulation for various drugs. Among these drugs are many peptides and proteins [2], since the oral route cannot be used for these molecules.

In relation to systemic absorption of drugs, absorption in the lung can be described as the passage of a series of barriers by the drug in order to enter the systemic circulation. It is important to realize that physiological conditions in the lung differ widely from site to site.

Major physiological factors that affect pulmonary absorption are [10]:

- Mucociliary transport in the airways that constantly drains fluid and solid particles (bacteria) in a counter-current flow to the oral cavity. A drug that is deposited in the airways can be cleared by this mechanism within several hours. So, if systemic absorption over the bronchial membrane is required, this has to occur relatively fast.

- The epithelial cells in the alveoli are covered by a thin layer of so-called epithelial lining fluid. This fluid in turn is covered by a monolayer of lung surfactant, which also is present in large amounts in the alveolar lining fluids. Moreover, the lining fluid often contain enzymes that can metabolize drug substances.
- The epithelial cell layer forms the major barrier to absorption of drug molecules. In the large airways stratified epithelium occurs, whereas in the alveoli the epithelium is only one cell layer thick.
- After passing the alveolar epithelium, the molecule enters the interstitium being part of the extracellular space inside the tissue. In the alveoli this space is relatively small.
- Finally, for passage into the blood, the molecules have to pass the endothelial membrane of the capillaries, separating the interstitial space from the blood. The endothelial membrane is considered to be much 'leakier' than the epithelial membrane. Therefore it is not considered as a major barrier during drug absorption.
- Macrophages can also form a functional barrier for some particular drug substances during pulmonary absorption. Macrophages are able to ingest particulate material present in the alveoli or airways. After phagocytosis the macrophages migrate either to the ciliated bronchial airways or via the alveolar interstitial space to the lymphatic system. Infection or inflammation may increase their numbers significantly, and increased phagocytosis of particles may occur. If protein drugs, deposited in the lung as particles, are internalized by macrophages, the drug may be partly destroyed by the efficient proteolytic degradation mechanism of these cells.

For an efficient pulmonary absorption process, the alveolar membrane seems to be an optimal absorption site for a number of reasons.

- In contrast to the airways, there is hardly any mucociliary clearance from the alveoli.
- The alveolar membrane forms the largest surface area in the lung.
- The area of the alveoli is 43 to 102 m^2 which is large in comparison to the surface area of the airways which have a cumulative area of about 2.5 m^2 [10].
- The alveolar epithelium is thinner and leakier than the bronchial epithelium.

If a drug is deposited in the alveoli, it will, in first instance, come into contact with the alveolar lining fluids. Long chain phospholipids (often referred to as 'lung surfactant') are the major constituent of this fluid. These amphiphilic insoluble molecules form a molecular monolayer covering the epithelial surface fluids. Before any absorption of drug from the alveoli can occur, the drug will have to be dissolved. When dry powder inhalers are used dissolution of the drug in the alveolar lining fluids might be significantly affected by the presence of such phospholipids. The dissolution of lipophilic drug molecules in particular, is likely to be enhanced by their presence.

The fact that the volume of the epithelial lining fluids in the alveoli is small, implies that for many drugs this volume is insufficient to provide sink conditions for dissolution. The absorption rate will therefore be highly dependent on the dissolution rate of the inhaled product. The micronization of drugs for inhalation is therefore not only a requirement for deep lung deposition but is also useful for the rapid absorption of the drug through fast dissolution of the solid drug.

The alveolar epithelium consists of so-called Type I and Type II cells. Type I cells cover over 90% of the alveolar surface, have a large surface, and are thin. Type II cells are larger in numbers but are small. Therefore, they cover only about 7% of the surface of the alveoli. Type II cells produce the phospholipids that make up the surfactant layer.

It should be noted that the permeability per surface unit of alveolar epithelium *per se* is not particularly high. The significant absorption found for various substances after pulmonary administration is rather explained by a number of beneficial factors such as the large surface area of the alveoli, the low volume of the epithelial lining fluid, the relatively thin diffusion layer, the absence of mucociliary clearance from the alveoli as well as the limited enzymatic activity in the lining fluids.

Passage over the epithelial membrane from the apical to the basal site may occur via different routes. The fast absorption found for molecules smaller than 40 kDa is generally explained by paracellular transport through the tight junctions between the epithelial cells. Although rather incorrect from a physiological point of view, the estimation of 'pore sizes' of alveolar epithelium on the basis of transport rates of solutes, may help to predict whether or not a certain molecule can pass the alveolar epithelium. The value of 40 kDa is compatible with the presence of pore structures with a diameter of about 5 nm. The turnover of epithelial surface cells may be responsible for the transient existence of larger openings in the alveolar epithelium. However, whether these pores have any significance to drug absorption is unknown.

The absorption of molecules that are larger than 40 kDa is generally slow and incomplete. These molecules probably cannot pass through the tight junctions of the epithelial membrane, but have to be transported by a transcytotic mechanism in order to be absorbed. Receptor-mediated endocytosis is a crucial mechanism here. The subsequent transport through the epithelial cells may occur in coated and non-coated vesicles. The non-coated vesicles are called caveolae. Macromolecules (after receptor recognition) may be sequestered both in coated vesicles as well as in caveolae. In the alveolar Type I cells, large numbers of caveolae are found (about 1.7 million per cell). The caveolae have internal diameters of 50 to 100 nm which is large enough to contain macromolecules with sizes over 400 kDa. However, in spite of these well-defined physiological processes the evidence for massive transport of larger macromolecules via this pathway is scarce. In general it is doubtful whether transcytosis via caveolae may significantly contribute to the absorption of macromolecules [10,32].

In relation to the above it is obvious that passage of the pulmonary epithelium may depend on characteristics of a drug molecule. Not only the size, but also its solubility, overall charge, structural conformation and potential aggregation can have a significant effect on the absorption rate and bioavailability of the drug after pulmonary deposition.

3.4.1 Systemic Delivery of Peptides and Proteins

Many studies have been carried out regarding the absorption of peptides and proteins after pulmonary drug delivery. The perspectives of a non-parenteral route of administration for larger proteins led to studies on the pulmonary absorption of proteins of different size. To date, over 30 different proteins have been evaluated with regard to absorption rate and

Table 3.2. Absorption data after pulmonary administration of peptides and proteins. Data from reference [32].

Substance (species)	**Molecular weight**	**Administration**	**Formulation**	**Absorption parameters**[a]
DDAVP (rat)	1071 Da	Aerosol	Solution	84 % in adult rats; t_{max} 0.5–1 h; 17–41 % in young rats
Hu leuprolide acetate (human)	1209 Da	Aerosol (MDI)	Solution/ suspension	4–18 % independent of formulation
Dog leuprolide acetate (dog)	1209 Da	IT instillation	Solution	4.6–95% increasing with deposition distance from epiglottis
Insulin (human)	5786 Da	Aerosol (nebulized)	Solution pH 7	20–25 % vs. s.c. injection
Insulin (human)	5786 Da	Aerosol (nebulized)	Solution	75 % for smokers; 25 % in non-smokers; t_{max} 15–20 min
PTH–84 (rat)	9418 Da	IT instillation	Solution pH 5	> 20 %; t_{max} 15–90 min
G–CSF (hamster)	18.6 kDa	IT instillation	Solution pH 4	45 %; t_{max} 1–2 h
Interferon–α (human)	19 kDa	Aerosol (nebulized)	Solution	t_{max} 2–16 h; absorption < 5 %
Interferon a (rat)	19 kDa	IT instillation	Solution	> 56 %; t_{max} 3– 9 h
Growth hormone (rat)	22 kDa	IT instillation	Solution	36 %
Growth hormone (rat)	22 kDa	Aerosol (nebulized)	Solution	9–10 %; t_{max} 1–4 h
DNase I (rat)	32 kDa	Aerosol	Solution	15 % absorption in 24 h
DNase I (monkey)	32 kDa	Aerosol	Solution	< 2 %
Peroxidase (guinea pig)	40 kDa	IT instillation	Solution	0.1–4 % h^{-1} absorption rate
Antitrypsin (sheep)	52 kDa	Aerosol (nebulized)	Solution	50 % in 50 h lost from peripheral lung; bioavailability. 16 % via lymphatics
Albumin (human)	68 kDa	Aerosol (nebulized)	Solution	50 % in 20 h lost from lung
Albumin (sheep)	68 kDa	Bronchial catheter	Solution	1 % h^{-1} lost from lung
Albumin (rat)	68 kDa	IT instillation	Solution	4–5 % in 96 h; t_{max} 16–24 h.
IgG (rat)	150 kDa	IT instillation	Solution	1.5–1.8 % in 192 h, t_{max} 16 h

[a] the relative bioavailability is given as the percentage of the administered dose. It should be mentioned that some bioavailabilities are unrealistically high. For correct interpretation of data, it may be advised to calculate the bioavailability from the deposited fraction of the dose.
IT, intratracheal

bioavailability [2,16]. In Table 3.2 data selected by Adjei [32] are shown, which illustrate the kinetic behaviour of proteins of different size after pulmonary administration.

Significant variations occur between results of studies using the same protein. Such variations can be explained by differences in the experimental conditions used and the different ways in which, for example, the bioavailability can be expressed. Assuming that the alveoli are the major site of absorption for all proteins administered, it is obvious that the fraction of the protein actually reaching the alveoli will largely determine the amount of protein that can finally be absorbed. Unfortunately, the reported fraction of proteins actually reaching the alveoli after pulmonary administration varies significantly between the different studies. Therefore, conclusions cannot be made with regard to the relative bioavailability after pulmonary administration. Moreover, it is not possible to estimate the fraction that passes the epithelium in relation to the amount of protein that has entered the alveoli.

When nebulizers are used to generate an aerosol, the fraction of the drug reaching the alveoli will be low and variable (see Section 3.5.1). Particularly for low molecular weight proteins (< 20 kDa), the fraction deposited in the alveoli might be the limiting factor for drug absorption. For example, the low absorption found for aerosolized Leuprolide acetate in humans is more likely to be explained by the low and variable portion of drug being deposited in the alveoli, than by the limited passage of the Leuprolide over the epithelial membrane. In animal studies the determination of the fraction absorbed is further complicated by the fact that it is often not feasible to establish the amount of drug that is actually inhaled. To overcome this problem instillation is often used as a method of administration. However, using this method of administration there remains uncertainty regarding the fraction of the administered protein that reaches the alveoli, it is likely that a significant part of the fluid only reaches the bronchi or the bronchioles.

In a study performed in rabbits, rhG-CSF in a powder formulation (aerodynamic diameter < 4 μm) was insufflated via an intratracheal tube and compared to intratracheal instillation of a solution of the drug. In this study it was shown that a direct relation exists between the amount of protein that was deposited deep into the lung and the relative bioavailability [33].

An alternative method which could be used to establish the fraction of protein that actually reaches the alveoli is the so-called co-aerosolization. If a protein is aerosolized from a solution that also contains another low molecular weight substance (deposition marker), it can be assumed that the fractions of protein and deposition marker reaching the alveoli will be the same. The deposition marker should be a substance with a known alveolar epithelial membrane passage (e.g. tobramycin or a decapeptide) which does not undergo absorption after oral administration. The fraction of the deposition marker that is deposited in the alveoli can be established from plasma (and urine) measurements of the deposition marker. The maximum fraction of protein that can pass the alveolar membrane will then be known. The ratio between the deposited fraction and the fraction that has been absorbed into the systemic circulation (as can be established form plasma or urine analysis) will provide an estimation of the protein passage across the alveolar membrane.

Alternatively the membrane passage of human airway epithelial cell lines can be studied *in vitro*. A number of bronchial epithelial cell lines is available, such as the 16HBE14o- and Calu–3 cell lines. These cell lines can be installed in diffusion chambers to measure transport rates [34]. A major disadvantage of the currently used cell lines is that they provide information about bronchial epithelial transport only. Since bronchial epithelium is very different from alveolar epithelium, the information from these *in vitro* studies is of limited value for the prediction of the bioavailability of pulmonary administered proteins.

The isolation and characterization of alveolar Type II cells which transform into alveolar Type I cells has been described, as well as a monolayer culture of alveolar Type I cells [35,36].

Recently, a monolayer of human alveolar epithelial cells was used to study the bioadhesive properties of lectins [37]. Lectins are sugar-recognizing adhesive molecules (they bind to epithelial cells) and are thought to increase the bioavailability of larger molecules by triggering vesicular transport processes.

Proteases occurring in the epithelial lining fluids are another source of variability in protein absorption from the lung. In the epithelial lining fluids proteases and peptidases do occur. Inhibition of proteases and peptidases with substances such as bacitracin or aprotinin might improve the bioavailability of proteins. For example bacitracin was shown to increase insulin bioavailability by a factor of 6.8 [3].

The enzymatic degradation of insulin was also shown to occur in the cytosol of alveolar cells, the pH optimum of the proteases being 7.4 [38]. To what extent intracellular proteases play a significant role in limiting the absorption of insulin is not clear, since the size of insulin likely allows paracellular transport over the alveolar epithelium. However, for proteins of higher molecular weight, that require transcellular transport, these proteases might certainly limit bioavailability.

For proteins with higher molecular weights the extent and rate of absorption tend to decrease and variability in absorption increases even more (see Table 3.2). For these proteins molecular conformation, charge and self-aggregation can largely determine passage through the epithelial membrane. It is crucial that these molecules are presented in an optimal way to the absorbing membrane. Consequently, formulation as discussed in Section 3.6.3 is critical for these proteins.

The addition of absorption enhancers, like bile salts (glycocholate), fatty acids (linoleic acid), surfactants (lecithins, polyoxyethylene-9-lauryl ether or N-lauryl-β-D-maltopyranoside) and chelators (EDTA) can significantly increase the absorption of various proteins. However, the application of enhancers is limited by their toxicity. For example polyoxyethylene-9-lauryl ether and sodium glycocholate caused serious oedema, haemorrhage and inflammation of the lung after intratracheal instillation [39].

Since larger proteins are transported by the transcellular route it is important to investigate potential enzymatic degradation both in the coated and non-coated vesicles as well as in lysosomes. The alveolar endothelial cells were shown to contain various proteases [35,38], which (depending on the cellular routing of a particular protein) can influence its bioavailability. More basic knowledge concerning receptor-mediated, endocytotic and transcytotic processes should be acquired in order to utilize physiological transport systems for pulmonary absorption of macromolecules. In addition, it is necessary to study the influence of various diseases on this route of administration.

3.5 Devices for Therapeutic Aerosol Generation

As described in Section 3.3 in more detail, particles in the aerosol cloud should preferably have an aerodynamic diameter between 0.5 and 7.5 μm. Currently, three different types of devices are used to generate aerosol clouds for inhalation: nebulizers (jet or ultrasonic), (pressurized) metered dose inhalers (pMDIs) and dry powder inhalers (DPIs). The basic function of these three completely different devices is to generate a drug-containing aerosol cloud that contains the highest possible fraction of particles in the desired size range.

Pressurized metered dose inhalers are still the most frequently used systems and they have proven their value in therapy. However, their application in early phases of biopharmaceutical research and further development of dosage forms seems less convenient, since they require special components including propellants, special containers, metering valves, and controlled filling conditions (pressure-filling or cold-filling).

Nebulizers and dry powder inhalers seem more appropriate systems to be used in the early stages of development of drug products for pulmonary drug delivery. However, it should not be concluded from this that the development of formulations for nebulizers or DPIs is easier and exhibits fewer theoretical and practical problems.

Which system is the most suitable for a particular drug or therapy is determined by both the physicochemical properties of the drug as well as by patient condition in relation to the chosen therapy. Asthma and COPD treatment using drugs such as β_2-agonists or corticosteroids is carried out with MDIs and DPIs. For children, nebulizers seem to be preferred, but MDIs with spacers can also be used. For antibiotic (e.g. tobramycin or colistin) therapy in cystic fibrosis patients nebulizers still seem the device of choice. Probably the patient population in this case is too small to make the development of DPIs or MDIs containing antibiotic drugs economically feasible. When peptide or protein delivery is considered, newer and more advanced systems such as the 'AERx™ system' or dry powder generators such as the 'Inhale Therapeutic System (Innova™)' have been developed [40,41].

3.5.1 Nebulizers

Nebulizers are applied to aerosolize drug solutions or suspensions. There are two basic types: the air jet and ultrasonic nebulizer [42]. Jet nebulizers have a two-fluid nozzle for atomizing the drug solution. Compressed air passes through a narrow hole and entrains the drug solution from one or more capillaries mainly by momentum transfer. The liquid break-up process depends on the design of the nozzle, the air pressure and the physical properties of the drug solution. Droplets in the required size range are entrained by the airflow from the nozzle. Larger droplets impact on a baffle and are returned to the reservoir. Auxiliary airflows, generated by the patient, may pass through special vents to the nebulization cup in order to improve droplet entrainment from the nozzle area. In an ultrasonic nebulizer, droplets are produced by a piezoelectric crystal vibrating at a high frequency. The frequency and again the properties of the drug solution determine the droplet size distribution of the mist.

Many reviews on the relevant technical aspects for drug nebulization are available (e.g. [43–45]. The greatest disadvantages of nebulizers are their poor deposition efficiency (see Section 3.11) and low output rate (e.g. [46]). Several developments have been reported to improve their efficacy, like the use of open vents or breath-assisted open vents [47] and adapted aerosol delivery [48]. A renewed interest in nebulizer therapy may also come from the generation of special aerosols, such as liposomes [49].

The AERx™ pulmonary delivery system [40,41] can be regarded as a combination of a MDI and a nebulizer. This system forms an aerosol by extrusion of an aqueous drug-containing solution through a disposable nozzle containing an array of precisely micromachined holes. The droplets are entrained by the airflow passing over the blister. Control over the size distribution of the holes enables the formation of droplets having a narrow size distribution.

Moreover, the system will release the aerosol cloud only when the pre-programmed optimal inhalation flow is generated by the patient. These features enable a controlled and targeted delivery to the lung.

3.5.2 Metered Dose Inhalers

The metered dose inhalers consist of four basic functional elements, container, metering valve, actuator and mouthpiece.

The drug is dissolved or suspended in the liquefied propellant which might contain other excipients. The energy for atomization of the drug suspension (or solution) from a metered dose inhaler is supplied by a liquefied propellant. When after actuation a small amount of the suspension or solution is released from the metering valve connected to the pressurized container, the propellant starts evaporating rapidly, thereby disrupting the liquid into small droplets. Initial droplet size and droplet speed are too high for effective deposition in the lower respiratory tract (the target area), however. Evaporation and deceleration in the upper respiratory tract (mouth and throat) is essential. Consequently, the inhalation manoeuvre is extremely relevant for deposition efficacy (particularly the co-ordination between firing and inhalation of a dose), in spite of the fact that no energy from the inspiratory air (except heat for evaporation of the propellant) is required for fine droplet generation. If spacers are used, the inhalation manoeuvre becomes less critical. For the 3M Autohaler, no firing of a dose is necessary, because dose release is breath triggered.

With respect to the formulations used in MDIs, the development over recent years has focused on the replacement of chlorofluorocarbon (CFC) propellants by hydrofluoroalkane propellants. Recently, new developments have been reviewed in a number of papers [50–52].

3.5.3 Dry Powder Inhalers

Dry powder inhalers have initially found their application in inhalation therapy as a CFC-free alternative for the older MDIs. However, nowadays they seem to have a much larger potential [14,53], because of the high lung deposition that can be attained and also because they are suitable for the pulmonary delivery of therapeutic peptides and proteins [2,10,16].

Dry powder inhalers are generally described as 'breath actuated' devices, because the inspiratory airstream releases the dose from the dose system and supplies the energy for the generation of fine drug particles from the powder formulation. Because the efficiency of dose release and powder disintegration increases with increasing inspiratory flow rate for most DPIs, these devices would be better described as 'breath controlled' devices. In Section 3.9, the effect of resistance and clinical conditions on the flow curve and relevant flow parameters for DPIs are discussed.

Basically, devices used as dry powder inhalers contain four basic functional elements, i.e.

- Powder container. Dry powder inhalers may contain the dry powder formulation in many different forms. The first DPI, the Spinhaler™ contained single doses in capsules. Other systems, like the Diskus™ or Diskhaler™ may contain the metered dose in blisters, whereas systems like the Turbohaler™, or Novolizer™, have multi-dose containers.

- Dosing system.
- Disintegration principle. In general, the powders in the inhaler are not formulated as single particles, but as adhesive mixtures or spherical pellets (Figure 3.3). These mixtures or pellets are suitable for processing and metering. However, the particle size of these mixtures or pellets is far too large for lung deposition. Therefore, the pellet or mixture has to be disintegrated to make an aerosol cloud with the desired particle size (< 5 µm). Many different disintegration principles exist. They may vary from a simple screen (Rotahaler™) to twisted powder channels (Turbuhaler™) or a cyclone chamber as used in de Novolizer™ [54].
- A mouthpiece. The mouthpiece may be used to control the direction of the aerosol cloud in the mouth and throat, in order to reduce drug deposition in the oropharyngeal cavities. De Boer *et al.* [55] use a so-called sheath flow to reduce mouth deposition.

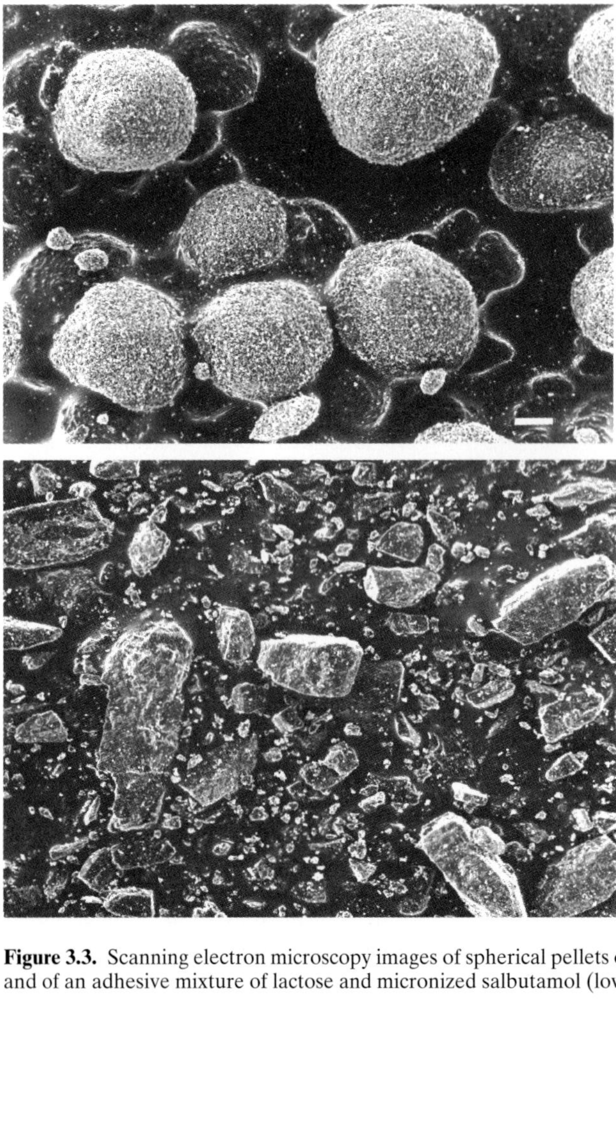

Figure 3.3. Scanning electron microscopy images of spherical pellets of budesonide (upper photograph) and of an adhesive mixture of lactose and micronized salbutamol (lower photograph).

In Table 3.3 some advantages and disadvantages of the use of dry powder inhalers are summarized.

Table 3.3. Advantages and disadvantages of dry powder inhalers versus metered dose inhalers, partly from reference [14].

Advantages of dry powder inhalers	Disadvantages of dry powder inhalers
• Propellant free • Less need for patient coordination • Less potential for formulation problems • Less potential problems with drug stability • Less potential for extractables from device components	• Performance depends on the patient´s inspiratory flow profile • Resistance of the device and other design parameters • Potential difficulties to obtain dose uniformity • Less protection from environmental effects and patient abuse • More expensive • Not available worldwide

3.6 Formulations for Inhalation Products

3.6.1 Formulations for Nebulizers

The physical characteristics of the solution or suspension that is used in a nebulizer may have a significant effect on both the generated droplet size as well as on the drug output rate. Theoretically, the viscosity of the solution is expected to influence the drug output rate (mass flow through the nozzle) and droplet size distribution of aerosols generated by jet nebulizers. Yet, conflicting experimental results were found [56–58]. It should be noted that, due to solvent evaporation, the concentration of drug in the reservoir increases during the nebulization process. This might result in an increased viscosity and affect the nebulizer performance.

For ultrasonic nebulizers the relation between viscosity and droplet size is more obvious. As could be expected on theoretical grounds, droplet size was found to be proportional to viscosity [57]. As a consequence, viscous solutions might not be aerosolized at all [58,59].

Droplet size increases with increasing surface tension of the drug solution [60]. However, surface tension should not become so low that foaming will occur, since this may prevent aerosol formation.

Because the relationship between the physical characteristics and the nebulizer performance is less straightforward than expected, it should be stressed once again that laboratory evaluation of the specific drug formulation in combination with the intended nebulizers is required, before their use *in vivo*.

3.6.2 Formulations for Dry Powder Inhalers

The physicochemical characteristics of the components (both drugs as well as excipients) used in dry powders are of significant importance for the performance of the inhalation sys-

Table 3.4. Major properties of powders to be determined during pre–formulation and some of the methods to be used.

Powder property	Method
• Particle size and size distribution	• Cascade impactor analysis Wet or dry laser diffraction analysis Microscopy Coulter counter analysis Sieve analysis Sedimentation analysis Time of flight measurements (Scanning electron) microscopy
• Particle surface area, shape and texture (morphology)	• Scanning electron microscopy Atomic force microscopy Particle flow (e.g. angle of repose) BET measurements
• Moisture sorption and desorption	• Dynamic gravimetric sorption (DVS)
• Surface energy	• Contact angle measurements Isothermal microcalorimetry Gravimetric sorption Inverse gas chromatography
• Crystallinity and crystal form	• Differential scanning calorimetry Thermogravimetric analysis Isothermal microcalorimetry Infra red analysis X–ray diffraction Sorption and desorption measurements (DVS)
• Solubility and dissolution rate	
• Partition coefficient	
• Stability in dry state and in solution	
• Impurities	

tem. Staniforth [61] gave an overview of the required pre-formulation tests for dry powder formulation development. The nature of the surface (e.g. surface morphology, crystallinity or surface energy) of the particles is of utmost importance and should be studied in detail. Two reviews [62,63] described a number of methods for characterizing particle morphology.

Table 3.4 summarizes the major properties of powders which need to be characterized in pre-formulation. The effect of micronization (or other high energy processes), which is often applied to the powder (surface), should also be investigated as this may alter the properties of the powder during processing [64]. This may also alter the performance of the formulation.

Different dry powder formulations for inhalation were recently reviewed [53,65]. Spherical pellets or adhesive mixtures are the most used formulation principles in dry powder inhaler systems (see Figure 3.3). Spherical pellets consist of the pure micronized drug or the micronized drug combined with a micronized excipient such as lactose or glucose. The small drug particles are formulated into the large spherical pellets to improve processing properties such as flowability and precision of metering. The pellets should be strong enough to withstand the filling process as well as normal handling and shock as may occur during use by the patient. On the other hand, it should be taken into account that optimal lung deposition is achieved through full disintegration of the spheres into the primary particle size. This contradiction makes optimum formulation of spherical pellets difficult. Spherical pellets are produced by low shear mixing procedures (e.g. tumbling or planetary mixers) with or without mixing aids such as small stainless steel balls, often followed by sieving procedures. Boerefijn *et al.* [66] investigated the effects of agglomerate size and humidity on breakage of lactose agglomerates. They showed that the extent of breakage was larger for the smaller particles, whereas humidity was found to decrease the propensity for breakage.

Adhesive mixtures require large carrier crystals to improve the handling properties of the powders. Dispersion of the small drug particles over the larger carrier material should assure dose uniformity. However, the small drug particles should be removed from the carrier material during inhalation, to render an aerosol cloud of respirable particles. If the particles remain on the carrier, mouth or throat deposition of the drug will occur, which might decrease therapeutic efficacy or cause serious side-effects.

The adhesion of the drug particles to the carrier is largely dependent on the surface properties of the carrier and drug as mentioned in Table 3.4. Variation in these properties can affect the dispersion of the drug over the carrier as well as the binding between drug and carrier. Both may have a significant influence on the drug delivery performance of the formulation. Concessio *et al.* [67] correlated powder flow and particle detachment from solid surfaces to *in vitro* disintegration efficacy (deposition) as well as to *in vivo* efficacy (bronchodilation) in guinea pigs. A direct correlation between powder flow on the one hand and ease of particle separation and aerosol dispersion on the other hand was observed. Furthermore it was found that formulations with a higher *in vitro* deposition had an increased *in vivo* efficacy. Lactose is the only carrier used in adhesive mixtures as yet; the carrier is added to the formulation to improve the processing and metering of the micronized drug. When increasing amounts of fine particles were used on the lactose, the drug delivery was found to increase [68]. Furthermore smoothing of the lactose surface increased the release of salbutamol sulphate from the surface [69]. On the other hand, increased surface roughness and enlarging the surface area were desirable to improve the release of pranlukast hydrate [70]. Obviously, detailed information concerning parameters determining the interaction between drug and carrier lactose is still lacking.

Staniforth and co-workers managed to reduce the effect of the lactose surface by co-processing the carrier (e.g. co-milling, mixing or surface modification) with up to 2% L-leucine. This process is called corrasion. This approach significantly increased the release of beclomethasone diproprionate from the carrier [14,53,61]. Another approach to modifying the surface properties of carrier and drug could be the use of super critical fluid crystallization. This technique gives precise control over the particle size, shape and crystallinity of the particles produced [71,72]. Lactose, trehalose and mannitol were also found to be suitable as drug carriers for the pulmonary delivery of proteins. Sucrose was less suitable due to its hygroscopicity [73].

3.6.3 Formulations for Peptides and Proteins

The formulation of small organic molecules in most cases uses established processes and only a limited number of excipients (mainly lactose or a small number of propellants). In contrast,

Table 3.5. Different issues to consider for peptide or protein inhalation formulations.

- Particle size morphology and surface characteristics
- Moisture sorption behaviour
- Stability in dry state and dissolved
- Tendency to form aggregates
- Charge of the molecule, isoelectric point
- Solubility and dissolution behaviour
- Crystallinity and crystal form

the formulation of peptide and protein powders for inhalation requires more advanced techniques and a wide variety of excipients and production processes [16,65,74]. The reason for this difference is found in the more complex nature of the problems and requirements related to peptide and protein formulations. Table 3.5 summarizes a number of issues that need to be considered when peptide or protein formulations for inhalation therapy are developed. Many of the characteristics mentioned in Table 3.5 can be affected by the processes used to prepare the protein or by the composition of the formulation used. Major formulation problems connected to peptides and proteins are their low stability, hygroscopic nature, and tendency to form aggregates, which are too large to cross the alveolar membrane.

If possible, adhesive mixtures or spherical pellets, prepared using simple excipients such as sugars are also preferred for protein formulations. For the preparation of dry peptide-containing formulations the most important techniques are lyophilization, spray freeze-drying, spray-drying, co-precipitation and super critical fluid extraction. When lyophilization is used as the drying method, milling to obtain the desired particle size can be used. For spray-drying or supercritical fluid extraction the desired particle size can be obtained immediately from the drying process. Lucas *et al.* [75] investigated different micronized bovine serum albumin–maltodextrin (50 : 50) mixtures. Improved aerosolization behaviour was found for adhesive mixtures based on carrier lactoses with surfaces that were modified by micronized lactose or micronized polyethylene glycol 6000. Maa *et al.* [76] compared particles prepared by spray freeze-drying with particles prepared by spray-drying. The particles contained recombinant human deoxyribonuclease-1, or anti-IgE monoclonal antibody and different sugars as excipient. The large size of the spray freeze-dried particles (about 8–10 μm) in combination with their high porosity, turned out to result in improved aerosol performance compared to the denser and smaller spray-dried particles. The lyophilization of proteins was recently reviewed by Wang [77].

Protein instability can either be of a physical or chemical nature. The major mechanisms underlying the degradation of proteins were recently extensively reviewed [78] Unfolding of the protein is the main cause of physical instability and may lead to denaturation, aggregation or surface adsorption. Excipients that preserve the protein in its preferred state of hydration may be used to stabilize the protein. Several studies described the role of different excipients (often in combination with production processes) in the stabilization of proteins [65,79–82]. The major excipients used for stabilization of proteins are classified in Table 3.6. The incorporation of the proteins in amorphous solid matrices of sugar (often referred to as sugar glasses), seems an effective method to stabilize the solid protein [83–85]. The stabiliza-

Table 3.6. Excipients used for protein formulations for inhalation.

Class	**Excipient**
• Carbohydrates	Sucrose Lactose Trehalose Inulin Dextrates Dextran Cyclodextrins
• Polyols	Sorbitol Mannitol
• Buffers	Sodium citrate Citric acid Sodium phosphate Sodium biphosphate Amino acids
• Surfactants	Polysorbate 80 Tween 20 Poloxamer 188 Dipalmitoyl phosphatidylcholine Alkylbenzene sulfonate
• Polymers	PEG PVP
• Amino acids	Glycine Lysine
• Proteins	Albumin
• Salts	Sodium chloride Calcium chloride Sodium sulfate
• Chelators	Disodium EDTA

tion is explained by the fact that in these amorphous sugar matrices hydrogen bonds between water and the protein in an aqueous environment are replaced by hydrogen bonds between the sugar and protein. This allows the protein to maintain its conformation and provides mechanical protection. Furthermore, inclusion of the protein in the matrix protects it from the environment thereby preventing degradation processes such as hydrolysis or oxidation. It is essential that the sugar in these systems remains amorphous and has a glass transition temperature above storage temperature. In the rubbery state, the glasses are not stable; crystallization may occur and the protection from environmental influences disappears. The glass transition temperatures of many sugars is above 50°C when the sugars are pure and completely free of water. However, both moisture and the included protein may reduce the glass transition temperature, which makes many sugars unsuitable for the formation of sugar glasses. The moisture content of the products is not only important because of the plasticizing effect, but also for their aerosol performance, since a high moisture content may increase powder cohesiveness. Compatibility of the sugar with the protein is necessary to obtain stable formulations. In this respect the use of reducing sugars such as sucrose or glucose is less satisfactory.

Trehalose is often referred to as the sugar of choice for preparing sugar glasses. It is a non-reducing disaccharide with a glass transition temperature of about 120°C in the anhydrous state. However, its glass transition temperature is rapidly decreased when the moisture content in the sugar increases. Considering the hygroscopic nature of trehalose this is a potential hazard and adequate moisture protection is essential. Furthermore, crystallization to the trehalose dihydrate occurs easily at a relative humidity above 60%. From this perspective, the use of a sugar polymer such as inulin (which is a fructose polymer terminating with a glucose unit) seems much more suitable. Inulin is also a non-reducing sugar. By changing the chain length (number of fructose units) of the molecule, physical characteristics like the glass transition temperature can be changed. Moreover, due to the polymeric character of inulin, crystallization is less likely to occur.

Sugar glasses are prepared by spray-drying, freeze-drying or vacuum-drying. Freeze-drying produces the lowest change in the sugar glass of degradation, whereas spray-drying may result in altering a large proportion of the particles to the preferred size range of 2 to 5 μm. In contrast to most other sugars or polyols that yield amorphous materials on spray-drying, mannitol was found to crystallize during spray-drying [73].

A final advantage of the use of sugar glasses is the fact that they include the proteins in a mono-molecular form. In the glassy state, mobility in the systems is insufficient to allow aggregation of the proteins. Upon dissolution of the sugar matrix the protein is released in its mono-molecular form which might enhance its passage through absorptive membranes.

Small amounts of surfactants may be used to prevent aggregation of proteins and may enhance the refolding process when the dried protein dissolves. Buffers may also help to prevent aggregation of the dissolved drug. Similarly, polymers may be used as aggregation inhibitors or to form matrices. Chan *et al.* [86] prepared crystalline powders of recombinant human deoxyribonuclease with high fractions of sodium chloride. These powders were formulated as adhesive mixtures on lactose and mannitol and showed improved aerosolization behaviour compared to the pure protein.

Preparation of high porosity particles may require special excipients, such as dipalmitoyl phosphatidylcholine or special drying techniques such as spray freeze-drying [76,87,88]. These large porous particles may combine the advantages that larger particle sizes contribute to the properties of powders with an improved aerosol performance. Furthermore, these large porous particles may be used to obtain sustained release of the incorporated drug [89,90].

Other techniques that have been used to obtain sustained release inhalation products are: the coating of the aerosol particles with paraffin wax or encapsulation or incorporation in biodegradable polymers such as poly(L-lactic acid) or poly(DL-lactide-co-glycolide) [91,92]. Talton *et al.* [93] described a new spray coating technique for applying ultra-thin coating layers on particles. Finally, some authors describe the use of liposomes or other phospholipid-containing systems to prolong drug release or lung retention [94]. Liposome vaccine formulations have also been used for immunization via the pulmonary route. These developments will not be discussed in detail here.

One of the major questions in relation to absorption enhancers such as surfactants or sustained release products is their safety. Whether damage to lung tissue is caused by the different excipients is not yet clear. The results obtained so far are not very promising for substances like surfactants [39]. What the effects of repetitive administration of insoluble or slowly (bio)degrading particles might be, remains to be established.

When nebulizers are used to produce the aerosol cloud, the proteins should be dissolved in aqueous solutions. The poor stability of many proteins in solution will make dissolution of the (freeze) dried protein just before nebulization necessary. The solvent may again contain stabilizers such as buffers or salts. During jet and ultrasonic nebulization, high shear forces might be exerted on the solutions. Both jet and ultrasonic nebulization may cause aggregation of the proteins in solution. Therefore, the resistance of the protein solution to nebulization should be investigated *in vitro* before use *in vivo*. Aggregation can be decreased by the addition of Tween 20, Tween 80 or polyethylene glycol 8000 or by cooling the solution (for ultrasonic nebulizers) [95,96].

A number of reports have described excipients and formulations for proteins used for MDIs [16,97,98].

3.7 Variables and Interactions in Dry Powder Inhalation

In Section 3.5.3, dry powder inhalers have been referred to as breath-controlled devices. The efficacy of dry powder inhalation is a function of many factors, influencing the delivered dose of fine particles and the deposition of these particles in the respiratory tract. Figure 3.4 shows that DPI performance is influenced both directly and indirectly by the design of the inhalation system. The powder formulation, the dose (measuring) system and the powder disintegration principle have to be designed correctly for release of sufficient fine drug particles in

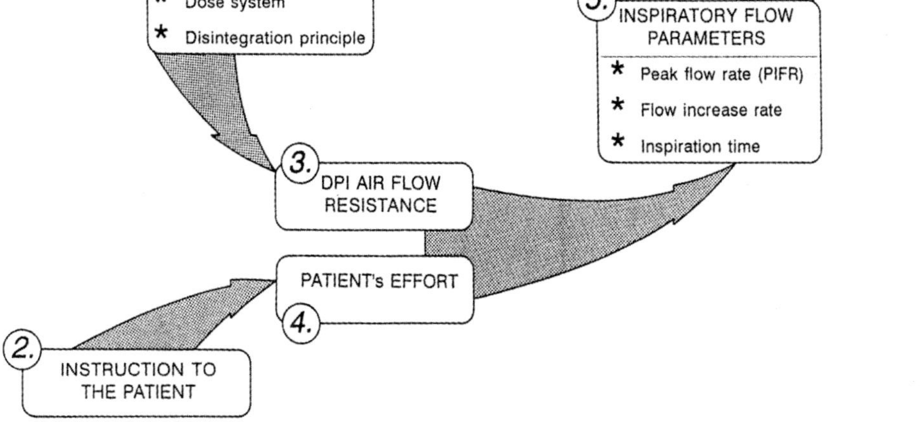

Figure 3.4. Relevant variables and interactions in dry powder inhalation therapy.

the correct size distribution during adequate inhalation. The inspiratory flow as the driving force for discharge of the dose system, fine particle generation during powder disintegration and particle deposition in the respiratory tract is one of the most important parameters. If the patient is unable to achieve the threshold values (for the relevant flow parameters) for good inhaler performance, the fine particle fraction will be too low for adequate efficacy. If the flow rate is too high, a substantial loss of the fine particle dose by inertial deposition in the oropharynx must be expected.

The nature of the flow curve achieved, depends on three different factors: the instructions given to the patient, the patient's effort following these instruction and the resistance to airflow of the DPI (Figure 3.4). Most instructions for use of DPIs prescribe forceful and deep inhalation. Patient interpretation of these instructions often vary considerably. Patient variables also include age, gender, condition and clinical assessment. In Section 3.9 the attainable pressure drops across external resistances as a function of clinical condition will be discussed more in detail. The DPI's resistance is a consequence of its design. Narrow channels in the dose system and the disintegration principle in addition to turbulent air zones, increase the resistance and reduce the attainable peak flow rate through the device. This is an advantage from the deposition point of view. High resistance DPIs generally have a high disintegration efficacy and do not require high flow rates to achieve an acceptable dose of fine particles. Recently, a multi-dose dry powder inhaler has been introduced (Sofotec Novolizer™) in which the resistance can be controlled over a certain range by means of a sheath flow around the aerosol cloud, without changing the fine particle output [55].

3.8 Airflow Resistance

The underpressure created in the respiratory tract is the driving force for the airflow through an inhalation device. The attainable underpressure and the rate of the airflow both depend on the total resistance in the airways and inhaler. The pressure drop achieved during inhalation is furthermore a function of the anatomy of the lungs, the effort made by the patient, pathological factors and the presence of exacerbations (e.g. in case of asthma).

A large proportion of the airflow resistance in the airways (internal resistance: R_i) is offered by the upper respiratory tract in which the airflow is already turbulent at relatively low flow rates of 30 to 40 l min^{-1} ($R_E > 2000$: see also Section 3.2.1). During quiet mouth breathing, the mouth, pharynx, larynx and trachea account for 20–30% of total airway resistance. The same region contributes as much as 50% to total resistance during heavy breathing however. In the small peripheral airways (those less than 2 mm in diameter), resistance is quite low and the contribution to R_i is not more than 10–20% [21].

For nebulizers and MDIs, the external resistance (R_E) is quite low. Different approaches have been made to describe the external airflow resistance of DPIs. Olsson and Asking [99] derived an empirical relationship between flow rate (Φ) and pressure drop (ΔP), $\Delta P = C.\Phi^{1.9}$, for a number of inhalers (such as Rotahaler™, Spinhaler™ and Turbuhaler™) in which they define the proportionality coefficient (C) as the airflow resistance. This relationship differs only slightly from the general (theoretical) equation for orifice types of flow constrictions:

$$\Phi_V = Fu(\mathrm{A}) \times (2\Delta P/\rho_A)^{0.5} \tag{3.2}$$

where Φ_V is the volumetric flow rate, ρ_A is the density of the air (upstream of the flow constriction) and *Fu*(A) is a function of the cross-section of the flow constriction. Several authors use a simplification of this formula, written in terms of driving force ($\sqrt{\Delta P}$), airflow resistance (R_E) and volumetric airflow (Φ_V) (e.g. references [22,100]).

$$\sqrt{\Delta P} = R_E \times \Phi_V \tag{3.2a}$$

In this equation, R_E is the reciprocal of $Fu(A) \times \sqrt{(2/\rho_A)}$. If there is no interaction between the flow through the inhaler and the flow in the respiratory tract, total resistance (R_{TOT}) may be written as $R_{TOT}^2 = R_I^2 + R_E^2$. It has been reported that a high external resistance influences the shape and width of the human pharynx and larynx in a manner favourable for deep lung deposition [101].

The resistance of dry powder inhalers can be calculated by measuring the volumetric flow rate and the pressure drop across the device simultaneously and applying Eq. 3.2a for the calculation. Values for some marketed DPIs are given in Table 3.7, showing that there are remarkable differences between devices.

Table 3.7. Airflow resistances of some marketed dry powder inhalers in $kPa^{0.5}$ min l^{-1}.

Reference	[100]	[144]	[145]	Mean
Glaxo Rotahaler	0.013	0.013	0.014	0.0133
Fisons Spinhaler	0.016	0.015	0.015	0.0153
Pharbita Cyclohaler[a]	0.017	0.018	0.018	0.0177
Sofotec Novolizer	–	–	0.026	0.0260
Glaxo Diskhaler[c]	0.021	0.018	0.030	0.0230
Glaxo Diskus	–	–	0.032	0.0320
Astra Turbuhaler	0.031	0.039	0.040	0.0367
Inhalator Ingelheim[b]	0.056	0.053	0.048–0.058	0.0540
Chiesi Inhaler	–	0.093	–	0.0930

[a] Also known as ISF inhaler and Novartis (Foradil) Inhaler.
[b] The range in reference [145] is a consequence of poor reproducibility in capsule piercing.
[c] The spread for the Glaxo Diskhaler may come from the use of two different devices: four-dose [145] and eight-dose [100, 144].
1 technical bar equals 1000 mbar = 10^5 Pa (N m^{-2} or kg s^{-2} m^{-1}) = 10^2 kPa
Resistances have been (re-)calculated for standard liters per minute (at 20 °C: L_S min^{-1} = 1,0733*L_N min^{-1}).

3.9 Inspiratory Pressure and Relevant Flow Parameters

Several authors measured attainable pressure drops as a function of the external resistance for different groups of volunteers. Healthy male subjects (during maximal inspiration) are able to create a pressure drop (on average) of 6.7 kPa through an airflow resistance of 0.038 $kPa^{0.5}$ min l^{-1}, (which is in the range of that of marketed DPIs, see Table 3.7), whereas females are able to create a pressure drop of 3.8 kPa under the same resistance [22]. Differences between different groups of patients depend on the degree to which pulmonary func-

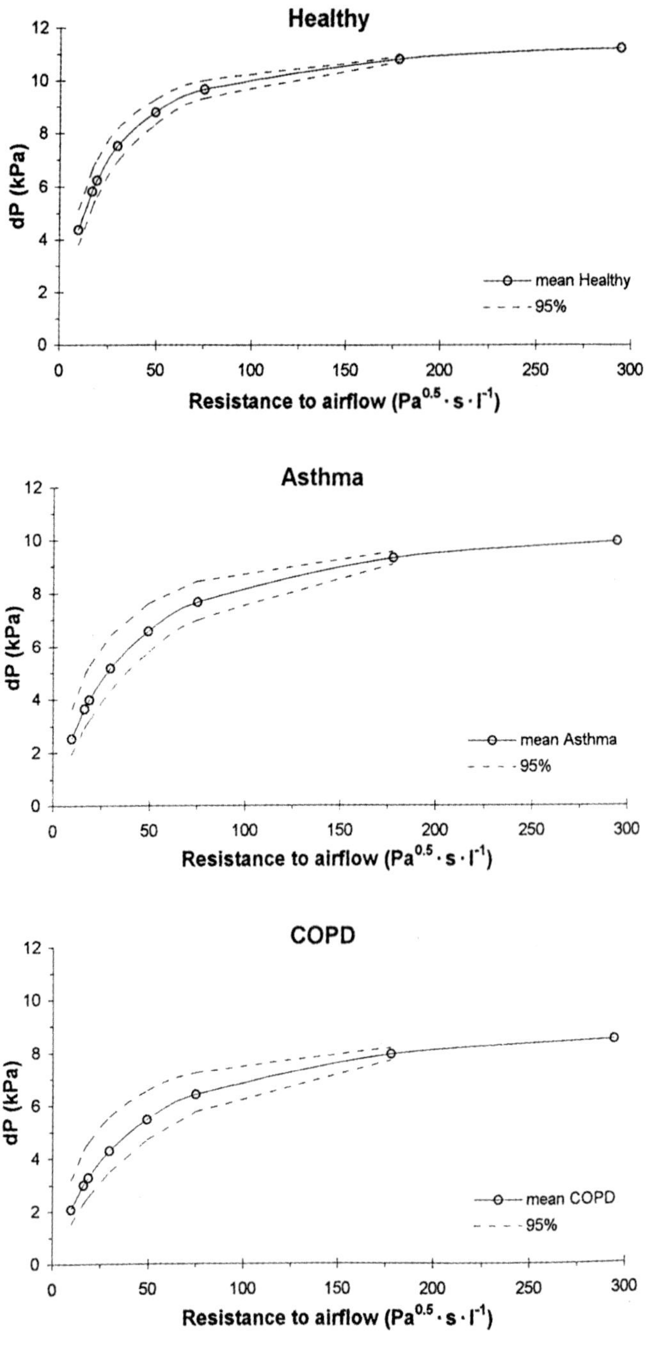

Figure 3.5. The effect of external resistance and clinical conditions on the pressure drop generated at maximum inhalation effort. The area between the dashed lines represents the 95% confidence interval.

tion has been deteriorated. However, pulmonary obstructions usually restrict the expiratory performance rather than the inhalation manoeuvre (e.g. [102,103]). Therefore, maximal inspiratory pressure (MIP) values of asthmatic and COPD patients can be of the same order of magnitude as that of healthy subjects [104]. Only severe COPD patients may not be able to generate MIPs higher than 1.5–2.0 kPa. The effect of airflow resistance and clinical condition on the generated pressure drop across an external airflow resistance is summarized in Figure 3.5 (data derived from reference [104]). The data show that *in vitro* testing of dry powder inhalers at only 4 kPa, as prescribed by various guidelines, is inadequate for the prediction of their performance in practice.

Attained flow parameters during inhalation can either be calculated (Eq. 3.1) from recorded pressure drop curves or measured directly, using the equipment described in Section 3.9.1. Various studies report peak inspiratory flow rate (PIFR) values for healthy and diseased adult subjects calculated from data on inhalation without external resistance (so-called 'control values')(e.g. [22,100,105,106]). As for the attainable pressure drop, asthma, COPD and cystic fibrosis may decrease PIFR, depending on the severity of the disease. The addition of an external airflow resistance, such as a DPI strongly affects the PIFR. Within the range of resistances for marketed DPIs, the average PIFR at maximal inhalation by healthy adults may decrease from 159 l min^{-1} (for $R = 0.015$ $kPa^{0.5}$ min l^{-1}, equals the Rhone Poulenc Spinhaler™) to only 62 l min^{-1} (for $R = 0.040$ $kPa^{0.5}$ min l^{-1}, equals the Astra Turbuhaler™) [22]. A maximal average flow rate of approximately 60 l min^{-1} through the Turbuhaler has also been reported for asthmatics (e.g. references [103,107,108]).

Various flow parameters may be relevant to good DPI performance. The effect of PIFR on the *in vitro* dose of fine particles from DPIs has been the subject of many studies (e.g. [109–112]. The actual dose of the fine particle fraction produced varies with the ranges of flow rates applied. The flow rate being the measuring principle used for the study and definition of the fine fraction. Most studies show a strong increase in fine particle output with increasing PIFR for the ASTRA Turbuhaler™ [110–112]. For the Glaxo Diskus™ and Diskhaler™, the effect of PIFR is less extreme: in some cases an more or less constant, but also a much lower output of approximately 15 to 25% of the label claim, has been obtained [109,110]. Some recent studies refer to the fact that the Pulmicort Turbuhaler™ also has a constant fine particle yield if the acceleration towards peak flow (flow increase rate: FIR) is high enough [113,114]. For a FIR > 8 l s^{-2}, the fine particle output is already maximal at flow rates above 40 l min^{-1}. For capsule inhalers, the inhalation time has to be long enough for all the particles to pass rapidly through the narrow holes pierced in the capsule ends before inhalation stops.

3.9.1 Measurement of the Inspiratory Flow Curve

Many different techniques are available for flow measurement and for recording of respiratory functions or flow parameters in particular (e.g. [115,116]). However, not all methods are appropriate for measurement of inhalation flows, either because they have low frequency responses or they influence the shape of the inspiratory flow curve by a large volume or by the inertia of the measuring instrument (e.g. rotameters). They may also interfere with the aerosol cloud from the inhalation device during drug deposition studies.

Electronic equivalents of traditional spirometers (pneumotachographs, pneumotachometers or anemometers) are often used in clinical practice. They integrate expiratory flow rates in order to compute flow–volume curves, from which expiratory parameters such as PEF (peak expiratory flow rate) and FEV_1 (forced expiratory volume in 1 s) are derived (e.g. [22,117]). They are often laminar flow meters (e.g. Vitalograph 2100 Spirometer, Jaeger Masterscreen IOS), measuring the volumetric flow rate (Φ_v) at the upstream pressure of the flow head. They can also be used for measuring the inspiratory flow curve, but the flow rate has to be corrected for pressure and air density when they are used with an add-on resistance (e.g. during simulation of inhalation through dry powder inhalers) [118]. Similar corrections, including discharge and correction coefficients, are necessary for head meters (e.g. venturi or orifice meters) [119]. Thermal mass flow meters, also referred to as 'hot-wire pneumotachographs' do not require these corrections, but they have a high internal resistance, which makes them inappropriate for patient characterization studies. Another great disadvantage of the techniques mentioned so far is that they are in-line components of the total flow scheme, which makes them difficult to use for *in vitro* drug deposition studies. For this reason, on-line pressure drop measurement across a flow constriction in the flow set-up is often recommended. The flow constriction can be the add-on resistance (during patient characterization) or the inhalation device (during drug deposition measurement). Once the airflow resistance of the flow constriction is known from previous calibration (as in Table 3.7 for DPIs), the flow curve can be monitored (or adjusted) on the differential pressure signal without interfering with the aerosol cloud or adding additional resistance to the flow set-up using Eq. 3.2a for the calculations.

3.10 *In Vitro* Particle Size Analysis and Deposition Measurements

Methods for analysis of the particle size distribution in the aerosol cloud include techniques such as time of flight measurement (TOF), inertial impaction and laser diffraction. Dynamic light scattering (photon correlation spectroscopy) is confined to particles (in suspension) in the submicron range. In addition to the size distribution, the particle velocity distribution can be measured with the Phase Doppler technique.

Inertial impaction is most widely applied for the characterization of inhalation systems. The principles of particle separation on the basis of inertial and drag forces have been well described for many different applications. Theoretical cut-off diameters (for particles with 50% collection efficiency) of impactors can be calculated on the basis of Stokes numbers for nozzles of a particular design [8,120]. Many different designs are available, but only a few are described in the United States and European Pharmacopoeia [121,122].

Inertial impaction has many inaccuracies and limitations and there are also some relevant differences between deposition *in vitro* (impactor) and *in vivo* (respiratory tract) which cause poor correlation between impactor data and lung deposition data. The most important difference is that deposition *in vitro* is by inertial impaction only, whereas deposition *in vivo* is by sedimentation and diffusional deposition as well. Except for the possible passage of the final stage (by the finest particles), particle collection *in vitro* is almost 100% efficient. In contrast, collection efficiency *in vivo* decreases with decreasing particle size (deposition mechanism) to a minimum of 20% for particles with an aerodynamic diameter of 0.5 μm (e.g. [8]). The classification by impactors into a small number of classes for the relevant particle sizes (one to seven) may provide insufficient discrimination. For example, the theoretical cut-off (with 50% collection efficiency) for the third stage of the multistage liquid impinger (MSLI) is 3.1 μm at 60 l min^{-1} for particles with a true density of 1.5 g cm^{-3}. So, the fraction retained from the fourth impactor stage plus filter is smaller than 3.1 μm. If the MMAD of this fraction is 2 μm, deposition efficiency in the respiratory tract can be nearly 40%. If however, the MMAD of this fraction is only 1 μm, the deposition efficiency is only about 20%, which is twice as low. This difference cannot be judged from the cascade impactor result.

Usually, the inspiratory flow through impactors is an on–off function: a (solenoid) valve is opened to allow suction at a pre-set flow rate through the inhalation device over a certain period. Some recent guidelines prescribe a flow rate corresponding with 4 kPa pressure drop for DPIs. Asking and Olsson derived a useful relationship for the effective cut-off diameters for the different stages of the MSLI as function of the flow rate for the range between 30 and 100 l min^{-1} [123]. Impactors have relatively large volumes and high airflow resistances which confine the range of adjustable PIFR and FIR. Probably the major drawback of cascade impactor analysis is that the technique is extremely laborious and time consuming.

Laser diffraction is a fast alternative for analysis of the size distribution of particles in an aerosol cloud. The theory of laser diffraction is well understood [124,125]) but this technique requires special measures to test inhalation devices and to interpret the results correctly. One of the major problems is that flow adjustment through the inhaler is not possible. Furthermore, the presence of carrier particles from adhesive mixtures may disturb the measurement of the fine drug particles and the size distribution obtained is of an unknown delivered mass fraction of the dose. These practical problems and limitations have been solved by the design of a new modular inhaler adapter for the Sympatec™ laser diffraction apparatus (Figure 3.6).

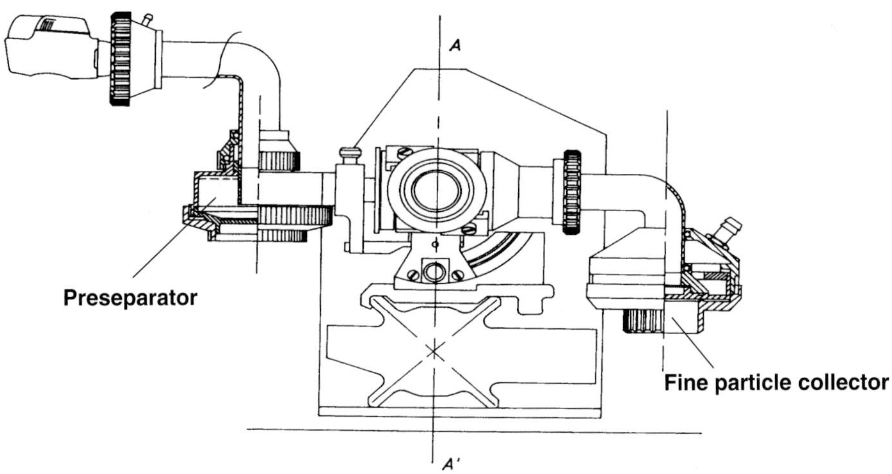

Figure 3.6. Schematic representation of a new inhaler adapter for laser diffraction characterization of the aerosol cloud.

The adapter for DPI testing consists of a closed central housing with a pre-separator for large (carrier) particles and a fine particle collector for analysis of the fine particle mass fraction [126]. The adapter for nebulizer testing can be tilted in order to fit the angled mouthpieces of this type of inhalation device. The closed system allows accurate control of the inspiratory flow rate through the inhaler without limitations regarding PIFR and FIR.

The application of the laser diffraction technique is sometimes questioned because it measures geometric instead of aerodynamic particle diameters. However, the aerodynamic diameter can be calculated when the dynamic shape factor and density are known. Moreover, the dynamic shape factor (χ) of micronized particles will often be only slightly higher than 1 and so is the true particle density ($1.0 < \rho_P < 1.4$ g cm^{-3}). As a consequence, the aerodynamic diameter differs only slightly from the equivalent volume diameter (see Eq. 3.1).

3.11 *In Vitro* and *In Vivo* Deposition Efficacy of Inhalation Systems

Many studies present and discuss *in vitro* and *in vivo* drug deposition results obtained with inhalation systems. It is often difficult to compare *in vitro* results from different studies, because different testing equipment and different definitions for the fine particle dose may have been used.

The aerosol clouds from nebulizers have been investigated both with the laser diffraction technique (e.g.[48,60]) and with cascade impactors [45,49]. Droplet size distributions generated by 14 different devices (including eight jet and six ultrasonic) from laser diffraction analysis were presented by Le Brun *et al.* for an aqueous 10% tobramycin solution [127]. They found volume median diameters ranging from 1.3 to 3.3 μm at an inspiratory flow rate of 40 l min^{-1} (X_{90}-values ranging from 2.3 to 7.9 μm). Other factors that may influence the droplet size (distribution) are surface tension and viscosity of the drug solution, nozzle pressure and inspiratory flow rate (e.g. [45,57,60]). The *in vivo* deposition from standard nebulizers is furthermore influenced by the patient's breathing pattern (waste of aerosol during exhalation) and the output (rate) of the device (e.g. [47,48]). As a result of all these critical factors, lung deposition from nebulizers is generally low, between 2 and 12% of the dose for jet nebulizers [47] and between 1 and 32% for ultrasonic devices [23].

Many studies refer to the particle size (distribution) of droplets in the aerosol from MDIs (e.g. [51,128]). Droplet sizes from conventional CFC-MDIs are often not optimal for deep lung deposition: the MMADs range from approximately 2.7 to 4.8 μm [51,52,128,129]. That is why *in vitro* fine particle (< 5 μm) fractions may be as low as 3 to 11% of the delivered dose for MDIs with the drug in solution (e.g. reference [130]) versus 20 to 35% for suspension MDIs [130,131]. Because initial particle velocities are high (exceeding 30 m s^{-1} at the actuator orifice), considerable losses occur in the oropharynx [23]. Under optimal conditions, including the inhalation technique, no more than 15 to 20% of the dose is deposited in the lungs (e.g. reference [29,131,132]). Newly formulated HFA(hydrofluoroalhane)-MDIs may be better, with MMADs between 1 and 2 μm [51,128,129], although it has been shown that the mean particle diameter depends on the type of HFA and increases with increasing diameter of the actuator orifice as well as increasing the concentration of non-volatile components in

lations that not only stabilize the protein, but also deliver it in the adequate physicochemical state to the absorbing membrane. In addition, it will be necessary to monitor closely the dissolution rate of drug in order to control the rate of absorption. Only if these requirements are met, will it be possible for a significant number of potential therapeutic proteins to be administered by the pulmonary route.

With regard to delivery to the lung, local therapies for asthma or COPD have an established position. New developments will focus on inhalation devices and formulations that allow a more reproducible and easy generation of the aerosol cloud and a less critical inhalation procedure. Generally, improved deposition of the drug in the airways is desirable to improve dosing accuracy and decrease side-effects. Together with the introduction of new drug substances technical improvements can still significantly improve the therapy of pulmonary diseases. For example, a major improvement in the treatment of cystic fibrosis can be achieved in the coming 5 years due to the development of new inhalation therapies for antibiotic drugs. Currently, the pulmonary use of about 10 antibiotic drugs has been reported but only two (tobramycin and colistin) have found a place in regular prophylactic use and therapy. Moreover, these drugs are still administered by quite inefficient nebulizers that provide only a deep lung deposition of less than 10% of the administered dose. Both the development of effective antibiotics against microorganisms such as *Pseudomonas aeruginosa* and *Burkholderia cepacia* as well as the development of innovative formulations and devices to improve lung deposition can largely optimize antibiotic therapy. Among others, improvements in this field can lead to better treatment of cystic fibrosis and increase life expectations. On the longer term, pulmonary administered gene therapy might even further improve cystic fibrosis therapy. Yet, significant technical hurdles have to be overcome before widespread use in therapy can be achieved.

In conclusion, it can be stated that the pulmonary administration of drugs is likely to expand rapidly in the coming years. Yet many questions still exist and extensive basic research is required before its therapeutic potential can be fully exploited in daily therapeutic practice.

Acknowledgements

Dirk K. F. Meijer is acknowledged for critically reading the manuscript. Bert Stok is thanked for his help in preparing the reference list.

References

[1] *Pharmaceutical Inhalation Aerosol Technology* (Ed. Hickey AJ), Marcel Dekker Inc., New York, **1992.**

[2] *Inhalation Delivery of Therapeutic Peptides and Proteins* (Eds Adjei AL, Gupta PK), Marcel Dekker Inc., New York, **1997.**

[3] Patton JS, Bukar J, Nagarajan S, *Adv. Drug Deliv. Rev.* **1999,** *35*, 235–247.

[4] Crook K, Porteous DJ, In: *Inhalation Delivery of Therapeutic Peptides and Proteins* (Eds Adjei AL, Gupta PK), pp. 555–585. Marcel Dekker Inc., New York, **1997.**

[5] Johnson LG, Boucher RC, In: *Inhalation Delivery of Therapeutic Peptides and Proteins* (Eds Adjei AL, Gupta PK), pp. 515–553. Marcel Dekker Inc., New York, **1997.**

[6] *Inhalation Aerosols: Physical and Biological Basis for Therapy* (Ed. Hickey AJ). Marcel Dekker Inc., New York, **1996.**
[7] In: *Proceedings of the International Workshop on Aerosol Inhalation, Lung Transport, Deposition and the Relation to the Environment: Recent Research Frontiers, Warsaw, Poland, 14–16 September 1995* (Eds Marijnissen JCM, Grado'n L). Kluwer Academic Publishers, Dordrecht, **1996.**
[8] *Aerosol Technology: Properties, Behaviour, and Measurement of Airborne Particles* (Ed. Hinds WC) John Wiley & Sons, New York, **1982.**
[9] Ganderton D, Jones T, *Drug Delivery to the Respiratory Tract*. Ellis Horwood Ltd., Chichester, **1987.**
[10] Patton JS, *Adv. Drug Deliv. Rev.* **1996,** *19*, 3–36.
[11] Wu-Pong S, Byron PR, *Adv. Drug Deliv. Rev.* **1996,** *19*, 47–71.
[12] Schulz H, *Pharm. Sci. Technol. Today* **1998,** *1*, 336–344.
[13] Li W-I, Perzl M, Ferron GA, Batycky R, Heyder J, Edwards DA, *J. Aerosol Sci.* **1998,** *29*, 995–1010.
[14] Ashurst I, Malton A, Prime D, Sumby B, *Pharm. Sci. Technol. Today* **2000,** *3*, 246–256.
[15] Yu J, Chien Y, *Crit. Rev. Ther. Drug Carrier Syst.* **1997,** *14*, 395–453.
[16] Niven RW, *Crit. Rev. Ther. Drug Carrier Syst.* **1995,** *12*, 151–231.
[17] Li W-I, Edwards DA, *Adv. Drug Deliv. Rev.* **1997,** *26*, 41–49.
[18] Morrow PE, Yu CP, In: *Aerosols in Medicine: Principles, Diagnosis and Therapy* (Eds Morén F, Newhouse M, Dolovich M), pp. 149–191. Elsevier Scientific Publishers, Biomedical Division, London, **1985.**
[19] Martonen TB, Yang Y, Hwang D, Fleming JS, *Comput. Biol. Med.* **1995,** *25*, 431–446.
[20] Weibel ER, *Morphometry of the Human Lung*. Springer Verlag, Berlin, **1963.**
[21] Netter FH, In: *Respiratory System: A Compilation of Paintings Depicting Anatomy and Embryology, Physiology, Pathology, Pathophysiology, and Clinical Features and Treatment of Diseases* (Eds Divertie MB, Brass A), pp. 46–59. CIBA Pharmaceutical Company, Summit, NJ, **1980.**
[22] de Boer AH, Winter HMI, Lerk CF, *Int. J. Pharm.* **1996,** *130*, 231–244.
[23] Newman SP, *Deposition and Effects of Inhalation Aerosols*, ISBN: 91-86058-03-07, AB DRACO, Lund, **1983.**
[24] Staniforth JN, Int. Patent Publication: WO97/03649, **1997.**
[25] Timsina MP, Martin GP, Marriott C, Ganderton D, Yianneskis M, *Int. J. Pharm.* **1994,** *101*, 1–13.
[26] Kirk WF, *Pharm. Int.* **1986,** *7*, 150–154.
[27] Byron PR, *Drug Develop. Ind. Pharm.* **1986,** *12*, 993–1015.
[28] Moren F, *Drug Develop. Ind. Pharm.* **1987,** *13*, 695–728.
[29] Jackson WF, *Inhalers in Asthma: The New Perspective*, ISBN: 1-899520-02-3, Astra Draco AB, Lund, **1995.**
[30] Newman SP, Pavia D, In: *Aerosols in Medicine: Principles, Diagnosis and Therapy* (Eds Morén F, Newhouse MT, Dolovich MB), pp. 193–218. Elsevier Scientific Publishers, Biomedical Division, London, **1985.**
[31] Morén F, Andersson J, *Int. J. Pharm.* **1980,** *6*, 295–300.
[32] Adjei AL, In: *Inhalation Delivery of Therapeutic Peptides and Proteins* (Eds Adjei AL, Gupta PK), pp. 771–814. Marcel Dekker Inc., New York, **1997.**
[33] Niven RW, Lott FD, Yp AY, Cribbs JM, *Pharm. Res.* **1994,** *11*, 1101–1109.
[34] Forbes B, *Pharm. Sci. Technol. Today* **2000,** *3*, 18–27.
[35] Forbes B, Wilson CG, Gumbleton M, *Int. J. Pharm.* **1999,** *180*, 225–234.
[36] Elbert KJ, Schäfer UF, Schäfers H-J, Kim K-J, Lee VHL, Lehr C-M, *Pharm. Res.* **1999,** *16*, 601–608.
[37] Lehr C-M, *J. Control. Rel.* **2000,** *65*, 19–29.
[38] Shen Z, Zhang Q, Wei S, Nagai T, *Int. J. Pharm.* **1999,** *192*, 115–121.
[39] Suzuki M, Machida M, Adachi K, Otabe K, Sugimoto T, Hayashi M, Awazu S, *J. Toxicol. Sci.* **2000,** *25*, 49–55.
[40] Gonda I, Schuster JA, Rubsamen RM, Lloyd P, Cipolla D, Farr SJ, *J. Control. Rel.* **1998,** *53*, 269–274.
[41] Rubsamen R, In: *Inhalation Delivery of Therapeutic Peptides and Proteins* (Eds Adjei AL, Gupta PK), pp. 703–731. Marcel Dekker Inc., New York, **1997.**
[42] Newman SP, *Nebulizer Therapy: Scientific and Technical Aspects*. AB DRACO, Lund, **1989.**
[43] Le Brun PPH, de Boer AH, Heijerman HGM, Frijlink HW, *Pharm. World Sci.* **2000,** *22*, 75–81.
[44] McCallion ONM, Taylor KMG, Bridges PA, Thomas M, Taylor AJ, *Int. J. Pharm.* **1996,** *130*, 1–11.

[45] Niven RW, Brain JD, *Int. J. Pharm.* **1994,** *104*, 73–85.
[46] Le Brun PPH, de Boer AH, Gjaltema D, Hagedoorn P, Heijerman HGM, Frijlink HW, *Int. J. Pharm.* **1999,** *189*, 205–214.
[47] O'Callaghan C, Barry PW, *Thorax* **1997,** *52*, S31–S44.
[48] Denyer J, Nikander K, In: *Proceedings of Respiratory Drug Delivery VI (Hilton Head, SC)*, pp. 311–314. Interpharm Press, Buffalo Grove, IL, **1998.**
[49] Bridges PA, Taylor KMG, *Int. J. Pharm.* **1998,** *173*, 117–125.
[50] Vervaet C, Byron PR, *Int. J. Pharm.* **1999,** *186*, 13–30.
[51] Brambilla G, Ganderton D, Garzia R, Lewis D, Meakin B, Ventura P, *Int. J. Pharm.* **1999,** *186*, 53–61.
[52] Keller M, *Int. J. Pharm.* **1999,** *186*, 81–90.
[53] Malcolmson RJ, Embleton JK, *Pharm. Sci. Technol. Today* **1998,** *1*, 394–398.
[54] Reiners M, Stang N, Wolf-Heuss E, In: *Respiratory Drug Delivery VII* (Eds Dalby RN, Byron PR, Farr SJ, Peart J), pp. 333–336. Serentec Press, Raleigh, **2000.**
[55] de Boer AH, Hagedoorn P, Gjaltema D, *Eur. Respir. J.* **1997,** *10*, 236S–236S.
[56] Newman SP, Pellow PGD, Clarke SW, *Atomiz. Spray Technol.* **1987,** *3*, 1–11.
[57] McCallion ONM, Taylor KMG, Thomas M, Taylor AJ, *Pharm. Res.* **1995,** *12*, 1682–1688.
[58] Le Brun PPH, de Boer AH, Gjaltema D, Hagedoorn P, Heijerman HGM, Frijlink HW, *Int. J. Pharm.* **1999,** *189*, 215–225.
[59] Boucher RMG, Kreuter J, *Ann. Allergy* **1968,** *26*, 591–600.
[60] McCallion ONM, Taylor KMG, Thomas M, Taylor AJ, *Int. J. Pharm.* **1996,** *129*, 123–136.
[61] Staniforth JN, In: *Respiratory Drug Delivery V: Program and Proceedings. Proceedings of the Symposium on Respiratory Drug Delivery, Phoenix, Arizona, 28 April–2 May 1996* (Eds Dalby RN, Byron PR, Farr SJ), pp. 65–73. Interpharm Press, Buffalo Grove, **1996.**
[62] Hickey AJ, Concessio NM, *Adv. Drug Deliv. Rev.* **1997,** *26*, 29–40.
[63] Buckton G, *Adv. Drug Deliv. Rev.* **1997,** *26*, 17–27.
[64] Buckton G, Choularton A, Breezer AE, Chatham SM, *Int. J. Pharm.* **1988,** *47*, 121–128.
[65] Johnson KA, *Adv. Drug Deliv. Rev.* **1997,** *26,* 3–15.
[66] Boerefijn R, Ning Z, Ghadiri M, *Int. J. Pharm.* **1998,** *172*, 199–209.
[67] Concessio NM, VanOort MM, Knowles MR, Hickey AJ, *Pharm. Res.* **1999,** *16*, 828–834.
[68] Larhrib H, Zeng XM, Martin GP, Marriott C, Pritchard J, *Int. J. Pharm.* **1999,** *191*, 1–14.
[69] Zeng XM, Martin GP, Marriott C, Pritchard J, *Int. J. Pharm.* **2000,** *200*, 93–106.
[70] Kawashima Y, Serigano T, Hino T, Yamamoto H, Takeuchi H, *Int. J. Pharm.* **1998,** *172*, 179–188.
[71] Shekunov BY, York P, *J. Crystal Growth* **2000,** *211*, 122–136.
[72] York P, Hanna M, In: *Respiratory Drug Delivery V: Program and Proceedings. Proceedings of the Symposium on Respiratory Drug Delivery, Phoenix, Arizona, 28 April–2 May 1996* (Eds Dalby RN, Byron PR, Farr SJ), pp. 231–239. Interpharm Press, Buffallo Grove, **1996.**
[73] Byron PR, Naini V, Philips EM, In: *Respiratory Drug Delivery V: Program and Proceedings. Proceedings of the Symposium on Respiratory Drug Delivery, Phoenix, Arizona, 28 April–2 May 1996* (Eds Dalby RN, Byron PR, Farr SJ), pp. 103–111. Interpharm Press, Buffalo Grove, **1996.**
[74] Adjei AL and Gupta PK, In: *Inhalation Delivery of Therapeutic Peptides and Proteins* (Eds Adjei AL, Gupta PK), pp. 625–665. Marcel Dekker Inc., New York, **1997.**
[75] Lucas P, Anderson K, Staniforth JN, *Pharm. Res.* **1998,** *15*, 562–569.
[76] Maa Y-F, Nguyen P-A, Sweeney T, Shire SJ, Hsu CC, *Pharm. Res.* **1999,** *16*, 249–254.
[77] Wang W, *Int. J. Pharm.* **2000,** *203*, 1–60.
[78] Wang W, *Int. J. Pharm.* **1999,** *185,* 129–188.
[79] Wong D, Parasrampuria J, *Pharm. Technol.* **1997,** *21*, 34–50.
[80] Andya JD, Maa Y-F, Costantino HR, Nguyen P-A, Dasovich N, Sweeney TD, Hsu CC, Shire SJ, *Pharm. Res.* **1999,** *16*, 350–358.
[81] Costantino HR, Carrasquillo KG, Cordero RA, Mumenthaler M, Hsu CC, Griebenow K, *J. Pharm. Sci.* **1998,** *87*, 1412–1420.
[82]Won CM, Molnar TE, McKean RE, Spenlehauer GA, *Int. J. Pharm.* **1998,** *167*, 25–36.
[83] Fox KC, *Science* **1995,** *267*, 1922–1923.
[84] Levine H, Slade L, *Cryo-letters* **1988,** *9*, 21–63.
[85] Angell CA, *Science* **1995,** *267*, 1924–1935.
[86] Chan H-K, Clark A, Gonda I, Mumenthaler M, Hsu C, *Pharm. Res.* **1997,** *14*, 431–437.
[87] Vanbever R, Mintzes JD, Wang J, Nice J, Chen D, Batycky R, Langer R, Edwards DA, *Pharm. Res.* **1999,** *16*, 1735–1742.
[88] Edwards DA, Ben-Jebria A, Langer R, *J. Appl. Physiol.* **1998,** *85,* 379–385.

[89] Vanbever R, Ben-Jebria A, Mintzes JD, Langer R, Edwards DA, *Drug Develop. Res.* **1999,** *48*, 178–185.
[90] Ben-Jebria A, Chen D, Eskew ML, Vanbever R, Langer R, Edwards DA, *Pharm. Res.* **1999,** *16*, 555–561.
[91] Pillai RS, Yeates DB, Miller IF, Hickey AJ, *J. Appl. Physiol.* **1998,** *84*, 717–725.
[92] El-Baseir MM, Kellaway IW, *Int. J. Pharm.* **1998,** *175*, 135–145.
[93] Talton J, Fitz-Gerald J, Rajiv Singh, Hochhaus G, In: *Respiratory Drug Delivery VII* (Eds Dalby RN, Byron PR, Farr SJ, Peart J), pp. 67–81. Serentec Press, Raleigh, **2000.**
[94] McAllister SM, Alpar HO, Teitelbaum Z, Bennet DB, *Adv. Drug Deliv. Rev.* **1996,** *19*, 89–110.
[95] Niven RW, Prestrelski SJ, Treuheit MJ, Ip AY, Arakawa T, *Int. J. Pharm.* **1996,** *127*, 191–201.
[96] Ip AY, Arakawa T, Silvers H, Ransone CM, Niven RW, *J. Pharm. Sci.* **1995,** *84*, 1210–1214.
[97] Williams RO, III, Liu J, *Pharm. Res.* **1997,** *14*, S634.
[98] Quinn EA, Forbes RT, Williams AC, Oliver MJ, McKenzie L, Purewal TS, *Int. J. Pharm.* **1999,** *186*, 31–41.
[99] Olsson B, Asking L, *J. Aerosol Med. Deposition Clear. Eff. Lung* **1994,** *7*, 201–204.
[100] Clark AR, Hollingworth AM, *J. Aerosol Med.* **1993,** *6*, 99–110.
[101] Svartengren K, Lindestad P, Svartengren M, Philipson K, Bylin G, Camner P, *Am. J. Respir. Crit. Care Med.* **1995,** *152*, 32–37.
[102] Brown PH, Ning ACWS, Greening AP, McLean A, Crompton GK, *Eur. Respir. J.* **1995,** *8*, 1940–1941.
[103] Meijer RJ, van der Mark ThW, Aalders BJ, Postma DS, Koëter GH, *Thorax* **1996,** *51*, 433–434.
[104] de Koning JP, van der Mark ThW, Rutgers SR, Coenegracht PMJ, Lerk CF, *Am. J. Respir. Crit. Care Med.* **1998,** *157*, A635.
[105] Timsina M, Martin GP, Marriott C, Ganderton D, Lee KC, Suen KO, Yianneskis M, *J. Aerosol Med.* **1993,** *6*, 41.
[106] Sarinas PSA, Robinson TE, Clark AR, Canfield J, Chitkara RK, Fick Jr RB, *Chest* **1998,** *114*, 988–992.
[107] Engel T, Heinig JH, Madsen F, Nikander K, *Eur. Respir. J.* **1990,** *3*, 1037–1041.
[108] Newman SP, Morén F, Trofast E, Talaee N, Clarke SW, *Eur. Respir. J.* **1989,** *2*, 247–252.
[109] de Boer AH, Hagedoorn P, Gjaltema D, *Eur. Respir. J.* **1996,** *23 (Suppl.)*, 206S.
[110] Olsson B, *J. Aerosol Med. Deposition Clear. Eff. Lung* **1995,** *8*, S13–S19.
[111] Ross DL, Schultz RK, *J. Aerosol Med. Deposition Clear. Eff. Lung* **1996,** *9*, 215–226.
[112] Prime D, Grant AC, Slater AL, Woodhouse RN, *J. Aerosol Med. Deposition Clear. Eff. Lung* **1999,** *12,* 75–84.
[113] de Boer AH, Bolhuis GK, Gjaltema D, Hagedoorn P, *Int. J. Pharm.* **1997,** *153*, 67–77.
[114] Everard ML, Devadason SG, Le Souëf PN, *Respir. Med.* **1997,** *91*, 624–628.
[115] Bonfig KW, *Technische Durchflussmessung: mit Besonderer Berücksichtigung Neuartiger Durchflussmessverfahren.* Vulkan-Verlag, Essen, **1977.**
[116] Sullivan WJ, Peters GM, Enright PL, *Respir. Care* **1984,** *29*, 736–749.
[117] Holmann JCh, Schibler A, Kraemer R, *Schweiz. Med. Wochenschr.* **1989,** *119*, 1713–1718.
[118] Kreit JW, Sciurba FC, *Am. J. Respir. Crit. Care Med.* **1996,** *154*, 913–917.
[119] Perry RH, Chilton CH, *Chemical Engineers' Handbook*. McGraw-Hill, New York, **1973.**
[120] John W, *J. Aerosol Sci.* **1999,** *30*, 1317–1320.
[121] Physical tests and determinations on aerosols. In: *United States Pharmacopoeia XXII, NF XVII, Ninth Supplement*, pp. 3584–3591, **1993.**
[122] Preparations for inhalation: aerodynamic assessment of fine particles. In: *European Pharmacopoeia*, pp. 143–152. European Pharmacopoeia, Strasbourg, **1997.**
[123] Asking L, Olsson B, *Aerosol Sci. Technol.* **1997,** *27*, 39–49.
[124] Müller RH, Schumann R, *Teilchengrössenmessung in der Labor Praxis* (Paperback APV). Wissenschaftliche Verlagsgesellschaft GmbH, Stuttgart, **1996.**
[125] Heuer M, Leschonski K, *Part. Charact.* **1985,** *2*, 7–13.
[126] de Boer AH, Gjaltema D, Witt W, Frijlink HW, In: *Respiratory Drug Delivery VII* (Eds Dalby RN, Byron PR, Farr SJ, Peart J), pp. 585–587. Serentec Press, Raleigh, **2000.**
[127] Le Brun PPH, de Boer AH, Gjaltema D, Hagedoorn P, Heijerman HGM, Frijlink HW, *Int. J. Pharm.* **1999,** *189*, 205–214.
[128]Tzou T-Z, *Int. J. Pharm.* **1999,** *186*, 71–79.
[129] Stein SW, *Int. J. Pharm.* **1999,** *186*, 43–52.
[130] Bell JH, Brown K, Glasby J, *J. Pharm. Pharmacol.* **1973,** *25*, 32P–36P.
[131] Newman SP, Clarke SW, *Chest* **1993,** *103*, 1442–1446.

[132] Newman SP, Weisz AWB, Talaee N, Clarke SW, *Thorax* **1991,** *46*, 712–716.
[133] Leach CL, *Respir. Med.* **1998,** *92*, 3–8.
[134] Conway JH, Walker P, *Eur. Respir. J.* **1999,** *14*, 196s.
[135] Pauwels R, Newman S, Borgström L, *Eur. Respir. J.* **1997,** *10*, 2127–2138.
[136] Selroos O, Pietinalho A, Riska H, *Clin. Immunother.* **1996,** *6*, 273–299.
[137] Newman SP, Pitcairn GR, Hirst PH, Bacon RE, O'Keefe E, Reiners M, Hermann R, *Eur. Respir. J.* **2000,** *16*, 178–183.
[138] Griesenbach U, Chonn A, Cassady R, Hannam V, Ackerley C, Post M, Tanswell AK, Olek K, O'Brodovich H, Tsui L-C, *Gene Ther.* **1998,** *5*, 181–188.
[139] Kukowska–Latallo JF, Raczka E, Quintana A, Chen C, Rymaszewski M, Baker Jr JR, *Hum Gene Ther.* **2000,** *11*, 1385–1395.
[140] Orson FM, Kinsey BM, Hua PJ, Bhogal BS, Densmore CL, Barry MA, *J. Immunol.* **2000,** *164*, 6313–6321.
[141] Tyler RC, Fagan KA, Unfer RC, Gorman C, McClarrion M, Bullock C, Rodman DM, *Am. J. Physiol.* **1999,** *277*, L1199–L1204.
[142] Li S, Tan Y, Viroonchatapan E, Pitt BR, Huang L, *Am. J. Physiol. Lung Cell. Mol. Physiol.* **2000,** *278*, L504–L511.
[143] Christofidou-Solomidou M, Pietra GG, Solomides CC, Arguiris E, Harshaw D, Fitzgerald GA, Albelda SM, Muzykantov VR, *Am. J. Physiol. Lung Cell. Mol. Physiol.* **2000,** *278*, L794–L805.
[144] Olsson B, Asking L, *J. Aerosol Med.* **1994,** *7*, S43–S47.
[145] de Boer AH, In: *Proceedings of The Second Annual Henry Stewart Update on Understanding Equivalence Testing for Inhaled Drugs, London, 22–23 June 1998*, Russell House Publ., London, **1998.**

4 Cell Specific Delivery of Anti-Inflammatory Drugs to Hepatic Endothelial and Kupffer Cells for the Treatment of Inflammatory Liver Diseases

Barbro N. Melgert, Leonie Beljaars, Dirk K. F. Meijer, Klaas Poelstra

4.1 Introduction

Fibrosis or scarring of the liver occurs after damage to liver tissue. Most chronic liver diseases eventually result in excess scarring leading to liver cirrhosis. This fatal disease, to date, can only be effectively treated with a liver transplantation. Since this is a costly procedure, hampered by the lack of donor organs among other technical factors, much effort has been put into developing new drugs. The drugs available are not sufficiently effective and/or cause too many adverse side-effects. Therefore drug targeting is an option in trying to maximize efficacy and minimize adverse drug reactions.

Chronic liver diseases are characterized by an inflammatory and a fibrotic component, both of which can be targets for pharmacological intervention. This chapter focuses on the treatment of liver fibrosis through the targeting of anti-inflammatory drugs. The target cells within the liver for anti-inflammatory treatment and possible entry mechanisms in these target cells will be identified. In addition, the different drug carriers and drug targeting preparations will be reviewed.

Since drug targeting implies the manipulation of drug distribution in the whole body, emphasis should be put on *in vivo* studies. In contrast to *in vitro* studies, studies in the intact organism will provide more definite insight into the cell specificity of carrier systems, the potential toxicity, immunogenicity, and the ability of the carrier system to pass anatomical barriers en route to the target cells. Moreover, it is of the utmost importance that these parameters are also studied in the diseased state, since the targeting potential of carriers can change dramatically under pathological conditions. *In vitro* studies with various liver preparations can be used to study endocytosis, carrier degradation and intracellular release of the targeted drug in more detail. In addition, the concept of drug targeting should also be tested in human tissue. Possibilities to include early (kinetic) screening in human tissue will also be discussed in this chapter.

4.2 The Liver

At the crossroads between the digestive tract and the rest of the body resides the largest solid organ of the body: the liver. Because of its interposition, the liver has a dual blood supply.

Nutrient-rich blood arrives through the portal vein and oxygen-rich blood through the hepatic artery. Together these channels import a large variety of endobiotics and xenobiotics, ranging from nutrients to toxic substances derived from the digestive system. The main function of the liver, therefore, is to maintain the body's metabolic homeostasis. This includes the efficient uptake of amino acids, carbohydrates, lipids and vitamins and their subsequent storage, metabolic conversion, and release into blood and bile; synthesis of serum proteins; hepatic biotransformation of circulating compounds, a process which converts hydrophobic substances into water-soluble derivatives that can be secreted into bile or urine, as well as phagocytosis of foreign macromolecules and particles such as bacteria.

Classically the liver has been divided into hexagonal lobules centred around the terminal hepatic venules. Blood enters the liver through the portal tracts that are situated at the corners of the hexagon. The portal tracts are triads of a portal vein, an hepatic artery, and a common hepatic bile duct. The vast expanse of hepatic tissue, mostly consisting of parenchymal cells (PC) or hepatocytes, is serviced via terminal branches of the portal vein and hepatic artery, which enters the tissue at intervals. The hepatocytes are organized into cords of cells radially disposed about the central hepatic venule. Between these cords are vascular sinusoids that transport the blood to the central hepatic venules. The blood is collected through the hepatic venules into the hepatic vein which exits the liver into the inferior vena cava (Figure 4.1).

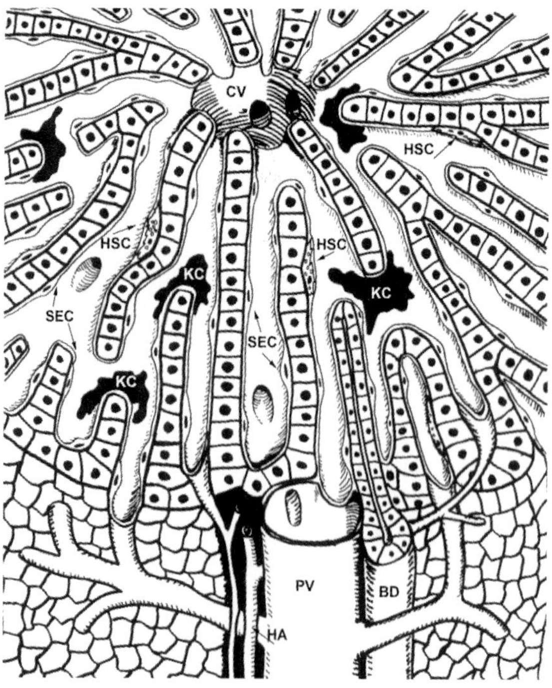

Figure 4.1. Schematic representation of the architecture of the liver. Blood enters the liver through the portal vein (PV) and hepatic arteries (HA), flows through the sinusoids, and leaves the liver again via the central vein (CV). KC, Kupffer cells; SEC, sinusoidal endothelial cells; HSC, hepatic stellate cells; BD, bile duct. Modified from reference 98.

The sinusoids are lined by the discontinuous and fenestrated sinusoidal endothelial cells (SEC) that demarcate the extrasinusoidal space of Disse. The abundant microvilli of the hepatocytes protrude into this space, which also contains the fat-containing lipocyte or hepatic stellate cell (HSC). At a strategic position along the luminal side of the endothelial cells are the resident tissue macrophages, the Kupffer cells (KC). Also located on the endothelial lining are the Pit cells, that correspond to large granular lymphocytes with natural killer activity. Between the abutting hepatocytes are bile canaliculi: channels in between the plasma membranes of facing hepatocytes, that are delineated from the vascular space by tight junctions. These intercellular spaces constitute the outermost reaches of the biliary tree. The canaliculi emanate from the centrilobular regions, progressively drain into the canals of Hering at the fringes of the portal tracts, and biliary fluid finally collects in the interlobular bile ducts.

4.2.1 The Parenchymal Cell (PC)

The liver consists mainly of parenchymal cells, or hepatocytes. Most drug-targeting preparations designed for liver targeting of therapeutic compounds are directed towards this cell type, generally aiming at the asialoglycoprotein receptor using galactose residues coupled to a core molecule for binding. This chapter, however, will not discuss this type of targeting, but further information can be found in several reviews [1–3].

Hepatocytes make up 60–70% of the total number of liver cells. They have a well-organized intracellular structure with huge numbers of cell organelles to maintain the high metabolic profile. At the apical side or canalicular membrane the cell is specialized for the secretion of bile components. There are several ATP-dependent transport carriers located on this side of the membrane, which transport bile salts, lipids and xenobiotics into the canaliculus. On the sinusoidal side, the cells specialize in uptake and secretion of a wide variety of components. To increase the surface of the membrane for this exchange with the bloodstream, the sinusoidal domain of the membrane is equipped with irregular microvilli. The microvilli are embedded into the fluid and matrix components of the space of Disse and are in close contact with the sinusoidal blood because of the discontinuous and fenestrated SECs. To facilitate its metabolic functions numerous membrane transport mechanisms and receptors are situated in the membrane.

4.2.2 The Sinusoidal Endothelial Cell (SEC)

The endothelial lining of the sinusoids in the liver differs from the other capillaries in the body and is adapted to form a selective barrier between blood and hepatocytes. The basement membrane is composed of non-fibril-forming collagens including types IV, VI and XIV, glycoproteins and proteoglycans. The lining is discontinuous and the SECs are perforated by numerous fenestrae that lack diaphragms. This allows direct contact of the hepatocytes with most plasma proteins in the space of Disse, but prevents direct contact with blood cells, large chylomicrons, bacteria and viruses. SECs play an important role in the pathogenesis of several acute and chronic inflammatory liver diseases. Consequently they are attractive target cells for anti-inflammatory therapies.

The SECs account for 20% of all liver cells and are the first cells, together with the KCs, to encounter potentially harmful materials present in the portal blood. They are therefore equipped with scavenger capabilities and certain defence mechanisms to prevent damage to other cell types. The SECs have an active scavenging system for the majority of physiological and foreign soluble (waste) macromolecules [4,5]. Clearance mechanisms include receptor-mediated endocytosis, transcytosis, and phagocytosis. To regain local homeostasis after ingestion of injurious substances and after other detrimental events, the SECs can also produce cytokines, eicosanoids, and adhesion molecules for the mobilization of other hepatic cell types and cells of the immune system.

4.2.2.1 Receptor-mediated Endocytosis

Targeting to SECs should be directed at specific receptors present on this cell type. A wide range of proteins and other molecules can be taken up by SECs through receptor-mediated endocytosis. For example, SECs play an important role in the uptake of degradation products of the extracellular matrix. For this purpose they have hyaluronan [6], (pro)collagen, and fibronectin receptors [7]. The first two receptors are uniquely located on SECs. Elevated levels of serum hyaluronan and fibronectin, that are often found in liver disease [8], are usually the result of dysfunction of the clearance capacity of SECs combined with an increased production by HSCs [9].

Scavenger receptors on the SECs are instrumental in another important endocytic mechanism. They recognize and endocytose modified proteins that have a high net negative charge [9]. SECs predominantly express two types of scavenger receptors: the class AI and the class AII scavenger receptor [10]. Physiological substrates for these receptors were found to be the N-terminal propeptides of types I and III procollagen [11] and the lipid A moiety of endotoxin [12]. Most studies, however, have used non-physiological substrates such as negatively-charged albumins [13] and acetylated low-density lipoproteins (LDL) [14] to characterize these receptors. Yet, the binding of both physiological and non-physiological substrates is Ca^{2+}-independent and is followed by rapid endocytosis and degradation in lysosomes.
The SECs are further equipped with a receptor that specifically interacts with mannose- and N-acetylglucosamine-terminated glycoproteins. Unlike the scavenger receptor, binding of ligands to this so-called mannose receptor is Ca^{2+}-dependent, but is also followed by rapid endocytosis and degradation in lysosomes [15]. The receptor is thought to be involved in the uptake of micoorganisms like yeasts, bacteria, and parasites [16], but has also been shown to be involved in the uptake of tissue-type plasminogen activator [17]. In addition, the receptor is involved in antigen uptake for subsequent antigen presentation [18]. This indicates that SECs may also be involved in cell-mediated immune responses in the liver.

Other uptake-linked receptors found on the SECs are the Fc receptor for the uptake of immunoglobulins [19], the CD14 receptor for the binding of lipopolysaccharide (LPS) bound to LPS binding protein [20], the platelet derived growth factor AA receptor [21] and the glucagon receptor [22].

4.2.2.2 Phagocytosis and Transcytosis

SECs are normally able to internalize only small particles (up to 0.23 μm). In conditions of impaired KC function, however, they have also been found to phagocytose larger particles [23]. They are also responsible for the receptor-mediated transcytosis of several compounds, such as insulin [24] and transferrin [25].

4.2.2.3 Regulation of the Inflammatory Process by SECs

Exposure of the SECs to pathogens or cytokines produced by other cells during stress induces activation of the SECs and subsequent production of cytokines, eicosanoids, and/or adhesion molecules. For instance, after activation with LPS, a main component of the walls of gram-negative bacteria and a major inducer of inflammation and non-specific immune functions [20], SECs produce a number of pro- and anti-inflammatory cytokines. Pro-inflammatory cytokines shown to be produced were: tumour necrosis factor alpha (TNFα) [26]; interleukin-1 alpha/beta(IL-1α/β) [27]; the major inducer of acute phase proteins interleukin-6 (IL-6) [28]; and the neutrophil chemo-attractant interleukin-8 (IL-8) [29]. Anti-inflammatory cytokines shown to be produced were: interleukin-10 (IL-10) [27] and hepatocyte growth factor (HGF) [30].

Eicosanoids are the oxidative metabolites derived from the cell membrane component arachidonic acid. Arachidonic acid is released from the cell membrane by phospholipase A_2 and enzymatically converted to either prostaglandins (PGs) by cyclo-oxygenase or leukotrienes (LTs) by lipoxygenase. Eicosanoids is the collective name of prostaglandins and leukotrienes. SECs and KCs are the major sources of eicosanoids, whereas the PCs are considered to be the most important target cells for them. The main eicosanoid produced by SECs was found to be PGE_2 [31], although PGD_2 has also been reported to be a major product [32]. The type of PG released may be a result of the difference in the induction stimulus used. Eicosanoid production is induced by many circulating substances; LPS, interferon gamma (IFNγ), TNFα, and platelet activating factor (PAF). PGE_2 is postulated to be involved in liver regeneration [33] and inhibition of hormone-stimulated glycogenolysis [31], PGD_2 was found to induce glycogenolysis [34].

SECs, like the vascular endothelium, play an active part in the control of leucocyte recruitment in cases of acute and chronic inflammatory conditions. Leucocyte recruitment from the blood compartment is a crucial determinant for the induction of immunity and inflammation. SECs control this process by producing cytokines that activate leucocytes and by expressing adhesion molecules. Under inflammatory conditions upregulation of intracellular adhesion molecule 1 (ICAM-1) and vascular cell adhesion molecule 1 (VCAM-1) was found [35;36], as well as expression of E-selectin and P-selectin [37]. Together with the expression of CD4 on SECs it has been postulated that these adhesion molecules might also be involved in the adhesion of KC cells to the sinusoidal wall [20].

4.2.3 The Kupffer Cell (KC)

Kupffer cells are the largest reservoir of fixed-tissue macrophages and are quantitatively the most important cell type for the removal of circulating microorganisms, LPS, tumour cells, immune complexes, and other circulating tissue and microbial debris [38]. They account for about 15% of the liver cell population in number and they are preferentially located in the periportal areas [39].

4.2.3.1 Receptor-mediated Endocytosis

Similar to the targeting of compounds to SECs, drug targeting preparations designed to modify KC functions have to be directed at KC-specific receptors. KCs are able to remove numerous soluble and particulate substances from the circulation and they possess many receptor systems that mediate this clearance, some of which have also been described for SECs. Like SECs, they possess fibronectin receptors, mannose receptors, Fc receptors, CD14 receptors, and the scavenger receptors class AI and AII [40]. In addition to these receptors, KCs also possess the novel member of the class A scavenger receptor family, the macrophage receptor with collagenous structure (MARCO) [41]. Besides these types of scavenger receptors, they also have macrosialin scavenger receptors for the uptake of oxidized LDL [10] and scavenger receptors class BI for the removal of high-density lipoproteins (HDL) [42]. For the uptake of unmodified LDL, KCs also have special LDL receptors [43].

Mannose receptors on KCs essentially recognize the same molecules as the mannose receptors present on SECs, but they exhibit different kinetics [44]. Besides the mannose receptors, KCs have two other carbohydrate-specific receptors. One is the galactose particle receptor, recognizing galactose-terminated oligosaccharides on particles and mediating endocytosis of desialylated erythrocytes [45]. The other is the fucose receptor which interacts not only with fucose-terminated glycoproteins, but also with galactose-exposing neoglycoproteins [46].

KCs also possess receptors for the complement components C1q and C3b [47;48]. The complement system is one of the main defence mechanisms of the body against invading pathogens. It is composed of a group of serum proteins that are part of a multienzymatic cascade. Activation of complement generates membranolytic components and protein fragments that enhance phagocytosis and mediate immune responses. KCs have the optimal capacity to remove complexes coated with complement from the circulation.

4.2.3.2 Phagocytosis

Not all KCs are phagocytic to the same extent; periportal KCs generally have a higher level of phagocytic activity than those in other regions of the liver [49]. Prior to phagocytosis, particulate material like viruses, bacteria and erythrocytes may be opsonized and bound by specific receptors, but this is not essential for phagocytosis [50].

4.2.3.3 Regulation of the Inflammatory Process by the KC

As is the case for SECs, endocytosis of substances represents more than just circulatory clearance mechanisms. The uptake of potentially toxic material can activate KCs to function as either inflammatory cells or accessory cells. As accessory cells they express major histocompatibility complex (MHC) class II molecules on their surface , synthesize IL-1β and present antigens to T cells [51]. As inflammatory cells, KCs enhance chemotaxis, phagocytosis, and oxidative metabolism of inflammatory cells [52] by producing cytokines, eicosanoids and reactive oxygen species (ROS). After LPS stimulation, KCs produce chemokines such as monocyte chemotactic protein (MCP-1), macrophage inflammatory protein-1α/β (MIP-1α/β), RANTES (Regulated upon Activation, Normal T-cell Expressed and Secreted) and IL-8 [53,54], in addition to TNFα, IL-1α/β, interferon alpha/beta (IFNα/β), and IL-6 [55]. Release of these mediators will lead to activation and local infiltration of inflammatory cells and activation of other resident hepatic cells. KCs also produce transforming growth factor beta (TGFβ), which stimulates collagen synthesis by HSCs, while inhibiting proliferation of these cells [56].

Besides LPS, other particulate and soluble agents are known to stimulate the formation of eicosanoids, e.g. PGE_2, PGD_2, and thromboxane [57]. These agents also elicit nitric oxide and superoxide anion formation, which may help to destroy phagocytosed microorganisms or particles [58].

4.2.4 The Hepatic Stellate Cell (HSC)

Another resident hepatic cell that is important in the pathogenesis of chronic liver diseases is the hepatic stellate cell (also known as fat-storing cell, Ito cell, lipocyte, perisinusoidal cell). They are located in the space of Disse and represent 5–8% of all liver cells. With cytoplasmatic extensions encircling the sinusoid they regulate blood flow through the sinusoidal lumen, in response to endothelin-1, nitric oxide, angiotensin-II, thromboxane A2, and the prostaglandins $F_{2\alpha}$, I_2, and E_2 [59]. They also contain many vitamin A-rich lipid droplets which account for 75% of the total amount of retinoids stored in the body. As well as controlling the uptake, storage, and release of retinoids, HSCs are the major regulators of the extracellular matrix composition after activation. They produce and secrete matrix proteins such as collagens I, III, IV, V and VI, fibronectin, laminin, tenascin, undulin, hyaluronic acid and proteoglycans [60], as well as extracellular matrix degrading metalloproteinases and their inhibitors, the tissue inhibitors of metalloproteinases (TIMPs) [61].

HSCs have a dual phenotype. In healthy livers they have the quiescent phenotype, regulating retinoid storage and blood flow. In response to liver injury, however, they acquire an activated myofibroblast-like phenotype. During the transition to the activated phenotype there is a gradual loss of lipid droplets and an increased expression of α-smooth muscle actin (αSMA). In rat livers this is also accompanied by a loss of desmin expression [62]. A consequence of HSC activation is the change in synthetic activity towards production of excess collagen I and III molecules and other matrix molecules. These matrix proteins are deposited in the space of Disse obstructing efficient exchange of proteins and reducing the diameter of the sinusoids, thereby impeding blood flow. This process is called capillarization. It is also accompanied by a loss of fenestration of the sinusoidal endothelial lining, which further hampers the diffusion of proteins between plasma and hepatic cells. The alterations in the appearance of the sinusoid are the hallmark of fibrosis.

The transdifferentiation of HSCs to myofibroblasts, producing extracellular matrix constituents is characterized by an increased expression of several receptors, including the platelet derived growth factor (PDGF) receptor, the collagen type VI receptor, and the insulin-like growth factor II/mannose-6-phophate receptor (IGFII/M6P). For reviews on this subject see Beljaars *et al.* [63], Li *et al.* [64], and Bissell [65].

4.3 Hepatic Inflammation and Fibrosis

Virtually any insult to the liver can cause hepatocyte destruction and parenchymal inflammation. If the insult is minor and occurs only once, local restoration mechanisms will suffice to repair the damage. If, however, the insult is major or persistent, an inflammatory response will be generated. This inflammation is the result of cytokine-mediated activation of sinusoidal cells, their subsequent release of pro-inflammatory cytokines and their expression of adhesion molecules for the recruitment of circulating leucocytes. Once the damage is under control and the inciting insult has been eliminated, the inflammatory process will end and local mechanisms will proceed until the damage is repaired. Usually little scar tissue will be detectable, because of extracellular matrix remodelling. During conditions of chronic liver injury, however, the repair process does lead to scar tissue formation, which is deposited within the liver until impairment of liver function occurs. This process is called liver fibrogenesis and the end stage, or irreversible stage, is referred to as liver cirrhosis (Figure 4.2).

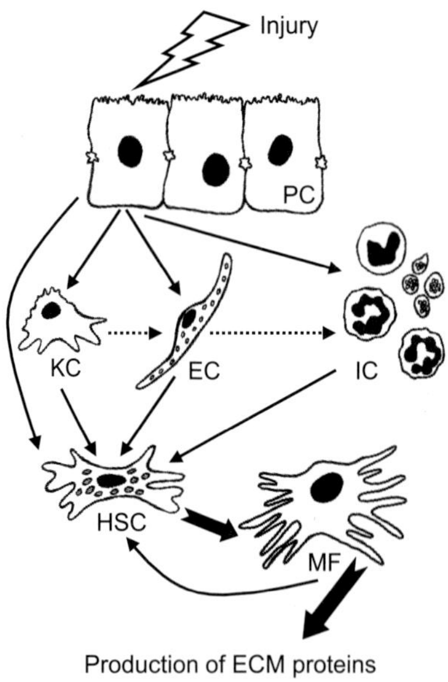

Figure 4.2. Diagram outlining the pathogenesis of liver fibrosis. Injury to parenchymal cells (PC) results in the activation of Kupffer cells (KC) and sinusoidal endothelial cells (SEC) and the recruitment of inflammatory cells (IC). These cells release cytokines, growth factors and reactive oxygen species that induce activation and proliferation of hepatic stellate cells (HSC). HSCs gradually transform into myofibroblasts (MF), the major producers of extracellular matrix (ECM) proteins.

After damage or infection, monocytes and KCs in the area detect the damaged cells or infectious agent and respond with release of primary mediators such as TNFα, IL-1 and some IL-6. These cytokines activate the surrounding cells, that respond with a secondary, amplified release of cytokines. This second wave includes large amounts of IL-6, which induce the synthesis of acute phase proteins in hepatocytes and chemoattractants such as IL-8 and MCP-1. These events will then lead to the typical inflammatory reactions. Both IL-1 and TNFα activate the central regulatory protein of many reactions involved in immunity and inflammation, nuclear factor kappa B (NFκB). These cytokines cause dissociation of NFκB from its inhibitor IκB, which makes translocation of NFκB to the nucleus possible. In the nucleus active NFκB induces the transcription of the 'second wave' cytokines (see also Chapter 7 for the molecular mechanisms of cytokine-mediated cell activation).

The release of TNFα and IL-1 also upregulates adhesion molecules like ICAM-1 and VCAM-1 on SECs, that are subsequently responsible for the adhesion and recruitment of circulating neutrophils. KCs and PCs release IL-8, which is a potent neutrophil chemoattractant. The attracted neutrophils and KCs are stimulated to release large amounts of reactive oxygen species (ROS: hydrogen peroxide, superoxide anion and nitric oxide (NO) radicals). The production of NO is also mediated through the NFκB pathway. The enzyme responsible for the increased synthesis of NO, inducible NO synthetase (i-NOS), is increasingly expressed through NFκB-mediated stimulation of the i-NOS promotor region.

TGFβ and TNFα produced by KCs and PDGF produced by SECs subsequently play an important role in the activation of HSCs. TGFβ appears to be the most important cytokine in stimulating the production of scar tissue components like collagens by HSCs. The mechanism of activation is probably via the IGF-II/M6P receptor, which is also increasingly expressed on activated HSCs. As yet unknown factors produced by KCs [66] stimulate expression of PDGF receptors on the surface of HSCs. In the presence of PDGF the HSC will now proliferate as well. On chronic stimulation, HSC stimulation and proliferation will result in production of excess extracellular matrix and the onset of fibrosis. KC-produced mediators appear to be important for HSC stimulation, but substances directly released by PCs are also found to be mitogenic [67].

Since not every insult necessarily results in liver fibrosis, counter-regulatory mechanisms must also exist. During inflammation, elimination of ROS by SECs and KCs is enhanced via increased expression of radical scavengers like superoxide dismutases and glutathione peroxidase. The radical nitric oxide itself also has an anti-inflammatory role. It has been described to prevent leucocyte adhesion to the endothelium [68] and to block an activation pathway of thrombocytes by stimulating guanylyl cyclase [69]. Furthermore, both PGE_2 and IL-10 can downregulate cytokine production by macrophages [70,71] and can also inhibit the antigen-presenting properties of SECs and KCs [18,72]. HGF produced by KCs, SECs, and quiescent HSCs is a potent mitogen for PCs and stimulates liver regeneration. It is probably aided by PGE_2 which also stimulates DNA synthesis in PCs [73]. Finally, scar tissue formation is not only regulated by production of extracellular matrix components, but also by the degradation of matrix components. Activated and quiescent HSCs, KCs, and SECs produce matrix metalloproteinases that are responsible for matrix degradation [74].

Whether liver regeneration will dominate over scar tissue formation depends on many factors, including the nature and the duration of the injury and the genetic background of the individual. It is still unclear at which point liver regeneration is no longer possible and fibrogenesis will progress to cirrhosis. When fibrogenesis takes over, however, collagens type I and III which are normally concentrated in the portal tracts and around central veins, are deposited throughout the liver. Collagen IV and VI and other components of the extracellular matrix are also increasingly expressed. The normal liver only contains 1–2% connective tissue, but in patients with cirrhosis this can increase up to a maximum of 50% [75]. The increased amount of extracellular matrix results in severe disruption of blood flow and impaired diffusion of solutes between PCs and plasma, which may have implications for drug targeting preparations to hepatic cells. Deposition of collagens in the space of Disse is also accompanied by the loss of fenestrations in SECs, which further impairs the movement of proteins between PCs and plasma. The subsequent resistance to portal flow induces portal hypertension and together with the reduced metabolic capacity of the liver this leads to four major clinical consequences: development of ascites, the formation of portosystemic venous shunts leading to dangerous esophagogastric varices, congestive splenomegaly causing haematologic abnormalities, and hepatic encephalopathy because of the exposure of the brain to an altered metabolic milieu. Other complications arising from the progressing fibrosis of the liver are the appearance of renal failure (hepatorenal syndrome), endotoxemia and hepatic failure. When loss of the hepatic functional capacity exceeds 80–90%, liver transplantation is usually the only option for survival. Many new pharmacological approaches to the therapy of fibrosis are being explored, but lack of effectiveness or a small therapeutic window remain major obstacles. These approaches may therefore benefit from drug targeting strategies [3].

4.4 Liver Cirrhosis: Causes and Therapy

Cirrhosis is among the top 10 causes of death in the Western World. This is largely the result of alcohol abuse, viral hepatitis and biliary diseases [75]. The causes for cirrhosis can be roughly divided into six categories:

1. Chronic exposure to toxins such as alcohol, drugs or chemicals,
2. Viral hepatitis resulting from infection with the hepatitis B, C or D viruses,
3. Metabolic disorders such as Wilson's disease (copper storage disease) and haemochromatosis (iron overload disease),
4. Autoimmune diseases such as primary biliary cirrhosis (PBC), primary sclerosing cholangitis (PSC) and autoimmune hepatitis,
5. Venous outflow obstruction,
6. Cirrhosis of unknown causes.

Obviously the best treatment for cirrhosis is removal of the injurious event. In the case of viral hepatitis, viral load can at least be temporarily reduced with anti-viral agents such as lamivudine, ribavirin and/or IFNα [76]. Unfortunately, complete removal of the injurious event is frequently not possible. Moreover, by the time cirrhosis is diagnosed the fibrotic process has usually progressed beyond 'the point of no return' and removal of the injurious event will have little effect. Successful pharmacological treatment to reverse the fibrotic process is not yet available. Several drugs have been tested in clinical trials, but the most effective treatment remains a liver transplantation.

The bile acid ursodeoxycholic acid has shown some promise in slowing down the fibrotic process in cholestatic patients, especially those suffering from PBC and PSC [77,78]. Its mechanism of action, however, is still a matter of debate.

Penicillamine, an inhibitor of collagen crosslinking, was evaluated in PBC, but failed to demonstrate any efficacy [79]. More promising results were found for colchicine, which inhibits collagen synthesis and secretion and enhances collagenase activity. Long-term use of colchicine prolonged survival in patients with mild to moderate cirrhosis, regardless of the cause [77,80]. Other types of collagen synthesis inhibitors, like the prolyl hydroxylase inhibitors, have been studied in experimental animal models [81], but have not yet found their way into the clinic.

Several types of immunosuppression have also been tried. Azathioprine alone was found to have no effect on PBC [82], but additional benificial effects were found in combination with ursodeoxycholic acid and corticosteroids [78]. Cyclosporin showed some success, especially in corticosteroid-resistant autoimmune hepatitis [83], but its use is generally considerably limited by severe side-effects. Corticosteroids were effective in the management of several types of autoimmune chronic active hepatitis [84,85] and in the management of acute alcoholic hepatitis [86]. Their use, however, has to be brief in order to minimize side-effects. In the treatment of PBC, corticosteroids alone were found to be toxic and had only limited efficacy [77].

A promising new development in drug therapy is the endothelin-antagonists [87,88]. Though not yet clinically tested, these compounds show potential in the management of portal hypertension, a hallmark of cirrhosis. Again, uptake of these antagonists by other parts of the body hampers their applicability [89], which might be circumvented by drug targeting.

4.5 Drug Targeting to the Liver

With no effective drugs available and the unacceptable side-effect profile of those drugs which have been studied so far, liver cirrhosis might benefit from the targeting of drugs to cells within the liver. There are several ways to intervene in the fibrotic process. One way is the targeting of drugs to SECs and KCs to modulate their release of pro-inflammatory mediators. This may arrest the inflammatory process leading to cirrhosis. Another way is the delivery of drugs to HSCs to inhibit collagen production or to enhance their extracellular matrix degrading capabilities. This chapter will focus on targeting to KCs and SECs to influence the inflammatory process that is the basis of most forms of liver cirrhosis. As mentioned before these cells have a number of specific entry mechanisms that could be used for cell-specific delivery of drugs. By either enclosing drugs in particles or by coupling drugs to macromolecular carriers with high affinity for certain uptake mechanisms, drugs can be concentrated in the target cells without causing side-effects elsewhere in the body. The choice for a type of carrier is determined by a number of considerations, depending on the specificity of the carrier, the potency of the drug and the entry mechanism during pathological conditions. The possible carriers directed to KCs and SECs show a considerable overlap, because these cells share many receptor-mediated endocytotic uptake mechanisms, such as uptake mediated by scavenger receptors or mannose receptors. Most of the carriers directed to SECs and KCs are designed for these receptors and are reviewed below.

4.5.1 Carriers Directed at SECs and KCs

4.5.1.1 Albumins

Albumin is one of the soluble macromolecular carriers available for drug targeting purposes. With a molecular weight of approximately 67 kDa, it is small in size as compared to other potential carriers. It can be derivatized with molecules that will determine its cell specificity, and with drug molecules. A maximum of about 60 molecules can be coupled to albumin through the ε-NH_2 of the lysine residues. Table 4.1 shows the modified albumins that have been used for targeting to SECs and KCs.

Albumin modified with negatively charged groups like succinic acid (Suc-HSA) and aconitic acid (Aco-HSA) are avidly taken up by SECs via the scavenger receptors type A. These receptors are also present on KCs, but have a slightly different substrate specificity. This was elegantly shown with formaldehyde-treated HSA (Form-HSA). The scavenger receptors on SECs take up the monomeric negatively-charged Form-HSA, whereas these receptors on KCs take up the polymeric Form-HSA [90]. Scavenging receptors are also involved in the uptake of maleylated albumin (Mal-BSA), which was designed for the targeting of chemotherapeutics to macrophages. It was found to be taken up by the non-parenchymal cells of the liver and by peritoneal macrophages, uptake by SECs was not determined [91].

The subtle differences in substrate specificity were found for the mannose receptor as well. Both SECs and KCs have mannose receptors, but mannosylated albumin (Man_{10}-HSA) is almost exclusively taken up by KCs. The relatively low mannose substitution (10 molecules of mannose per HSA molecule) combined with the extra negative charge added to the albumin molecule by the coupling of the mannose molecules to the lysine-residues of HSA, directs

Table 4.1. Albumin carriers designed for targeting to SEC and KC.

Carrier	SEC	KC
Suc-HSA	+++	–
Aco-HSA	+++	–
Form-HSA	++	++
Man_{10}-HSA	+	+++
Mal-BSA	ND	+++
Nap_{20}-HSA	+++	+
$Dexa_{10}$-HSA	+++	+

HAS, human serum albumin; BSA, bovine serum albumin; Suc, succinic acid; Aco, aconitic acid; Form, formaldehyde; Man, mannose; Mal, maltose; Nap, naproxen; Dexa, dexamethasone. ND, not done; –, no uptake; +, small uptake; ++, moderate uptake; +++, abundant uptake.

this carrier to the KCs. When the mannose substitution is increased, the uptake by SECs also increases [92]. For a long time this carrier has been assumed to be inert. However, recent studies from our laboratory indicate that the carrier Man_{10}-HSA may activate KCs and induce an immunological response [93]. Whether this limits the use of this carrier remains to be established. The subsequent coupling of dexamethasone to Man_{10}-HSA attenuated this immunological response [93].

Direct modification of albumin with drugs like naproxen (Nap_{20}-HSA) and dexamethasone ($Dexa_{10}$-HSA) changes the protein into a substrate for the scavenging receptors type A. These drugs are coupled to the free ε-NH_2-groups of the lysine residues in albumin. Normally these NH_2-groups are positively charged through protonation. Coupling of a drug molecule inhibits this protonation. The albumin molecule is left with a relative negative charge and becomes a substrate for the scavenger receptors. Apart from the net negative charge, it has been postulated that the added hydrophobicity of these drug molecules is an important feature in determining their affinity for the scavenger receptors [94].

After interaction of the aforementioned carriers with specific receptors, the carrier is then taken up by endocytosis and transported intracellularly to acidified endosomes and lysosomes. The carrier is proteolytically degraded in the lysosomes and if a drug is coupled to the carrier, it is then released to diffuse into the cytoplasmic compartment.

4.5.1.2 Liposomes

Liposomes are small vesicles composed of unilamellar or multilamellar phospholipid bilayers enclosing an aqueous space. Soluble drugs can readily be incorporated into this aqueous space and lipophilic drugs can be incorporated into the lipid bilayer. The loading capacity for drugs is therefore much greater than that of the modified albumins. Elimination from the circulation is dependent on the lipid composition, charge, and size of the liposomes. Common liposomes such as neutral and negatively-charged liposomes, are however, primarily cleared by the phagocytotic processes of the cells of the reticuloendothelial system (RES), the KCs having the greatest responsibility for this process. This feature of liposomes can seriously limit the use of liposomes in targeting other sites in the body [95]. It has been shown for instance that the targeting of cytostatic agents such as adriamycine to tumours is associated with loss of KC function [96], thereby contributing to the immuno-suppressed status of patients. The high KC uptake has been suprisingly under-exploited in drug targeting approaches to treat liver diseases. Liposomes have been used for the targeting of anti-Leishmania drugs [97,98] and immunomodulators [99] and have greatly increased the efficacy of these drugs in Leishmania infections and metastatic tumour growth, respectively. However, intervening in the fibrotic process by modulating KC or SEC functions with liposome-encapsuled drugs has not yet been attempted.

The exact mechanism responsible for the uptake of liposomes by KCs and SECs is not clear. Most studies confirm internalization of whole liposomes in an energy-dependent phagocytic process in which the liposomes are delivered to the lysosomes. The liposomal lipids are completely degraded and the encapsulated solutes released. Neutral liposomes consisting of lipids such as cholesterol and phosphatidylcholine are probably cleared by receptor-mediated mechanisms, due to the adsorption of opsonizing proteins onto the lipid bilayer. Some of the opsonizing proteins that have been found to play a role are complement factors [100] and fibronectin [101]. The opsonization of liposomes by plasma proteins, in particular complement factors C3bi and C5a, affects the cell selectivity of this carrier, causing uptake by the RES and by neutrophils [102]. The uptake of liposomes containing the negatively-charged phospholipid phophatidylserine (PS) is still a matter of debate. These liposomes may be taken up by the RES via specific PS receptors or via scavenger receptors, but here too uptake appears to be mediated predominantly by plasma proteins that bind in a PS-specific manner to liposomes [103]. The influence of plasma proteins on the uptake route of PS-containing liposomes was shown by Kamps *et al. In vitro* studies with liposomes containing 30% PS showed scavenger receptor-mediated uptake in KCs and SECs, but subsequent *in vivo* studies did not reveal a significant contribution of scavenger receptors to the KC uptake of these liposomes [104]. Specific scavenger-mediated uptake of liposomes by SECs was achieved by coating liposomes with negatively-charged albumins [105].

Lipoproteins are endogenous carriers for the transport of cholesterol and other lipids in the blood circulation and can be regarded as 'natural liposomes'. Because they are endogenous, they are not immunogenic and escape recognition by the RES. They are cleared from the circulation by specific lipoprotein receptors that recognize the apolipoproteins [106]. They can be directed to non-lipoprotein receptors as well, by chemical modification of the apolipoprotein moiety. Specific scavenger receptor-mediated uptake by SECs was achieved by the acetylation of LDL, whereby oxidized LDL was specifically taken up by KCs via the scavenger receptors and lactosylated LDL via the galactose-particle receptors [106–108].

The lipid core can be used to incorporate lipophilic drugs, whereas more hydrophilic drugs have to be provided with a lipophilic anchor to enable incorporation. Oleyl, retinyl and cholesteryl residues have been used for this purpose [109]. Chemical derivatization will however, alter the pharmacological activity of the parent drug in most cases. The anchors should therefore be easily removable inside the cell, yielding the original drugs. These liposomes have not as yet been used much to target drugs to KCs and SECs. Just one study described the enhancement of the tumouricidal activity of KCs with the immunomodulator muramyldipeptide incorporated in lactosylated LDL [110].

4.5.1.3 Carriers with Intrinsic Anti-inflammatory Activity

Another approach to drug targeting is the use of carriers with an intrinsic pharmacological activity. In this 'dual targeting' strategy a beneficial effect is achieved both from the carrier itself and the drug it carries. The negatively-charged HSA carriers, for instance, developed for the targeting of drugs to HIV-infected cells, exert strong antiviral activity themselves [111]. Possible carriers with intrinsic anti-inflammatory activity are superoxide dismutase (SOD) and alkaline phosphatase (AP).

SOD is a major oxygen free radical scavenging enzyme, which may therefore have beneficial effects in liver fibrosis. Through mannosylation or coupling to the polyanion DIVEMA, SOD was made more liver specific. Both conjugates showed superior inhibition of intrahepatic ROS production in fibrotic rats as compared to unmodified SOD. DIVEMA-SOD, however, exhibited the most potent inhibitory effects [112]. Although their mode of action is most probably *extracellular* free radical scavenging, Man-SOD and DIVEMA-SOD are likely to be taken up rapidly by mannose receptors and scavenger receptors, respectively. However, depending on the dose, a considerable fraction may be present on the cell surface, either bound to the receptor or through re-exposure via retroendocytosis after prior internalization [113]. Therefore, sufficient enzymatic activity might still be obtained in the extracellular space.

AP is a membrane-anchored protein, that can be shed into the general circulation, which was shown to be able to detoxify LPS *in vivo* through dephosphorylation [114]. This dephosphorylating activity could be enhanced by increasing the negative charge of the enzyme through succinylation [115]. Using AP as a carrier for anti-inflammatory drugs to KCs, the main site of LPS uptake, it could intrinsically contribute to therapeutic efficacy in cirrhosis through detoxicification of LPS. The LPS-detoxifying activity of KCs in cirrhotic livers is impaired and consequently LPS may promote the fibrotic process [116].

4.5.2 Targeting to other Hepatic Cells

Selective delivery of anti-fibrotic drugs to HSCs would be an elegant option in the design of effective anti-fibrotic therapy. Only recently the first carriers targeted to HSCs were developed: albumin modified with mannose 6-phosphate groups for uptake via mannose 6-phosphate/insulin-like growth factor II receptor and albumin derivatized with cyclic peptides containing amino acid sequences that mimic the binding site of either collagen type VI or PDGF to their receptors [117,118]. In addition to being used as drug carriers, these carriers could also be intrinsically active. The mannose 6-phosphate/insulin-like growth factor II receptor is involved in the activation of the fibrogenic mediator TGFβ [119,120], which, in theory, could be competively inhibited by the mannose 6-phosphate-modified albumin. The same competition between carrier and endogenous ligands can be anticipated for collagen type VI and PDGF receptors. The approach of using cyclic peptides with the receptor-recognizing domains of various cytokines or growth factors, that will mediate binding to their respective receptors, can also be exploited for the design of other dual active carriers to cell types such as SEC and KC.

4.6 Anti-inflammatory Drugs

Several classes of drugs can be potentially used to reduce the release of pro-inflammatory mediators by SECs and KCs in the fibrotic process.

4.6.1 Nonsteroidal Anti-inflammatory Drugs (NSAIDs)

NSAIDs are drugs related to acetylsalicylic acid which inhibit cyclooxygenase (COX), the enzyme in the synthesis of PGs and thromboxanes from arachidonic acid. There are two isoforms of cyclooxygenase, COX-1 and COX-2 [121]. The former is constitutively expressed in blood vessels, stomach and kidney, while COX-2 is normally not present at these sites. It can, however, be induced under inflammatory conditions by certain serum factors, cytokines, and growth factors [122]. Most of the currently used NSAIDs non-selectively inhibit both COX-1 and COX-2. They are widely used in inflammatory disorders of the joints such as arthritis, of the tendons and of the bursae. The side-effects are to some extent the result of the non-selective inhibition of the constitutive COX-1 production by PGs in other tissues. In the kidneys this may lead to renal insufficiency and in the gastrointestinal tract to the formation of ulcers and bleeding [123,124].

Acetylsalicylic acid was shown to prevent cirrhosis under certain experimental conditions [125]. Naproxen and indomethacin partially protected against LPS and D-galactosamine-induced hepatotoxicity [126] Acetylsalicylic acid and ibuprofen were also protective in endotoxic shock [127]. Endotoxaemia is one of the complications in cirrhotic patients [128] and is probably caused by an impaired ability of the liver to take up and detoxify gut-derived LPS [116]. The presence of portosystemic shunts in cirrhotic patients may also contribute to this spill-over of LPS into the systemic circulation [129]. NSAIDs, however, are also reported to provoke deleterious effects on renal function in cirrhosis [130], and can therefore not be used in cirrhotic patients. Cell-specific delivery of NSAIDs to SECs and/or KCs may make application of these drugs in cirrhosis feasible by circumventing the renal side-effects.

4.6.2 Glucocorticosteroids

Glucocorticosteroids are the synthetic derivatives of the adrenal gland hormone cortisol. At pharmacological doses they prevent or suppress inflammation and other immunologically mediated processes. These drugs are therefore used for a variety of inflammatory diseases such as allergic diseases, rheumatic disorders, renal diseases, bronchial asthma, skin and gastrointestinal diseases [122]. The anti-inflammatory and immunosuppressive activities of glucocorticosteroids are most likely due to the inhibition of the production of a wide range of cytokines, chemokines, eicosanoids, and metalloproteinases in many cell types. In macrophages they block the release of numerous cytokines (IL-1, IL-6, TNFα), inhibit the expression of the MHC class II antigens, depress production and release of pro-inflammatory PGs and LTs, and depress tumouricidal and microbicidal activities of activated macrophages [131]. In the case of neutrophils they inhibit neutrophil adhesion to endothelial cells, thereby reducing the infiltration of neutrophils at inflamed sites. At pharmacological doses they only modestly block neutrophil functions such as lysosomal enzyme release and respiratory burst [132]. Glucocorticosteroids also have profound effects on the activation and subsequent function of endothelial cells. Besides inhibiting cytokine and eicosanoid release, they depress vascular permeability, LPS-induced upregulation of adhesion molecules ICAM-1 and endothelial leucocyte adhesion molecule 1 or E-selectine, and expression of MHC class II antigens [133,134]. Moreover, they inhibit the secretion of the complement pathway proteins C3 and factor B [135].

The molecular mechanisms underlying glucocorticosteroid inhibition of inflammatory responses are slowly being unravelled (see also Chapter 7). After entering the cell, glucocorticosteroids bind to the glucocorticoid receptor (GR) present in the cytoplasm. Following ligand binding, the GR is redirected to the nucleus where it can interact with specific DNA sequences. The expressions of many proteins involved in inflammatory reactions are regulated by the transcriptional regulatory proteins Activator Protein-1 (AP-1) and NFκB. The ligand–GR complex decreases the AP-1-dependent activation of some pro-inflammatory genes by interacting directly with AP-1, thereby sequestering AP-1 away from its binding site [136]. The NFκB-dependent activation of pro-inflammatory genes is inhibited in a different way. Glucocorticosteroids stimulate the synthesis of IκBα inhibitory protein, which traps the activated NFκB in inactive cytoplasmic complexes [137]. Recently it has been shown that more complex mechanisms like nuclear competition for limiting amounts of coactivators between the GR and the p65 component of NFκB, also contribute to the effect of glucocorticosteroids [138].

The continued use of glucocorticosteroids at supraphysiological doses will lead to several side-effects, some of them potentially life-threatening. These include increased susceptibility to infections, osteoporosis, hyperglycaemia, myopathy, behavioural disturbances and hypertension [122]. The severity of these side-effects limits the use of glucocorticosteroids, but would also justify a drug targeting approach. Beneficial effects of targeted glucocorticosteroids that inhibit many of the harmful mediators of the fibrotic process are therefore anticipated.

4.6.3 Other Anti-inflammatory Drugs

Most NSAIDs decrease COX activity without decreasing the generation of lipoxygenase-produced LTs. These substances also contribute to the inflammatory response through a variety of effects, such as that on smooth muscle contractility (LTC_4, LTD_4, LTE_4); neutrophil aggregation, degranulation and chemotaxis (LTB_4); vascular permeability (LTC_4, LTD_4, LTE_4); and on lymphocytes (LTB_4). In recent years, a large number of drugs have been developed that act either as lipoxygenase inhibitors or as LT receptor antagonists. Studies so far have shown only limited toxicity of these drugs [139].

Several other agents are under study that are designed to produce combined blockade of COX and lipoxygenase. One such example is tenidap sodium, a novel antiarthritic agent, which also appears to block IL-1 formation and action [140].

Another drug that has been found to have anticytokine activity is pentoxifylline. It was initially characterized as a haemorheologic agent for the treatment of peripheral vascular diseases [141]. In addition, it was also found to be capable of inhibiting the pro-inflammatory actions of IL-1 and TNFα on neutrophil function and cytokine production by monocytic cells [142]. Its mechanism of action is the inhibition of phosphodiesterases, leading to increased intracellular levels of cyclic adenosine monophosphate [143]. Besides its effects on the cytokine network, pentoxifylline also exerted an anti-fibrogenic action in cultures of fibroblasts and in animal models of fibrosis [144] and could therefore be an attractive candidate for targeting hepatic inflammation.

4.7 Anti-fibrotic Drugs

HSCs are the major contributors to the deposition of extracellular matrix in fibrotic livers and should therefore be the target for anti-fibrotic therapy. With the recent development of carriers for this cell type [117,118] targeting of anti-fibrotic drugs has become a realistic option. Using the carriers that are internalized by activated HSCs, potential anti-fibrotic drugs include collagen synthesis inhibitors, e.g. the prolyl hydroxylase inhibitors [81,145], inhibitors of HSC activation, e.g. NFκB inhibitors or histone deacetylase inhibitors (trichostatin A) [146], and inhibitors of portal hypertension, the endothelin antagonists [87,88]. With the carriers that stay at the outer surface of activated HSCs it is possible to deliver anti-fibrotic drugs to the extracellular microenvironment of HSCs. Interesting candidates would be metalloproteinase activators [145], TGFβ-neutralizing compounds, and PDGF-binding molecules.

4.8 Testing Liver Targeting Preparations

We have developed albumin-conjugates of dexamethasone and naproxen for targeting to SECs and KCs. These conjugates have been extensively tested in vitro and in vivo in rats with bile duct ligation-induced liver fibrosis [93,147–150].

4.8.1 Distribution

Biodistribution is one of the first things that should be tested after construction of a drug targeting preparation for the liver. Is the drug delivered to the intended site within the liver and is the drug released from the carrier? In our opinion only *in vivo* studies can give a correct answer to the first question. Once the success of targeting is established, *in vitro* studies can be used to provide insights into the cellular handling of the preparations.

There are several ways to investigate the biodistribution of drug targeting preparations. Uptake in the target organ as compared to other organs in the body can be studied by radiolabelling the drug targeting preparation and measuring radioactivity in the organs either by organ harvesting or with positron emission tomography or gamma camera studies. Ideally, the labelling should not influence the distribution of the carrier, but this cannot be excluded in all cases [151]. Alternatively, a drug that is easily detectable can be measured in homogenates of the organs without prior radiolabelling. In our laboratory, we developed antibodies against dexamethasone coupled to an albumin carrier allowing the detection of unlabelled dexamethasone-containing drug targeting preparations in human and rat tissue [152]. The initial distribution of the compound can be assessed at 10 min after injection, whereas at later time points degradation, metabolism and redistribution can be studied. The pharmacokinetic behaviour of the drug targeting preparations can be investigated using ELISA methods.

Since the structure of the liver is completely changed during fibrosis, distribution of a liver targeting compound has to be tested in the pathological state as well. Parts of the liver may become inaccessible by portosystemic shunts and individual cells may be hampered in the uptake of compounds by the excess extracellular matrix deposited around them. Uptake processes themselves may also be impaired. The phagocytic activity of KCs has been reported to be depressed in some forms of fibrosis [153,154] and some receptors, such as the hyaluronic acid receptor, are found to be downregulated [155], whereas others can be upregulated [156,157]. There are several animal models for liver fibrosis available. Those that are most frequently used are fibrosis induced by bile duct ligation or occlusion (BDL or BDO) and fibrosis induced by carbon tetrachloride [158–160].

Once specific liver uptake is established, the intrahepatic distribution needs to be addressed. The importance of this issue is exemplified by the apparently high uptake of untargeted Dexa by the liver. Dexa itself was taken up exclusively by the hepatocytes, whereas targeted Dexa was taken up by SECs and KCs [152], the target cells for anti-inflammatory therapy.

Qualitative analysis of intrahepatic distribution is possible with immunohistochemistry. With antibodies against the carrier or the carrier-bound drug this compound can be localized in liver sections [152]. To identify the cell type(s) involved in the uptake, the sections can subsequently be double stained with markers for the different cell types. In the rat liver, the monoclonal antibodies HIS52 (anti-rat endothelial cell antigen-1 or anti-RECA-1), ED2, the combination of anti-desmin and anti-glial fibrillary acidic protein, and anti-αSMA are generally used to identify SECs, KCs, quiescent HSCs, and activated HSCs, respectively.

To determine the uptake by the different cell types quantitatively, the cells can be isolated from the liver after injection of the drug targeting preparation. The amount of targeting conjugate can be detected in the subsequent cell fractions by determining the amount of drug present or counting the amount of radioactivity when the compound is labelled. Rat and human liver cells can be isolated after perfusion of the liver with collagenase and/or pronase [161,162]. Separation of the different cell types is performed by centrifugal elutriation, by density gradients (using Percoll, Nycodenz, stractan, or sucrose), or by magnetic retention of the cells with specific antibodies attached to insoluble magnetic beads [163]. There are several drawbacks to using these methods: the cell fractions obtained are usually not 100% pure, which makes interpretation of the results problematic [162]; selection of normal or diseased cells from fibrotic livers cannot be excluded; and isolation of cells from fibrotic livers is very difficult with respect to viability and purity of the fractions; pronase, which is needed for the isolation of HSCs, affects the viability of hepatocytes [164]. Isolation of all cell types from one liver is therefore impossible.

For obvious reasons it is impossible to determine the biodistribution of liver targeting preparations in humans. It is, however, possible to determine the intrahepatic distribution in human liver tissue using two *in vitro* methods. The first makes use of pieces of human liver, both non-diseased and cirrhotic, in a perfusion set-up, the so-called human liver lobe perfusion [165]. These pieces of liver can be perfused with a liver targeting preparation for at least 90 min, after which time the intrahepatic distribution can be determined by subsequent cell isolation or immunohistochemical analysis. The liver targeting compound and the cell types involved are again identified with specific antibodies. Human SECs, KCs and HSCs can be detected with anti-GP96, anti-CD68 and anti-α-SMA, respectively. The second method uses slices of human liver tissue, as described in Chapter 12. After incubation of precision-cut liver slices with a drug targeting compound, the latter can be localized by immunohistochemical analysis of cryostat-cut sections.

4.8.2 Cellular Processing

To study the uptake mechanism of drug targeting preparations and the release of a drug from the carrier several *in vitro* techniques are available in addition to *in vivo* studies. Besides liver slices and isolated perfused liver tissue of both rat and human origin [166], cultured cells derived from cell lines or from cell isolations of rat or human liver tissue may be used. The primary cultures generally reflect the *in vivo* situation better than the immortalized cells of a cell line culture. However, one should bear in mind that during the isolation procedure target receptors may be destroyed or damaged by the enzymes collagenase and/or pronase. Liver slices and perfused liver tissue, therefore, seem to be the most attractive alternatives for *in vitro* studies. Both methods use liver tissue which is unaffected by isolation and culture procedures in which the hepatic cells still have their normal cell–cell contacts. This allows the study of the cellular processing of liver targeting preparations within the organ, in the presence of all other resident cell types and their secreted mediators.

The receptors involved in the uptake and the route of internalization can be assessed using specific ligands and inhibitors. The binding of a targeting compound to a certain receptor is generally studied at 4°C when internalization is low or absent. The preparation under study is incubated with increasing amounts of specific ligand for a potential receptor. Inhibition of the binding of the preparation indicates that the two compounds compete for the same receptor. Ligands used for the scavenger receptor are Form-HSA, Suc-HSA, polyinosinic acid, and acetylated LDL. For the mannose receptor mannan is mostly used.

With inhibitors affecting different levels of the endocytotic pathway, the route of internalization can be clarified. The vesicles formed after internalization of receptor-bound liver targeting preparation are endosomes that are subsequently acidified through an ATP-dependent proton pump [167]. This acidification is necessary to dissociate the receptor and carrier. Vesicles containing receptor molecules often recycled to the membrane, while vesicles containing targeting molecules are trafficked to the lysosomes [168]. This acidic organelle contains a variety of aggressive enzymes to degrade proteins, lipids and oligosaccharides [169]. Chloroquine and ammonium chloride neutralize the acidic pH of the endosomes and lysomes [170]. The ligand can no longer dissociate from the receptor and a used receptor is recycled to the cell membrane and this subsequently blocks the uptake of new ligands. Colchicine is a microtubule depolymerizing drug. Microtubules are important for both the intracellular organization of vesicles and their routing between compartments. Consequently, colchicine inhibits endocytosis of the liver targeting preparation, the movement of endocytotic vesicles, and receptor recycling [171]. Monensin also prevents acidification of endosomes and lysosomes, but additionally it inhibits the release of the receptor–ligand-containing vesicles from the microtubules [172]. Although all of these compounds block endocytosis, none of them inhibit the initial binding of the targeting preparation to its specific receptor.

4.8.3 Efficacy and Toxicity

Initially, testing whether drug targeting conjugates have pharmacological or toxic effects can best be done *in vitro*. A large variety of drug targeting preparations can be tested under several experimental conditions, using only few experimental animals. The most promising compounds can then be studied *in vivo*. In addition to using individual cell cultures, the method of precision-cut liver slices seems to be an excellent procedure for the rapid testing of liver targeting preparations. The effect of anti-inflammatory targeting preparations can be studied in liver slices by the inhibition of mediator release after activation of resident cells with LPS. $Dexa_{10}$-HSA, for instance, was shown to inhibit LPS-induced TNFα release more effectively than an equimolar quantity of free dexamethasone [152]. Another method is to study the effects on LPS-induced cell activation in a liver perfusion set-up system. Isolated perfused livers of rats pretreated with *Corynebacterium parvum* and subsequently challenged with LPS are a well-established model for hepatic inflammation. LPS-induced cholestasis and hepatic damage in this model were successfully prevented by the administration of Nap_{23}-HSA, whereas an equimolar dose of uncoupled naproxen did not have any effect on these parameters [173]. Currently, we are also investigating whether LPS-induced cellular activation in the human liver lobe perfusion set-up can be exploited for efficacy studies in human tissue [165]. With the information gathered from experiments concerning uptake mechanisms, cellular processing and *in vitro* efficacy, combined with *in vivo* pharmacokinetic data of the targeting conjugate and the free drug, it may be possible to make estimations about treatment efficacy. The liver targeting compounds can then be studied in models of acute hepatic inflammation and chronic hepatic inflammation or fibrosis. Acute inflammation is usually induced by i.v. LPS injection, after which plasma levels of relevant cytokines such as TNFα and IL-1β can be measured. This acute hepatic inflammatory response can be induced in either normal or fibrotic rats [147,152]. Livers of these rats may be analysed (immuno)histochemically for KC activation, infiltration of inflammatory cells and other signs of inflammation. This model of acute inflammation was used to study the effects of $Dexa_{10}$-HSA or Nap_{23}-HSA. Fibrotic rats receiving LPS after $Dexa_{10}$-HSA or Nap_{23}-HSA administration showed less signs of toxicity and a prolonged survival as compared to vehicle-treated controls [147,152].

Chronic inflammatory processes like fibrosis can be induced physically or chemically. As stated before, bile duct ligation (BDL)/bile duct occlusion (BDO) and tetrachloride-induced models of fibrosis are the most frequently used models. BDL/BDO-induced fibrosis is characterized by portal proliferation of bile ducts with collagen deposition and neutrophil infiltration, and periportal hepatocellular necrosis [158,174,175]. The combination of increasing intrabiliary pressure, anoxia, release of inflammatory mediators, and the cytotoxic effect of bile acids induce inflammatory processes leading to HSC proliferation and transformation to myofibroblasts. These cells produce large quantities of extracellular matrix molecules which are characteristic of hepatic fibrosis. In the experimental model of carbon tetrachloride-induced fibrosis on the other hand, hepatotoxicity is mediated by free radicals. The ensuing oxidative stress leads to lipid peroxidation and mitochondrial dysfunction which perpetuate cell damage leading to micronodular fibrosis [176].

Unfortunately, there are few serum parameters available to determine the degree of fibrosis [177]. Those that best show the correlation with the degree of fibrosis are serum laminin, hyaluronic acid, and procollagen type III peptide levels. However, these parameters are not sensitive enough to detect small differences in the fibrotic process. Therefore, the effect of targeted anti-inflammatory treatment has to be examined invasively, either by taking a liver biopsy or examining the whole liver. Again, (immuno)histochemical methods can be used to examine collagen deposition, KC/SEC activation, HSC proliferation and transformation, and infiltration of inflammatory cells. In addition to these methods, deposition of collagen and other matrix molecules can be measured quantitatively after careful tissue extraction [178,179].

4.9 Targeting of Anti-inflammatory Drugs for the Treatment of Liver Fibrosis

As described above there are several carriers available for targeting the key cells in the hepatic inflammatory process. The method of loading a carrier with anti-inflammatory drugs largely depends on the proposed entry mechanism of the carrier into the cell. Most drugs are not active when coupled to albumin or incorporated in liposomes and have to be released from the carrier first. Ideally, the carrier should be stable enough in the bloodstream for the drug to be released only within the target cell. In the case of drug-filled liposomes and drug molecules covalently linked to albumin this means the carrier must be degraded in the target cell for the drug to be released. Most of the receptors described above which are responsible for the uptake of carriers are linked to a lysosomal degradation route. After receptor-mediated uptake most of the carrier is thus lysosomally degraded and a pharmacologically active drug can be released. For covalently attached drugs this means enzymatic (lysosomal hydrolases or reductases) or hydrolytic (acid environment) degradation of the chemical bond between the drug and the carrier.

The advantage of using liposomal carriers is that drugs can be easily incorporated by dissolving them in the aqueous phase of the liposome core or in the lipid phase. The drawback of chronic treatment with a particle-based carrier system for KC or SEC targeting, however, is the possible blockade of cell functions if the carrier is slowly degraded and accumulates in the cells [180,181]. Moreover, because of their size liposomes might be impeded in reaching their target cells during fibrosis when extensive ECM deposition is present in the liver. In contrast, the soluble macromolecular carriers are readily able to penetrate fibrotic tissue and reach their target cells without hinderance [152,174]. Other complications with the use of liposomes are related to the adsorption of plasma proteins to the surface of the liposomes. This may lead to the loss of cell-selectivity due to opsonization by the RES and the activation of the complement system (see Section 4.5.1.2).

In the case of soluble macromolecular carriers, like the albumins, drugs have to be covalently attached to the carrier protein. Drugs can be coupled directly to the carrier or via an acid-labile or enzymatically-sensitive spacer (see also Chapter 11) [182]. The free amino groups of the lysine residues in the protein are attractive candidates for coupling drugs and homing devices. In HSA up to 60 lysine molecules can be derivatized. The carrier Man-HSA has fewer lysine residues available, because some of the lysine amino groups have been used to attach the mannose molecules [183]. However, the higher the loading the more deviations from the normal charge distribution of the protein will occur. This may lead to immunogenicity and loss of cell specificity. Therefore, the number of drugs and homing devices coupled to the carrier protein should not exceed a threshold beyond which the surface charge and tertiary structure of the protein are likely to be significantly affected.

4.9.1 Targeting of NSAIDs

For targeting with a soluble macromolecular carrier, the NSAID naproxen (Nap) was coupled via its carboxyl groups to the free amino groups of the lysine residues in the (Man-)HSA molecule, resulting in a direct amide linkage (Figure 4.3). This type of bond is not very sensitive to proteolytic degradation and incubation with lysosomal lysates showed release of a lysine conjugate of Nap [184]. This Nap-lysine, however, was equipotent to Nap itself with respect to inhibition of PGE_2 synthesis in sensitized guinea-pig trachea challenged with antigen [185].

Distribution, pharmacokinetics and efficacy of a conjugate consisting of approximately 20 molecules of Nap coupled to one molecule of HSA were studied in detail. As compared to free Nap, Nap coupled to HSA was preferentially taken up by the liver, mainly by SECs, but to a lesser extent also by KCs. Scavenger receptors were responsible for this uptake [148,184]. Liver fibrosis induced significant alterations in the pharmacokinetic behaviour of Nap_{20}-HSA. The initial plasma concentration of Nap_{20}-HSA was markedly lower in fibrotic rats and was accompanied by an increase in the volume of distribution during the terminal elimination phase [149]. LPS-induced acute inflammation did not significantly change the pharmacokinetics of Nap_{20}-HSA. Positive effects of treatment with Nap_{20}-HSA were observed in two separate models of hepatic inflammation. In isolated perfused livers of rats pretreated with *Corynebacterium parvum* and subsequently challenged with LPS, Nap_{20}-HSA treatment was protective in a dose-dependent manner at concentrations 30 times lower than conventional doses [173]. As well as inhibiting alanine aminotransferase release, Nap_{20}-HSA also prevented cholestasis and increased vascular resistance in this experimental set-up. Treatment of endotoxaemic fibrotic rats with Nap_{20}-HSA was found to significantly increase survival and markedly reduce toxic effects to the kidneys [147].

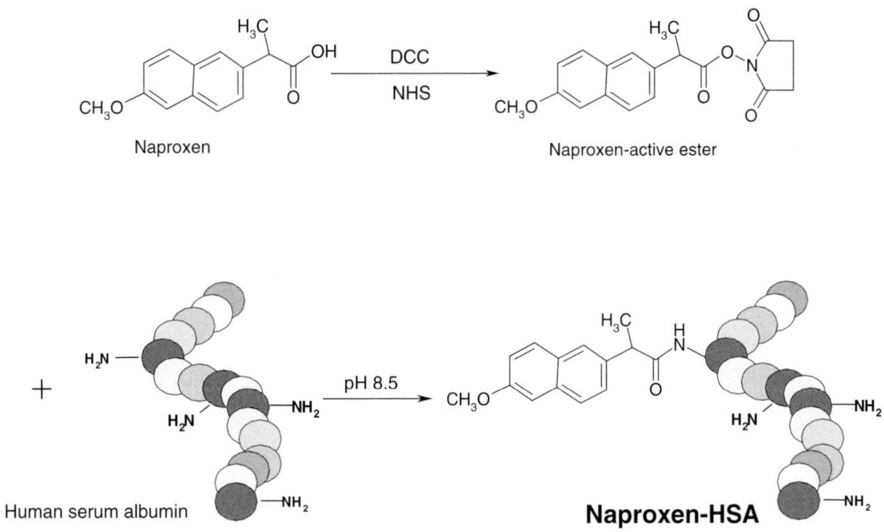

Figure 4.3. The chemical synthesis of naproxen-HSA. Naproxen is first converted to an ester and is then coupled to the free ε-NH_2 of the lysine residues in human serum albumin (HSA). NHS: N-hydroxysuccinimide, DCC: dicyclohexylcarbodiimide.

A pronounced alteration in the intrahepatic distribution of Nap was observed when Nap was coupled to mannosylated HSA as compared to Nap coupled to HSA. Coupling to Man_{10}-HSA resulted in a major shift in intrahepatic distribution from mainly SECs to mainly KCs [184].

One study has described the use of NSAID-loaded liposomes for the targeting of inflammatory lesion sites for the treatment of postoperative pain and pain related to various types of cancer [186]. In this study they showed strong and immediate analgesic effects in relieving pain with NSAID-loaded liposomes, but did not compare this with the analgesic effects of free NSAID.

4.9.2 Targeting of Glucocorticosteroids

The glucocorticosteroid dexamethasone (Dexa) was coupled to HSA and Man_{10}-HSA for targeting to SECs and KCs. Dexa itself could not be coupled directly to the protein, and

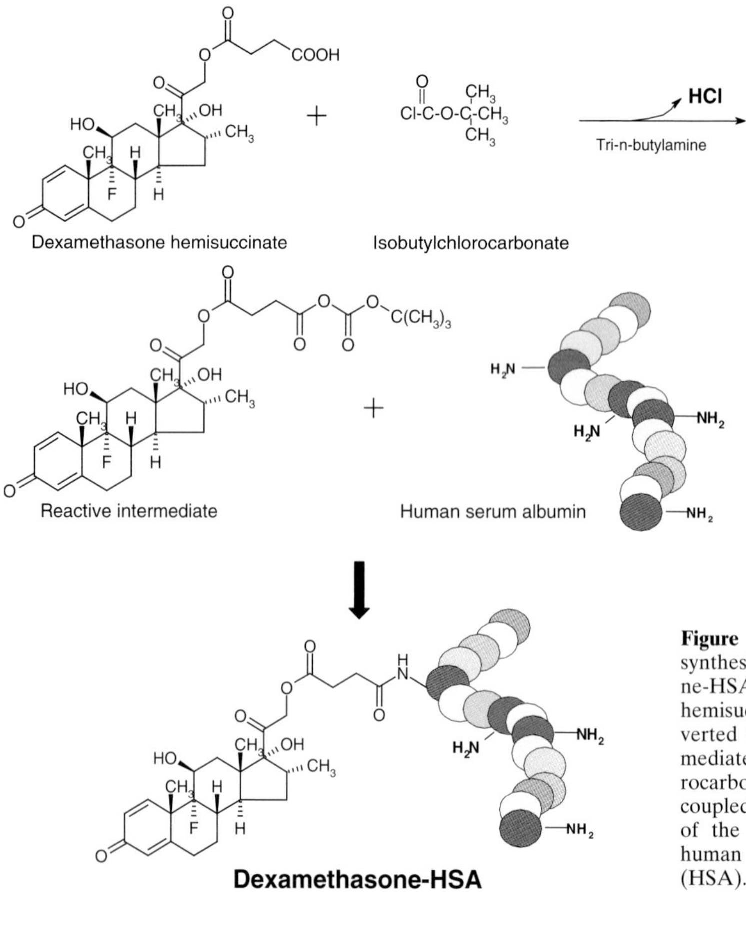

Figure 4.4. The chemical synthesis of dexamethasone-HSA. Dexamethasone hemisuccinate is first converted to a reactive intermediate with isobutylchlorocarbonate and is then coupled to the free ε-NH_2 of the lysine residues in human serum albumin (HSA).

therefore had to be derivatized to create a reactive compound. As described by Fiume *et al.* [187] succinic acid was coupled to the alcohol group on C21 yielding Dexa hemisuccinate. The introduced carboxyl group could then easily be coupled to the free amino groups of the lysine residues in the HSA molecule yielding Dexa$_{10}$-HSA and Dexa$_5$-Man$_{10}$-HSA (Figure 4.4) [152]. The ester bond between native Dexa and the succinate spacer proved to be more sensitive to proteolytic enzymes than the amide bond between the succinate spacer and the protein. Lysosomal degradation of the Dexa-HSA conjugate, therefore, yielded the native Dexa.

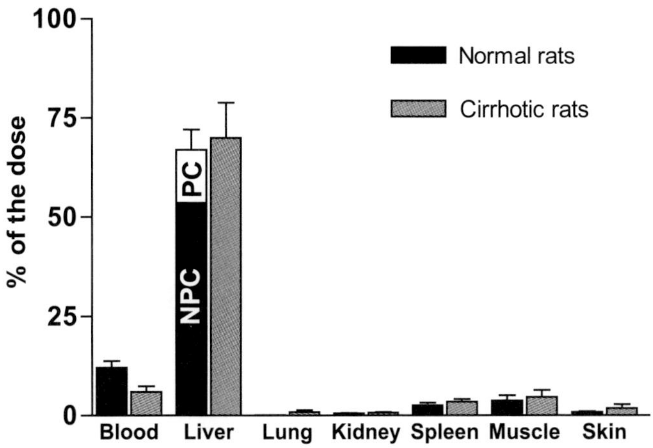

Figure 4.5. Tissue and intrahepatic distribution of 125I-Dexa$_5$-Man$_{10}$-HSA 10 min after injection in normal ($n = 4$) and cirrhotic rats ($n = 5$). NPC, non-parenchymal cells; PC, parenchymal cells.

In rats, both Dexa$_{10}$-HSA and Dexa$_5$-Man$_{10}$-HSA were mostly taken up by the liver, in healthy as well as fibrotic livers (Figure 4.5). Intrahepatic distribution studies showed that Dexa$_{10}$-HSA was taken up by SECs and KCs, whereas Dexa$_5$-Man$_{10}$-HSA was taken up by KCs [93,152]. Interestingly, in human livers Dexa$_{10}$-HSA was found to be taken up by SECs and KCs of healthy livers, but in cirrhotic livers only by KCs [165]. In liver slices, these two conjugates showed superior inhibition of LPS-induced release of TNFα as compared to untargeted Dexa, indicating specific inhibition of KC and SEC function.

The efficacy of Dexa$_{10}$-HSA *in vivo* was shown in a model of acute inflammation. Fibrotic rats pretreated with conjugated and non-conjugated Dexa showed increased survival after LPS injection. Although the anti-inflammatory effects of Dexa$_{10}$-HSA could be demonstrated, superiority of conjugated Dexa as compared to free Dexa could not be established in this model.

When Dexa$_5$-Man$_{10}$-HSA was administered chronically to rats with BDL, reduced infiltration of ROS-producing cells and an increased glycogen content in hepatocytes was found, suggesting more efficient liver function, this was not the case for non-conjugated Dexa [93]. The specific inhibition of KC function in this model, however, also accelerated the development of fibrosis. The mechanism causing this acceleration is currently under investigation, but is not related to the immunogenicity of the conjugate [93]. These studies may indicate that in addition to the pro-fibrotic actions of KCs, these cells are also endowed with anti-fibrotic capacities that are downregulated by dexamethasone.

Dexa has also been incorporated into several particle-type carriers, although most of them were not designed for the specific targeting of KCs or SECs. Yokoyama and Watanabe incorporated Dexa-21-palmitate into lipid microspheres for targeting inflammatory cells and macrophages in the treatment of rheumatoid arthritis [188]. They demonstrated high uptake of lipid-incorporated Dexa by macrophages, up to five times higher anti-inflammatory activity of lipid-incorporated Dexa as opposed to free Dexa and a significantly higher rate of improvement in patients with rheumatoid arthritis treated with encapsulated Dexa. Magnani *et al.* used Dexa-21-phosphate encapsulated in human erythrocytes for this same purpose [189]. They also showed high concentrations of corticosteroids in macrophages and effective inhibition of the respiratory burst of stimulated macrophages by encapsulated Dexa , which was not found for free Dexa. Recently, we incorporated Dexa disodium phosphate into liposomes especially designed for KC targeting [190]. These liposomes are currently used to study their effects on the development of fibrosis in rats subjected to bile duct ligation.

4.10 Selective Drug Delivery for the Treatment of Other Hepatic Disorders

Drug targeting preparations have been given to patients with various infectious diseases. For the treatment of Leishmaniasis, liposomes as well as mannosylated HSA have been used to deliver antiparasitic drugs, such as methotrexate, amphotericin B, doxorubicin and muramyl dipeptide to KCs [97,98,191–193]. These conjugates all inhibited the growth of Leishmania parasites in Kupffer cells as well as in splenic macrophages in mice infected with this parasite. Liposomal amphotericin B was also tested in immunocompetent patients with visceral Leishmaniasis and was proven to be an effective treatment. Side-effects typical for amphotericin B (hypokalaemia, nephrotoxicity) occurred significantly less frequently after treatment with the liposomal formulation as compared to the convential formulation of the drug [98].

Drug targeting preparations based on lactosylated HSA have been used for the treatment of chronic viral hepatitis, because these viruses reside in hepatocytes. Fiume *et al.* coupled several anti-viral nucleoside analogues to this carrier for the treatment of hepatitis [194,195]. The conjugate of adenine arabinoside monophosphate and lactosylated HSA (araAMP-lacHSA) has been studied in animals as well as in humans [1,196]. From the clinical trials it was concluded that use of this conjugate allowed more prolonged treatment of chronic hepatitis B than free araAMP, because of the lack of side-effects after chronic application of the conjugate, which enhanced its chemotherapeutical index. To date, however, no follow-up has been published.

The treatment of tumours in the liver with drug targeting preparations is hampered by the lack of tumour specifity of most preparations. Liposomes incorporating the immunomodulator muramyl tripeptide phosphatidylethanolamine have been used as an 'aspecific' approach to increasing the number of tumouricidal macrophages in the liver in order to prevent the development of metastases [99]. To date, the greatest tumour cell specificity has been obtained with monoclonal and bispecific antibody carriers [197]. In addition to these tumour-directed strategies, targeting is also aimed at the tumour vasculature. This is extensively reviewed in Chapter 9.

4.11 Conclusions

Hepatic inflammation and fibrosis of the liver are multifactorial processes that cannot be treated successfully with drugs currently on the market. These drugs either lack suitable efficacy or cause too many side-effects. New directions for therapy include the targeting of anti-inflammatory drugs to the key players in the chronic inflammatory process: the Kupffer cells and liver endothelial cells. Several studies have shown that targeting of, for instance, the anti-inflammatory drugs dexamethasone and naproxen to these cell types is feasible, but few *in vivo* studies have been conducted to investigate the effects of targeted therapeutic intervention. Initial studies in our laboratory with dexamethasone targeted to KCs revealed that these cells may have important anti-fibrotic abilities and that KC-delivered dexamethasone yields not only anti-inflammatory, but also pro-fibrotic effects. These results show that in drug targeting research more emphasis should be placed on studying therapeutic interventions in pathological models. However, it should be noted that therapeutic intervention aimed at one cell type may not be sufficient to treat the disease, since all hepatic cell types contribute to the development of liver fibrosis. It may therefore be necessary to target several drugs to different cell types simultaneously.

It should also be emphasized that results obtained from studies with experimental animals can not be directly translated to the human situation. We have shown that the characteristic intercellular distribution of the drug targeting preparation $Dexa_{10}$-HSA is different in human and rat cirrhotic liver tissue [165]. Careful evaluation of animal experimental data should therefore be combined with studies in human test models. Cell culture systems are a valuable technique, but for drug targeting purposes a more integrated system is preferred. In the case of the liver these could include precision-cut liver slices or a liver lobe perfusion of human liver tissue. In contrast to liver cell cultures, liver slices and liver lobes contain all the different liver cell types, and the complex cell–cell contacts and interactions that exist *in vivo* are maintained. The major challenge in the near future will be to establish the relevance of the concept of drug targeting in experimental models of disease, in human tissue *in vitro*, and finally in patients with liver diseases.

References

[1] Fiume L, Di Stefano G, Busi C, Mattioli A, Bonino F, Torrani-Cerenzia M, Verme G, Rapicetta M, Bertini M, Gervasi GB, *J. Viral. Hepat.* **1997,** *4*, 363–370.
[2] Smith RM, Wu GY, *Semin. Liver. Dis.* **1999,** *19*, 83–92.
[3] Meijer DKF, Molema G, *Semin. Liver Dis.* **1995,** *15*, 202–256.
[4] Martinez I, Sveinbjornsson B, Vidal Vanaclocha F, Asumendi A, Smedsrod B, Porcheron J, Margonari J, Gorry F, Boillot O, *Biochem. Biophys. Res. Commun.* **1995,** *27*, 1658–1641.
[5] Melkko J, Hellevik T, Risteli L, Risteli J, Smedsrod B, *J. Exp. Med.* **1994,** *179*, 405–412.
[6] Mcgary CT, Yannariello-Brown J, Kim DW, Stinson TC, Weigel PH, *Hepatology* **1993,** *18*, 1465–1476.

[7] Smedsrod B, De Bleser PJ, Braet F, Lovisetti P, Vanderkerken K, Wisse E, Geerts A, *Gut* **1994,** *35*, 1509–1516.
[8] Ichida T, Sugitani S, Satoh T, Matsuda Y, Sugiyama M, Yonekura K, Ishikawa T, Asakura H, *Liver* **1996,** *16*, 365–371.
[9] Smedsrod B, Pertoft H, Gustafson S, Laurent TC, *Biochem. J.* **1990,** *266*, 313–327.
[10] van Oosten M, van de Bilt E, Van Berkel TJC, Kuiper J, *Infect. Immun.* **1998,** *66*, 5107–5112.
[11] Smedsrod B, Melkko J, Araki N, Sano H, Horiuchi S, *Biochem. J.* **1997,** *322*, 567–573.
[12] Hampton RY, Golenbock DT, Penman M, Krieger M, Raetz CR, *Nature* **1991,** *352*, 342–344.
[13] Yoshioka T, Yamamoto K, Kobashi H, Tomita M, Tsuji T, *Liver* **1994,** *14*, 129–137.
[14] Van Berkel TJC, Van Velzen A, Kruijt JK, Suzuki H, Kodama T, *Biochem. J.* **1998,** *331*, 29–35.
[15] Magnusson S, Berg T, Biochem. J. 1989, 257, 651–656.
[16] Ezekowitz RAB, Williams DJ, Koziel H, Armstrong MYK, Warner A, Richards FF, Rose RM, *Nature* **1991,** *351*, 155–158.
[17] Noorman F, Rijken DC, *Fibrin. Proteol.* **1997,** *11*, 173-186.
[18] Knolle PA, Uhrig A, Hegenbarth S, Loser E, Schmitt E, Gerken G, Lohse AW, *Clin. Exp. Immunol.* **1998,** *114*, 427–433.
[19] Iwamura S, Enzan H, Saibara T, Onishi S, Yamamoto Y, *Hepatology* **1995,** *22*, 1456–1461.
[20] Scoazes JY, Feldman G, *Hepatology* **1991,** *14*, 789–797.
[21] Gressner AM ,Bachem MG, *Digestion* **1995,** *56*, 335–346.
[22] Watanabe J, Kanai K, Kanamura S, *J. Histochem. Cytochem.* **1988,** *36*, 1081–1089.
[23] Kanai M, Murata Y, Mabuchi Y, Kawahashi N, Tanaka M, Ogawa T, Doi M, Soji T, Herbert DC, *Anat. Rec.* **1996,** *244*, 175–181.
[24] Bergman RN, *Recent. Prog. Horm. Res.* **1997,** *52*, 359–385.
[25] Omoto E, Minguell JJ, Tavassoli M, *Pathobiology* **1992,** *60*, 284–288.
[26] Nagano T, Kita T, Tanaka N, *Int. J. Exp. Pathol.* **1992,** *73*, 675–683.
[27] Rieder H, Meyer zum Büschenfelde K-H, Ramadori G, *J. Hepatol.* **1992,** *15*, 237–250.
[28] Knolle PA, Loser E, Protzer U, Duchmann R, Schmitt E, Zumbuschenfelde KHM, Rosejohn S, Gerken G, *Clin. Exp. Immunol.* **1997,** *107*, 555–561.
[29] Ohkubo K, Masumoto T, Horiike N, Onji M, *J. Gastroenterol. Hepatol.* **1998,** *13*, 696–702.
[30] Noji S, Tashiro K, Koyama E, Nohno T, Ohyama K, Taniguchi S, Nakamura T, *Biochem. Biophys. Res. Commun.* **1990,** *173*, 42–47.
[31] Hashimoto N, Watanabe T, Shiratori Y, Ikeda Y, Kato H, Han K, Yamada H, Toda G, Kurokawa K, *Hepatology* **1995,** *21*, 1713–1718.
[32] Kuiper J, Zijlstra FJ, Kamps JA, Van Berkel TJ, *Biochim. Biophys. Acta* **1988,** *959,* 143–152.
[33] Tsujii H, Okamoto Y, Kikuchi E, Matsumoto M, Nakano H, *Gastroenterology* **1993,** *105*, 495–499.
[34] Kuiper J, De Rijke YB, Zijlstra FJ, Van Waas MP, Van Berkel TJ, *Biochem. Biophys. Res. Commun.* **1988,** *157*, 1288–1295.
[35] van Oosten M, van de Bilt E, de Vries HE, Van Berkel TJC, Kuiper J, *Hepatology* **1995,** *22*, 1538–1546.
[36] Steinhoff G, Behrend M, Schrader B, Duijvestijn AM, Wonigeit K, *Am. J. Pathol.* **1993,** *142*, 481–488.
[37] Lopez S, Prats N, Marco AJ, *Am. J. Pathol.* **1999,** *155*, 1391–1397.
[38] West MA, Heaney ML, In: *Hepatocyte and Kupffer Cell Interactions* (Eds Billiar TR, Curran RD), pp. 209–241. CRC Press, London, Tokyo, **1992.**
[39] Bioulac Sage P, Kuiper J, Van Berkel TJ, Balabaud C, *Hepatogastroenterology* **1996,** *43*, 4–14.
[40] Toth CA, Thomas P, *Hepatology* **1992,** *16*, 255–266.
[41] van der Laan LJ, Dopp EA, Haworth R, Pikkarainen T, Kangas M, Elomaa O, Dijkstra CD, Gordon S, Tryggvason K, Kraal G, *J. Immunol.* **1999,** *162*, 939–947.
[42] Fluiter K, vanderWesthuijzen DR, Vanberkel TJC, *J. Biol. Chem.* **1998,** *273*, 8434–8438.
[43] Kamps JA, Kruijt JK, Kuiper J, Van Berkel TJ, *Biochem. J.* **1991,** *276 (Pt 1)*, 135–140.
[44] Sano A, Taylor ME, Leaning MS, Summerfield JA, *J. Hepatol.* **1990,** *10*, 211–216.
[45] Shimada K, Kamps JAAM, Regts J, Ikeda K, Shiozawa T, Hirota S, Scherphof GL, *BBA–Biomembranes* **1997,** *1326*, 329–341.
[46] Biessen EA, Bakkeren HF, Beuting DM, Kuiper J, Van Berkel TJ, *Biochem. J.* **1994,** *299 (Pt 1)*, 291–296.
[47] Armbrust T, Nordmann B, Kreissig M, Ramadori G, *Hepatology* **1997,** *26*, 98–106.
[48] Maruiwa M, Mizoguchi A, Russell GJ, Narula N, Stronska M, Mizoguchi E, Rabb H, Arnaout MA, Bhan AK, *J. Immunol.* **1993,** *150*, 4019–4030.
[49] Bouwens L, Baekeland M, De Zanger RB, Wisse E, *Hepatology* **1986,** *6*, 718–722.

[50] Coleman DL, Eur. J. Clin. Microbiol. 1986, 5, 1–5.
[51] Knolle P, Lohr H, Treichel U, Dienes HP, Lohse A, Schlaack J, Gerken G, *Z. Gastroenterology* **1995,** *33*, 613–620.
[52] Crawford JM, *Curr. Opin. Gastroenterol.* **1997,** *13*, 175–185.
[53] Kopydlowski KM, Salkowski CA, Cody MJ, Van Rooijen N, Major J, Hamilton TA, Vogel SN, *J. Immunol.* **1999,** *163*, 1537–1544.
[54] Yamaguchi Y, Matsumura F, Takeya M, Ichiguchi O, Kuratsu J-I, Horiuchi T, Akizuki E, Matsuda T, Okabe K, Ohshiro H, Liang J, Mori K, Yamada S, Takahashi K, Ogawa M, *Hepatology* **1998,** *27*, 727–734.
[55] Luster MI, Germolec DR, Yoshida T, Kayama F, Thompson M, *Hepatology* **1994,** *19*, 480–488.
[56] Roth S, Gong W, Gressner AM, *J. Hepatol.* **1998,** *29*, 915–922.
[57] Zhang F, Warskulat U, Wettstein M, Schreiber R, Henninger HP, Decker K, Haussinger D, *Biochem. J.* **1995,** *312 (Pt 1)*, 135–143.
[58] Vogl S, Petermann H, Dargel R, *Liver* **1996,** *16*, 313–320.
[59] Rockey DC, Weisiger RA, *Hepatology* **1996,** *24*, 233–240.
[60] Ramadori G, Knittel T, Saile B, *Digestion* **1998,** *59*, 372–375.
[61] Arthur MJP, *Digestion* **1998,** *59,* 376–380.
[62] Kawada N, *Histol. Histopathol.* **1997,** *12*, 1069–1080.
[63] Beljaars E, Melgert BN, Meijer DKF, Molema G, Poelstra K, In: *Drug Targeting Technology: A Critical Analysis of Physical, Chemical and Biological Methods* (Ed. Schreier H). Marcel Dekker Inc., New York, **2001.**
[64] Li D, Friedman SL, *J. Gastroenterol. Hepatol.* **1999,** *14*, 618–633.
[65] Bissell DM, *J. Gastroenterol.* **1998,** *33*, 295–302.
[66] Friedman SL, Arthur MJ, *J. Clin. Invest.* **1989,** *84*, 1780–1785.
[67] Gressner AM, *J. Hepatol.* **1995,** *22*, 28–36.
[68] Kanwar S, Kubes P, *New. Horiz.* **1995,** *3*, 93–104.
[69] Radziszewski W, Chopra M, Zembowicz A, Gryglewski R, Ignarro LJ, Chaudhuri G, *Int. J. Cardiol.* **1995,** *51*, 211–220.
[70] Goss JA, Mangino MJ, Callery MP, Flye MW, *Am. J. Physiol.* **1993,** *264*, G601–G608.
[71] Goss JA, Mangino MJ, Flye MW, *J. Surg. Res.* **1992,** *52*, 422–428.
[72] Fennekohl A, Schieferdecker HL, Jungermann K, Puschel GP, *J. Hepatol.* **1999,** *30*, 38–47.
[73] Callery MP, Mangino MJ, Flye MW, *Hepatology* **1991,** *14*, 368–372.
[74] Alcolado R, Arthur MJP, Iredale JP, *Clin. Sci.* **1997,** *92*, 103–112s.
[75] Scholmerich J, Holstege A, *Drugs.* **1990,** *40 (Suppl. 3)*, 3–22.
[76] Hoofnagle JH, *Digestion* **1998,** *59*, 563–578.
[77] Kaplan MM, *Semin. Liver. Dis.* **1997,** *17*, 129–136.
[78] Wolfhagen FH, van Hoogstraten HJ, van Buuren HR, Berge-Henegouwen GP, Ten Kate FJ, Hop WC, van der Hoek EW, Kerbert MJ, van Lijf HH, den Ouden JW, Smit AM, de Vries RA, van Zanten RA, Schalm SW, *J. Hepatol.* **1998,** *29*, 736–742.
[79] Messner M, Brissot P, *Drugs* **1990,** *40 (Suppl. 3)*, 45–57.
[80] Kershenobich D, Vargas F, Garcia-Tsao G, Perez TR, Gent M, Rojkind M, *N. Engl. J. Med.* **1988,** *318*, 1709–1713.
[81] Bickel M, Baringhaus K-H, Gerl M, Günzler V, Kanta J, Schmidts L, Stapf M, Tschank G, Weidmann K, Werner U, Baringhaus KH, Gunzler V, *Hepatology* **1998,** *28*, 404-411.
[82] Crowe J, Christensen E, Smith M, Cochrane M, Ranek L, Watkinson G, Doniach D, Popper H, Tygstrup N, Williams R, *Gastroenterology* **1980**, *78*, 1005–1010.
[83] Fernandes NF, Redeker AG, Vierling JM, Villamil FG, Fong TL, *Am. J. Gastroenterol.* **1999,** *94*, 241–248.
[84] Chazouilleres O, Wendum D, Serfaty L, Montembault S, Rosmorduc O, Poupon R, *Hepatology* **1998,** *28*, 296–301.
[85] Czaja AJ, *Ann. Intern. Med.* **1996,** *125*, 588–598.
[86] Batey RG, *Alcohol Suppl.* **1994,** *2*, 327–333.
[87] Kawada N, Seki S, Kuroki T, Kaneda K, *Biochem. Biophys. Res. Commun.* **1999,** *266,* 296–300.
[88] Baveja R, Kresge N, Bian X, Yokoyama Y, Sonin NV, Clemens MG, Zhang JX, *Hepatology* **1999,** *30*, 232A.
[89] Poo JL, Jimenez W, Maria MR, Bosch-Marce M, Bordas N, Morales-Ruiz M, Perez M, Deulofeu R, Sole M, Arroyo V, Rodes J, *Gastroenterology* **1999,** *116*, 161–167.
[90] Jansen RW, Molema G, Harms G, Kruijt JK, Van Berkel TJC, Hardonk MJ, Meijer DKF, *Biochem. Biophys. Res. Commun.* **1991,** *180*, 23–32.

[91] Chaudhuri G, Mukhopadhyay A, Basu SK, *Biochem. Pharmacol.* **1989,** *38,* 2995–3002.
[92] Ogawara K, Hasegawa S, Nishikawa M, Takakura Y, Hashida M, *J. Drug. Target.* **1999,** *6,* 349–360.
[93] Melgert BN, Weert B, Molema G, Meijer DKF, Poelstra K, *Hepatology* **1999,** *30,* 329A.
[94] Skogh T, Stendahl O, Sundqvist T, Edebo L, *Int. Arch. Allergy Appl. Immunol.* **1983,** *70,* 238–244.
[95] Rooijen Nv, Sanders A, *Int. J. Pharm.* **1998,** *162,* 45–50.
[96] Daemen T, Hofstede G, Ten Kate MT, Bakker-Woudenberg IAJM, Scherphof GL, *Int. J. Cancer* **1995,** *61,* 716–721.
[97] Kole L, Das L, *J. Infect. Dis.* **1999,** *180,* 811–820.
[98] Coukell AJ, Brogden RN, *Drugs* **1998,** *55,* 585–612.
[99] Thomas K, Nijenhuis AM, Dontje BH, Daemen T, Scherphof GL, *Clin. Exp. Metast.* **1995,** *13,* 328–336.
[100] Harashima H, Sakata K, Funato K, Kiwada H, *Pharm. Res.* **1994,** *11,* 402–406.
[101] Rossi JD, Wallace BA, *J. Biol. Chem.* **1983,** *258,* 3327–3331.
[102] Liu DX, *Adv. Drug. Deliv. Rev.* **1997,** *24,* 201–213.
[103] Scherphof GL, Kamps JAAM, *Adv. Drug. Deliv. Rev.* **1998,** *32,* 81–97.
[104] Kamps JA, Morselt HW, Scherphof GL, *Biochem. Biophys. Res. Commun.* **1999,** *256,* 57–62.
[105] Kamps JAAM, Morselt HWM, Swart PJ, Meijer DKF, Scherphof GL, *Proc. Natl Acad. Sci. USA* **1997,** *94,* 11681–11685.
[106] Bijsterbosch MK, Van Berkel TJC, *Adv. Drug. Deliv. Rev.* **1990,** *5,* 231–251.
[107] De Rijke YB, Biessen EA, Vogelezang CJ, Van Berkel TJ, *Biochem. J.* **1994,** *304,* 69–73.
[108] Kuiper J, Bakkeren HF, Biessen EAL, Van Berkel TJC, *Biochem. J.* **1994,** *299,* 285–290.
[109] Firestone RA, Pisano JM, Falck JR, McPhaul MM, Krieger M, *J. Med. Chem.* **1984,** *27,* 1037–1043.
[110] van de Water B, Van Berkel TJC, Kuiper J, *Mol. Pharmacol.* **1994,** *45,* 971–977.
[111] Kuipers ME, Van den Berg M, Swart PJ, Laman JD, Meijer DKF, Koppelman MHGM, Huisman H, *Biochem. Pharmacol.* **1999,** *57,* 889–898.
[112] Swart PJ, Hirano T, Kuipers ME, Ito Y, Smit C, Hashida M, Nishikawa M, Beljaars L, Meijer DKF, Poelstra K, *J. Hepatol.* **1999,** *30,* 1034–1043.
[113] Takakura Y, Masuda S, Tokuda H, Nishikawa M, Hashida M, *Biochem. Pharmacol.* **1994,** *47,* 853-858.
[114] Poelstra K, Bakker WW, Klok PA, Kamps JA, Hardonk MJ, Meijer DKF, *Am. J. Pathol.* **1997,** *151,* 1163–1169.
[115] Poelstra K, Grasmeijer R, Meijer DKF, *Hepatology* **1996,** *24,* 327A.
[116] Nolan JP, *Hepatology* **1989,** *10,* 887–891.
[117] Beljaars L, Molema G, Schuppan D, Geerts A, De Bleser PJ, Weert B, Meijer DKF, Poelstra K, *J. Biol. Chem.* **2000,** *275,* 12743–12751.
[118] Beljaars L, Molema G, Weert B, Bonnema H, Olinga P, Groothuis GMM, Meijer DKF, Poelstra K, *Hepatology* **1999,** *29,* 1486–1493.
[119] De Bleser PJ, Jannes P, Van Buul-Offers SC, Hoogerbrugge CM, Van Schravendijk CFH, Niki T, Rogiers V, Van den Brande JL, Wisse E, Geerts A, *Hepatology* **1995,** *21,* 1429–1437.
[120] Weiner JK, Chen AP, Davis BH, *J. Biol. Chem.* **1998,** *273,* 15913–15919.
[121] Smith WL, *Am. J. Physiol.* **1992,** *263,* F181–F191.
[122] Goodman A, Gilman AZ, In: *Goodman and Gilman's the pharmacological basis of therapeutics.* McGraw-Hill, Inc., New York, **1993.**
[123] Langman MJ, Weil J, Wainwright P, Lawson DH, Rawlins MD, Logan RF, Murphy M, Vessey MP, Colin-Jones DG, *Lancet* **1994,** *343,* 1075–1078.
[124] Gentilini P, Cirrhosis, *J. Hepatol.* **1993,** *19,* 200–203.
[125] Endoh T, Tang Q, Denda A, Noguchi O, Kobayashi E, Tamura K, Horiguchi K, Ogasawara H, Tsujiuchi T, Nakae D, Sugimura M, Konishi Y, *Carcinogenesis* **1996,** *17,* 467–475.
[126] Badger AM, Olivera D, Talmadge JE, Hanna N, *Circ. Shock* **1989,** *27,* 51–61.
[127] Wise WC, Cook JA, Eller T, Halushka PV, *J. Pharmacol. Exp. Ther.* **1980,** *215,* 160–164.
[128] Lin RS, Lee FY, Lee SD, Tsai YT, Lin HC, Lu RH, Hsu WC, Huang CC, Wang SS, Lo KJ, *J. Hepatol.* **1995,** *22,* 165–172.
[129] Nolan JP, *Hepatology* **1981,** 1, 458–465.
[130] Boyer TD, Zia P, Reynolds TB, *Gastroenterolgy* **1979,** *77,* 215–222.
[131] Boumpas DT, Chrousos GP, Wilder RL, Cupps TR, Balow JE, *Ann. Intern. Med.* **1993,** 119, 1198–1208.
[132] Sternberg EM, Wilder RL, In: *Arthritis and Allied Conditions* (Eds McCarthy DJ, Koopman WJ), pp. 665–682. Lea & Febiger, Philadelphia, **1993.**

[133] Cronstein BN, Kimmel SC, Levin RI, Martiniuk F, Weissmann G, *Proc. Natl Acad. Sci. USA* **1992,** *89*, 9991–9995.
[134] Zimmerman GA, Prescott SM, McIntyre TM, *Immunol. Today* **1992,** *13*, 93–100.
[135] Coulpier M, Andreev S, Lemercier C, Dauchel H, Lees O, Fontaine M, Ripoche J, *Clin. Exp. Immunol.* **1995,** *101*, 142–149.
[136] Didonato JA, Saatcioglu F, Karin M, *Am. J. Respir. Crit. Care Med.* **1996,** *154*, S11–5.
[137] Auphan N, Didonato JA, Rosette C, Helmberg A, Karin M, *Science* **1995,** *270*, 286–290.
[138] Sheppard KA, Phelps KM, Williams AJ, Thanos D, Glass CK, Rosenfeld MG, Gerritsen ME, Collins T, *J. Biol. Chem.* **1998,** *273*, 29291–29294.
[139] Chanarin N, Johnston SL, *Drugs* **1994**, *47*, 12–24.
[140] Brooks PM, *Agents Actions (Suppl.)*, **1993,** *44*, 161–163.
[141] Muller R, *J. Med.* **1981,** *12*, 209–235.
[142] Sullivan GW, Carper HT, Novick Jr WJ, Mandell GL, *Infect. Immun.* **1988,** *56*, 1722–1729.
[143] Ward A, Clissold SP, *Drugs* **1987,** *34,* 50–97.
[144] Windmeier C, Gressner AM, *Gen. Pharmacol.* **1997,** *29*, 181–196.
[145] Schuppan D, Strobel D, Hahn EG, *Digestion* **1998,** *59*, 385–39
[146] Niki T, Rombouts K, De Bleser PJ, de Smet K, Rogiers V, Schuppan D, Yoshida M, Gabbiani G, Geerts A, *Hepatology* **1999,** *29*, 858–867.
[147] Albrecht C, Meijer DKF, Lebbe C, Sägesser H, Melgert BN, Poelstra K, Reichen J, *Hepatology.* **1997,** *26*, 1553–1559.
[148] Melgert BN, Wartna E, Lebbe C, Albrecht C, Molema G, Poelstra K, Reichen J, Meijer DKF, *J. Drug Target.* **1998,** *5*, 329–342.
[149] Albrecht C, Melgert BN, Reichen J, Poelstra K, Meijer DKF, *J. Drug Target.* **1998,** *6*, 105–117.
[150] Melgert BN, Poelstra K, Jack VK, Molema G, Meijer DKF, *Hepatology* **1998,** *28,* 440A.
[151] Rosenmund A, Kuyas C, Heberli A, *Biochem. J.* **1986,** *240*, 239–245.
[152] Melgert BN, Olinga P, Jack VK, Molema G, Meijer DKF, Poelstra K, *J. Hepatol.* **2000,** *32*, 603–611.
[153] Noda T, Mimura H, Orita K, *Hepatogastroenterology* **1990,** *37*, 319–323.
[154] Petermann H, Heymann S, Vogl S, Dargel R, *J. Hepatol.* **1996,** *24*, 468–477.
[155] Laurent TC, Laurent UB, Fraser JR, *Ann. Med.* **1996,** *28*, 241–253.
[156] Weiner JA, Chen A, Davis BH, *J. Biol. Chem.* **1998,** *273*, 15913–15919.
[157] Wong L, Yamasaki G, Johnson RJ, Friedman SL, *J. Clin. Invest.* **1994**, *94*, 1563–1569.
[158] Kountouras J, Billing BH, Scheuer PJ, *Br. J. Exp. Pathol.* **1984,** *65*, 305–311.
[159] Fischer-Nielsen A, Poulsen HE, Hansen BA, Hage E, Keiding S, *J. Hepatol.* **1991,** *12*, 110–117.
[160] Boigk G, Stroedter L, Herbst H, Waldschmidt J, Riecken EO, Schuppan D, *Hepatology* **1997,** *26*, 643–649.
[161] Braet F, De Zanger R, Sasaoki T, Baekeland M, Janssens P, Smedsrod B, Wisse E, *Lab. Invest.* **1994,** *70*, 944–952.
[162] Geerts A, Niki T, Hellemans K, De Craemer D, Van den Berg K, Lazou J-M, Stange G, Van de Winkel M, De Bleser PJ, *Hepatology* **1998,** *27*, 590–598.
[163] Dong QG, Bernasconi S, Lostaglio S, De Calmanovici RW, Martin Padura I, Breviario F, Garlanda C, Ramponi S, Mantovani A, and Vecchi A, *Arterioscler. Thromb. Vasc. Biol.* **1997,** *17*, 1599–1604.
[164] Kuiper J, Brouwer A, Knook DL, Van Berkel TJC, In: *The Liver: Biology and Pathobiology* (Eds Arias IM, Boyer JL, Fausto N, Jakoby WB, Schachter DA, Shafritz DA), pp. 791–818. Raven Press Ltd., New York, **1994.**
[165] Melgert BN, Olinga P, Weert B, Slooff MJH, Meijer DKF, Poelstra K, Groothuis GMM, *Drug Metab. Dispos.* **2001,** *29*, 361–367.
[166] Olinga P, Meijer DKF, Slooff MJH, Groothuis GMM, *Toxicol. In Vitro* **1997,** *12*, 77–100.
[167] Yamashiro DJ, Maxfield FR, *J. Cell Biochem.* **1984,** *26,* 231–246.
[168] Geuze HJ, Slot JW, Strous GJAM, Lodish HF, Schwartz AL, *Cell* **1983,** *32*, 1277–1287.
[169] De Duve C, Wattiaux R, *Annu. Rev. Physiol.* **1966,** *28*, 435–492.
[170] Thorens B, Vassalli P, *Nature* **1986,** *321*, 618–620.
[171] Clarke BL, Weigel PH, *Biochem. J.* **1989,** *262*, 277–284.
[172] Goltz JS, Wolkoff AW, Novikoff PM, Stockert RJ, Satir P, *Proc. Natl Acad. Sci. USA* **1992,** *89*, 7026–7030.
[173] Lebbe C, Reichen J, Wartna E, Sägesser H, Poelstra K, and Meijer DKF, *J. Drug Target.* **1997,** *4*, 303–310.
[174] Beljaars L, Poelstra K, Molema G, Meijer DKF, *J. Hepatol.* **1998,** *29*, 579–588.

[175] Hinz S, Franke H, Machnik G, Muller A, Dargel R, *Exp. Toxicol. Pathol.* **1997,** *49*, 281–288.
[176] Castilla-Cortazar I, Garcia M, Muguerza B, Quiroga J, Perez R, Santidrian S, Prieto J, *Gastroenterology* **1997,** *113*, 1682–1691.
[177] Schuppan D, Stolzel U, Oesterling C, Somasundaram R, *J. Hepatol.* **1995,** *22*, 82–88.
[178] Gerling B, Becker M, Waldschmidt J, Rehmann M, Schuppan D, *J. Hepatol.* **1996,** *25*, 79–84.
[179] Creemers LB, Jansen DC, Veen-Reurings A, van den BT, Everts V, *Biotechniques* **1997,** *22*, 656–658.
[180] Allen TM, Murray L, MacKeigan S, Shah M, *J. Pharmacol. Exp. Ther.* **1984,** *229*, 267–275.
[181] Storm G, Oussoren C, Peeters PAM, Barenholz Y, In: *Liposome Technology*, (Ed. Gregoriadis G), pp. 345–383. CRC Press, Boca Raton, **1993.**
[182] Soyez H, Schacht E, Van Der Kerken S, *Adv. Drug Deliv. Rev.* **1996,** *21*, 81–106.
[183] Kataoka H, Tavassoli M, Blood 1985, 65, 1165–1171.
[184] Franssen EJF, Jansen RW, Vaalburg M, Meijer DKF, *Biochem. Pharmacol.* **1993,** *45*, 1215–1226.
[185] Franssen EJF, Van Amsterdam RGM, Visser J, Moolenaar F, De Zeeuw D, Meijer DKF, *Pharm. Res.* **1991,** *8*, 1223–1230.
[186] Ohmukai O, *Adv. Drug Deliv. Rev.* **1996,** *20*, 203–207.
[187] Fiume L, Chiricolo M., Busi C., Mattioli A., Spinelli C., Spinosa G., Bartolini G., Minghetti L., Orlandi M., Tomasi V. European Community Patent no.: 91117891.1 (EP0482540A1), **1992.**
[188] Yokoyama K, Watanabe M, *Adv. Drug Deliv. Rev.* **1996,** *20*, 195–201.
[189] Magnani M, D'Ascenzo M, Chiarantini L, Antonelli A, Naftalin R, *Drug Deliv.* **1995,** *2*, 151–155.
[190] Kamps JAAM, Morselt HWM, Scherphof GL, *Prog. Drug Deliv. Syst.* **1996,** *5*, 89–92.
[191] Sett R, Sarkar HS, Das PK, *J. Antimicrob. Chemother.* **1993,** *31*, 151–159.
[192] Sett R, Sarkar K, Das PK, *J. Infect. Dis.* **1993,** *168*, 994–999.
[193] Sarkar K, Das PK, *J. Immunol.* **1997,** *158*, 5357–5365.
[194] Ponzetto A, Fiume L, Forzani B, Song SY, Busi C, Mattioli A, Spinelli C, Marinelli M, Smedile A, Chiaberge E, Bonino F, Gervasi GB, Rapicetta M, Verme G, *Hepatology* **1991,** *14*, 16–24.
[195] Zahm FE, d'Urso N, Bonino F, Ponzetto A, *Liver* **1996,** *16*, 88–93.
[196] Cerenzia MT, Fiume L, Venon WD, Lavezzo B, Brunetto MR, Ponzetto A, Di Stefano G, Busi C, Mattioli A, Gervasi GB, Bonino F, Verme G, *Hepatology* **1996,** *23*, 657–661.
[197] Fanger MW, Morganelli PM, Guyre PM, *Crit. Rev. Immunol.* **1992,** *12*, 101–124.
[198] Muto M, *Arch. Histol. Jpn.* **1975,** *37*, 369–386.

5 Delivery of Drugs and Antisense Oligonucleotides to the Proximal Tubular Cell of the Kidney Using Macromolecular and Pro-drug Approaches

Marijke Haas, Yukio Kato, R. Folgert G. Haverdings, Frits Moolenaar, Kokichi Suzuki, Dick de Zeeuw, Yuichi Sugiyama, Dirk K.F. Meijer

5.1 Introduction

The need for specific delivery of drugs to their site(s) of action is evident in the case of extremely toxic agents that have to be administered in high doses such as anti-tumour drugs. But what would be the rational for specific drug targeting to the kidney? Clearly, the kidney is one of the organs with the highest exposure to drugs circulating in the body. Around 25% of the cardiac output flows through the two kidneys. In addition, many compounds are concentrated in the proximal tubule by active transport processes while often high luminal concentrations of drugs are reached due to water reabsorption. So, at first glance, one may argue that little extra would be gained by renal selective targeting. However, we believe that there can be several good reasons for renal-specific drug delivery. First, although many drugs used in the treatment of renal diseases do reach the kidney in sufficient quantities, they may cause undesirable extra-renal effects. Second, the intra-renal transport of a drug may not be optimal in relation to the target cell within the organ. Third, some drugs are largely inactivated before they reach the site of action in the kidneys. Finally, pathological conditions such as abnormalities in glomerular filtration, tubular secretion, or the occurrence of proteinuria can affect the normal renal distribution of a drug. Renal-specific drug targeting therefore can be an attractive option to overcome such problems and to improve the therapeutic index of a drug. Furthermore, cell-specific drug targeting within the kidney may provide an interesting tool in understanding the mechanisms of drug action and to manipulate renal physiology.

5.1.1 Kidneys and their Functions

The major function of kidneys is to filter the redundant nutrients and metabolites out of the blood, including those that come from the natural breakdown of tissues as well as those that we ingest with food intake. In this way, the kidneys maintain the homeostatic balance with respect to water and electrolytes as well as nutrients and metabolites. In addition, the actions of the kidneys also regulate blood pressure and erythropoiesis.

To do this, the kidney is equipped with nephrons, the basic units of the kidney. The nephron consists of a glomerulus and a tubule (Figure 5.1). The tubule is subdivided into the proximal

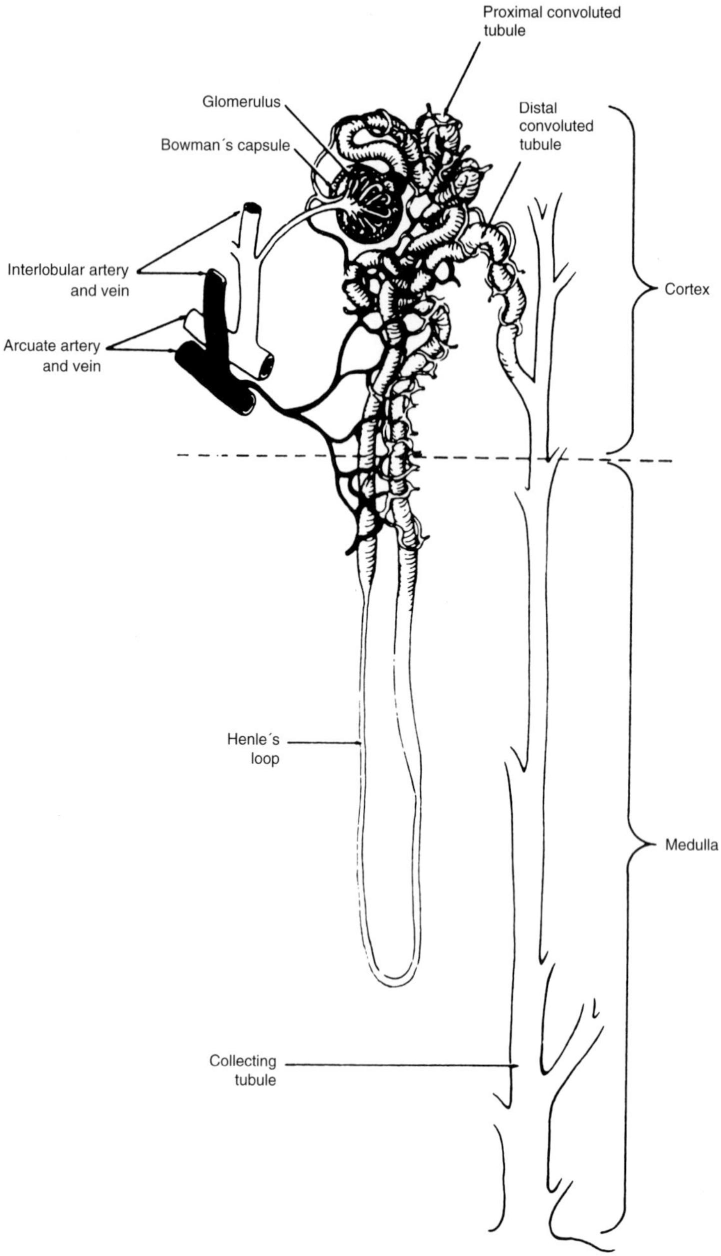

Figure 5.1. The functional nephron with representative blood supply. Reprinted with permission from reference [154].

convoluted tubule, proximal straight tubule, Henle's loop, the distal tubule and the collecting duct. In the tubule, a monolayer of epithelial cells separates the tubular lumen from the blood. A close network of arterial and venous capillaries provides close contact between the blood circulation and the tubular cells.

The blood first reaches the glomerulus, the filter unit of the nephron. The glomerular filtrate, i.e. blood deprived of macromolecules and blood cells, passes through the tubular lumen. The blood which is not filtered, flows through the efferent arteriole into the network of capillaries around the tubules supplying the proximal and distal tubules with blood.

5.1.2 Proximal Tubular Cells and their Functions

The proximal tubular cell plays a major role in the elimination of both inorganic and organic substrates. The cells have two distinct membrane domains. The basolateral membrane is in contact with the blood, and the apical brush-border membrane lines the tubular lumen.

Methods of traversing the basolateral membrane include uptake systems for organic cations and anions via facilitated diffusion and/or active transport [1]. Organic anions and cations cross the basolateral membrane via ATP-driven or secondary active processes (H^+-antiport) [2]. Basolateral uptake processes include the gamma-glutamyl transport system [3] and those for glycoproteins [4]. Certain proteins (insulin, epidermal growth factor (EGF)) are transcytosed across the tubular cells from the blood to the tubular lumen via receptor-mediated uptake [5].

In healthy individuals, useful endogenous compounds that are freely filtered by the glomerulus, only appear in the urine in small quantities. These compounds are 'rescued' by tubular reabsorption. These 'rescue mechanisms' consist of a variety of, mostly, carrier-mediated processes at the luminal site of the tubular cell. Substances transported by reabsorptive systems include sugars [6], amino acids [7], dipeptides [8], urate [9], folate [10], nucleosides [11] and proteins [12].

Apart from the elimination function, the kidney disposes of many endogenous and exogenous substances through metabolic conversion. Many compounds are highly concentrated in the proximal tubular lumen before being eliminated in the urine [13]. Therefore the driving force for metabolic conversion can be high. For instance metabolic clearance of indomethacin occurs predominantly by renal glucuronidation due to efficient enterohepatic recycling/deconjugation processes followed by carrier-mediated accumulation in the tubular cells [14].

For exogenous compounds such as drugs, various enzymes involved in both phase I and phase II metabolic routes are present, e.g. various isoforms of cytochrome p450, cytochrome b5, glucuronyl transferase and sulfotransferase [15].

In addition, renal tubular cells contain various proteases for the degradation of proteins and oligopeptides. These enzymes are located predominantly in the lysosomes and microsomes of these cells, but some have been reported on the brush-border membranes [16]. Degradative enzymes include various endopeptidases, exopeptidases and esterases [17].

In principle, the above-mentioned transport and metabolic functions of the tubule can be used for renal delivery and (re-)activation of (pro-)drugs and macromolecular drug targeting preparations.

5.1.3 Cellular Targets for Drug Delivery in the Kidney

The renal glomerulus consists of endothelial cells, glomerular epithelial cells and mesangial cells.

The mesangial cells of the glomerulus and the proximal tubular cells are the first choice targets for renal drug delivery. Both cell types play a central role in many disease processes in the kidney.

The mesangium is a specialized pericapillary tissue. It contains predominantly mesangial cells constituting contractile endocytic capillary pericytes embedded in the extracellular matrix. There is a continuous flow of blood plasma into the mesangium through mesangial fenestrations including the sieving of even relatively large particles. The mesangial cells are particularly highly reactive to foreign substances and pathogenic agents. As a consequence of such noxious triggers, mesangial cells respond with the synthesis of a host of inflammatory factors [18]. Consequently this cell type is an interesting target for renal drug delivery in the case of acute and chronic inflammatory conditions.

Several factors have been identified that trigger activity of the proximal tubular cell. Glomerular and systemically-derived cytokines and growth factors reach the tubular cells by filtration, peritubular secretion or diffusion through the interstitium [19]. Hypoxia, ischaemia, nephron loss and luminal obstruction cause tubular cell activation in an adaptive response to compensate for loss of function. Furthermore, tubular protein overload as a result of glomerular proteinuria and high tubular delivery of glucose in the diabetic state are considered to be important factors causing tubular activation (Figure 5.2).

As a consequence of such noxious triggers, proximal tubular cells respond with the synthesis of a host of inflammatory mediators [20]. Because of this, the proximal tubular cell is a central target for drug delivery.

To date, only a limited number of studies have focused on drug delivery to the mesangium cell and only a modest degree of selectivity has been obtained in this respect [21,22]. More extensive studies have been performed on targeting drugs to the proximal tubular cell. Therefore, in this chapter, only targeting to the proximal tubular cell will be addressed.

5.1.4 Renal Pathology and the Proximal Tubular Cell for Therapeutic Intervention

Targeting of anti-inflammatory and anti-fibrotic drugs to the proximal tubular cell may prevent tubulointerstitial inflammation and scarring secondary to systemic and glomerular infection and proteinuria. Furthermore, tubular drug delivery may be beneficial during shock, renal transplantation, ureter obstruction, diabetes, proteinuria, renal carcinoma and some tubular defect diseases such as Fanconi and Bartter's syndrome.

An argument against the concept of cell-specific drug delivery to the kidney is that, in most cases, the aforementioned diseases are not associated with only one cell-type in the kidney. However, after being released into the proximal tubular cell, the targeted drug may be redistributed locally through diffusion out of the cell, after which it becomes active in interstitium and downstream cells. Furthermore, cell-specific drug delivery will allow more aggressive

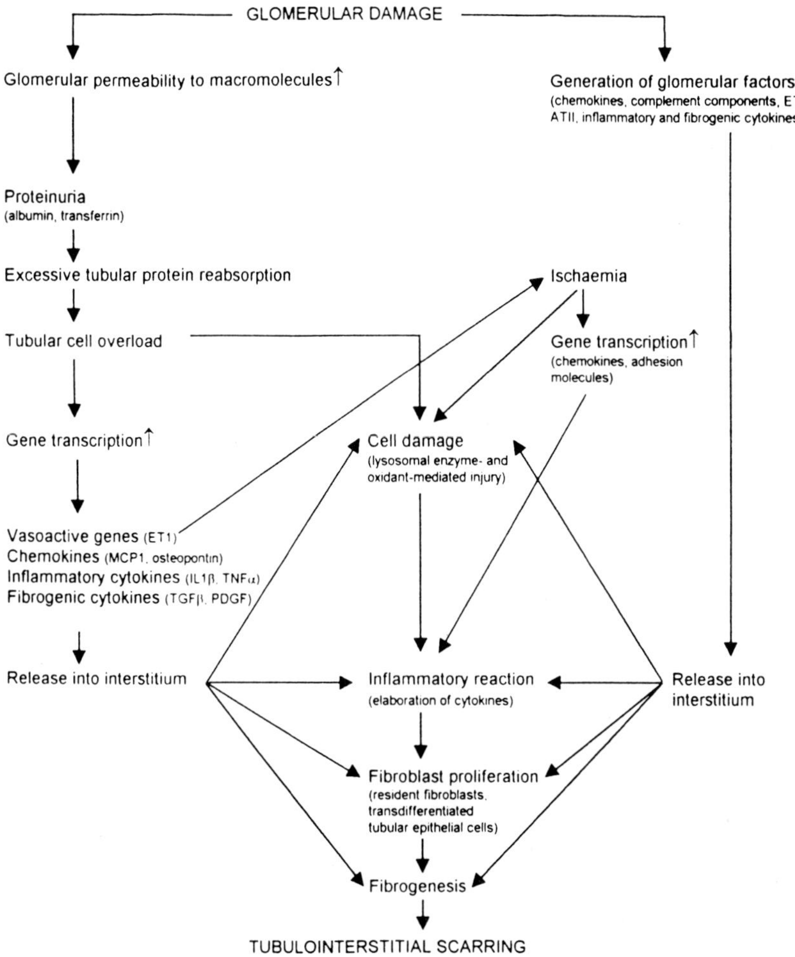

Figure 5.2. Pathogenic mechanisms that are potentially involved in tubulointerstitial fibrogenesis in glomerular kidney disorders. Reprinted with permission from reference [19].

treatment of the targeted cell and because of that, may improve the therapy when given in combination with the conventional treatment.

5.1.5 Targeting to the Proximal Tubular Cell

In this chapter, three aspects of drug targeting to the proximal tubular cell will be discussed in the light of recent advances in this field. First, various pro-drug concepts designed for selective renal delivery with emphasis on the use of alkylglycoside vectors, will be described. Subsequently, the use of low-molecular weight proteins as potential drug carriers will be discussed and finally, the delivery of antisense oligonucleotides to the proximal tubules is reviewed.

5.2 Renal Delivery Using Pro-Drugs

5.2.1 The Alkylglycoside Approach

5.2.1.1 Introduction

The alkylglycoside vector is a kidney-specific delivery system that has recently been established [23–25]. This vector is efficiently taken up from the basal side of the renal epithelium in a blood flow-limited manner and it can be used with several types of therapeutic molecules. The following sections summarize and discuss, first, how the novel kidney-specific alkylglycoside vector was identified, second, its structural and size requirements and third, the potential limitations of delivery to the kidney and the characterization of its binding sites on kidney cell membranes.

5.2.1.2 Concept of the Alkylglycoside Approach

To identify a novel target for tissue-specific drug delivery, Suzuki *et al.* [23–25] focused on sugars as probes. The reason for this was that sugar-recognition has been reported to play a key role in cell–cell, cell–matrix, and cell–molecule interactions including receptor-mediated endocytosis [26]. For instance, it has been established that the galactose moiety in carbohydrate chains determines the systemic clearance of glycoproteins [27]. Most of the carbohydrate receptors, such as the galactose- and mannose-specific receptors, are located in specific cell types in the liver (see Chapter 4 for more details on this subject). Some studies have described sugar moieties as sites for drug delivery in organs other than the liver [28]. In the tubular cells of the kidney various carrier-mediated processes for basolateral and apical membrane transport have been described, including sugar transport. Until recently, membrane transporters for sugars and glycoprotein receptors have not been studied as targets for drug delivery to the kidney. Therefore, studies were undertaken in which several types of sugars were introduced into a model peptide drug (arginine vasopressin, AVP) via different lengths of alkyl-chain spacer (Figure 5.3a). Subsequently, their tissue distribution characteristics were examined [23,24].

5.2.1.3 Distribution of Alkylglycoside-derivatized AVP In Vivo

Because tissue-specific vectors are aimed at increasing the influx of a drug into the target, assessment of unidirectional transport from the circulating plasma into the target organ is essential. In this context, integration plot analysis is a convenient *in vivo* method in which a tracer amount of vector is injected intravenously and the plasma (C_p) and tissue (C_T) con-

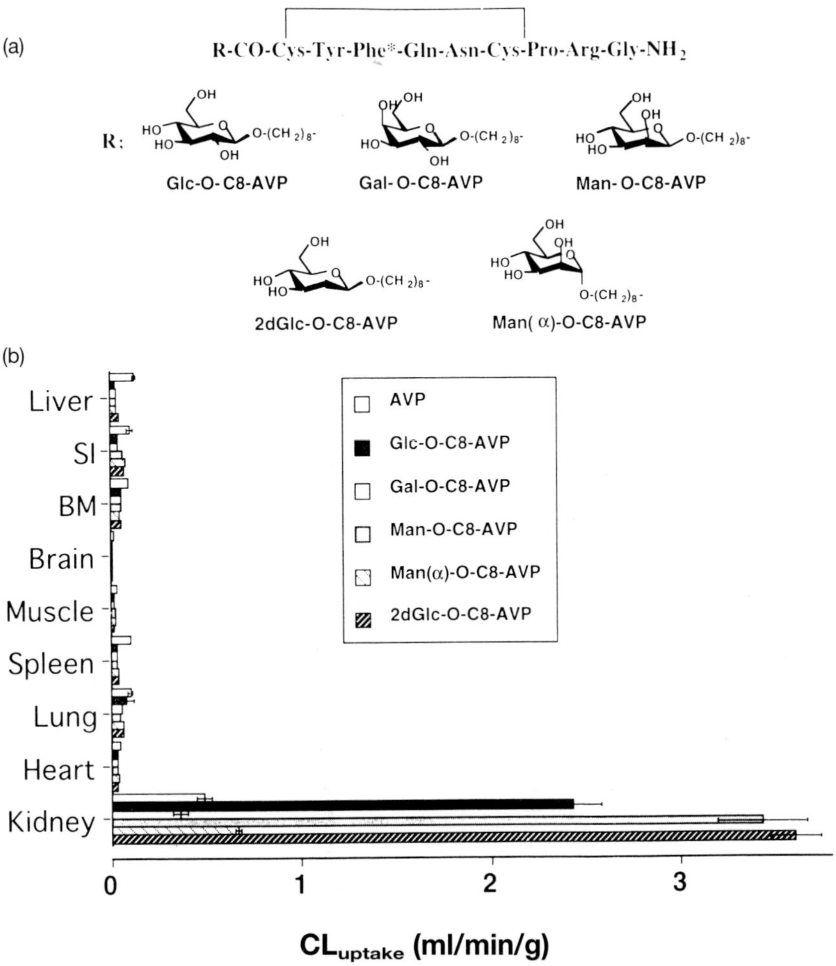

Figure 5.3. (a) Structure and (b) tissue uptake clearance of alkylglycoside-derivatized AVP. *Position of ^{3}H-label. Adapted from reference [23].

centration profiles are monitored (Figure 5.4a) [29]. Figure 5.3b shows the CL_{uptake} for several sugar-modified AVPs. It shows that the tissue-distribution profile largely depends on the sugar moiety, and that glucose- (Glc), mannose- (Man) and 2dGlc-O-C8-AVP exhibit kidney-specific distribution [23].

In the kidney, the CL_{uptake} includes glomerular filtration as well as renal binding and or uptake from the basal site. Hence, such CL_{uptake} should give the apparent value ($CL_{\text{uptake,app}}$):

$$CL_{\text{uptake,app}} = f_p GFR + CL_{\text{uptake,kidney}} \tag{5.1}$$

where f_p, GFR, and $CL_{\text{uptake,kidney}}$ represent respectively, the unbound plasma fraction, the glomerular filtration rate, and the CL_{uptake} value representing the unidirectional drug associ-

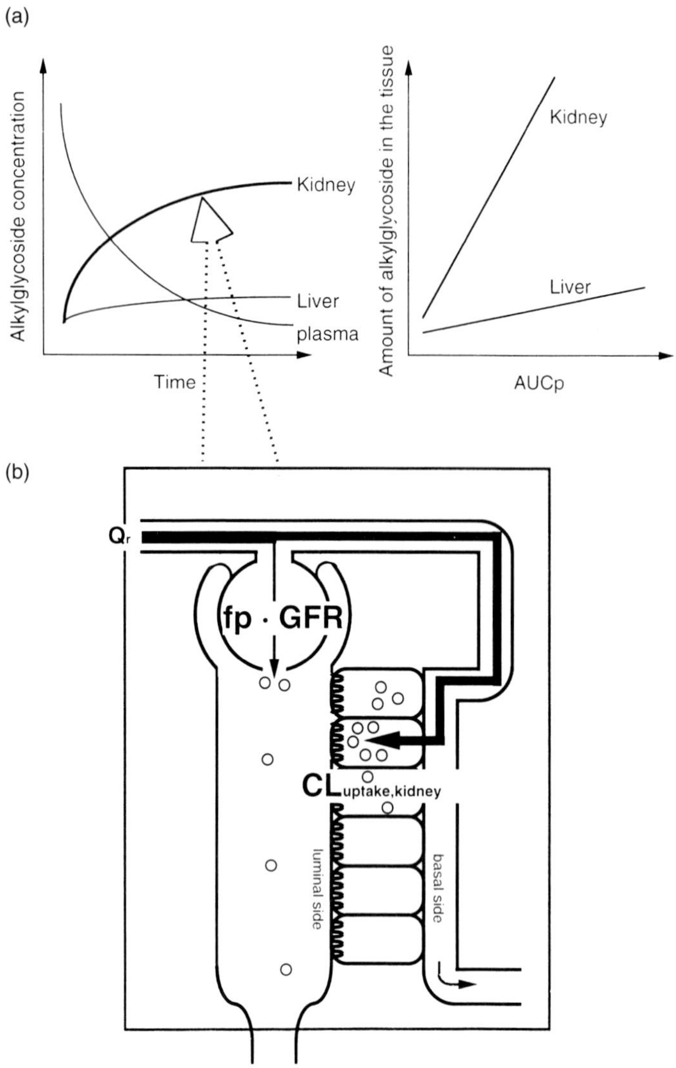

Figure 5.4. Schematic diagram for (a) the integration plot analysis and (b) renal processing of alkylglycoside.

ation from the basal site of the kidney (Figure 5.4b). Based on this equation, the targeting efficiency from the basal site can be estimated as $CL_{\text{uptake,kidney}}$. A similar analysis for p-aminohippurate and inulin enabled us to estimate the renal plasma flow rate (Q_r) and GFR, respectively. The $CL_{\text{uptake,app}}$ for Glc-O-C8-AVP was close to Q_r (~2.4 ml min^{-1} g^{-1}) and much higher than f_pGFR (~0.13 ml min^{-1} g^{-1}), suggesting that renal accumulation of Glc-O-C8-AVP occurs mainly from the blood ($f_pGFR << CL_{\text{uptake,kidney}}$) [23]. Semimicro- and microautoradiography revealed that this renal uptake occurs in the cortex, and Glc-O-C8-AVP is distributed in the proximal convoluted tubules rather than in the glomeruli [23].

5.2.1.4 Specific Binding of Alkylglycoside-derivatized AVP in Kidney Plasma Membranes

Possible explanations for a blood flow-limited uptake in kidney include the existence of specific uptake mechanisms, such as receptor-mediated endocytosis and carrier-mediated transport. Since the former mechanism is initiated by binding of the ligand to the cell-surface receptor, the specific binding of alkylglycoside compounds to isolated tubular plasma membranes was examined [23,24].

Scatchard analysis revealed specific binding of Glc-, Man- and 2dGlc-O-C8-AVP exhibiting kidney-specific distribution *in vivo* (Figure 5.3b), with a dissociation constant (*K*d) of 10–60 nM. This did not occur with Gal- and Man(α)-O-C8-AVP. Saturation of the CL_{uptake} of Glc-O-C8-AVP in the kidney *in vivo* occurred at a similar concentration (40–80 nM) of unbound ligand in the renal capillary space [23]. These results suggest that specific binding site(s) are involved in the renal distribution of alkylglycoside conjugates of AVP.

5.2.1.5 Structure–Kinetic Relationship Studies

To develop alkylglycoside moieties as drug delivery vectors, a systematic analysis was performed to identify the structural requirements for both vectors and drugs. This allowed us to understand the spectrum and limitations of compounds that can be delivered by this system. A binding study using isolated tubular membranes enabled the investigation of such struc-

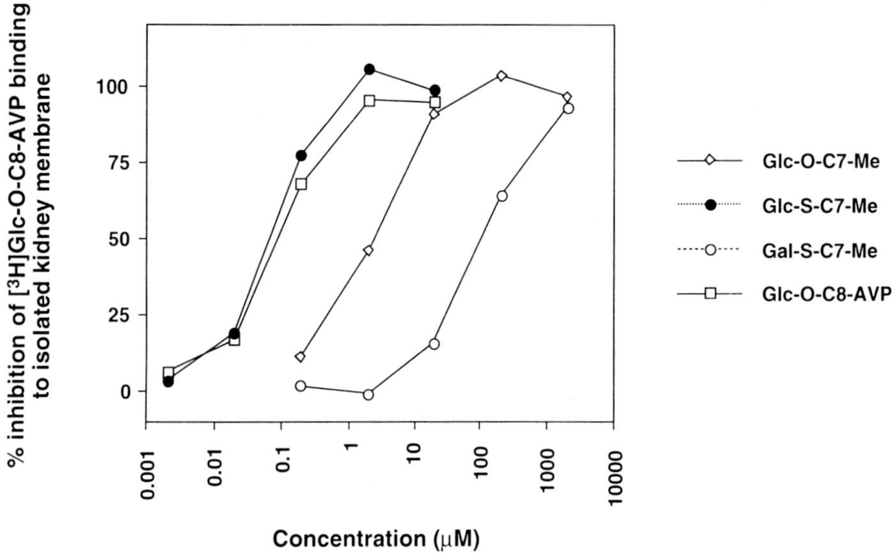

Figure 5.5. Inhibition of specific binding of Glc-O-C8-AVP to rat kidney membrane fraction. Adapted from reference [24].

tural requirements. When the drug moiety (AVP) was removed from Glc-O-C8-AVP, its affinity constant was two orders of magnitude lower than the original (IC_{50} of Glc-O-C7-Me ~ 3 μM) [24]. In addition, removal of the alkyl-chain almost abolished binding (IC_{50} of Glc-O-Me > 1 mM) [24]. Thus, sugar, alkyl, and drug moieties seem to be essential for renal targeting. However, S-glycoside alone exhibited a much higher affinity, the IC_{50} of Glc-S-C7-Me (~ 70 nM) being almost comparable with that of Glc-O-C8-AVP (Figure 5.5) [24]. The CL_{uptake} of Glc-S-C7-Me in the kidney was close to the renal plasma flow rate and much higher than the CL_{uptake} in other organs. Scatchard analysis of Glc-S-C7-Me and Glc-S-C8-AVP revealed a *K*d of 10–20 nM [24]. These results imply that Glc-S-C7-Me (octyl β-D-thioglucoside) may be a suitable delivery vector for the kidney. It is noteworthy that this moiety has been widely used as a detergent, but the specific binding observed occurs at a much lower concentration than that at which it exerts a detergent effect (~ mM range) [30].

To optimize the alkyl-chain length, the effect of different numbers of methylene groups on the CL_{uptake} *in vivo* and specific binding to kidney membranes was examined. Glc-S-C5-AVP showed a much lower CL_{uptake} whereas Glc-S-C11-AVP had a higher CL_{uptake} and specific binding than Glc-S-C8-AVP [24]. By screening the inhibition potential of various types of sugars and/or alkyl moieties, it was found that an equatorial OH group in the 4 position is essential, while the inhibitory properties were not affected by the orientation of the OH group at the 2 position or by its absence (Figure 5.6) [25]. The length, branching and charge in the region of the glycoside bond within the methylene group also appeared to be important for specific binding [25]. Thus, the alkylglycoside moiety appears to be essential for kidney targeting.

The types of therapeutic compounds that, in principle, can be delivered by conjugation with this vector remains to be studied in detail. Until now, a marked increase in renal CL_{uptake} has been found only for low-molecular-weight compounds such as alkylglycoside-derivatized AVP, oxytocin, tryptamine, and 4-nitrobenz-2-oxa-1,3- diazole (NBD) [23–25]. To elucidate the size limitation of compounds to be delivered, acylated polylysines (APL) with a range of molecular weights (Mw) were conjugated to the Glc-S-C8 moiety. The CLuptake in kidney of Glc-S-C8-APL with a mean Mw of 17 kDa was much larger than that in other organs, half the renal clearance being accounted for by f_pGFR. Thus, molecular weight limitation seems to be critical for the renal targeting.

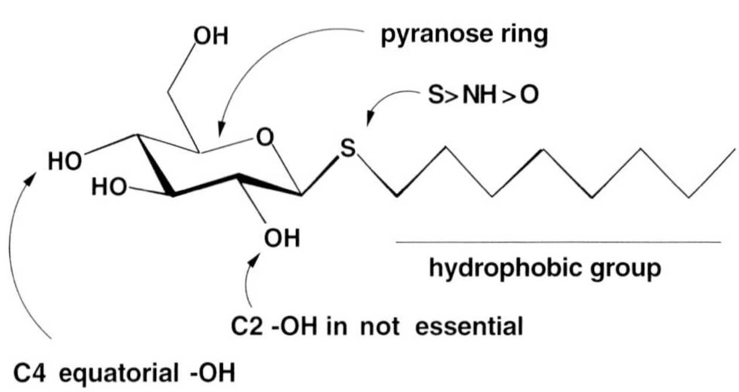

Figure 5.6. Structural requirements for a renal targeting vector.

5.2.1.6 Identification of Target Molecules for Alkylglycosides

So far, the target molecule that binds alkylglycosides in the kidney has not been conclusively identified. It is known that several types of transporters are localized on the proximal tubules. These include organic anion transporters, organic cation transporters, and oligopeptide transporters [31,32]. However, since the renal distribution of the alkylglycoside vector is highly dependent on the structure of the sugar moieties, and both basic and neutral peptides (AVP and oxytocin, respectively) are recognized, the involvement of these transporters seems to be minor. Sugar transporters are classified as facilitated sugar transporters and Na^+/glucose co-transporters [33,34]. The facilitated transporters do not concentrate the ligand inside the cells whereas the secondary active Na^+/glucose co-transporter that is located on the brush-border membrane can transport molecules through the existing Na^+-gradient [34]. Since the degree of inhibition of alkylglycoside binding to the kidney membrane by glucose (10–100 mM) is minor [25], involvement of this transporter is also unlikely. In addition, kidney lectin [35] which recognizes acidic sugars, does not seem to be a potential transporter. Further studies are therefore needed to clarify the major transport system.

Downregulation of CL_{uptake} is one of the methods used to discriminate between carrier-mediated transport and receptor-mediated endocytosis [29,36]. This concept is based on the general finding that an excess concentration of a ligand can induce internalization and subsequent degradation of cell-surface receptors, resulting in a reduction of the cell-surface receptor density. After intravenous administration of excess unlabelled Glc-O-C8-AVP, the CL_{uptake} of tracer Glc-O-C8-AVP initially declined followed by a gradual recovery, suggesting that Glc-O-C8-AVP is taken up predominantly via receptor-mediated endocytosis [23]. However, it is possible that the temporary reduction of CLuptake can also be explained by competitive inhibition by the unlabelled ligand instead of a downregulation of the receptor [23]. Recently, cross-linking ^{125}I-labelled alkylglycoside to the renal plasma membrane revealed the presence of a binding protein with a Mw of 62 kDa. This band disappeared in the presence of excess unlabelled ligand and was clearly located at the basolateral membranes, and not in the brush-border membranes [37].

5.2.1.7 Perspectives of Renal Delivery with Alkylglycoside Vectors

The most notable feature of the alkylglycoside vector system is the highly efficient uptake via the basal site of the renal plasma membranes. This may provide a higher rate of target delivery compared to the targeting methods that exploit glomerular filtration with subsequent reabsorption from the apical position, such as the low-molecular weight protein method described below in this chapter [38,39]. The CL_{uptake} in the latter case cannot exceed the $f_{\mathrm{p}}GFR$ based on Eq. 5.1 (Figure 5.4b). Furthermore, delivery via the basal site will not be negatively affected by proteinuria.

Concentration as a result of uptake and subsequent retention of the ligand in the proximal tubules may be suitable for certain types of therapeutic agents. The plasma concentration of Glc-S-C8-NBD fell rapidly while the kidney concentration of the intact ligand remained almost constant, the kidney-to-plasma concentration ratio being ~ 200 at 30 min after injection

[24]. Uptake which has the effect of concentrating the therapeutic agent is one of the advantages of using receptor- and/or transporter-mediated drug delivery.

The critical factors that need to be addressed include the limited range of drugs that can be delivered by this system. The present findings suggest that the system cannot be applied to macromolecules such as genes and proteins. For application to low-molecular-weight compounds, the therapeutic activity of the drug needs to be regained after release from its vector in the kidney. The pharmacological activity of AVP was affected by derivatization [40,41] and our recent findings suggest that the kidney-targeting potential is low for certain types of anionic drugs.

It should be noted also that distribution may occur to organs other than the kidney. For example, oxytocin derivatives also exhibited CL_{uptake} in the small intestine. A similar phenomenon was observed for Glc-S-C8-tryptamine where the CL_{uptake} in small intestine and liver also increased by derivatization [24]. These findings cannot be explained simply by the existence of a single binding site for alkylglycoside vectors in the different organs. The presence of multiple binding sites is supported by the finding that inhibition of the specific binding of Glc-S-C8-tyrosine by Glc-S-C8-AVP cannot be fitted to a single site kinetic model [37]. To clarify the renal and extra-renal transport mechanisms, kinetic analysis performed by changing the structure of the ligand may not be sufficient and molecular biological analysis may be helpful, for example by characterizing the target binding protein. This should reveal the scope and limitations of this alkylglycoside strategy in clinical and pathological situations.

5.2.2 The Amino Acid Pro-drug Approach

5.2.2.1 Introduction

Most research on tubular cell-specific drug delivery has been focused on the development of pro-drugs that should be activated by more or less kidney-selective enzymes. The relevant literature will be reviewed briefly and discussed with regard to the benefits and limitations compared to the alkylglycoside and macromolecular approaches of renal drug targeting. The 'soft drug' concept will be discussed as a potential method by which targeted drugs are inactivated efficiently after reentering the circulation.

5.2.2.2 The Concept of the Amino Acid Pro-drug

In the design of drugs, the usefulness of renal-specific enzymes which enable the site-specific release of the active drug, should be taken into account. The design of kidney-selective pro-drugs is based upon the relatively higher amounts of certain enzymes in the proximal tubular cells than elsewhere in the body.

These strategies are aimed at either cytosolic enzymes, such as L-amino acid decarboxylase, β-lyase and N-acetyl transferase, or enzymes that are expressed at the brush border of the proximal tubule and to a lesser extent on the basolateral membrane, such as γ-glutamyl transpeptidase (GGT).

The technology involves one or more chemical modifications of the parent compound using chemical moieties that, with regard to size, are comparable to or even smaller than the parent drug. This type of pro-drug may be degraded intracellularly into the active drug, resulting in its release and subsequent secretion into the tubular lumen or via the interstitium back into the circulation.

Alternatively, the pro-drug may be a substrate for brush-border enzymes of the proximal tubular cell, resulting in release of the active drug in the tubular lumen and subsequent reabsorption at distal sites or elimination in the urine.

5.2.2.3 Renal Specificity of Amino Acid Pro-drugs and their Effects

Several drugs have been coupled to gamma-glutamyl transferase, γ-glutamyl. The γ-glutamyl pro-drug of l-dopa (gludopa) showed a higher renal specificity [42–44] than the pro-drug of dopamine [45]. Gludopa induced renal-specific effects such as increases in renal blood flow and salt excretion while systemic blood pressure remained unaffected [42,43].

Another example is the pro-drug γ-glutamyl-sulphamethoxazole. This pro-drug did not show renal selectivity, either because of its rapid removal from the kidney, or due to cleavage in non-target tissues containing a low concentration of the enzyme. On the other hand the N-acetyl-γ-glutamyl derivative showed pronounced renal specificity [46,47]. In this respect, pro-drug accumulation in the kidney was due to carrier-mediated transport at the basolateral membrane site, which is sensitive to buthionine sulfoximine and probenecid. Other N-acetyl-γ-glutamyl pro-drugs have been developed and tested on the basis of the same principle. Of the derivatives tested, N-chloroacetyl-γ-glutamyl-sulfamethoxazole appeared to have the highest renal selectivity [46]. N-acetyl-γ-glutamyl-aminowarfarin was not a successful pro-drug since it was not rapidly secreted via the tubule and therefore did not reach the enzyme site [48]. In fact, this pro-drug was selectively excreted in the bile.

The N-acetyl-γ-glutamyl pro-drug of the hydralazine-like vasodilator CGP 18137 showed a higher renal-selective activity than the parent compound, CGP 18137. In contrast to the parent drug, the pro-drug caused a decrease in renal resistance without any effect on blood pressure [49].

Because of the high phosphatase activity in the kidney, dopamine-phosphate ester pro-drugs have been synthesized, e.g. SIM 2055 (N-methyl-dopamine-4-O-phosphate) [50]. Although the mechanism of renal selectivity of this compound is not yet understood in detail, it is thought to be due to the high renal blood flow and the high renal phosphatase activity together with the high affinity of the kidney for the released drug.

Cysteine-S-conjugates have also been proposed as kidney-selective pro-drugs. Renal metabolism of S-6-(purinyl)-L-cysteine resulted in the formation of 6-mercaptopurine by the action of β-lyase [51]. However, besides formation of the intended parent compound, other S-conjugates may be formed by various radical reactions, which may induce renal toxicity.

5.2.2.4 Benefits and Limitations of the Amino Acid Pro-drug

A potential benefit of the pro-drug approach is that the compounds can, in principle, be designed for oral administration. Furthermore, immunogenicity which results from using protein conjugates as drug carriers, will not be a problem. However, in contrast to the LMWP approach, in which the kinetics of the protein carrier overrule the intrinsic kinetics of the drug to be targeted by accumulation after the absorptive process, conjugation with amino acids or small peptides does not necessarily lead to higher specificity for renal uptake. Therefore, for each drug, different derivatives should be synthesized and tested for the desired kinetic profile.

5.2.2.5 The Soft Drug Concept

When drugs that are activated in the kidney are transported back into the circulation they may be deposited elsewhere in the body. This undesirable consequence may be overcome by using the so-called soft drug approach [52]. After administration of N-acetyl-γ-glutamyl-CGP 18137, active CGP 18137 is specifically generated within the kidney but partly diffuses back into the circulation. However, in the circulation, this vasodilating agent is rapidly inactivated by a chemical reaction [49]. For reasons unknown to us, this innovative product was not developed further. However, this example illustrates the potential of soft drugs in the field of drug targeting. Irrespective of the fact that the drug is generated from a pro-drug or released from a carrier, inactivation of the active compound after being released from the kidney into the blood circulation, could evidently add to the renal selectivity and therapeutic safety.

5.2.3 The Folate Pro-drug Approach

5.2.3.1 Introduction

The kidney has an important role in conserving folate to counteract a potential deficiency of this essential vitamin. Circulating folate, in the form of 5-methyltetrahydrofolate, is filtered through the glomeruli and extensively reabsorbed within the nephron into the renal vascular circulation. The kidney contains a high affinity folate-binding protein (FBP) that is concentrated in the proximal tubule cells [53,54]. Immunocytochemical studies have located FBP to the brush-border membrane, endocytic vacuoles and dense apical tubules, indicating a reabsorption of folate through endocytosis of the FBP–folate complex followed by dissociation and recycling of FBP [55]. In this study no significant labelling was found in lysosomes at any time, implying that there is no transport of FBP to lysosomes for degradation.

Recently it was shown that folate transport from the basolateral site occurs as readily as that from the luminal site, indicating that changes in secretion can mediate excess urinary folate excretion [56].

5.2.3.2 Potential Renal Selectivity of Folate Constructs

It has been hypothesized that folate receptor-mediated endocytosis can be exploited for the selective delivery of drugs by covalent attachment to folate via its γ-carboxyl group. This concept was primarily designed for the targeting of various biomolecules to solid tumours. For a number of human tumours, having a high over-expression of a membrane-associated folate receptor, *in-vitro* studies have shown that folic acid derivatization allowed selective delivery to cancer cells in the presence of normal cells. Thus high tumour selectivity was achieved with folate-targeted imaging agents [57], antineoplastic drugs [58,59], protein toxins [60], liposomes [61] and antisense oligonucleotides [62].

Interestingly, after intravenous administration of a radiolabelled folate conjugate (^{111}-Indium-diethylenetriaminepenta acid (DTPA)-folate) in the rat, the conjugate was rapidly excreted in the urine. Moreover, after intravenous administration to athymic mice with a human tumour cell implant, the radiotracer was not only taken up by the subcutaneous tumour but was also taken up by the kidneys in significant quantities [63], indicating substantial renal selectivity of the folate conjugate. In addition to the kidney, the liver also has a high concentration of the folate-receptor [64].

5.2.3.3 Benefits and Limitations of Folate

To date, the possibility of using folate binding for the purpose of renal drug targeting has not been studied. Since the kidney is not the only organ containing folate-receptors, the physicochemical properties of the conjugate may be important determinants of the success of targeting.

5.3 Renal Delivery Using Macromolecular Carriers: The Low Molecular Weight Protein Approach

5.3.1 Introduction

Low molecular weight proteins (LMWP) are freely filtered proteins with a molecular weight of less than 30 000 Dalton and are considered to be suitable as renal-specific drug carriers. The concept is based on four principles:

- The carrier has functional groups allowing drug attachment.
- The LMWP accumulates specifically in the kidney, in particular in the tubular cells through a reabsorption mechanism.
- The physicochemical properties of the LMWP overrule those of the linked drug.
- The drug–LMWP conjugate is stable in the circulation but after arrival in the kidney, the active drug is released in the catabolically-active lysosomes of the proximal tubular cells (Figure 5.7).

As reviewed by Franssen *et al.* [65], drugs can be directly coupled to LMWPs via the lysine amino group of the protein to form an amide bond. Alternatively, the drug can be coupled to the protein via different spacers such oligopeptides (amide bond), (poly)-alpha-hydroxy acids (ester bond), pH-sensitive cis-aconityl spacers (acid-sensitive amide bond) and SPDP spacers (disulfide bond) (see Chapter 11). The ability of the kidney to release the parent drug from such drug-spacer derivatives and drug–LMWP conjugates by enzymatic or chemical hy-

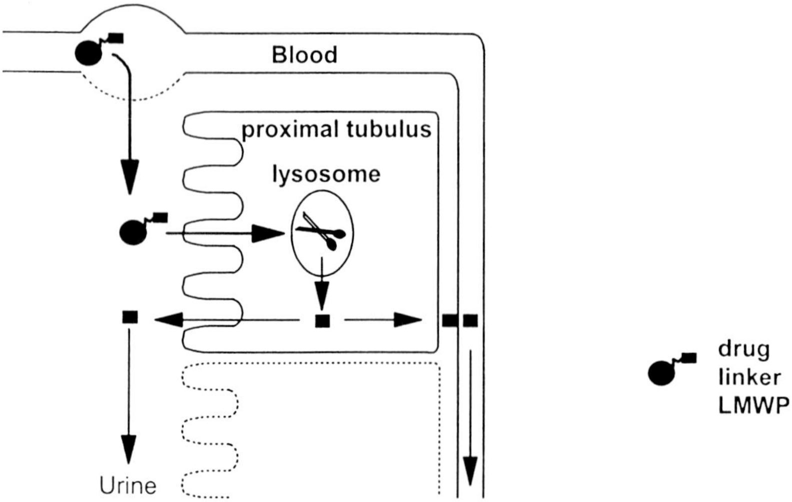

Figure 5.7. Schematic representation of the mechanism by which drug targeting to the proximal tubular cell of the kidney might be achieved using a low molecular weight protein (LMWP) as a carrier.

drolysis of the bond, have been tested in renal cortex homogenates and lysosomal lysates as well as in *in vivo* studies. It was found that lysosmal proteases can cleave the peptide bond between the carboxylic acid group of a drug and an α-amino group of an amino acid. However, the bond between the carboxylic acid group of the drug and the ε-amino group of lysine could not be cleaved. Since the conjugation of drugs to amino groups of a protein will predominantly occur at the ε-lysine residues and only to a small extent at the N-terminal α-amino group, direct conjugation of a drug via its carboxylic acid group will not result in the quantitative regeneration of the parent compound [66]. Drugs with a terminal carboxyl group, such as naproxen [67], can be released as the parent drug from LMWP conjugates using ester spacers such as L-lactic acid. Increasing spacer length by intercalating a tetra (L-lactic acid) moiety between the drug and the protein further increases the rate of drug release, indicating increased accessibility of the bond to the enzymes.

Drugs that have primary amino groups available for conjugation, for instance dopamine and doxorubicin, can in principle be coupled to LMWPs via oligopeptides. In contrast to the carboxypeptidases, the aminopeptidases appear to possess a broader specificity. To allow the release of terminal amino group-containing drugs in the acid environment of the lysosomes without the requirement of enzymes, an acid-sensitive spacer can be used.

Drugs coupled via a disulfide bond like, captopril, are rapidly released from the protein-spacer moiety of the conjugate, enzymatically by β-lyase and/or non-enzymatically by thiol-disulfide exchange with endogenous thiols [68].

The different aspects of drug targeting using LMWPs that have been studied to date are discussed below. As an example, we use the data of two conjugates, naproxen–lysozyme and captopril–lysozyme.

5.3.2 Renal uptake of LMWP Conjugates

5.3.2.1 Renal Uptake of Native LMWPs

Comparison of the kinetic features of different LMWPs revealed that all LMWPs tested so far (such as lysozyme, cytochrome-c and aprotinin) are quickly cleared from the circulation and accumulate rapidly in the kidney [38]. The fractions of the injected LMWP that are reported to be taken up by the kidney vary between 40–80 % of the injected dose. In our studies, using external counting of radioactivity, at least 80 % of the intravenously injected LMWPs was finally taken up by the kidneys, which is in agreement with renal extraction studies [69,70]. However, studies in which the actual amount of LMWP in the kidney was measured directly in the tissue, indicated a lower, but still substantial accumulation of 40% of the injected dose [71,72]. Apart from the kidney, LMWPs do not seem to accumulate elsewhere in the body (Figure 5.8).

From this we concluded that LMWPs are potentially suitable to serve as renal-specific drug carriers: a drug–LMWP conjugate will be rapidly removed from the circulation and the drug can be intra-renally released. Consequently, major distribution to extra-renal tissue and related unwanted effects elsewhere in the body can, in principle, be avoided. It is assumed that secondary redistribution of the generated drug from the kidney is relatively slow so that systemic concentrations remain below the therapeutic window for extra-renal effects.

Figure 5.8. Renal specificity of a radiolabelled LMWP. Gamma-camera imaging after an intravenous injection of a radiolabelled low molecular weight protein (LMWP) in the rat, showing the predominant uptake of the LMWP by the kidneys.

5.3.2.2 Renal Delivery of Naproxen–Lysozyme

Targeting of nonsteroidal anti-inflammatory drugs (NSAIDs) such as naproxen could be of interest for the treatment of proteinuria and tubular defects such as Fanconi syndrome and Bartter's syndrome [73,74]. Although a conjugate with an ester spacer is preferred to a conjugate with a direct peptide linkage [66,67], we continued our research using naproxen directly conjugated to lysozyme. The synthesis of the conjugate with an ester spacer (naproxen–L-lactic acid–lysozyme) is cumbersome, but fortunately the catabolite of the conjugate with the direct peptide linkage (naproxen–lysine) appeared to have an inhibitory effect on prostaglandin synthesis *in vitro* which was equivalent to that of the parent drug [66].

The coupling of 2 moles of naproxen to 1 mole of lysozyme did not affect the renal uptake of lysozyme in the rat: like native lysozyme, the conjugate rapidly accumulated in the kidney [75]. Focusing on the drug moiety of the conjugate, it was shown that conjugation of naproxen to lysozyme distinctly altered the kinetics of the drug. Conjugation to lysozyme resulted in a 70-fold increase in naproxen concentrations in the kidney (Figure 5.9a) [76].

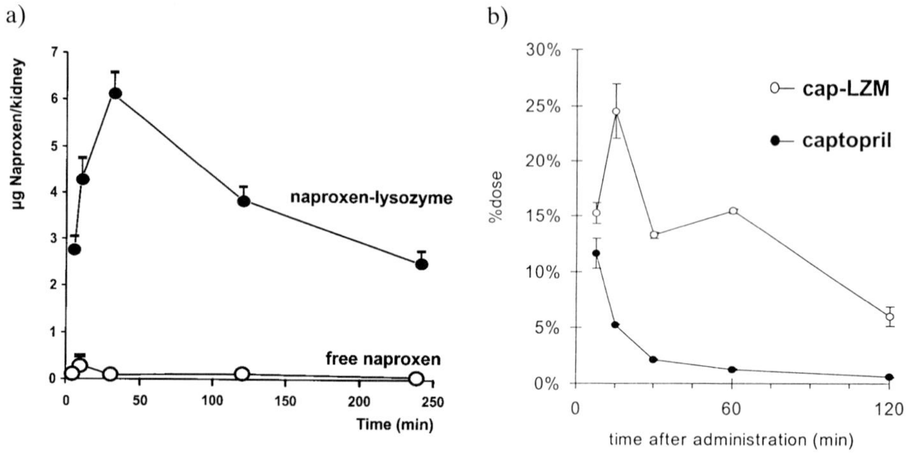

Figure 5.9. The concentration–time course of (a) naproxen and (b) captopril in the kidney after intravenous injection of the parent drug or the drug–lysozyme (LZM) conjugate. Values are given as means + SEM.

5.3.2.3 Renal Delivery of Captopril–Lysozyme

Angiotensin-converting enzyme (ACE) inhibitors such as captopril exert a long-term renoprotective effect. Among other effects, they lower systemic blood pressure and renal plasma flow and effectively reduce urinary protein excretion. Renal delivery of ACE-inhibitors may increase this efficacy and reduce extra-renal side-effects. Renal targeting of an ACE-inhibitor can also be useful in clarifying the contribution of local ACE inhibition to these renoprotective effects.

A spacer was used to link captopril via a disulfide bond to the LMWP lysozyme. Conjugation of captopril to lysozyme resulted in a 6-fold increase in captopril accumulation in the rat kidney (Figure 5.9b) [77]. This modest enrichment, as compared to that achieved with naproxen–lysozyme, was due to fact that, in contrast to naproxen, free captopril is cleared very efficiently by the kidney itself. Thus, delivery via lysozyme reabsorption only leads to a limited improvement of renal accumulation of captopril.

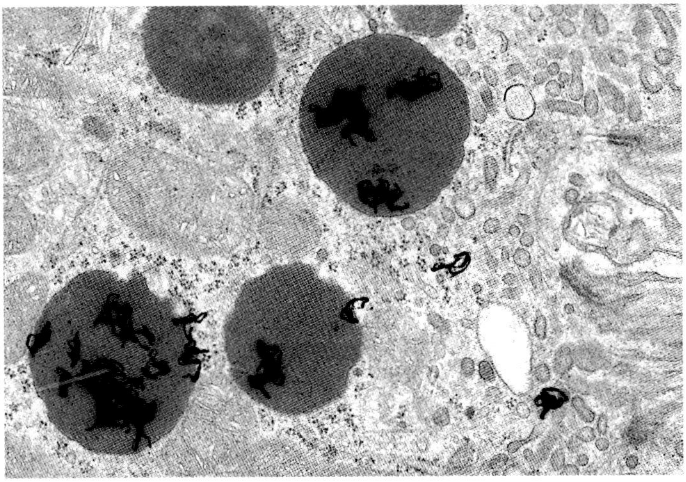

Figure 5.10. Accumulation of a radiolabelled LMWP in the lysosomes of the proximal tubular cell. Electron microscope autoradiography of renal proximal tubular cells from a rat injected i.v. with [125I]-tyramine-cellobiose-labelled cytochrome-c, 4 h prior to fixation through the abdominal aorta. An intense lysosomal accumulation of the protein is observed in three dark electron-dense lysosomes . A few grains are seen over the apical endocytic apparatus. Part of the luminal brush border is found in the upper right hand corner. Magnification, x 25 000. Unpublished data from E. I. Christensen, Arhus, Denmark, and M. Haas, Groningen, Netherlands.

5.3.3 Renal Catabolism of LMWP-conjugates

5.3.3.1 Renal Catabolism of Native LMWPs

Morphological (Figure 5.10) and biochemical studies have established that after endocytosis by the proximal tubular cell, LMWPs migrate via endosomes to the proteolytically active lysosomes [78,79]. Within the lysosomes the LMWPs are degraded into small peptides and single amino acids. Whereas the renal uptake rate of various LMWPs appeared to be similar, LMWPs are catabolized with distinct individual differences in their catabolic rate as indicated from the difference in the rate of decline of radioactivity in the kidney (Figure 5.11). The rate of catabolism seemed unrelated to the size or charge of the protein alone [80,81]. Probably multiple structural factors play a role in this process. A crucial factor may be the different endosomal migration times of LMWPs from the tubular lumen to the lysosomes. Whereas cytochrome-c accumulated in the lysosomes within 3 min, lysozyme seemed to migrate for 20 min before the commencement of degradation [72, 82]. Also the intrinsic activity of the reabsorbed protein may play a role. For instance, the long renal half-life of aprotinin, an inhibitor of proteolytic enzymes, may be explained by an inhibition of its own degradation, as suggested by Bianchi [71]. These studies suggest that the LMWP method of renal drug targeting results in cell-selective delivery followed by controlled drug release which can be manipulated at various stages of the renal deposition process. The lysosomes are stacked with a variety of proteolytic enzymes in an acidic environment. Programmed drug release from a drug–carrier conjugate may therefore be achieved using peptide, ester or acid-labile bonds

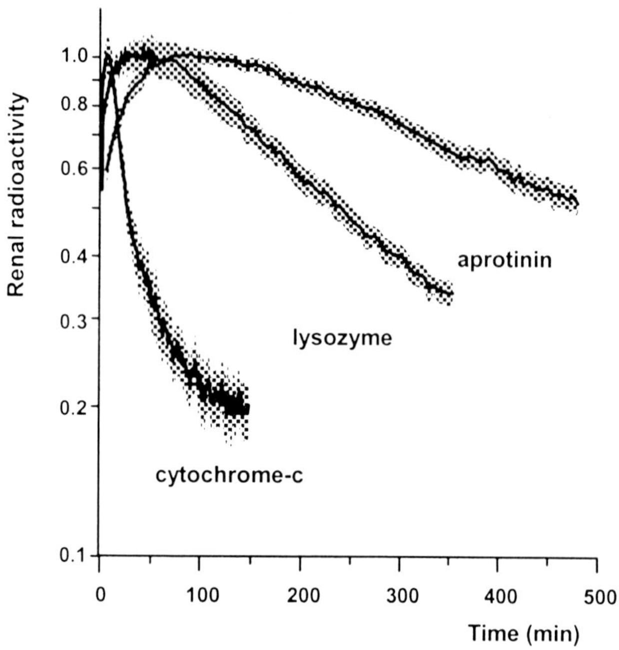

Figure 5.11. Time course of clearance from the kidney of radiolabelled LMWPs after intravenous injection. After renal uptake, the radiolabelled protein is gradually catabolized and the radioactive breakdown products released from the kidney, as shown by the decline of renal radioactivity over time.

between the drug and protein carrier. Consequently both the differences in rate of catabolism between LMWPs as well as the rate of hydrolysis of the bond between the drug and carrier may be used to manipulate the rate of drug release in the kidney. The variable migration times of different LMWPs and their conjugates after endocytosis may have consequences for the intracellular concentration profiles. For instance, in order to achieve relatively constant cellular levels of the drug, an LMWP which is only slowly degraded might be preferred as a drug carrier. In contrast, if short-term peak levels of the drug are preferred, treatment with a rapidly processed protein (with a short migration time) may be a more appropriate choice. Certain drugs (e.g. peptides and nucleotides) should be released before entering lysosomes to prevent inactivation by degradative enzymes. For such drugs, a prolonged endosomal migration time combined with simple hydrolysis of the drug–protein linkage in the acidic environment of the endosomes, will be preferred to achieve adequate drug release and prevent an abortive route to the lysosomes.

5.3.3.2 Renal Catabolism of Naproxen–Lysozyme

The coupling of 2 moles of naproxen to 1 mole of lysozyme did not affect the catabolism of lysozyme in rat kidney [66,75]. After delivery to the kidney, naproxen in the form of naproxen–lysine was gradually released from the conjugate. This catabolite was subsequently eliminated from the kidney and after a single injection, drug levels in the renal tissue gradually decreased with a $t_{1/2}$ of 160 min (Figure 5.9a).

No detectable amounts of naproxen or its lysine conjugates were found in the plasma after administration of the conjugate and it can be inferred that excretion into the urine is the crucial process which determines the elimination rate $t_{1/2}$. The lack of diffusion into the bloodstream is a favourable property in relation to unwanted extra-renal effects.

5.3.3.3 Renal Catabolism of Captopril–Lysozyme

After renal uptake, captopril was rapidly released from the conjugate as indicated by the rapid decrease in renal captopril levels with time (Figure 5.9b). The difference in renal $t_{1/2}$ of naproxen and captopril after delivery with lysozyme is likely to be due to an unequal rate of release from the lysozyme conjugates. Whereas naproxen–lysozyme requires a peptidase for cleavage, captopril is released from the conjugate enzymatically by β-lyase and/or non-enzymaticaly by thiol-disulfide exchange with endogenous thiols. To reduce the rate of captopril–lysozyme breakdown, two different cross-linking reagents, SPDP and SMPT, were tested. Although an SMPT link between two proteins is in principle less susceptible to disulfide reduction [83], no difference in degradation rate was found between the SPDP and the SMPT captopril–lysozyme conjugates (Kok *et al.*, unpublished data).

5.3.4 Effects of Targeted Drugs Using an LMWP as Carrier

5.3.4.1 Renal Effects of Naproxen–Lysozyme

Having obtained promising kinetic profiles, the potential renal effects of naproxen–lysozyme in the rat were investigated [84]. Naproxen, as an inhibitor of cyclooxygenase, blocks prostaglandin synthesis. Among other effects, naproxen reduced furosemide-stimulated urinary excretion of prostaglandin E_2 (PGE_2) as well as the natriuretic and diuretic effects of furosemide. Studies with the conjugate showed that naproxen–lysozyme treatment clearly prevents furosemide-induced excretion of PGE_2. This occurred with a dose of naproxen that was not effective in the unconjugated form. Surprisingly, this effect occurred in the absence of a change in natriuretic and diuretic response to furosemide. In this respect the pharmacological effect differed from treatment with a high dose of free naproxen. An explanation for these differences remains to be found. One possibility is that there is a difference in the intra-renal kinetics of the NSAID compared with the parent drug. Free naproxen is extensively reabsorbed in the distal tubule of the kidney via which route it may effectively inhibit prostaglandin synthesis in the medullary interstitial cells. On the other hand, naproxen–lysine is more hydrophilic and may be unable to reach the sites of prostaglandin synthesis involved in the furosemide-induced excretion of sodium and water. These data shows that renal drug targeting preparations can also be used as a tool to unravel the mechanisms of renal therapeutic effects.

5.3.4.2 Renal and Systemic Effects of Captopril–Lysozyme

With regard to the pharmacological effects of the captopril–lysozyme conjugate, the following observations were made (Kok *et al.*, unpublished data). The extent of ACE-inhibition in the plasma and kidney tissue was measured after i.v. administration of captopril–lysozyme and an equimolar dose of free captopril. It was shown that conjugation to lysozyme caused a similar though more sustained inhibition of renal ACE-activity by captopril. The inhibition of plasma ACE-activity was clearly reduced but not entirely prevented by conjugation of captopril to lysozyme. Possibly, the S-S linked drug conjugate is partly degraded in the circulation. It is also possible that after degradation of the conjugate in the kidney, captopril was transported back into the bloodstream. The rapid intracellular release may provide a sufficient driving force for transport across the basolateral membranes.

Captopril–lysozyme did not significantly affect systemic blood pressure whereas an equimolar dose of captopril alone decreased blood pressure significantly. Whereas free captopril (5 mg kg^{-1}) completely prevented an angiotensin-I-induced blood pressure increase, an equimolar amount of captopril–lysozyme did not. However, in line with the direct ACE activity measurements in renal tissue and plasma, in captopril–lysozyme-treated rats the angiotensin-I-induced blood pressure increase was lower than in untreated rats, suggesting that systemic activity was not fully prevented.

Neither free nor conjugated captopril affected glomerular filtration. Renal plasma flow increased to the same degree after treatment with free or conjugated captopril (1 mg kg^{-1}). Although the complete dose–effect relationship was not studied, we can conclude that conjugation of captopril to lysozyme did not prevent the drug from acting on the renal plasma flow. Whether this effect is determined by intra-renal or systemic ACE-inhibition remains to be investigated.

At present, the synthesis of lysozyme conjugates with ACE-inhibitors other than captopril is under investigation. Some of these ACE-inhibitors may be advantageous for renal delivery. The amount of conjugate required for therapy can be reduced when using an ACE-inhibitor with a higher affinity for ACE (e.g. lisinopril). Furthermore, the stability of the conjugate in plasma may be increased by using an ACE-inhibitor which is conjugated to lysozyme via a linkage that is highly stable in plasma (e.g. lisinopril can in principle be coupled via an acid-sensitive spacer).

5.3.5 Renal Disease and LMWP Processing

Proteinuria is one of the most prominent abnormalities found in renal disease and is one of the factors held responsible for the progressive loss of renal function. As a consequence of the glomerular leakage of proteins, the proximal tubular cells are exposed to increasing amounts of protein. This pathological condition can be anticipated to influence the deposition and metabolism of protein-linked drugs. It is likely, in such a situation, that drug–LMWP conjugates will have to compete with the overload of protein for tubular uptake as well as for catabolism. The effect of proteinuria on the renal processing of LMWPs has been examined in a number of studies [85–92]. Collectively, these studies clearly indicate that the effect of proteinuria on renal uptake and degradation of LMWPs depends on the severity and duration of the protein leakage. However, it should be noted that tubular reabsorption of LMW-Ps is only slightly reduced during adriamycin-induced chronic proteinuria [92]. With respect to LMWP catabolism, the data suggest that protein overload will lead to reduced proteolytic degradation. In that case, an acid labile spacer or a disulfide bond should be chosen to guarantee an adequate rate of drug release.

We found a difference in susceptibility to proteinuria between cationic LMWP cytochrome-c and neutral LMWP myoglobulin with respect to their catabolism. This may indicate that the effect of proteinuria on LMWP catabolism is determined by the proximal tubular segment in which the LMWPs and the protein overload are processed [88,89,93,94]. We speculate that, through coupling to a specific LMWP, drugs can be delivered specifically to those proximal tubular cells that are predominantly affected by proteinuria. This might be essential for drugs chosen to protect the tubular cell from further damage by proteinuria. In addition, it may be possible to use certain LMWPs as drug carriers to circumvent the proteinuria-affected cells. In that case, treatment of diseases unrelated to proteinuria will not be hindered by the severity of proteinuria.

5.3.6 Renal Delivery of High Doses of LMWPs

The renal cell responsible for the uptake of LMWPs is the proximal tubular cell. LMWPs are relatively freely filtered by the glomerulus and subsequently reabsorbed by the proximal tubular cell by megalin/gp330 receptor-mediated endocytosis [95]. In healthy individuals, the relatively moderate amounts of endogenous LMWPs are completely reabsorbed by the proximal tubular cells. However, for drug targeting purposes, larger doses of LMWP may be required. We compared the urinary loss of intact LMWP after intravenous administration of different doses of LMWP by either single dose injections or by continuous infusions in healthy rats. From these studies, we concluded that after a continuous low-dose infusion the non-reabsorbed fraction is considerably less than that after single high-dose injections. However, infusion could not entirely prevent the loss of intact LMWP into the urine (the loss was 8% of the dose after 100 mg lysozyme kg^{-1} over 6 h and rose to 33% following 1000 mg lysozyme kg^{-1} over 6 h).

Cojocel *et al.* demonstrated clear adverse effects after relatively high doses of lysozyme [96]. We studied these aspects in more detail and concluded that lysozyme should be given in a dose of less than 100 mg^{-1} kg^{-1} over 6 h to minimize the negative effects on systemic blood pressure, glomerular filtration and renal blood flow. From these data, it emerged that LMWPs are suitable to serve as drug carriers to the proximal tubular cell of the kidney. However, the conjugate should preferably be administered in low-dose by constant infusion to limit the systemic and renal toxicity and to reduce the urinary loss of the intact conjugate (unpublished data).

5.3.7 Limitations of the LMWP Strategy of Drug Delivery to the Kidney

Among the disadvantages of the LMWP strategy for the treatment of chronic renal disease are the requirement for parenteral administration and the possible immunogenicity of the drug conjugate. With respect to the administration route, the conjugate could possibly be administered subcutaneously or intramuscularly. This is a common administration route for polypeptide drugs such as insulin. If immunogenicity appears to be a serious limitation for chronic treatment, a synthetic polymer may be used as the 'reabsorptive' carrier instead [97,98].

For short-term clinical interventions with the aim of protecting the kidney during acute reperfusion or preventing allograft rejection after transplantation, the prerequisite of parenteral administration does not constitute a serious limitation.

5.4 Renal Delivery of Antisense Oligodeoxynucleotides

5.4.1 Introduction

Various macromolecular and pro-drug technologies designed to achieve selective renal drug accumulation and action have been discussed in the previous sections of this chapter. In these approaches, traditional drugs have been modified through coupling to carrier molecules. It is generally accepted that, at least in theory, antisense oligodeoxynucleotides (AS-ODN) offer a new approach for selective treatment [99,100].

In view of the preferential distribution of some AS-ODNs to the kidney the oligonucleotide backbone could even be employed for renal-specific drug delivery because of both their intrinsic activity and the potential of coupling of other agents to them.

Antisense refers to the use of single-stranded synthetic oligonucleotides to inhibit gene expression [99,100].The striking advantage of the antisense approach in comparison to traditional drugs is its potential for specificity. The binding affinity between the oligonucleotide and its target receptor is many orders of magnitude higher compared to that at other binding sites, as a result of the multiple interaction sites that exist on the target receptors [101]. Since affinity is proportional to the number of interactions between a drug and its receptor, the specificity of an AS-ODN depends on its length. The base pairing specificity of an AS-ODN of about 15-17 nucleotides in length appeared to be sufficient to inhibit only one target gene within the entire human genome [99]. For successful inhibition *in vivo*, the plasma and intracellular stability and the pharmacokinetic profile of the antisense molecule along with the turnover time of the inhibited gene are important determinants.

First, we will briefly review the different aspects that are of importance in the use of antisense for *in vivo* therapy. Second, we will describe the effects of antisense targeting to the proximal tubule of the kidney that have been obtained so far.

5.4.2 Mechanism of Action of Antisense Oligodeoxynucleotides

AS-ODN are designed to be complementary to the coding (sense) sequence of the mRNA in the cell. After hybridization to target sequences, translational arrest occurs via one of several putative mechanisms. The first mechanism is inhibition of transcription. Secondly, AS-ODN can prevent the synthesis of fully mature mRNA in the cytosol at the level of splicing, processing and transport across the nuclear membrane. The third mechanism is inhibition of translation by hybridization of the AS-ODN to the sense sequence and thereby preventing the ribosome from reading the mRNA code. Translation can be inhibited by AS-ODN which bind to important sites for translation such as translation initiation sites, poly(A) signals, and protein-binding regulatory sites. Finally, AS-ODN hybridization to the mRNA initiates specific cleavage of the RNA strand by activated RNase H and this cleavage results in destruction of the coding sequence and inhibition of mRNA translation [101].

5.4.3 Stabilization of Antisense Oligodeoxynucleotides

Phosphodiester AS-ODN are poor candidates for use as therapeutic agents *in vivo* due to their sensitivity to 3′- and 5′- exo/endonucleases. Because of this, various chemical modifications to the oligonucleotide backbone have been introduced to improve enzymatic stability while preserving their ability to hybridize cognate targets. Most common examples include the phosphorothioated and methylphosphonated analogues which have a sulfur atom and a methyl group, respectively, substituted for a non-bridging oxygen atom (Figure 5.12).

Phosphorothioated AS-ODN retain their negatively charged groups in the phosphodiester backbone and have the ability to induce mRNA degradation via RNase H. However, these compounds have a somewhat lower binding affinity to the target sequence. Moreover, non-sequence-specific activity has been reported for phosphorothioated AS-ODN, probably due to their stronger protein binding capacity [102,103].

Unlike phosphorothioates, methylphosphonated AS-ODN are uncharged compounds with a higher cellular uptake than unmodified AS-ODN. Unfortunately, these compounds appeared to be ineffective in some cell lines. This might be explained by the formation of

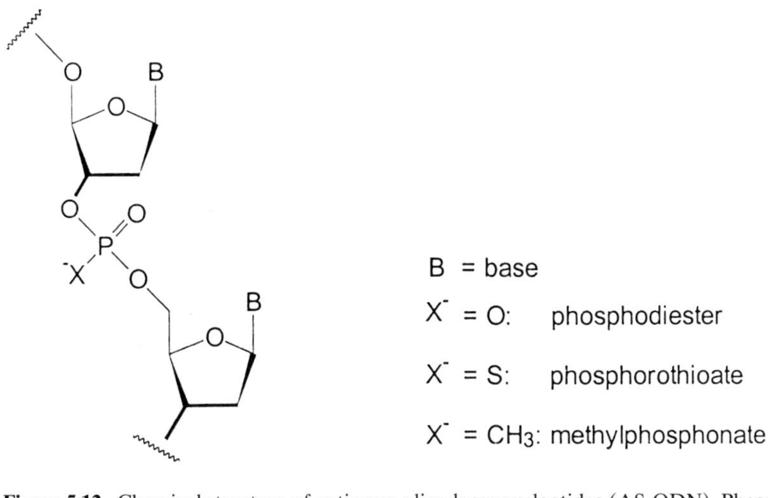

Figure 5.12. Chemical structure of antisense oligodeoxynucleotides (AS-ODN). Phosphorothioate and methylphosphonate AS-ODN have a sulfur atom and a methyl group respectively, substituted for a non-bridging oxygen atom to increase stability to nucleases.

diastereomers or the inability of methylphosphonates to induce mRNA degradation via RNase H [104].

To avoid the problem of chirality and to improve the potency and limit the non-specific actions of AS-ODN, new compounds are required. Synthesis of new AS-ODNs has further improved their nuclease stability, enhanced of cellular uptake and affinity through modification of the base, sugar and phosphate moieties of the oligonucleotides [105–108].

5.4.4 Pharmacokinetic Aspects of Antisense Oligodeoxynucleotides and Renal Distribution

The tissue distribution of AS-ODN after a single intravenous injection has been studied extensively in many species including mouse [109], rat [110], monkey [111] and man [112]. The majority of pharmacokinetic studies have been performed using phosphorothioated AS-ODNs. In general, the pharmacokinetic profiles of AS-ODNs of varying lengths (up to 20mer) and base compositions are remarkably similar in all species.

In plasma, most of the phosphorothioated AS-ODNs are protein bound [113,114]. Cossum and co-workers revealed that albumin and α_2-macroglobulin are responsible for this binding [114]. The protein binding capacity in rat was elevated after administration of doses higher than 15–20 mg kg^{-1} resulting in a dose-dependent increase in distribution volume and an increase in plasma clearance [115–117].

The rapid elimination from plasma following intravenous administration of phosphorothioated AS-ODN can be explained by a two compartment model in all species, i.e. an initial plasma half-life of less than 1 h [111,113,114] and a slower elimination half-life ranging between 20 and 50 h [111,113].

The kidneys and the liver primarily take up phosphorothioated AS-ODN after parenteral administration, accumulating more than 10% each, while the rest of the organs all accumulate less than 1% of the injected dose [110,111,114,116]. It is noteworthy that renal AS-ODN tissue levels exceed that of any other organ [110,113,114], as confirmed by the tissue to plasma ratios of approximately 85 and 20 for kidney and liver, respectively [113,118].

Autoradiographic studies of the kidney have shown the accumulation of AS-ODN to occur almost exclusively in the proximal tubular cells [110,119]. Oberbauer *et al.* reported that intravenously injected AS-ODN accumulated in proximal tubular cells, and electron microscopy revealed that AS-ODN did accumulate only in the brush border or lysosomal compartment. This implies that the AS-ODNs were not completely degraded after being taken up by the proximal tubule [110].

In the last 2 years, several AS-ODNs with modified backbone structures and sugar moieties have been developed and these are characterized by a significanty increased stability in plasma [107,108]. Chimeric AS-ODNs, consisting of a mixture of phosphorothioate and methylphosphonate nucleotides, also exhibited increased stability in plasma [106]. It is worth noting that these AS-ODNs also appeared to be more stable in various tissues including the kidney [106,107]. Agrawal *et al.* [105] and Crooke *et al.* [108] have shown that changes in the sugar moieties can further improve the tissue distribution of AS-ODNs in favour of the kidney.

5.4.5 Cellular Uptake of Antisense Oligonucleotides

Cellular uptake of AS-ODNs is restricted because of their large molecular mass as well as their polyanionic character. When added directly to cells in culture, only 1–2% of the AS-ODNs will be cell-associated. Therefore, enhanced AS-ODN uptake is a critical consideration in developing these agents for therapeutic applications.

The cellular uptake of AS-ODN is an energy-dependent process and takes place in a saturable and sequence-independent manner [120,121]. The exact mechanism of uptake remains controversial. From *in vitro* experiments, some authors have proposed that the uptake is endocytic and mediated by membrane receptor proteins. The receptor responsible for the cellular uptake of AS-ODNs was reported to consist of both a 30-kDa protein [122] and an 80-kDa membrane protein [121]. However, other workers have argued that AS-ODN binding to membrane proteins is relatively non-specific and is mostly charge associated, consistent with adsorptive endocytosis or fluid-phase pinocytosis [101]. As a result of these conflicting reports, it is unlikely that *in vitro* data can be safely extrapolated to what occurs in the intact organism.

In the kidney, AS-ODNs are filtered and subsequently reabsorbed by the proximal tubular cells. The AS-ODNs most likely accumulate in the proximal tubular cells via a receptor-dependent mechanism [110,123]. This hypothesis supports the apparent saturation of AS-ODN uptake in the kidney as reflected by a reduction of degree of renal uptake with increasing AS-ODN dose [110,116,117]. Moreover, Rappaport *et al.* described the existence of 40 and 97-kDa binding proteins for 18mer phosphorothioates in the renal brush border membrane [123]. In another study, a protein with a molecular weight of approximately 50 kDa which may serve as a transmembrane channel transporting AS-ODN into the tubular cell was described [124]. These channels have previously been reported for the uptake of proteins and phage DNA. The presence of such channels might explain why uptake in the proximal tubular cells is dependent on the nucleotide length as was demonstrated by Loke and co-workers [121]. It is noteworthy that scavenger receptors located at the basolateral site may also be responsible for additional tubular accumulation of AS-ODN [125].

5.4.6 Metabolism and Elimination of Antisense Oligodeoxynucleotides

A prerequisite to acquire an antisense effect is the maintenance of AS-ODN within the target cells. Several studies have reported that the majority of phosphorothioated AS-ODNs taken up by the kidney remains intact for several hours [110,113]. In fact, 4 days after administration, 3% of the infused dose was still present in the kidney intactly [110]. Although several studies have confirmed the presence of intact AS-ODN in the kidney, concomitant metabolism in the kidney of 20% after 6 h [113], 50% after 48 h [118,126] and 50% after 4 days [114] has also been reported.

In spite of the improved stability to nucleases, achieved through chemical modification, AS-ODN degradation in plasma still occurs, predominantly from the 3′-terminus. In the liver and kidney, the major sites of metabolism, AS-ODNs are degraded from the 5′-terminus as well [127,128].

Elimination of phosphorothioated AS-ODN takes place primarily via the urine. Approximately 30% of the injected dose is found in the urine within 24 h [110,113]. Althought in most cases only metabolites of AS-ODN could be demonstrated in the urine [110,118], Agrawal and co-workers described the excretion of intact AS-ODN in the urine after a dose of 30 mg kg^{-1} [126]. The saturation of plasma protein binding and proximal tubular uptake could explain this observation [114,116].

Excretion via faeces is a minor route of elimination, accounting for less than 10% of the administered dose [113, 126].

5.4.7 Effects of Antisense Targeting to the Proximal Tubule

Noiri *et al.* used AS-ODN to inhibit production of inducible nitric oxide synthase (iNOS) in an attempt to prevent NO production in an ischaemic kidney. A single intravenous injection of iNOS AS-ODN attenuated acute renal failure and reduced the morphological abnormalities [129].

Oberbauer *et al.* reported inhibition of a sodium/phosphate (Na/Pi-2) co-transporter by phosphorothioated AS-ODN. A single intravenous injection of the AS-ODN inhibited both the mRNA and the protein for the Na/Pi-2 co-transporter, and consequently suppressed luminal uptake of phosphate by the proximal tubules [130].

Wang *et al.* injected a Texas-red-labelled phosphorothioated AS-ODN into the dopamine 1A receptor in the rat renal interstitium. Fluorescence was detected after 24 h in both tubular epithelium and intra-renal vasculature. Treatment resulted in a 35% decrease in the dopamine 1A receptor protein, causing a reduction in urinary sodium excretion and urine output [131].

Rat kidneys were perfused *ex situ* with phosphorothioate intercellular adhesion molecule (ICAM)-1 AS-ODN and exposed to 30 min cold or warm ischaemia. After this time the kidneys were transplanted to syngeneic nephrectomized rats. Treatment with 10 mg antisense reduced the harmful effect of transplantation on renal function [132]

Cheng *et al.* showed that intravenous infusion of intracellular adhesion molecule (ICAM) AS-ODN markedly reduced ICAM-1 expression, alleviated infiltration of inflammatory cells and accumulation of extracellular matrix in the obstructed kidney of mice with unilateral obstruction of the ureter [133].

Repeated intravenous injection of osteopontin AS-ODN to Goodpasture syndrome rats blocked tubular osteopontin expression, attenuated monocyte infiltration and preserved renal plasma flow. No changes were found in osteopontin mRNA level, glomerular histology or proteinuria. The data suggest that interstitial inflammation as a consequence of glomerular disease can be prevented through a selective inhibition of tubular osteopontin expression using osteopontin AS-ODN [134].

Continuous infusion of transforming growth factor-β (TGF-β) AS-ODN in diabetic mice, decreased kidney TGF-β levels and attenuated the increase in kidney weight, and decreased levels of α1(IV)collagen and fibronectin mRNAs.

The above described studies show that the renal proximal tubular cell is a good target for antisense therapy [135].

5.4.8 Benefits and Limitations of Antisense Oligodeoxynucleotides

The introduction of therapy through the delivery of antisense oligodeoxynucleotides holds promise for the treatment of several diseases. It is more specific than conventional drugs while, in contrast with gene targeting, the effect is temporary so that therapy can be terminated when desired. Pharmacokinetic studies have revealed that the proximal tubular cell of the kidney is a suitable target for antisense therapy. Although recent studies have shown antisense oligodeoxynucleotides to be effective in the treatment of renal diseases, antisense targeting is however, a new approach to therapy and all the risks associated with it are not yet known.

5.5 Drugs for Renal Targeting

For the treatment of kidney diseases, several kinds of drugs are currently used. At present, angiotensin-converting enzyme inhibitors are the first choice drugs for the treatment of chronic kidney diseases that are characterized by loss of renal function and proteinuria [136]. These drugs exhibit only moderate side-effects. However, renal targeting of an ACE-inhibitor may improve the therapy in certain cases. For example, when proteinuria is accompanied by normal blood pressure, hypotension due to ACE-inhibition limits the amount of drug that can be given. Renal inflammation such as glomerulonephritis and tubulointerstitial inflammation are treated with corticosteroids [137]. These drugs have serious side-effects. Renal targeting of these drugs may allow a more aggressive treatment of the inflammation. Also, local suppression of the immune system may be useful to prevent transplant rejection. However, it is as yet unknown whether suppression of the local immune system is sufficient or whether the systemic system should also be suppressed to prevent rejection [138]. Renal tumours are characterized by insensitivity to the common anti-tumour drugs [139]. This is probably due to the unfavourable kinetic profile of these drugs. By renal targeting, an anti-tumour drug may reach the renal tumour while the extra-renal side-effects will be reduced.

5.6 *In-Vitro* and *In-Vivo* Models for Renal Targeting

5.6.1 *In-vitro* Models

In the isolated perfused kidney model, the artery of the kidney is perfused and urinary samples as well as venous blood samples can be collected to determine the drug concentration. A serious drawback of the model is that isolation and artificial perfusion greatly affect the function of the organ as shown by a dramatic drop in the glomerular filtration rate. Another *in-vitro* model is the isolated tubule in which samples can be taken from both the luminal and basolateral sites of the tubule [140,141]. The disadvantage of this technique as well as of the isolated kidney model, is that they require specific equipment and expertise and therefore can only be performed in rather specialized laboratories. Experiments using freshly isolated or cultured cells are more simple to carry out [142,143]. Tubular cells can be grown in a polarized fashion enabling the addition of drugs and removal of samples from both the luminal and basolateral sites of the cell.

5.6.2 *In vivo* Models

Obviously, none of the existing animal models of renal diseases are perfect reflections of the human situation. The natural model of progressive loss of renal function is the 5/6 nephrectomy. Drawbacks of this model are the large wound in the remaining kidney and the limited amount of tissue available for analysis. The two-kidney, one-clip Goldblatt model is a good model of renal vascular hypertension. Progressive loss of renal function due to essential hypertension can be studied using the spontaneously hypertensive rat (SHR). Several animal models for diabetic nephropathy exist [144]. Streptozotocin induces diabetes, resulting in a mild proteinuria and tubular dysfunction during the progression of the disease [145]. Also, animal models of spontaneous diabetes have been described [146]. The diabetic nephropathy that develops in these models is likely to be a good reflection of the human situation since it is a consequence of the same initial disease.

Several models of toxic nephritis have been developed. After an intravenous injection of adriamycin or puromycin, a chronic nephropathy develops which is characterized by severe proteinuria and glomerular sclerosis [147,148]. The severity of the proteinuria is much higher compared to human proteinuria, while the reduction in the glomerular filtration rate is limited. The progression of proteinuria after puromycin injection occurs in two phases, while adriamycin causes a gradual continuous increase in proteinuria. Overload proteinuria is a model in which bovine albumin is repeatedly injected into rats in large quantities [149]. The proteinuria is less aggressive than in the adriamycin and puromycin models and the model seems to be a better reflection of the human situation. However, this model is more difficult to set up. Several toxic agents, like cyclosporin and cadmium, accumulate in the proximal tubular cell, causing severe tubular damage. These models are a good reflection of tubular damage by toxic agents in humans.

For glomerulonephritis, several immunological models are available [150]. For example, injection of an antibody against thymocytes (anti-Thy 1.1 nephritis) causes a rapid mesangiolysis followed by proliferation [151,152].

In addition to other models [153], tubulointerstitial inflammation and fibrosis can be obtained by ureter obstruction. The inflammation develops very rapidly and is severe. The model is a good reflection of ureter obstruction in humans. However, a serious drawback in using this model for tubular drug delivery studies is the fact that glomerular filtration is absent.

5.7 Concluding Remarks

In this chapter, macromolecular and pro-drug approaches for cell-selective therapeutic intervention in the proximal tubular cell have been described. Using a low-molecular weight protein as a drug carrier, the drug is delivered to the lysosomes of the proximal tubular cell via reabsorption from the luminal site. Lysosomal delivery allows drug attachment via an acid-sensitive spacer or via biodegradable peptide or ester moieties. Using the alkylglycoside vector as a drug carrier, the drug is taken up via the basolateral site into the proximal tubular cell. It is as yet unknown to which compartment of the proximal tubular cell the drug is delivered using this carrier, and the subsequent stages such as drug release as well as the kinetics and dynamics during renal diseases remain to be studied. Yet, a basolateral delivery may be advantageous during severe reduction of glomerular filtration and presence of proteinuria. On the other hand, with low-molecular weight proteins a broader range of drugs (with respect to their physicochemical properties) can be delivered to the proximal tubular cells.

Oligonucleotide targeting to the kidney is more feasible than to many other tissues as a result of the glomerular filtration and tubular reabsorption of these poly-anionic agents. The effect is temporary allowing the therapy to be terminated when desired. Up until now, data has only been available on the kinetics and some renal and extra-renal effects of oligonucleotides in healthy animals.

The most relevant studies examining the effects of drug targeting in experimental disease, are yet to come. These studies may provide clues to the role of the proximal tubular cell in the various renal diseases and may determine whether treatment of renal disease can be accomplished by drug targeting to the proximal tubular cell. A further goal of renal targeting is the specific delivery of drugs to the filtration unit of the kidney, the glomerulus, which is also believed to play an important role in the progression of renal disease. Until recently only limited research has been focused on this target in the kidney [76].

References

[1] Ito S, *Pediatr. Nephrol.* **1999,** *13*, 980–988.
[2] Pritchard JB, Miller DS, *Am. J. Physiol.* **1991,** *261*, R1329–R1340.
[3] Welbourne TC, Matthews JC, *Am. J. Physiol.* **1999,** *277*, F501–F505.
[4] Suzuki K, Susaki H, Okuno S, Yamada H, Watanabe HK, Sugiyama Y, *J. Pharmacol. Exp. Ther.* **1999,** *288*, 888–897.
[5] Orlando RA, Rader K, Authier F, Yamazaki H, Posner BI, Bergeron JJ, Farquhar MG, *J. Am. Soc. Nephrol.* **1998,** *9*, 1759–1766.
[6] Hoffman BB, Ziyadeh FN, *Curr. Opin. Nephrol. Hypertens.* **1995,** *4*, 406–411.
[7] Murer H, *Contrib. Nephrol.* **1982,** *33*, 14–28.
[8] Adibi SA, *Am. J. Physiol.* **1997,** *272*, E723–E736.
[9] Maesaka JK, Fishbane S, *Am. J. Kidney Dis.* **1998,** *32*, 917–933.
[10] Christensen EI, Birn H, Verroust P, Moestrup SK, *Int. Rev. Cytol.* **1998,** *180*, 237–284.
[11] Leung S, Bendayan R, *Can. J. Physiol. Pharmacol.* **1999,** *77*, 625–630
[12] Birn H, Vorum H, Verroust PJ, Moestrup SK, Christensen EI, *J. Am. Soc. Nephrol.* **2000,** *11*, 191–202.
[13] van Ginneken CA, Russel FG, *Clin. Pharmacokin.* **1989,** *16*, 38–54.
[14] Moolenaar F, Cancrinius S, Visser J, de Zeeuw D, Meijer DKF, *Pharm. Weekbl. (Sci).* **1992,** *14*, 191–195.
[15] Pacifici GM, Viani A, Franchi M, Gervasi PG, Longo V, Di Simplicio P, Temellini A, Romiti P, Santerini S, Vanucci L, Mosca F, *Pharmacology* **1989,** *39*, 299–308.
[16] Minard FN, Grant DS, Cain JC, Jones PH, Kyncl J, *Biochem. Pharmacol.* **1980,** *29*, 69–75.
[17] Haga HJ, *Int. J. Biochem.* **1989,** *21*, 343–345.
[18] Kitamura M, Fine LG, *Kidney Int.* **1999,** *55*, 1639–1671.
[19] Vleming LJ, Bruijn JA, van Es LA, *Neth. J. Med.* **1999,** *54*, 114–128.
[20] Johnson RJ, *Kidney Int.* **1997,** *63 (Suppl.),* S2–S6.

[21] Manil L, Davin JC, Duchenne C, Kubiak C, Foidart J, Couvreur P, Mahieu P, *Pharm. Res.* **1994,** *11*, 1160–1165.
[22] Kitamura M, Taylor S, Unwin R, Burton S, Shimizu F, Fine LG, *J. Clin. Invest.* **1994,** *94*, 497–505.
[23] Suzuki K, Susaki H, Okuno S, Yamada H, Watanabe HK, Sugiyama Y, *J. Pharmacol. Exp. Ther.* **1999,** *288*, 888–897.
[24] Suzuki K, Susaki H, Okuno S, Sugiyama Y, *J. Pharmacol. Exp. Ther.* **1999,** *288*, 57–64.
[25] Suzuki K, Ando T, Susaki H, Mimori K, Nakabayashi S, Sugiyama Y, *Pharm. Res.* **1999,** *16*, 1026–1034.
[26] Varki A, *Glycobiology.* **1993,** *3*, 97–130.
[27] Ashwell G, Harford J, *Annu. Rev. Biochem.* **1982,** *51*, 531–554.
[28] Takakura Y, Hashida M, *Pharm. Res.* **1996,** *13*, 820–831.
[29] Kato Y, Sugiyama Y, *Crit. Rev. Ther. Drug Carrier Syst.* **1997,** *14*, 287–331.
[30] Tsuchiya T, Saito S, *J. Biochem (Tokyo)* **1984,** *96*, 1593–1597.
[31] Sweet DH, Pritchard JB, *Cell. Biochem. Biophys.* **1999,** *31*, 89–118.
[32] Endou H, *Toxicol. Lett.* **1998,** *102–103*, 29–33.
[33] Hediger MA, Rhoads DB, *Physiol. Rev.* **1994,** *74*, 993–1026.
[34] Thorens B, *Am. J. Physiol.* **1996,** *270*, G541–G553.
[35] Matsumoto I, Kitagaki H, Iida N, Saito Y, Seno N, *Carbohydr. Res.* **1986,** *151*, 261–270.
[36] Liu KX, Kato Y, Kino I, Nakamura T, Sugiyama Y, *Am. J. Physiol.* **1998,** *275*, E835–E842.
[37] Watanabe Y, Suzuki H, Suzuki K, Ando T, Nakabayashi S, Sugiyama Y, *Pharm. Res.* **2000,** *17*, 49–54.
[38] Haas M, De Zeeuw D, van Zanten A, Meijer DKF, *Kidney Int.* **1993,** *43*, 949–954.
[39] Franssen EJF, Moolenaar F, De Zeeuw D, Meijer DKF, *Adv. Drug Deliv. Rev.* **1994,** *14*, 67–88.
[40] Susaki H, Suzuki K, Ikeda M, Yamada H, Watanabe HK, *Chem. Pharm. Bull. (Tokyo)* **1994,** *42*, 2090–2096.
[41] Susaki H, Suzuki K, Ikeda M, Yamada H, Watanabe HK, *Chem. Pharm. Bull. (Tokyo)* **1998,** *46*, 1530–1537.
[42] Lee MR, *J. Auton. Pharmacol.* **1990,** *10*, s103–s108.
[43] Barthelmebs M, Caillette A, Ehrhardt JD, Velly J, Imbs JL, *Kidney Int.* **1990,** *37*, 1414–1422.
[44] Drieman JC, van Kan FJ, Thijssen HH, van Essen H, Smits JF, Struyker-Boudier HA, Br. *J. Pharmacol.* **1994,** *111*, 1117–1122.
[45] Minard FN, Grant DS, Cain JC, Jones PH, *Kyncl J, Biochem. Pharmacol.* **1980,** *29*, 69–75.
[46] Orlowski M, Mizoguchi H, Wilk S, *J. Pharmacol. Exp. Ther.* **1980,** *212,* 167–172.
[47] Drieman JC, Thijssen HH, Struyker-Boudier HA, *J. Pharmacol. Exp. Ther.* **1990,** *252*, 1255–1260.
[48] Drieman JC, Thijssen HH, *J. Pharmacol. Exp. Ther.* **1991,** *259*, 766–771.
[49] Drieman JC, Thijssen HH, Zeegers HH, Smits JF, Struyker-Boudier HA, *Br. J. Pharmacol.* **1990,** *99*, 15–20.
[50] Elfarra AA, Hwang IY, *Drug Metab. Dispos.* **1993,** *21*, 841–845.
[51] Casagrande C, Merlo L, Ferrini R, Miragoli G, Semeraro C, *J. Cardiovasc. Pharmacol.* **1989,** *14,* S40–S59.
[52] Hofbauer KG, Sonnenburg C, Stalder R, Criscione L, Kraetz J, Fuhrer W, Habicht E, *J. Pharmacol. Exp. Ther.* **1985,** *232*, 838–844.
[53] Selhub J, Nakamura S, Carone FA, *Am. J. Physiol.* **1987,** *252*, F757–F760.
[54] Christensen EI, Birn H, Verroust P, Moestrup SK, *Int. Rev. Cytol.* **1998,** *180*, 237–284.
[55] Birn H, Selhub J, Christensen EI, *Am. J. Physiol.* **1993,** *264*, C302–C310.
[56] Morshed KM, Ross DM, McMartin KE, *J. Nutr.* **1997,** *127*, 1137–1147.
[57] Mathias CJ, Wang S, Waters DJ, Low PS, Green MA, *J. Nucl. Med.* **1996,** *37*, 347P–348P.
[58] Turner FB, Andreassi JL, Ferguson J, Titus S, Tse A, Taylor SM, Moran RG, *Cancer Res.* **1999,** *59*, 6074–6079.
[59] Wang S, Low PS, *J. Control. Release* **1998,** *53,* 39–48.
[60] Leamon CP Low PS, *J. Drug Target.* **1994,** *2*, 101–112.
[61] Lee RJ, Low PS, *J. Biol. Chem.* **1994,** *269*, 3198–3204.
[62] Wang S, Lee RJ, Cauchon G, Gorenstein DG, Low PS, *Proc. Natl Acad. Sci. USA.* **1995,** *92*, 3318–3322.
[63] Wang S, Luo J, Lantrip DA, Waters DJ, Mathias CJ, Green MA, Fuchs PL, Low PS, *Bioconjug. Chem.* **1997,** *8*, 673–679.
[64] Clifford AJ, Heid MK, Muller HG, Bills ND, *J. Nutr.* **1990,** *120*, 1633–1639.
[65] Franssen EJF, Koiter J, Kuipers CAM, Bruins AP, Moolenaar F, De Zeeuw D, Kruizinga WH, Kellogg RM, Meijer DKF, *J. Med. Chem.* **1992,** *35*, 1246–1259.

[66] Franssen EJF, Moolenaar F, De Zeeuw D, Meijer DKF, *Pharm. Res.* **1993,** *10*, 963–969.
[67] Franssen EJF, van Amsterdam RGM, Visser J, Moolenaar F, De Zeeuw D, Meijer DKF, *Pharm. Res.* **1991,** *8*, 1223–1230.
[68] Bertani T, Poggi A, Pozzoni R, Delaini F, Sacchi G, Thoua Y, Mecca G, Remuzzi G, Donati MB, *Lab. Invest* **1982,** *46*, 16–23.
[69] Kau ST, Maack T, *Am. J. Physiol.* **1977,** *233*, F445–F454.
[70] Wallace AL, Stacy BD, Thorburn GD, *Pflugers Arch.* **1972,** *331*, 25–37.
[71] Bianchi C, Donadio C, Tramonti G, Auner I, Lorusso P, Deleide G, Lunghi F, Salvadori P, *Contrib. Nephrol.* **1988,** *68*, 37–44.
[72] Hysing J, Tolleshaug H, Curthoys NP, *Acta Physiol. Scand.* **1990,** *140*, 419–427.
[73] Vriesendorp R, Donker AJ, De Zeeuw D, De Jong PE, van der Hem GK, Brentjens JR, *Am. J. Med.* **1986,** *81*, 84–94.
[74] Usberti M, Pecoraro C, Federico S, Cianciaruso B, Guida B, Romano A, Grumetto L, Carbonaro L, *Pediatrics* **1985,** *75*, 501–507.
[75] Haas M, Kluppel ACA, Wartna ES, Moolenaar F, Meijer DKF, De Jong PE, De Zeeuw D, *Kidney Int.* **1997,** *52*, 1693–1699.
[76] Haas M, Meijer DKF, Moolenaar F, De Jong PE, De Zeeuw D, In: *International Yearbook of Nephrology 1996* (Eds Andreucci VE, Fine LG), pp. 3–11. Oxford University Press, Oxford, **1996.**
[77] Kok RJ, Grijpstra F, Walthuis RB, Moolenaar F, De Zeeuw D, Meijer DKF, *J. Pharmacol. Exp. Ther.* **1999,** *288*, 281–285.
[78] Maack T, Park CH, Camargo MJF, In: *The Kidney, Physiology and Pathophysiology* (Eds Seldin DW, Giebisch G), pp. 1773–1803. Raven Press, New York, **1985.**
[79] Christensen EI, Nielsen S, *Semin. Nephrol.* **1991,** *11*, 414–439.
[80] Haas M, De Zeeuw D, Meijer DKF, *Contrib. Nephrol.* **1993,** *101*, 78–84.
[81] Bianchi C, Donadio C, Tramonti G, Auner I, Lorusso P, Deleide G, Lunghi F, Vannucci C, Vitali S, Ricchiuti V, *Contrib. Nephrol.* **1990,** *83*, 39–46.
[82] Christensen EI, *Kidney Int.* **1976,** *10*, 301–310.
[83] Thorpe PE, Wallace PM, Knowles PP, Relf MG, Brown AN, Watson GJ, Knyba RE, Wawrzynczak EJ, Blakey DC, *Cancer Res.* **1987,** *47*, 5924–5931.
[84] Haas M, Moolenaar F, Meijer DKF, De Jong PE, De Zeeuw D, Clin. Sci. 1998, 95, 603–609.
[85] Thielemans N, Lauwerys R, Bernard A, *Nephron* **1994,** *66*, 453–458.
[86] Rustom R, Grime JS, Maltby P, Stockdale HR, Jackson MJ, Critchley M, Bone JM, *Clin. Sci.* **1993,** *85*, 733–736.
[87] Rustom R, Jackson MJ, Critchley M, Bone JM, *Miner. Electrolyte. Metab* **1992,** *18*, 108–112.
[88] Olbricht CJ, Cannon JK, Tisher CC, *Kidney Int.* **1987,** *32*, 354–361.
[89] Eisenberger U, Fels LM, Olbricht CJ, Stolte H, *Renal Physiol. Biochem.* **1995,** *18*, 89–96.
[90] Olbricht CJ, Cannon JK, Garg LC, Tisher CC, *Am. J. Physiol.* **1986,** *250*, F1055–F1062.
[91] Schaefer L, Gilge U, Heidland A, Schaefer RM, *Kidney Int.* **1994,** *47 (Suppl.)*, S64–S67.
[92] Haas M, De Boer E., Navis G., Meijer D. K. F., De Jong P. E., De Zeeuw D. *J. Am. Soc. Nephrol.* **1997,** *8*, A2876.
[93] Olbricht CJ, Gutjahr E, Cannon JK, Irmler H, Tisher CC, *Kidney Int.* **1990,** *37*, 918–926.
[94] Tojo A, Endou H, *Am. J. Physiol.* **1992,** *263*, F601–F606.
[95] Cui S, Verroust PJ, Moestrup SK, Christensen EI, *Am. J. Physiol.* **1996,** *271*, F900–F907.
[96] Cojocel C, Franzen-Sieveking M, Berndt W, Baumann K, *Pflugers. Arch.* **1984,** *402*, 34–38.
[97] Haas M, De Boer E, Moolenaar F, Navis G, Meijer DKF, De Jong PE, De Zeeuw D, *J. Am. Soc. Nephrol.* **1997,** *8*, A2877.
[98] van Kooten C, Langers AM, Bruijn JA, Daha MR, *Kidney Blood Press. Res.* **1999,** *22*, 53–61.
[99] Stein CA, Cheng YC, *Science* **1993,** *261*, 1004–1012.
[100] Wagner RW, *Nature* **1994,** *372*, 333–335.
[101] Rojanasakul Y, *Adv. Drug. Deliv. Rev.* **1996,** *18*, 115–131.
[102] Agrawal S, Zhao Q, *Curr. Opin. Chem. Biol.* **1998,** *2*, 519–528.
[103] Lipkowitz MS, Klotman ME, Bruggeman LA, Nicklin P, Hanss B, Rappaport J, Klotman PE, *Am. J. Kidney Dis.* **1996,** *28*, 475–492.
[104] Hélène C, Toulmé J-J, *Biochim. Biophys. Acta* **1990,** *1049*, 99–125.
[105] Agrawal S, Jiang Z, Zhao Q, Shaw D, Cai Q, Roskey A, Channavajjala L, Saxinger C, Zhang R, *Proc. Natl Acad. Sci. USA.* **1997,** *94*, 2620–2625.
[106] Zhang R, Iyer RP, Yu D, Tan W, Zhang X, Lu Z, Zhao H, Agrawal S, *J. Pharmacol. Exp. Ther.* **1996,** *278*, 971–979.

[107] Zhang R, Lu Z, Zhao H, Zhang X, Diasio RB, Habus I, Jiang Z, Iyer RP, Yu D, Agrawal S, *Biochem. Pharmacol.* **1995,** *50*, 545–556.
[108] Crooke ST, Graham MJ, Zuckerman JE, Brooks D, Conklin BS, Cummins LL, Greig MJ, Guinosso CJ, Kornbrust D, Manoharan M, Sasmor HM, Schleich T, Tivel KL, Griffey RH, *J. Pharmacol. Exp. Ther.* **1996,** *277*, 923–937.
[109] Agrawal S, Temsamani J, Galbraith W, Tang J, *Clin. Pharmacokinet.* **1995,** *28*, 7–16.
[110] Oberbauer R, Schreiner GF, Meyer TW, *Kidney Int.* **1995,** *48*, 1226–1232.
[111] Grindel JM, Musick TJ, Jiang Z, Roskey A, Agrawal S, *Antisense Nucleic Acid Drug Dev.* **1998,** *8*, 43–52.
[112] Maeshima Y, Kashihara N, Sugiyama H, Sekikawa T, Okamoto K, Kanao K, Morita Y, Yamasaki Y, Makino H, Ota Z, *J. Am. Soc. Nephrol.* **1994,** *5*, 835–835.
[113] Raynaud FI, Orr RM, Goddard PM, Lacey HA, Lancashire H, Judson IR, Beck T, Bryan B, Cotter FE, *J. Pharmacol. Exp. Ther.* **1997,** *281*, 420–427.
[114] Cossum PA, Truong L, Owens SR, Markham PM, Shea JP, Crooke ST, *J. Pharmacol. Exp. Ther.* **1993,** *267*, 1181–1190.
[115] Nicklin PL, Craig SJ, Phillips JA, In: *Antisense Research and Application* (Ed. Crooke ST), pp. 141–168. Springer Verlag, Berlin, **1998.**
[116] Geary RS, Leeds JM, Henry SP, Monteith DK, Levin AA, *Anticancer Drug Des.* **1997,** *12*, 383–393.
[117] Rifai A, Brysch W, Fadden K, Clark J, Schlingensiepen KH, *Am. J. Pathol.* **1996,** *149*, 717–725.
[118] Sands H, Gorey Feret LJ, Cocuzza AJ, Hobbs FW, Chidester D, Trainor GL, *Mol. Pharmacol.* **1994,** *45*, 932–943.
[119] Carome MA, Kang YH, Bohen EM, Nicholson DE, Carr FE, Kiandoli LC, Brummel SE, Yuan CM, *Nephron* **1997,** *75*, 82–87.
[120] Stein CA, Mori K, Loke SL, Subasinghe C, Cohen JS, Neckers LM, *Gene* **1988,** *72*, 333–341.
[121] Loke SL, Stein CA, Zhang XH, Mori K, Nakanishi M, Subasinghe C, Cohen JS, Neckers LM, *Proc. Natl Acad. Sci. USA.* **1989,** *86*, 3474–3484.
[122] Bennett RM, Gabor GT, Morritt MJ, *J. Clin. Invest.* **1985,** *76*, 2182–2190.
[123] Rappaport J, Hanss B, Kopp JB, Copeland TD, Bruggeman LA, Coffman TM, Klotman PE, *Kidney Int.* **1995,** *47*, 1462–1469.
[124] Hanss B, Leal-Pinto E, and Klotman PE, *J. Am. Soc. Nephrol.* **1994,** *6*, 362–362.
[125] Sawai K, Mahato RI, Oka Y, Takakura Y, Hashida M, *J. Pharmacol. Exp. Ther.* **1996,** *279*, 284–290.
[126] Agrawal S, Temsamani J, Tang JY, *Proc. Natl Acad. Sci. USA* **1991,** *88*, 7595–7599.
[127] Temsamani J, Tang JY, Padmapriya A, Kubert M, Agrawal S, *Antisense Res. Dev.* **1993,** *3*, 277–284.
[128] Temsamani J, Roskey A, Chaix C, Agrawal S, *Antisense Nucleic Acid Drug Dev.* **1997,** *7*, 159–165.
[129] Noiri E, Peresleni T, Miller F, Goligorsky MS, *J. Clin. Invest.* **1996,** *97*, 2377–2383.
[130] Oberbauer R, Schreiner GF, Biber J, Murer H, Meyer TW, *Proc. Natl Acad. Sci. USA* **1996,** *93*, 4903–4906.
[131] Wang ZQ, Felder RA, Carey RM, *Hypertension* **1999,** *33*, 504–510.
[132] Chen W, Bennett CF, Wang ME, Dragun D, Tian L, Stecker K, Clark JH, Kahan BD, Stepkowski SM, *Transplantation* **1999,** *68*, 880–887.
[133] Cheng QL, Chen XM, Li F, Lin HL, Ye YZ, Fu B, *Kidney Int.* **2000,** *57*, 183–190.
[134] Okada H, Moriwaki K, Kalluri R, Takenaka T, Imai H, Ban S, Takahama M, Suzuki H, *Am. J. Physiol.* **2000,** *278*, F110–F121.
[135] Han DC, Hoffman BB, Hong SW, Guo J, Ziyadeh FN, *Am. J. Physiol.* **2000,** *278*, F628–F634.
[136] Weir MR, *Am. J. Hypertens.* **1999,** *12*, 195S–203S.
[137] Briggs WA, Gimenez LF, Samaniego-Picota M, Choi MJ, Nadasdy T, Eustace J, Scheel PJ, Jr, *J. Clin. Pharmacol.* **2000,** *40*, 115–123.
[138] Gruber SA, *Immunol. Rev.* **1992,** *129*, 5–30.
[139] Motzer RJ, Russo P, *J. Urol.* **2000,** *163*, 408–417.
[140] Edwards RM, Stack EJ, Trizna W, *J. Pharmacol. Exp. Ther.* **1999,** *290*, 38–42.
[141] Muller-Berger S, Nesterov VV, Fromter E, *Pflugers Arch.* **1997,** *434,* 383–391.
[142] Takakura Y, Morita T, Fujikawa M, Hayashi M, Sezaki H, Hashida M, Borchardt RT, *Pharm. Res.* **1995,** *12*, 1968–1972.
[143] Baer PC, Tunn UW, Nunez G, Scherberich JE, Geiger H, *Exp. Nephrol.* **1999,** *7*, 306–313.
[144] Janssen U, Phillips AO, Floege J, *J. Nephrol.* **1999,** *12*, 159–172.

[145] Van Zwieten PA, Kam KL, Pijl AJ, Hendriks MG, Beenen OH, Pfaffendorf M, *Pharmacol. Res.* **1996,** *33*, 95–105.
[146] Velasquez MT, Kimmel PL, Michaelis OE, *FASEB J.* **1990,** *4*, 2850–2859.
[147] Wapstra FH, Van Goor H, Navis G, De Jong PE, de Zeeuw D, *Clin. Sci.* **1996,** *90*, 393–401.
[148] Marinides GN, Groggel GC, Cohen AH, Border WA, *Kidney Int.* **1990,** *37*, 749–757.
[149] Rollason TP, Brewer DB, *J. Pathol.* **1981,** *134*, 39–56.
[150] Kluth DC, Rees AJ, *J. Nephrol.* **1999,** *12*, 66–75.
[151] Bagchus WM, Hoedemaeker PJ, Rozing J, Bakker WW, *Lab. Invest.* **1986,** *55*, 680–687.
[152] Jefferson JA, Johnson RJ, *J. Nephrol.* **1999,** *12*, 297–307.
[153] Wilson CB, *Kidney Int.* **1989,** *35*, 938–953.
[154] Jungquiera LC, Carneiro J, Coutopoulos AN, In: *Basic Histology,* pp. 354–371. Lange Medical Publications, Los Altos, **1975.**

6 A Practical Approach in the Design of Colon-specific Drug Delivery Systems

Claudia S. Leopold

6.1 Introduction

Drug delivery to the large intestine has become attractive to researchers whose main interest is the treatment of colonic disorders and the delivery of peptide drugs to the colon. In contrast to colon-specific drug delivery, drug delivery to the small intestine can be easily achieved by using enteric coating polymers that are soluble in the neutral environment of the small intestine. The development and the design of colon-specific drug formulations represents a technological challenge as these dosage forms must pass through the upper gastrointestinal (GI) tract in intact form before delivering the drug to the colon.

Colon-specific drug delivery does not appear to make much sense at first because of the small area of absorption and the strong barrier properties of the colonic epithelium. However, the colon has some unique features, which make this organ attractive for site-specific drug delivery. On the one hand, the peptidase activity in the large intestine is significantly lower than that in the stomach and the small intestine and the colonic transit time is much longer than that of the upper GI tract. This allows the delivery of unstable peptide drugs and drugs with a low permeability to this lower intestinal region. On the other hand, the topical treatment of colonic disorders may lead to the reduction of both drug dose and side effects.

There are currently four strategies that are pursued to achieve colon specificity: first, by relying on the pH difference between the small and the large intestine (pH-controlled drug release); second, by exploiting the enzymatic activity of the colonic microflora (enzyme-controlled drug release); third, by relying on the relatively constant small intestinal transit time (time-controlled drug release) and fourth, by taking advantage of the increase in the luminal pressure in the colon due to strong peristaltic waves (pressure-controlled drug release). This chapter gives an overview of the delivery concepts in colon-specific drug delivery which are currently employed (see also Table 6.1).

6.2 Physiological Characteristics of the Colon

Drug delivery to the colon has become attractive to researchers interested in the delivery of peptide drugs to the large intestine and the topical treatment of colonic disorders. Because of the unique physiological characteristics of the large intestine, drug delivery to the colon can be achieved in different ways. One such feature is the colonic microflora (bacterial count: 10^{11}–10^{12} cfu ml^{-1}), which consist mainly of anaerobic or facultative anaerobic microorganisms [1] (Figure 6.1) that produce a variety of enzymes [2]. The ability of the colon to support

Table 6.1. Overview of potential possibilities for achieving drug delivery to the colon.

Delivery method	Principle	Achieved with	Examples
pH-controlled drug release	Difference in pH between the small and large intestine	pH-dependent dissolution of polymeric coatings	Enteric coatings, basic polymers
		pH-dependent polymer swelling of hydrogels	Acrylic polymers
		pH-dependent drug release from drug/ion exchange resin complexes	Insulin + gelatin B Olsalazine + anion exchange resin
Enzyme-controlled drug release	Degradation of dosage form components by the enzymes of the colonic microflora	Degradable pro-drugs	Mono-, oligo- or polymers with degradable drug–carrier bonds
		Coating materials with degradable bonds including capsule shells	Azo polymers, polymers with glycosidic bonds
		Hydrogels and matrices consisting of cross-linked, degradable polymers	Cross-linked guar, pectin, dextran, inulin, azo polymers
		Sustained release coating materials with degradable domains (pore formers)	Ethylcellulose or Eudragit® RS with galactomannans, β-cyclodextrin, glassy amylose, inulin
Time-controlled drug release	Relatively constant transit time in the small intestine of about 3 h	Time-dependent swellable polymers	Cellulose ethers, Eudragit® sustained-release coatings
		Slow build-up of an osmotic pressure in the dosage form	COER-24™
		Polymer layers with time-dependent erosion or dissolution	Cellulose ethers; Eudragit® E, chitosan (in combination with an acid in the dosage form)
Pressure-controlled drug release	Disintegration of the dosage form in the colon by intra-luminal pressure resulting from strong peristaltic waves	Thick coating consisting of water-insoluble, non-swellable polymers	Hard gelatin capsule with inner ethylcellulose coating

anaerobic bacterial flora is shown by its redox potential in the range of – 250 and – 480 mV [3,4]. Further characteristics are the slightly acidic environment in the proximal colon (pH 6.0–6.4), which results from the degradation of poly- and oligosaccharides to short-chain fatty acids, and a neutral or slightly alkaline environment in the distal colon (pH 7.0–7.4) [5] (Figure 6.2). Moreover, as a result of the strong and prolonged propulsive motility in the distal colon that occurs once or twice a day, the luminal pressure and thus the potentially destructive forces increase temporarily in this lower part of the large intestine, where solid faeces are formed [6]. Drug absorption from the colon is affected by the small effective surface area available for absorption and the tightly packed colon epithelium; however, colonic tran-

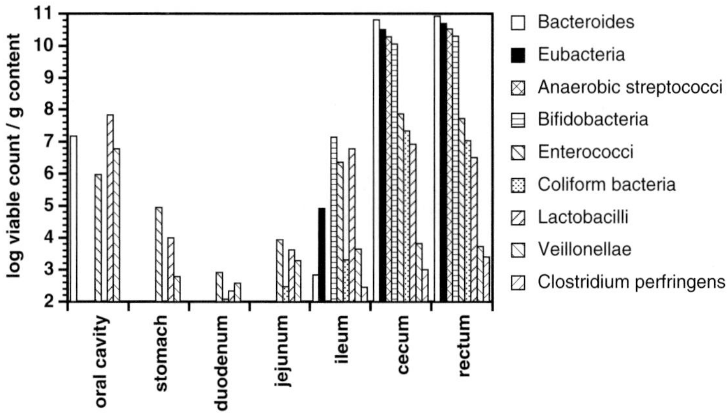

Figure 6.1. Bacterial flora of the human GI tract. Modified from reference [4].

sit time can last for up to 78 h, thereby allowing the absorption of drugs of even low permeability such as peptides.

6.3 Pathological Processes in the Colon

Ulcerative or inflammatory lesions may affect the physiology of the small and large intestine. Ulcer formation entails a circumscribed loss of tissue from the surface of an organ, which results from necrosis following cell destruction by chemicals and the like, or by restriction of the blood supply. Ulcers are among the most common and important lesions. Those that do not penetrate the muscularis mucosa are called erosions. Ulcerative conditions in humans must be differentiated from malignant ulcers, which are associated with neoplasia. Among

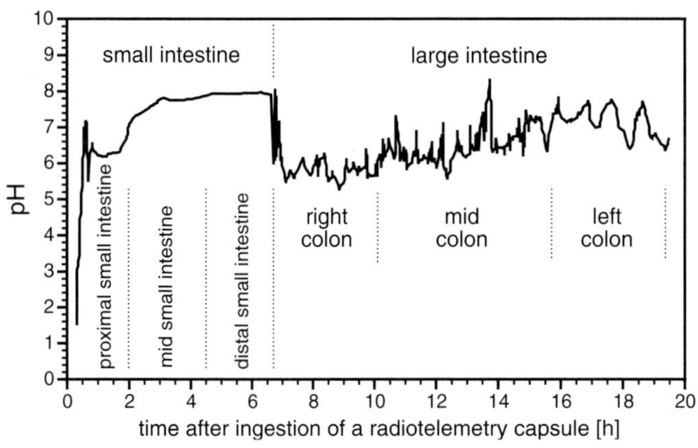

Figure 6.2. pH profile in the GI tract of a healthy subject, measured with a radiotelemetry capsule. Modified from reference [5].

the inflammatory bowel diseases of humans are regional enteritis, or Crohn's disease, and chronic ulcerative colitis [7]. These diseases are primarily treated with mesalazine, various corticosteroids and immunosuppressants.

Crohn's disease is granulomatous and in most cases it is a simultaneous disease of the ileum and colon. The primarily inflamed region is the distal ileum, and all intestinal layers are thickened. The mucosal surface is reddened, nodular, and cobblestone-like, with multiple linear ulcerations. The mucosal layer is thickened by inflammatory infiltrate, the submucosa and serosa by fibrosis, and the serosa by hypertrophy. Chronic ulcerative colitis is a systemic disease that starts at the rectum or the sigmoid colon and progresses proximally to involve the entire left side of the colon. The colonic crypts are the first sites of cell damage and death, and the disease primarily involves the mucosal layer of the intestine.

The aetiopathogenesis of inflammatory bowel disease is not yet known. Most authors agree that immunologic abnormalities in the local mucosa-associated immune system are of major importance. Under normal conditions this gut-associated immune system has to protect the host against invasion of potential pathogens or an inappropriate immune response to luminal antigens. Lymphocytes within the mucosal immune system differ in many respects from lymphocytes in other areas of the body. There are indications that the tissue-specific differentiation of mucosal T cells is disturbed in inflammatory bowel disease. An imbalance between helper and suppressor mechanisms in the intestinal mucosa could result in a sustained and overshooting inflammatory and immune reaction against antigens normally occurring in the intestinal lumen [8] (see Chapter 7 for a more detailed discussion on the pathophysiological processes in inflammatory bowel diseases). Further disease states of the large intestine that might be treated with colon-specific dosage forms in the future are diarrhoea, tropical sprue, coeliac disease, irritable bowel syndrome, and different types of cancer [7].

In those instances where a disease of the colon is to be treated locally through the use of a delivery system, testing in the appropriate animal model is extremely important. For example, the delivery of anti-inflammatory agents to the colon for treatment of inflammatory bowel disease must be evaluated in suitable animal models. A number of animal models for intestinal inflammation are available for the testing of colonic delivery systems. The methods employed include lymphatic obstruction, vascular changes, and neurogenic manipulation [9–11]. Intestinal inflammation in animals such as rodents may be produced by topical application or administration of irritant chemicals such as acetic acid, trinitrobenzenesulfonic acid, difluoromethyl ornithine, pepsin inhibitors, or degraded carrageenan [9,11]. Colon cancer may be induced by administration of carcinogens such as chanthrenes, aromatic amines, hydrazine derivatives, alkylnitrosamides, and aflatoxin [12]. In the future, transgenic animals will play an important role as models for various disease states.

6.4 Approaches to Colon-specific Drug Delivery

Four strategies are currently being pursued to achieve drug release specifically in the colon.

- The fact that the luminal pH of the healthy distal colon is slightly higher than that of the proximal small intestine has led to the development of oral dosage forms that are intended to release the drug at the colonic pH (pH-controlled drug release).

- Colonic microflora produce a variety of enzymes that are not present in the stomach or the small intestine and can therefore be used to deliver drugs to the colon after enzymatic cleavage of degradable formulation components or drug carrier bonds (enzyme-controlled drug release). It should be taken into consideration that because of the negative redox potential in the colon, enzymatic or chemical reduction reactions are favoured.
- The relatively constant transit time in the small intestine of approximately 3–4 h is another physiological characteristic which can be exploited to achieve colon specificity (time-controlled drug release). After gastric emptying, a time-controlled drug delivery system is intended to release the drug after a predetermined lag phase.
- Another strategy relies on the strong peristaltic waves in the colon that lead to a temporarily increased luminal pressure (pressure-controlled drug release). Pressure-sensitive drug formulations release the drug as soon as a certain pressure limit is attained, i.e. destruction force is exceeded.

Using mostly anti-inflammatory model drugs or drugs that are absorbable in the colon, many colon-specific dosage forms have been developed in the past, including pro-drugs, cross-linked hydrogels, matrices and coated dosage forms. However, whereas the synthesis of pro-drugs is possible only if the drug has suitable functional groups that can be bound to a carrier molecule, biodegradable hydrogels and matrices are problematic insofar as polymer degradation rates and thus drug release are often too slow. Most colon-specific drug delivery systems belong to the group of coated dosage forms because of the flexibility in the design of the latter and the improved coating procedures that have been developed in the past.
With regard to peptide and protein absorption poor membrane permeability, enzymatic instability, and large molecular size are three factors that have remained major hurdles for peptide formulations. Absorption-enhancing agents that have been effective, at least in research environments with smaller drug candidates, have also shown some limited efficacy in small animal models with certain peptides. In most cases, however, effective formulations have only achieved fairly low peptide absorption (< 10%) and have also resulted in significant alterations in the normal cellular morphology of the gastrointestinal tract, at least on a transient basis [13]. Current data suggest that the successful development of oral peptide formulations remains a significant challenge.

6.4.1 pH-Controlled Drug Release

Many of the marketed dosage forms developed for colon-specific drug delivery, such as the enteric coated formulations Asacolitin®, Azulfidine®, Claversal®, Salofalk®, Colo-Pleon®, Entocort® and Budenofalk® rely on the physiological pH difference between the small and the distal large intestine. In healthy subjects this pH difference amounts to about 0.5 pH units [4,5] (Figure 6.2). However, it has been shown that this difference in pH between the small and the large intestine is too small to guarantee reliable drug release in the colonic region [14–16]. Moreover, in patients with inflammatory bowel disease the luminal colonic pH drops to values between 2.5 and 4.7 [17–19], a fact that has been attributed to a failure of bicarbonate secretion rather than excessive bacterial fermentation [18].

Enteric *coating materials* not only protect a dosage form from the acidic environment in the stomach and allow drug delivery to the small intestine, they may also pass through the small intestine and dissolve only in the colon, depending on their dissolution pH and the thickness of the coating applied. Many of the oral drug preparations for the treatment of inflammatory bowel disease available on the market (e.g. Asacolitin®, Claversal®, Salofalk®) are coated with enteric polymers such as Eudragit® L or S, i.e. methacrylic acid copolymers with different degrees of substitution, which show pH-dependent dissolution behaviour. These polymers are supposed to induce drug release as soon as the luminal pH in the GI tract exceeds values of 6 or 7. However, studies with Eudragit® S-coated tablets in healthy subjects have shown, that drug release in the colon is not sufficiently reproducible [14,15]. Other studies verify the reliability of this delivery method. One reason for these inconsistent results is the decrease in the luminal pH after passage through the ileocaecal valve as a result of bacterial fermentation of non-absorbable oligo- and polysaccharides to short chain fatty acids. Only in the distal colon is a luminal pH of 7 attained, which differs only slightly from the average pH of the small intestine (6.5–6.8).

The OROS-CT™ delivery system (*O*ral *O*smotic *S*ystem for *C*olon *T*argeting) is an enteric formulation consisting of one drug compartment containing osmotically active excipients and a second compartment containing a swelling polymer (Figure 6.3). Both compartments are coated with a semi-permeable membrane and an outer enteric coating. After dissolution of the enteric coating the swelling polymer slowly pushes the liquid content of the osmotic compartment out of the micropore as a result of water penetration. This leads to combined pH-controlled and sustained drug release.

During an acute attack of inflammatory bowel disease the luminal pH of the large intestine which is normally 6.4–7.0, drops to values between 2.3 and 4.7 [17–19], which means that enteric coatings are unsuitable coating materials in this particular case. Coating materials that dissolve at a low pH or are degradable in an acidic environment may be used in such a case. Therefore, the basic polymers Eudragit® E, an aminoalkyl methacrylate copolymer, and polyvinylacetal diethylaminoacetate (AEA™) have been investigated *in vitro* as coating materials for oral dosage forms designed for the treatment of inflammatory bowel disease with dexamethasone as the model corticosteroid [20,21]. An *in vivo* study is planned.

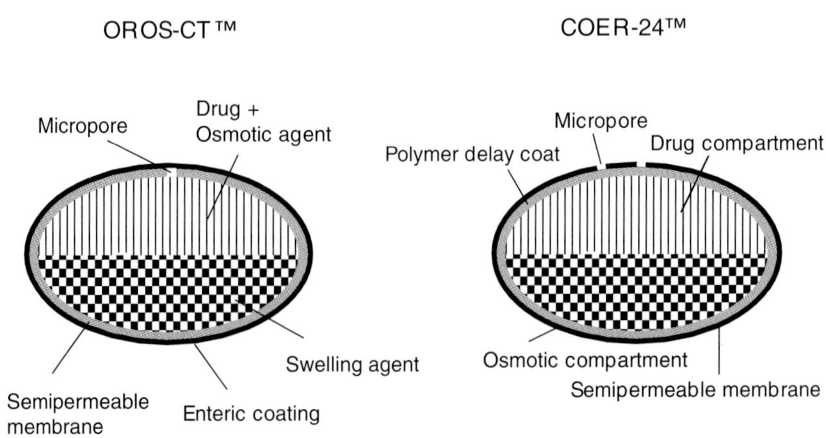

Figure 6.3. OROS-CT™ (*O*ral *O*smotic *S*ystem for *C*olon *T*argeting) and COER-24™ (*C*ontrolled *O*nset *E*xtended *R*elease) delivery systems.

pH-sensitive *ion exchange systems* represent another approach to how to achieve pH-controlled drug release in the colon. Drug ions, bound to an ion exchange resin, may be released pH-dependently into the large intestine, as in the case of insulin [22] or mesalazine [23]. In the latter case the drug is used in the form of its ionized pro-drug (olsalazine) and drug release occurs after microbial cleavage of the drug–carrier bonds. The ion exchange resin serves as polymer to prevent premature absorption of the pro-drug in the small intestine.

6.4.2 Enzyme-controlled Drug Release

Enzyme-controlled drug release takes advantage of the existence of enzyme-producing microorganisms in the colon (Figure 6.1). The colonic microflora produce a variety of enzymes, the activity of which may be exploited for colon-specific drug delivery. Among these enzymes are the azoreductase, various glycosidases, esterases and peptidases. Because of this physiological characteristic of the colon, biodegradable pro-drugs, coating materials, hydrogels and matrices have been developed (see below).

Pro-drugs are conjugates of drugs with carrier molecules mostly of inert nature. The microbial enzymes in the colon are responsible for the cleavage of the drug–carrier bond. A variety of pro-drugs have been synthesized, mainly azo compounds, glycosides, esters and amides [24].

Pro-drugs must not be cleaved by digestive enzymes of the upper GI tract and should not be susceptible to chemical hydrolysis. Moreover, pro-drug absorption in the small intestine should be negligible. Because of these requirements, the hydrophilicity, the molecular weight and the charge of the carrier molecules have to be regarded as critical parameters.

The azo pro-drug sulfasalazine (Azulfidine®, Colo-Pleon®), consisting of the drug mesalazine and the carrier molecule sulfapyridine, was the first pro-drug available for the treatment of inflammatory bowel disease. Because of side-effects caused by the pharmacologically active sulfapyridine carrier, other carrier molecules such as sulfanilic acid, p-aminobenzoic acid and its amino acid derivatives (benzalazine; ipsalazide; balsalazide, Colazide®) and mesalazine itself (olsalazine, Dipentum®) have been used in its place. In the case of glycoside pro-drugs, which were developed primarily for use with corticosteroids, monosaccharides have been intensively investigated as inert carrier molecules [25,26]. In addition, a variety of inert hydrophilic carriers of oligomeric as well as polymeric nature, some of them being enzymatically degradable themselves, have been used to prevent premature pro-drug absorption in the small intestine. Examples are β-cyclodextrin [27,28], dextran [29–34] and the polyanionic poly(L-aspartic acid) [35,36].

The Drug Delivery Index (*DDI*) allows a quantification of the reduction in the drug dose and the systemic exposure observed after drug release specifically to the colon [37]. It may be calculated using AUC (*A*rea *U*nder the plasma drug concentration–time *C*urve) data or drug concentrations in blood and colonic tissues under steady-state conditions:

$$DDI\text{ (Pro-drug vs. Drug)} = \frac{\dfrac{\text{AUC}^{\text{Tissue}}\text{ (Pro-drug)}}{\text{AUC}^{\text{Tissue}}\text{ (Drug)}}}{\dfrac{\text{AUC}^{\text{Blood}}\text{ (Pro-drug)}}{\text{AUC}^{\text{Blood}}\text{ (Drug)}}} \tag{6.1a}$$

$$DDI\,(\text{Pro-drug vs. Drug}) = \frac{\dfrac{Css^{\text{ Tissue}}\,(\text{Pro-drug})}{Css^{\text{ Tissue}}\,(\text{Drug})}}{\dfrac{Css^{\text{ Blood}}\,(\text{Pro-drug})}{Css^{\text{ Blood}}\,(\text{Drug})}} \tag{6.1b}$$

where *Css* is the steady-state drug concentration

The numerator of Eq. 6.1a describes to what extent drug concentrations are increased in the target tissue (colonic mucosa or muscle tissue) after pro-drug administration as opposed to drug administration. It may be regarded as the factor by which the pro-drug dose could be reduced in comparison to the drug dose. The denominator of Eq. 6.1a, which corresponds to the relative bioavailability of the drug released from the pro-drug, provides a measure of the reduction in the systemic exposure and thus the decrease of the systemic toxicity. In Eq. 6.1b the DDI is defined as the quotient of the tissue to blood concentration ratios after pro-drug and drug administration. With glucoside, glucuronide and dextran pro-drugs DDI values up to 9.7 have been reported in rats [25,38,39].

A disadvantage of the use of pro-drugs is the need for suitable functional groups such as amino-, hydroxy- or carboxy groups in both drug and carrier molecules (see Chapter 11 for more details on chemical synthesis routes applied in drug targeting/delivery strategies). Sometimes spacer molecules are necessary to link the drug to the carrier molecule, which often leads to complicated drug release kinetics. Another disadvantage of the pro-drug approach is the fact that for the approval of any newly synthesized pro-drug a toxicity study is required by regulatory agencies.

The group of enzymatically *degradable coating materials* comprises film-forming azo polymers and polymers with glycosidic bonds as well as conventional sustained release coating materials based on acrylates or ethylcellulose with biodegradable pore formers. The films must not be soluble or degradable in the upper GI tract. They should only be degradable in the colon and their degradation products need to be toxicologically harmless.

Cross-linked azo polymers were the first coating materials that were investigated with regard to their enzymatic degradability in the colon using insulin as the model drug [40,41]. A problem observed with these polymers was the rather slow degradation rate, probably a result of the lipophilicity of these compounds. Sufficiently hydrophilic linear azo polymers and pH-sensitive terpolymers based on acrylates were found to be more suitable coating materials [42–45].

In general, several problems have to be considered if azo polymers are used as an enzymatically degradable component of a colon-specific dosage form [46]. As a result of the enzymatic reduction to primary aromatic amines in the colon, the toxicity of these polymers is a critical issue. Moreover, the rate of microbial reduction of the polymeric azo compounds is often too slow. The reduction reaction may stop at the level of the hydrazo compounds instead of leading to the amines, which may influence the drug release mechanism and rate. In addition, the reduction reaction does not necessarily depend on the presence of the azoreductase, but may be induced by the negative redox potential in the colon. The latter also applies to the degradation of pro-drugs or polymers with disulfide bonds, which is the result of a chemical reduction step with no enzymatic reaction involved [47].

Biodegradable polymer coatings with glycosidic bonds, that have been developed in the past, are mainly based on galactomannans [48–50], chitosan [51] or the high molecular weight dextran fatty acid ester lauroyl dextran [52]. A special enzyme-controlled delivery system with pH-induced drug release has been developed by Watanabe *et al.* [53]. Drug release from this enteric tablet formulation begins after microbial degradation of an outer disaccharide layer (lactulose) to short-chain fatty acids in the colon and subsequent dissolution of the underlying basic Eudragit® E film. The colon-specific dosage form is available on the market as Codes™ and can incorporate any drug suitable for colonic drug delivery.

Degradable matrix films, consisting of a sustained release coating material and a poorly water-soluble but degradable pore former, are used if the pore former itself does not form an homogeneous coating film. These pore formers guarantee drug release in the colon after microbial degradation and the formation of pores in the film. As pore formers several oligo- and polysaccharides were investigated such as β-cyclodextrin [54], galactomannans [55], glassy amylose [56,57], pectin [58–60] and inulin [61]. A coated dosage form for colon-specific drug delivery with a matrix film consisting of ethylcellulose and the pore former glassy amylose is available on the market as Colal™.

If a biodegradable polymer does not exhibit satisfactory film-forming properties, it may also be used as a compression coat requiring a compaction process onto a drug-containing core [62–64].

Further examples of enzymatically degradable 'drug formulation wrappings' are capsule shells made of the polysaccharides chitosan [65,66] or cross-linked dextran [67].

Biodegradable polymers that cannot be used as coating materials for colon-specific drug delivery because they are readily water soluble and/or do not form films, may be used in the form of *hydrogels* and *matrices*. Hydrogels, consisting of cross-linked polymers, have been developed based on acrylates, polyether-esters and polysaccharides. In the case of acrylate- and polyether-ester-copolymers, azo-aromatic compounds have been used for cross-linking purposes and to guarantee degradability in the colon [68–70]. The higher the cross-linking density of these polymers, the lower their tendency to swell and the slower the degradation rate by azoreductase, ultimately resulting in slower drug release. Long chain azo-aromatic cross-linking compounds lead to faster polymer degradation and thus higher drug release rates. In general, the azo-aromatic cross-linking compounds that are used should have a rather small negative redox potential in order to ensure rapid degradation [71].

In the design of coatings, hydrogels and matrices based on azo polymers, the number of azo bonds in the polymers should not be too high, as this could lead to enzyme-saturated conditions slowing down the degradation process and thus drug release [72].

Polymer cross-linking leads to a decrease in the water solubility of many readily soluble polysaccharides, low water solubility being a requirement for colon-specific drug delivery. Dextrans [73,74], the mucopolysaccharide chondroitin sulfate [75,76], guar gum [77,78], pectin [79–82] and inulin [83–85] have all been investigated in cross-linked forms. Again, with a higher degree of cross-linking, the swelling properties of these polymers tend to be lower and this leads to a slower degradation rate and thus slower release of the drug. Poorly water-soluble drugs are usually released by an erosion-type mechanism.

6.4.3 Time-controlled Drug Release

Time-controlled drug release mechanisms rely on the fairly constant small intestinal transit time during which no measurable drug release occurs. Only after arrival of the dosage form in the colon may drug delivery begin. Such a delayed release mechanism can be achieved using compounds with swelling properties or osmotically active excipients, which leads to an increase in volume by water uptake resulting in a build-up of pressure inside the dosage form. After a predetermined lag phase the drug is released in a more or less pulsatile fashion, often accompanied by rupture of the outer coating layer. Such a lag phase may also be achieved with slowly eroding or dissolving coating layers [21]. In general, drug release from time-controlled delivery systems may be pH-induced, induced by swelling, pressure-induced or erosion-induced.

Time-controlled drug release with *pH-induced drug delivery* is a delivery method that does not depend on changes in the luminal pH of the GI tract but on a pH change within the dosage form itself.

Colon-specific drug formulations relying on the time-dependent dissolution of basic polymer layers such as Eudragit® E and chitosan under acidic conditions have been developed by Ishibashi *et al.* (CTDC: *C*olon-*T*argeted *D*elivery *C*apsule) [86–88] and Yamada *et al.* [89]. A solid organic acid incorporated in the dosage forms which dissolves as soon as it comes into contact with the penetrating intestinal fluid generates an acidic environment and induces the dissolution of the basic polymer layers and thus drug release. The onset of drug release depends on the thickness of the basic polymer layer or shell.

The organic acid-induced sigmoidal release system developed by Narisawa *et al.* consists of a drug core containing a solid organic acid which is coated with an ammonio methacrylate copolymer sustained-release coating (Eudragit® RS) [90,91]. During a lag phase, the drug permeability of the Eudragit® RS film increases drastically as a result of diffusion of the organic acid into the film thereby facilitating polymer hydration and inducing drug release.

Drug delivery induced by swelling may be achieved with swellable polymer layers based on cellulose ethers or acrylates, where with the latter pH-dependent swelling behaviour is feasible [92]. Examples are the oral Chronotopic® delivery system [93], the *T*ime-controlled *E*xplosion *S*ystem (TES) [94,95] and the TIME-CLOCK™-System [96–98] (Figure 6.4).

Swelling polymers may also be used as plugs, i.e. stoppers, to seal water-insoluble capsule bodies, as in the case of Pulsincap® [99,100]. The delayed drug release observed after plug ejection at the end of the lag phase, depends on the swelling properties of the polymer plug as well as on the geometry and chemical structure of the plug.

Swelling sustained-release coating polymers such as Eudragit® NE 30 D, i.e. poly(ethylacrylate-methylmethacrylate) [101,102] or Eudragit® RS [90], lead to a delay in drug release which is dependent on the thickness of the coating since these films have slow rates of swelling.

One dosage form available on the market which relies on *pressure-induced delivery* is the COER-24™ delivery system (*C*ontrolled *O*nset *E*xtended *R*elease). This formulation, developed for drugs that can be absorbed in the colon, is similar to the OROS-CT™ system (Figure 6.3). Here, a polymeric delay coat is responsible for the delayed drug release. During the lag phase the osmotic compartment swells resulting in pressure being applied to the drug

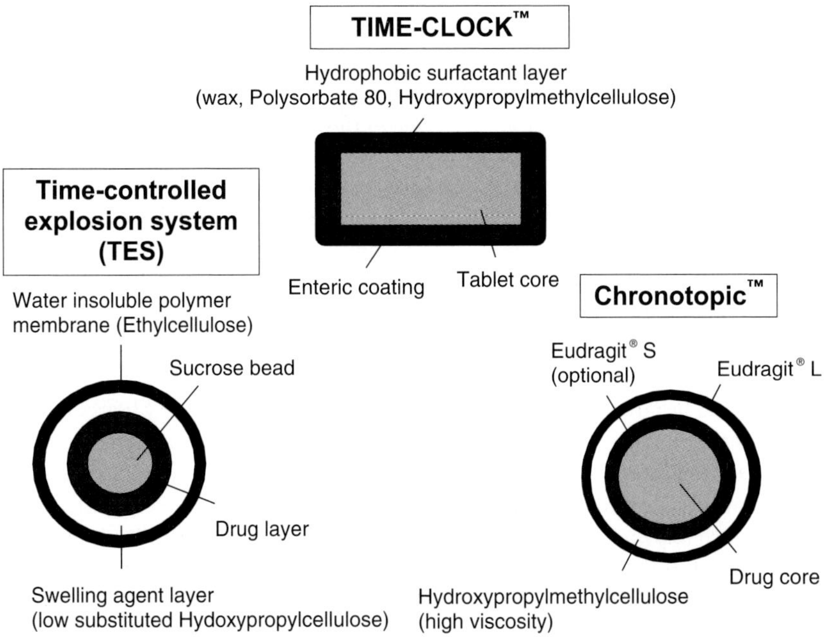

Figure 6.4. Swelling-induced time-controlled drug delivery systems.

compartment. Micropores in the outer semi-permeable film allow a sustained drug release after dissolution and extrusion of the underlying delay coat.

For an *erosion-induced drug delivery* system compactable cellulose ethers are suitable polymers [103]. Drug release, which is controlled by the erosion/dissolution of these polymeric layers, may be pH-dependent if an acid or basic polymer is used.

In summary, the lag time of drug release may be controlled by the rate at which water penetrates through the coating or shell, the rate of fluid absorption of the polymer layer, the osmotic activity of salts and osmopolymers, the erosion and dissolution rate of the polymer layers and the thickness of the layers or coatings.

6.4.4 Pressure-controlled Drug Release

Pressure-controlled delivery systems take advantage of the temporary increase in intra-luminal pressure in the colon in the event of a peristaltic wave. A drug formulation relying on the resulting destructive force has been developed and consists of a conventional hard gelatin capsule with a thick inner layer of ethylcellulose [104,105]. The rupture of the ethylcellulose shell with subsequent drug release is induced by an increase in the intra-luminal pressure and subsequent destructive force. Drug absorption from this formulation has been investigated *in vivo* in dogs and humans with mesalazine, carbamazepine [105,106] and caffeine [104] as the model drugs. The capsule contains the drug in solution because otherwise there may be insufficient fluid available in the distal portion of the colon for the drug to dissolve in.

6.5 Concluding Remarks

All the methods for colon-specific drug delivery presented in this overview are more or less susceptible to changes in the physiological environment (diet, disease state, medication, etc.). This may affect reliability and reproducibility of the delivery systems. However, most of the systems described have already been investigated *in vivo* with regard to colon specificity. Methods such as planar γ-scintigraphy [86,97,107] and three-dimensional magnetic marker monitoring [108,109] appear to be the methods of choice for dynamic visualization of the dosage form in the GI tract and the determination of the location and time of onset of drug release. In animal models, the measurement of blood and intestinal tissue concentrations of the perorally administered drug in order to quantify the reduction in the systemic exposure, is an additional approach [110]. However, most formulations are tested in healthy humans or animals in spite of the fact, that pathological conditions such as inflammatory bowel disease may have a significant influence on gut physiology. As yet, little is known regarding the effect of disease states on the intestinal transit time, GI tract motility, luminal pH, intra-luminal colonic pressure and composition of colonic microflora. It appears that time-controlled drug release systems, which rely on a constant transit time through the small intestine, are the most promising colon-specific delivery systems, as this transit time is only marginally influenced by pathological events in the GI tract. pH-controlled systems are reliable only if there is a significant difference in the luminal pH between the small and the large intestine. Under physiological conditions this pH difference is often too small. In certain pathological conditions such as inflammatory bowel disease however, a significant decrease in pH in the affected GI tract regions may be observed, and this has led to the development of novel pH-controlled delivery systems. Enzyme-controlled delivery systems are dependent on the activity of the microbial enzymes in the large intestine, which are susceptible to many environmental factors, particularly diet and medication. The reliability of pressure-controlled delivery systems is expected to be highly dependent on variations in intestinal peristalsis despite promising *in vivo* data.

References

[1] Simon GL, Gorbach SL, *Gastroenterology* **1984,** *86*, 174–193.
[2] Drasar BS, Hill MJ, In: *Human Intestinal Flora*, pp. 103-123. Academic Press, London, **1974.**
[3] Bragger JL, Yazaki E, Evans DF, *Pharm. Res.* **1997,** *14 (11 Suppl.)*, S656–S657.
[4] Mitsuoka T, *Intestinal Bacteria and Health*. Harcourt Brace Jovanovich, Tokyo, **1978.**
[5] Evans DF, Pye G, Bramley R, Clark AG, Dyson TJ, Hardcastle JD, *Gut* **1988,** *29*, 1035–1041.
[6] Shameem M, Katori N, Aoyagi N, Kojima S, *Pharm. Res.* **1995,** *12*, 1049–1054.
[7] Haeberlin B, Friend DR, In: *Oral Colon-Specific Drug Delivery* (Ed. Friend DR), pp. 1–43. CRC Press, Boca Raton, **1992.**
[8] Zeitz M, *Digestion* **1997,** *58 (Suppl. 1)*, 59–61.
[9] Bhatti MA, Hodgson HJF, *Int. J. Exp. Pathol.* **1995,** *76*, 309–315.
[10] Morteau O, Bueno L, *Gastroenterol. Clin. Biol.* **1995,** *19*, 204–214.
[11] Warren BF, Watkins PE, *J. Pathol.* **1994,** *172*, 313–316.
[12] Pories SE, Ramchurren N, Summerhayes I, Steele G, *Arch. Surg.* **1993,** *128*, 647–653.
[13] Fix JA, *J. Pharm. Sci.* **1996,** *85*, 1282–1285.
[14] Ashford M, Fell JT, Attwood D, Woodhead PJ, *Int. J. Pharm.* **1993,** *91*, 241–145.
[15] Ashford M, Fell JT, Attwood D, Sharma H, Woodhead PJ, *Int. J. Pharm.* **1993,** *95*, 193–199.
[16] Dew MJ, Ryder REJ, Evans N, Evans BK, Rhodes J, *Br. J. Clin. Pharmacol.* **1983,** *16*, 185–187.

[17] Fallingborg J, Christensen LA, Jacobsen BA, Rasmussen SN, *Dig. Dis. Sci.* **1993,** *38*, 1989–1993.
[18] Roediger W, Lawson MJ, Kwok V, Grant AK, Pannall PR, *J. Clin. Pathol.* **1984,** *37*, 704–707.
[19] Sasaki Y, Hada R, Nakajima H, Fukuda S, Munakata A, *Am. J. Gastroenterol.* **1997,** *92*, 114–118.
[20] Leopold CS, Eikeler D, *J. Drug Target.* **1998,** *6*, 85–94.
[21] Leopold CS, Eikeler D, *Drug Dev. Ind. Pharm.* **2000,** *26*, 1239–1246.
[22] Wunderlich J-C, Schick U, Werry J, Freidenreich J, International Patent WO 93/10767, **1993.**
[23] Pitzele BS, Jones PH, US Patent 4,374,932, **1983.**
[24] Leopold CS, *Pharmazie in unserer Zeit* **1997,** *26*, 289–298.
[25] Friend DR, Tozer TN, *J. Control. Rel.* **1992,** *19*, 109–120.
[26] Friend DR, *STP Pharma Sci.* **1995,** *5*, 70–76.
[27] Hirayama F, Minami K, Uekama K, *J. Pharm. Pharmacol.* **1996,** *48*, 27–31.
[28] Hirayama F, Minami K, Uekama K, *Proc. 2nd World Meeting APGI/APV* **1998,** *2*, 935–936.
[29] Harboe E, Larsen C, Johansen M, Olesen HP, *Pharm. Res.* **1989,** *6*, 919–923.
[30] Larsen C, Harboe E, Johansen M, Olesen HP, *Pharm. Res.* **1989,** *6*, 995–999.
[31] Larsen C, Jensen BH, Olesen HP, *Acta Pharm. Nord.* **1991,** *3*, 71–76.
[32] McLeod AD, In: *Oral Colon-specific Drug Delivery* (Ed. Friend DR), pp. 213–231. CRC Press, Boca Raton, **1992.**
[33] McLeod AD, Tolentino L, Tozer TN, *Biopharm. Drug Dispos.* **1994,** *15*, 151–161.
[34] McLeod AD, Fedorak RN, Friend DR, Tozer TN, Cui NR, *Gastroenterology* **1994,** *106*, 405–413.
[35] Leopold CS, Friend DR, *Int. J. Pharm.* **1995,** *126*, 139–145.
[36] Leopold CS, Friend DR, *J. Pharmacokinet. Biopharm.* **1995,** *23*, 397–406.
[37] McLeod AD, Tozer TN, In: *Oral Colon-specific Drug Delivery* (Ed. Friend DR), pp. 85–114. CRC Press, Boca Raton, **1992.**
[38] Fedorak RN, Haeberlin B, Empey LR, Cui N, Nolen H, III, Jewell LD, Friend DR, *Gastroenterology* **1995,** *108*, 1688–1699.
[39] Tozer TN, Rigod J, McLeod AD, Gungon R, Hoag MK, Friend DR, *Pharm. Res.* **1991,** *8*, 445–454.
[40] Saffran M, Kumar GS, Savariar C, Burnham JC, Williams F, Neckers DC, *Science* **1986,** *233*, 1081–1084.
[41] Saffran M, Kumar GS, Neckers DC, Peña J, Jones RH, Field JB, *Biochem. Soc. Trans.* **1990,** *18*, 752–754.
[42] Van den Mooter G, Samyn C, Kinget R, *Int. J. Pharm.* **1992,** *87*, 37–46.
[43] Van den Mooter G, Samyn C, Kinget R, *Pharm. Res.* **1994,** *11*, 1737–1741.
[44] Van den Mooter G, Samyn C, Kinget R, *Pharm. Res.* **1995,** *12*, 244–247.
[45] Van den Mooter G, Maris B, Samyn C, Augustijns P, Kinget R, *J. Pharm. Sci.* **1997,** *86*, 1321–1327.
[46] Lloyd AW, Martin GP, Soozandehfar SH, *Int. J. Pharm.* **1994,** *106*, 255–260.
[47] Schacht E, Wilding I, International Patent WO 91/11175, **1991.**
[48] Bauer KH, Deutsche Patentschrift 42 09 160, **1993.**
[49] Betzing J, Bauer KH, *Pharm. Ztg. Wiss.* **1992,** *137*, 131–134.
[50] Sarlikiotis AW, Bauer KH, *Pharm. Ind.* **1992,** *54*, 873–880.
[51] Sekigawa F, Onda Y, US Patent 5,217,720, **1993.**
[52] Kesselhut JF, Bauer KH, *Pharmazie* **1995,** *50*, 263–269.
[53] Watanabe S, Kawai H, Katsuma M, Fukui M, International Patent WO 95/28963, **1995.**
[54] Siefke V, Weckenmann H, Bauer K, *Proc. Int. Symp. Control. Rel. Bioact. Mater.* **1993,** *20*, 182–183.
[55] Lehmann K, Dreher KD, *Proc. Int. Symp. Control. Rel. Bioact. Mater.* **1991,** *18*, 331–332.
[56] Milojevic S, Newton JM, Cummings JH, Gibson GR, Bothman RL, Ring SG, Allwood MC, Stockham M, *STP Pharma Sci.* **1995,** *5*, 47–53.
[57] Siew LF, Newton JM, *Pharm. Res.* **1997,** *14 (11 Suppl.)*, S-658.
[58] Macleod GS, Fell JT, Collett JH, *Int. J. Pharm.* **1997,** *157*, 53–60.
[59] Wakerly Z, Fell JT, Attwood D, Parkins D, *Pharm. Res.* **1996,** *13,* 1210–1212.
[60] Wakerly Z, Fell JT, Attwood D, Parkins D, *Int. J. Pharm.* **1997,** *153*, 219–224.
[61] Vervoort L, Kinget R, *Int. J. Pharm.* **1996,** *129*, 185–190.
[62] Krishnaiah YSR, Satyanarayana S, Prasad YVR, Rao SN, *Int. J. Pharm.* **1998,** *171*, 137–146.
[63] Krishnaiah YSR, Satyanarayana S, Prasad YVR, *Drug Dev. Industr. Pharm.* **1999,** *25*, 651–657.
[64] Rubinstein A, Radai R, *Eur. J. Pharm. Biopharm.* **1995,** *41*, 291–295.
[65] Chen RH, Tsaih ML, Lin WC, *Carbohydr. Polymers* **1996,** *31*, 141–148.
[66] Tozaki H, Komoike J, Tada C, Maruyama T, Terabe A, Suzuki T, Yamamoto A, Muranishi S, *J. Pharm. Sci.* **1997,** *86*, 1016–1021.
[67] Brøndsted H, Andersen C, Hovgaard L, *J. Control. Rel.* **1998,** *53*, 7–13.

[68] Brøndsted H, Kopecek J, *Pharm. Res.* **1992,** *9*, 1540–1545.
[69] Kakoulides EP, Smart JD, Tsibouklis J, *J. Control. Rel.* **1998,** *54*, 95–109.
[70] Samyn C, Kalala W, Van den Mooter G, Kinget R, *Int. J. Pharm.* **1995,** *121*, 211–216.
[71] Bragger JL, Lloyd AW, Soozandehfar SH, Bloomfield SC, Marriott C, Martin GP, *Int. J. Pharm.* **1997,** *157*, 61–71.
[72] Van den Mooter G, Offringa M, Kalala W, Samyn C, Kinget R, *STP Pharma Sci.* **1995,** *5*, 36–40.
[73] Chiu HC, Hsiue GH, Lee YP, Huang LW, *J. Biomater. Sci.* **1999,** *10*, 591–608.
[74] Hovgaard L, Brøndsted H, *J. Control. Rel.* **1995,** *36*, 159–166.
[75] Rubinstein A, Nakar D, Sintov A, *Int. J. Pharm.* **1992,** *84*, 141–150.
[76] Rubinstein A, Nakar D, Sintov A, *Pharm. Res.* **1992,** *9*, 276–278.
[77] Rubinstein A, Gliko-Kabir I, *STP Pharma Sci.* **1995,** *5*, 41–46.
[78] Rubinstein A, Gliko-Kabir I, Penhasi A, Yagen B, *Proc. Int. Symp. Control. Rel. Bioact. Mater.* **1997,** *24*, 839–840.
[79] Ashford M, Fell J, Attwood D, Sharma H, Woodhead P, *J. Control. Rel.* **1993,** *26*, 213–220.
[80] Ashford M, Fell J, Attwood D, Sharma H, Woodhead P, *J. Control. Rel.* **1994,** *30*, 225–232.
[81] Rubinstein A, Radai R, Ezra M, Pathak S, Rokem JS, *Pharm. Res.* **1993,** *10*, 258–263.
[82] Wakerly Z, Fell J, Attwood D, Parkins D, *J. Pharm. Pharmacol.* **1997,** *49*, 622–625.
[83] Vervoort L, Van den Mooter G, Augustijns P, Bousson R, Toppet S, Kinget R, *Pharm. Res.* **1997,** *14*, 1730–1737.
[84] Vervoort L, Van den Mooter G, Augustijns P, Kinget R, *Int. J. Pharm.* **1998,** *172*, 127–135.
[85] Vervoort L, Rombaut P, Van den Mooter G, Augustijns P, Kinget R, *Int. J. Pharm.* **1998,** *172*, 137–145.
[86] Ishibashi T, Pitcairn GR, Yoshino H, Mizobe M, Wilding IR, *J. Pharm. Sci.* **1998,** *87*, 531–535.
[87] Ishibashi T, Hatano H, Kobayashi M, Mizobe M, Yoshino H, *Int. J. Pharm.* **1998,** *168*, 31–40.
[88] Ishibashi T, Hatano H, Kobayashi M, Mizobe M, Yoshino H, *J. Control. Rel.* **1999,** *57*, 45–53.
[89] Yamada A, Wato T, Uchida N, Fujisawa M, Takama S, Inamoto Y, US Patent 5,395,628, **1995.**
[90] Narisawa S, Nagata M, Danyoshi C, Yoshino H, Murata K, Hirakawa Y, Noda K, *Pharm. Res.* **1994,** *11*, 111–116.
[91] Narisawa S, Nagata M, Hirakawa Y, Kobayashi M, Yoshino H, *J. Pharm. Sci.* **1996,** *85*, 184–188.
[92] Abramowitz R, Ranadive S, Jain NB, Varia SA, *Pharm. Res.* **1997,** *14 (11 Suppl.)*, S656.
[93] Gazzaniga A, Sangalli ME, Giordano F, *Eur. J. Pharm. Biopharm.* **1994,** *40*, 246–250.
[94] Ueda S, Ibuki R, Kawamura A, Murata S, Takahashi T, Kimura S, Hata T, *J. Drug Target.* **1994,** *2*, 133–140.
[95] Ueda S, Hata T, Asakura S, Yamaguchi H, Kotani M, Ueda Y, *J. Drug Target.* **1994,** *2*, 35–44.
[96] Pozzi F, Furlani P, Gazzaniga A, Davis SS, Wilding IR, *J. Control. Rel.* **1994,** *31*, 99–108.
[97] Steed KP, Hooper G, Monti N, Strolin Benedetti M, Fornasini G, Wilding IR, *J. Control. Rel.* **1997,** *49*, 115–122.
[98] Wilding IR, Davis SS, Pozzi F, Furlani P, Gazzaniga A, *Int. J. Pharm.* **1994,** *111*, 99–102.
[99] Binns J, Stevens HNE, McEwen J, Pritchard G, Brewer FM, Clarke A, Johnson ES, McMillan I, *J. Control. Rel.* **1996,** *38*, 151–158.
[100] Hegarty M, Atkins G, International Patent WO 95/10263, **1995.**
[101] Kraeling MEK, Ritschel WA, *Methods Find. Exp. Clin. Pharmacol.* **1992,** *14*, 199–209.
[102] Rao SS, Ritschel WA, *Int. J. Pharm.* **1992,** *86*, 35–41.
[103] Shah NH, Phuapradit W, Railkar A, US Patent 5,482,718, **1996.**
[104] Muraoka M, Hu ZP, Shimokawa T, Sekino S, Kurogoshi R, Kuboi Y, Yoshikawa Y, Takada K, *J. Control. Rel.* **1998,** *52,* 119–129.
[105] Takaya T, Sawada K, Suzuki H, Funaoka A, Matsuda KI, Takada K, *J. Drug Target.* **1997,** *4*, 271–276.
[106] Takaya T, Niwa K, Muraoka M, Ogita I, Nagai N, Yano R, Kimura G, Yoshikawa Y, Yoshikawa H, Takada K, *J. Contr. Rel.* **1998,** *50*, 111–122.
[107] Adkin DA, Kenyon CJ, Lerner EI, Landau I, Strauss E, Caron D, Penhasi A, Rubinstein A, Wilding IR, *Pharm. Res.* **1997,** *14*, 103–107.
[108] Weitschies W, Kotitz R, Cordini D, Trahms L, *J. Pharm. Sci.* **1997,** *86*, 1218–1222.
[109] Weitschies W, Cardini D, Karaus M, Trahms L, Semmler W, *Pharmazie* **1999,** *54*, 426–430.
[110] Tozer TN, Friend DR, McLeod AD, *STP Pharma Sci.* **1995,** *5*, 5–12.

7 Vascular Endothelium in Inflamed Tissue as a Target for Site Selective Delivery of Drugs

Maaike Everts, Astrid J. Schraa, Lou F. M. H. de Leij, Dirk K. F. Meijer, Grietje Molema

7.1 Introduction

Chronic inflammatory disorders are characterized by an abundant leucocyte infiltration in the affected tissue due to the continuous recruitment of these inflammatory cells. Although these events are part of the body's defence mechanisms, excessive responses occur in the vicious circle of cell activation and cell destruction, leading to acute and chronic inflammatory disorders. Many current therapeutic approaches are based on attempts to control leucocyte activation. Recently there has been a growing interest in the role of the endothelial cells in leucocyte recruitment and the possibility of manipulating endothelial cell activation as a therapeutic approach [1,2]. The endothelium plays an important role in the inflammatory cascade, and has the advantage of being easily accessible for drug targeting preparations due to direct contact with the blood as well as the presence of endocytotic processes [3]. Therefore endothelial cells are attractive targets for cell-selective pharmacological intervention employing drug targeting strategies.

The aim of this chapter is to describe potential target epitopes on endothelial cells in chronic inflammation, the design of targeting devices on the basis of this knowledge and the use of these targeting devices in the intervention of endothelial cell activation. Several *in vitro* and *in vivo* test models for studying endothelial cell responses in inflammation are currently available and will be briefly described. Since drug targeting to the inflamed endothelium is a relatively new approach, the number of research reports dedicated to this topic is, at present, relatively small. Therefore, this chapter will include some general considerations regarding drug targeting to the activated endothelial cell. Adequate cell-selective delivery of potent anti-inflammatory agents may provide an important tool for increasing the efficacy and reducing the side-effects of such agents in the treatment of chronic inflammatory diseases.

7.2 Regulation of Immune Responses in (Chronic) Inflammation

7.2.1 Induction of an Immune Response

After an antigen (a 'non-self' molecular entity) has entered the body, it is recognized by the cells of the immune system. An important step in eliciting an efficient immune response to an

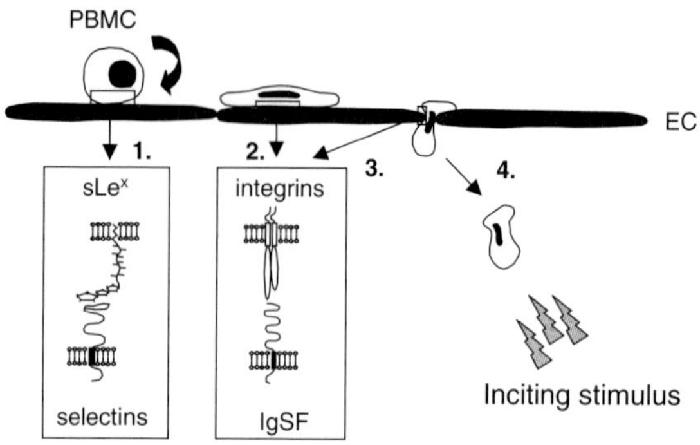

Figure 7.1. Leucocyte recruitment to sites of inflammation takes place via strictly regulated expression of adhesion molecules by the leucocytes (peripheral blood mononuclear cells, PBMC) and endothelial cells (EC). (1) Tethering of the leucocytes is mediated by interactions between members of the selectin family and their sialyl Lewis X (sLe^x) counterparts. Subsequent chemokine-mediated cellular activation leads to strong adhesion (2) and trans-endothelial migration (3) of the leucocytes into the underlying tissue. These processes are mediated by members of the integrin and immunoglobulin superfamily (IgSF) and homotypic interactions of the IgSF member CD31, among others. Cellular movement through the extracellular matrix (4) is facilitated by interactions between integrins and their extracellular matrix ligands, and a variety of chemokines and their respective receptors.

antigen is the recruitment of leucocyte subsets to the site of antigen presence or entry. The immune system is subsequently capable of efficiently eliminating the antigen.

The recruitment and migration of leucocytes into inflamed tissues is a carefully orchestrated process (Figure 7.1). It consists of sequential steps mediated by different families of adhesion molecules expressed by both the leucocytes and the endothelial cells at the site of inflammation [4]. Of these adhesion molecules, the selectin family mediates the initial contact and subsequent rolling of the leucocyte on the endothelium. It consists of three members, i.e. E- (endothelial), P- (platelet) and L- (leucocyte) selectin. Activated endothelial cells express E- and P-selectin. P-selectin is also expressed on platelets, whereas L-selectin is only expressed on subsets of leucocytes [5].

If during the rolling process the leucocyte is correctly activated, the affinity of the members of the integrin family of adhesion molecules on the leucocyte membrane increases. Examples of activating factors are cytokines such as interleukin (IL)-6 and IL-8, which can be produced by the activated endothelial cells, and chemokines such as monocyte chemotactic proteins (MCPs), growth related proteins (GROs) and interferon γ-inducible protein 10 (IP-10) [6]. The so-called counter receptors for integrins on the endothelium are members of the immunoglobulin superfamily (IgSF) and encompass Intercellular Adhesion Molecule-1 (ICAM-1) and Vascular Cell Adhesion Molecule-1 (VCAM-1). These molecules are highly expressed by activated endothelial cells in inflammatory sites. The interaction of integrins on the leucocyte with the immunoglobulin superfamily members on the endothelium mediates the firm attachment of the leucocyte, followed by transmigration into the tissue. In this latter process Platelet Endothelial Cell Adhesion Molecule-1 (PECAM-1, CD31) and a variety of matrix metalloproteases (MMPs) exert important functions. Although initially identified as an IgSF member with a main function in cell–cell contact, PECAM-1 was recently shown to be a modulator of vascular cell activation as well [7]. MMPs play a role in the degradation of the basal membrane and the migration of the leucocyte through the tissue in the direction of the antigen [8–11].

7.2.2 The Resolution of Inflammation

In normal, non-pathological inflammation, the resolution of an inflammatory reaction is strictly controlled. Elimination of the antigen by the immune system leads to the shutdown of the inflammatory reaction. Without the inciting stimulus being present, infiltrated leucocytes die by apoptosis as was demonstrated in a delayed-type hypersensitivity (DTH) response in healthy volunteers. This increase in apoptotic T cells which preceded the resolution of the reaction was probably due to an increased CD95 ligand (FasL) expression on the cells of the perivascular infiltrate and the deprivation of stimulatory cytokines such as IL-2 an IL-15. This process is furthermore associated with a decreased ratio of anti-apoptotic (e.g. Bcl-2) and pro-apoptotic (e.g. Bax) protein expression by the infiltrated leucocytes [12]. Collectively this indicates that the expression of pro- and anti-apoptotic proteins in leucocytes plays an important role in controlling inflammation. Similarly, these processes exert central regulatory functions in endothelial cell activation (see Section 7.3.3). In contrast to the above-described role in the resolution of inflammation, programmed cell death can also initiate inflammation. In an ischaemia–reperfusion model in mice, inhibition of apoptosis effectively prevented subsequent inflammation and tissue injury. Inhibition of early apoptosis of tissue parenchymal cells is most likely the mechanism behind this effect [13].
The events described above demonstrate the complexity and fine balance between various processes resulting in either resolution or enhancement of inflammation. Failure to control one or more of these, and possibly other presently unknown, regulatory mechanisms may lead to chronic activation of the immune system resulting in chronic inflammatory diseases.

7.3 Chronic Inflammatory Disorders

7.3.1 Pathophysiology of Chronic Inflammatory Disorders

In chronic inflammatory disorders there is a general over-activation of the immune system, the cause of which is often unknown. In all of these diseases, an increased number of infiltrated immune cells can be found in the inflamed tissues. Furthermore, increased expression of adhesion molecules on the endothelium can often be observed, although the adhesion molecules are expressed differentially in the various diseases.

7.3.1.1 Rheumatoid Arthritis

Rheumatoid arthritis (RA) is an inflammatory disease of the synovium which results in erosion, deformity and finally the destruction of joints. Inflammation of the joints is associated with a villous hypertrophy of the synovial membrane, which on microscopy shows proliferation of the lining layer with an inflammatory infiltrate. There is extensive expression of HLA-

DR on T cells, B cells and synovial lining cells, indicating strong immunological activity. RA is thought to be an autoimmune reaction, caused by an interaction between constitutional and environmental factors [14].

In human biopsies of patients with RA, an increased expression of the adhesion molecules E-selectin, VCAM-1, ICAM-1 and the CD11c integrin has been demonstrated [15,16]. Also, soluble mediators produced by perivascular cells (e.g. the chemokines Macrophage Inflammatory Protein (MIP)-1α and MIP-1β) or by endothelial cells themselves (e.g. IL-8 and IL-15) are important in the regulation of leucocyte infiltration in RA [17]. Tumour Necrosis Factor α (TNFα) is thought to be a particularly important inflammatory mediator contributing to the pathology of arthritis, as demonstrated by the beneficial effects of the TNFα-neutralizing therapies that are currently being explored [18]. However, present therapies for RA mainly aim at the inhibition of cyclooxygenase enzymes which are responsible for the overproduction of inflammatory mediators like prostaglandin E_2 in arthritis-affected joints [19].

7.3.1.2 Atherosclerosis

Atherosclerosis is a generalized degenerative disease that affects large and medium-sized arteries. The atherosclerotic plaque contains increased numbers of smooth muscle cells (which are morphologically abnormal), increased connective tissue and lipid, mostly cholesterol. Monocyte-derived macrophages and lymphocytes are also found in the plaques. Endothelial damage is believed to be the essential trigger for the development of atherosclerosis. Once injury of the endothelium has occurred, platelets and smooth muscle cells will adhere and aggregate. Subsequently smooth muscle cells will proliferate, collagen and elastin production will increase, and lipid is allowed to accumulate in the vessel wall through enhanced permeability at the site of injury [14].

Two important molecular participants in the atherosclerotic process are the transcription factor nuclear factor κB (NFκB) and the adhesion molecule CD40. Using immunohistochemical techniques the activated form of NFκB has been shown to be present in human atherosclerotic lesions in smooth muscle cells, macrophages and endothelial cells. In contrast, in vessels lacking atherosclerotic processes little or no activated NFκB was present [20]. Recently, there has been an increasing interest in the role of CD40 and CD40L (CD154), members of the TNF receptor and TNF family respectively, in chronic inflammation. Ligation of CD40 on vascular wall cells promotes upregulation of endothelial adhesion molecules such as E-selectin, ICAM-1 and VCAM-1 [21]. Furthermore, it stimulates mononuclear cell recruitment, participates in the weakening of the plaque and sets the stage for thrombosis by inducing tissue factor expression [22]. Additionally, CD40 ligation signals pro-angiogenic processes which are also prominent in atherosclerosis [23].

7.3.1.3 Inflammatory Bowel Disease

Inflammatory Bowel Disease (IBD) comprises several diseases, including ulcerative colitis and Crohn's disease. Ulcerative colitis is a disease of the colon, originating in the rectum and extending proximally to a variable extent. It frequently affects the entire colon but never causes significant involvement of the small intestine. Crohn's disease may affect any part of the gastrointestinal tract, although an ileocolitis is the most common localization. The aetiology of both diseases is unknown. Many humoral and immune phenomena are reputed to be involved, and it seems likely that immunological effector mechanisms are mainly responsible for causing chronic disease [14,24]. For instance, abnormal T cell responses to components of normal gut flora have been described [25], and it has been suggested that deficiencies in suppressive regulatory T cells are involved in the pathogenesis of IBD [26].

An important mechanism for the observed tissue accumulation of polymorphonuclear cells (PMNs) in IBD might be the high production of granulocyte colony-stimulating factor, resulting in a delay of neutrophil apoptosis [27]. In addition, PMNs of IBD patients have an increased capacity to produce cytokines TNFα and IL-1β, which is also reflected by the presence of TNFα in the affected tissues and in stool of IBD patients [28–30]. Furthermore, an increased expression of the chemokines IP-10, IL-8, MCP-1 and MCP-3 in infiltrated cells in the lamina propria in colitis patients was detected [31]. All these inflammatory mediators contribute to activation of endothelial cells, as shown by an increased expression of E- and P-selectin and ICAM-1 in biopsies of IBD patients, particularly at the base of ulcers in actively inflamed tissue [32]. The high expression of these adhesion molecules mediates the ongoing recruitment of inflammatory cells, resulting in a vicious circle of leucocyte infiltration and tissue damage (Figure 7.2).

7.3.1.4 Other Diseases

In a variety of other diseases inflammation is a pathological hallmark. These diseases include asthma, psoriasis and organ transplant rejection episodes [13]. Even Alzheimer's disease exhibits an inflammatory component [33]. Although the aetiology of these diseases differs to a large extent, they are all characterized by ongoing leucocyte recruitment and cytokine production. Therefore, they represent potential target diseases for treatment with endothelium-directed drug targeting constructs. Due to space limitations, these diseases will not however be discussed further.

7.3.2 Angiogenesis in Chronic Inflammation

A process that was more recently identified as being important in chronic inflammatory diseases is angiogenesis, the formation of new blood vessels from pre-existing capillaries or post-capillary venules. Angiogenesis is important in normal physiological conditions such as embryogenesis, wound healing and in the female reproductive cycle. It also plays a role in pathological conditions such as cancer, diabetic retinopathy and chronic inflammatory diseases [34].

The switch to the angiogenic phenotype involves a change in the local equilibrium between pro- and anti-angiogenic factors. The most extensively studied pro-angiogenic factors are vascular endothelial growth factor (VEGF) and basic fibroblast growth factor (bFGF). Following a pro-angiogenic stimulus, endothelial cells become activated and start to express proteolytic enzymes to break down the basement membrane and extracellular matrix. Sub-

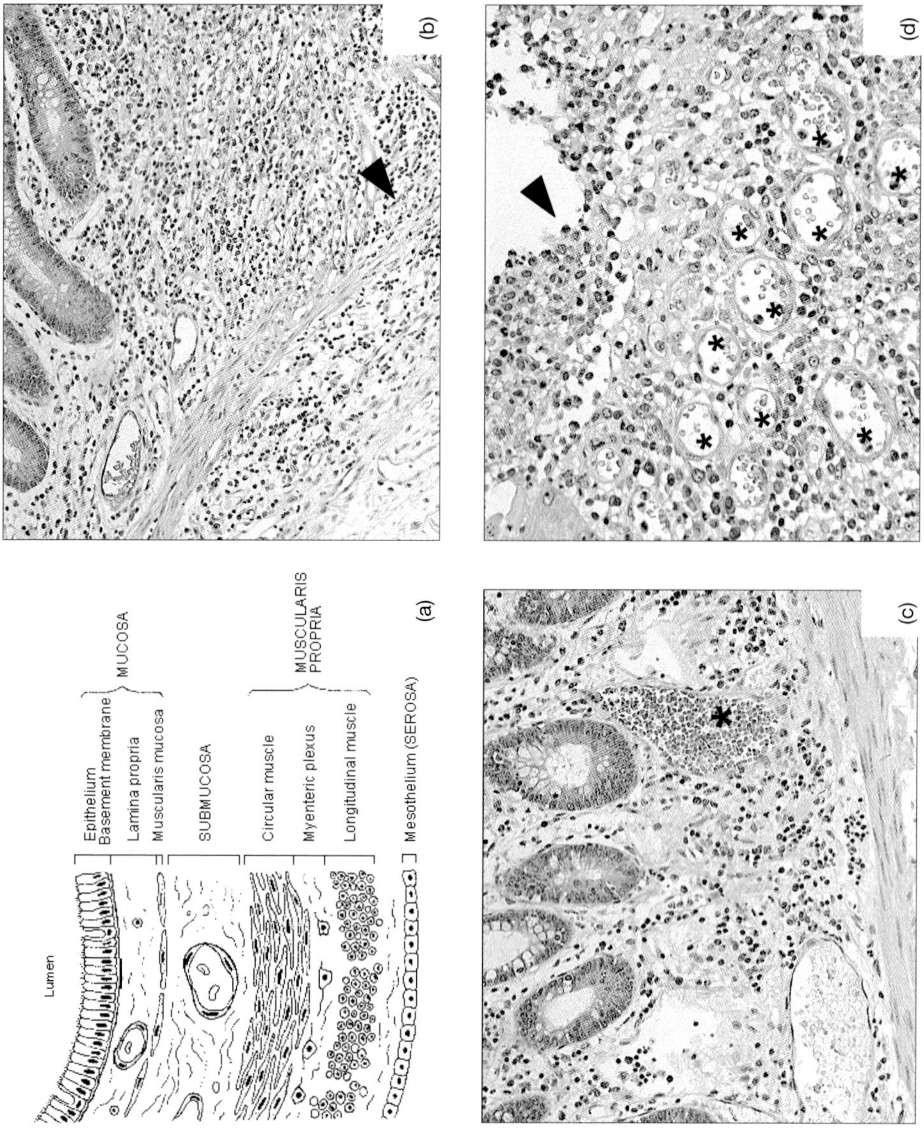

Figure 7.2. (a) Schematic representation of the different layers of the gut wall. From reference [153] (b–c) Haematoxylin/eosin staining of biopsies of a Crohn's disease lesion in the human colon. Biopsy orientation: intestinal lumen at the upper side of the photographs; (b) excessive leucocyte infiltration through the disintegrated muscle layer (muscularis mucosae; arrowhead); (c) active Crohn's disease is accompanied by blood coagulation within the blood vessels (asterisk) throughout the lesion; (d) the integrity of the epithelial layer of the intestinal tissue is lost (arrowhead) and excessive neovascularization can be observed (asterisks). Figure 7.2 (b to d) by courtesy of Dr G. Dijkstra, Department of Gastroenterology, University Hospital Groningen, The Netherlands.

sequently, they proliferate and migrate towards the tumour or inflammatory site. Finally they form a capillary lumen, after which vessel maturation takes place (for a more detailed description of the molecular regulation of angiogenesis, see Chapter 9) [34].

The putative role of angiogenesis in chronic inflammatory diseases is the maintenance of the inflammatory state by allowing ongoing recruitment of inflammatory cells and by supplying nutrients and oxygen to proliferating inflamed tissue. The increased endothelial surface creates an enormous capacity for the production of cytokines, adhesion molecules, and other inflammatory stimuli [35].

In many chronic inflammatory diseases, angiogenesis can be identified in the inflamed lesions. For example, in rheumatoid arthritis extensive neovascularization is present in the inflamed synovium where it is one of the earliest histopathological findings [36]. Since in RA synoviocytes exhibit characteristics of tumour cells, including somatic mutations in key regulatory genes such as H-ras and the p53 tumour suppressor, RA can be viewed as a multicentric tumour-like mass that invades and destroys its local environment [37]. Concurrent increased endothelial cell turnover may contribute to microvascular dysfunction and thereby facilitate persistent synovitis.

Although the identity of the factors that promote angiogenesis in RA specifically still remain unclear, both synovial tissue and fluid are enriched in angiogenesis-promoting factors. These include cytokines such as bFGF, VEGF, and IL-8, and soluble VCAM-1 and E-selectin [38].

In other chronic inflammatory diseases elevated levels of angiogenic factors were also found. In patients with psoriasis, skin lesions over-expressed IL-8 while the expression of one of the inhibitors of angiogenesis, thrombospondin-1, was downregulated [39].

Neovascularization in artherosclerotic lesions may be regulated by VEGF, as this factor is over-expressed by different cells in the plaque tissue [40–42]. The increased serum levels of VEGF that correlate with disease activity in patients with Crohn's disease and ulcerative colitis, indicate a role for this cytokine in promoting inflammation. Most likely, increased vascular permeability and/or wound healing via its pro-angiogenic activity are the basis for this effect [43].

There is at present no consensus on whether inflammation and angiogenesis can exist independently from each other. This is due to the lack of suitable marker-epitopes that cover all stages of angiogenesis, resulting in an underestimation or misinterpretation of the occurrence of angiogenesis. As cells of the immune system contribute significantly to the local production of pro-inflammatory and pro-angiogenic factors, it is likely though, that inflammation and angiogenesis affect each other to a considerable extent.

7.3.3 Activation Pathways of Endothelial Cells in Chronic Inflammation

Activation of endothelial cells leads to changes in endothelial cell properties such as loss of vascular integrity, expression of adhesion molecules, antithrombotic to prothrombotic phenotype changes, cytokine production and the upregulation of HLA molecules. All these diverse effects can be attributed to the activation of transcription factors [44]. Of the presently known transcription factors, NFκB is believed to be one of the most important in the regulation of endothelial cell activation. After a stimulus at the cell surface which is caused by e.g.

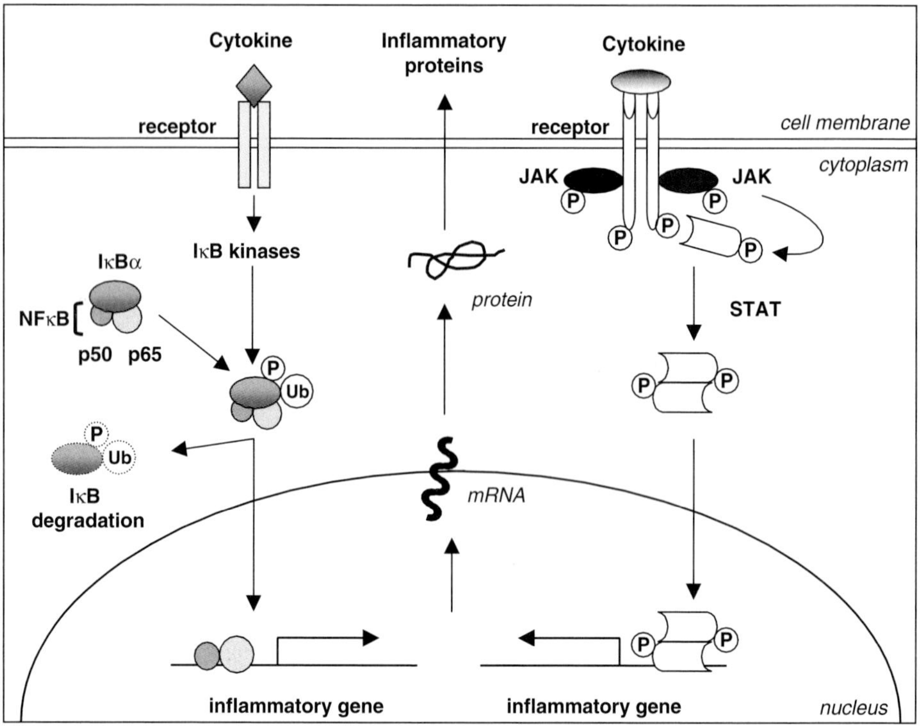

Figure 7.3. Schematic representation of some of the intracellular signal transduction pathways of cytokine receptor signalling. Interleukin (IL)-1 and TNFα, among others, induce cell activation via the NFκB pathway, although their respective signal transduction routes upstream of the NFκB system differ significantly. On interaction with their receptors, NFκB-inducible kinase (NIK) is activated resulting in IκB-kinase (IKK) phosphorylation. As a result, IκBα is phosphorylated, thereby becoming a substrate for ubiquitination and proteasome-mediated degradation. The released p50/p65 NFκB complex translocates to the nucleus, which results in the expression of inflammatory genes. The JAK/STAT route can be activated by cytokine (e.g. IL-6, IL-10 or IL-15) binding to their receptors. JAK phosphorylation is followed by phosphorylation of receptor subunits which function as a docking site for STAT molecules. Phosphorylated STAT molecules subsequently dimerize, translocate to the nucleus and modulate gene transcription.

interaction between a cytokine (IL-1, TNFα) and its receptor, UV radiation, lipopolysaccharide (LPS) or oxidized low density lipoproteins (oxLDL), NFκB-inducible kinase (NIK) is activated, which in turn phosphorylates IκB-kinase-1 (IKK-1), and perhaps IKK-2. The IKK molecules phosphorylate the inhibitor of κB (IκB) at serine-residues, resulting in ubiquitination and degradation of IκB by the proteasome machinery in the cell cytoplasm. The nuclear localization sequence (NLS) of the NFκB dimer then becomes exposed, after which the transcription factor travels to the nucleus and induces transcription of many pro-inflammatory genes (Figure 7.3) [45]. Since NFκB is a key component in the inflammation process, this important transcription factor is controlled by several autoregulatory loops. The expression of the inhibitory protein IκBα to which NFκB is bound in the cytoplasm, becomes upregulated when NFκB is activated, thereby repressing the transcription of VCAM-1 and E-selectin for instance [46]. Large quantities of NO produced by the inducible NO-synthase enzyme (iNOS) prevent further NFκB activation either by S-nitrosylation of cysteine 62 of the p50 subunit of NFκB or by stabilization of the inhibitory IκB-α protein [47,48]. As described above, the anti-apoptotic proteins Bcl-2, A20 and Bcl-x_L play an important role in controlling inflammation, as well as in endothelial cell activation. This is partly due to their ability to inhibit NFκB activation in endothelial cells. *In vitro*, these proteins block the induction of pro-inflammatory genes such as cytokines, pro-coagulant and adhesion molecules, and hence serve as a regulatory mechanism to restrain activation and injury [49,50].

Class I and II cytokine receptors that do not have direct tyrosine kinase activity mediate signal transduction in cells via JAK (janus kinase) and STAT (signal transduction and activator of transcription) molecules. Among these are receptors for IL-15 and GM-CSF, cytokines involved in T cell recruitment to rheumatoid arthritis lesions and pro-angiogenic responses of endothelium, respectively [17]. Each cytokine activates a pre-defined set of JAKs through interaction between one of the receptor subunits and a JAK molecule. The JAKs then become activated by reciprocal transphosphorylation and in turn phosphorylate substrates such as receptor subunits (Figure 7.3). This creates a docking site for signalling molecules leading to subsequent signal transduction cascades. Downstream of JAK activation lies a variety of targets, including the low molecular weight G protein Ras and its targets PI3-kinase and the serine/threonine kinase Akt or protein kinase B. In parallel, members of a family of STAT transcription factors sharing a central DNA-binding domain, can become phosphorylated. Their subsequent dimerization leads to nuclear localization and DNA-binding, although this is not sufficient for their transactivation. Through physical and functional interactions between STATs and other transcription factors gene transcription can take place to modulate cellular functions [51]. It has been reported that the c-Jun N-terminal kinase group of MAP kinases plays a role in endothelial cell signal transduction and activation upon exposure to inflammatory cytokines, in a manner similar to that described above. Due to space limitations, this will not be discussed further: a detailed description of the role of this regulatory pathway in various inflammatory conditions can be found in reference [52].

7.4 Targeting Drugs to the Endothelial Cell

In most, if not all, chronic inflammatory diseases endothelial cells are prominently involved in the disease process. This is demonstrated by an increased expression of adhesion molecules and production of cytokines, and their pro-angiogenic behaviour. This leads to continuous recruitment of leucocytes into the inflamed area, without (detectable) antigen present in the affected tissue, resulting in a vicious circle of tissue damage and leucocyte recruitment. Targeting inhibitory agents (in)to the endothelial cell may interrupt in this process by controlling the activation status of this cell type.

The advantage of endothelial cell targeting is the localization of the endothelial cells. Since they are in direct contact with the blood, the 'homing' devices do not have to cross the endothelial barrier to find their targets [53]. The expression of adhesion molecules and epitopes involved in angiogenesis are, at least in theory, suitable target epitopes for drug targeting purposes. In this respect it is important to distinguish between potential carriers that only bind to the target cells at the external surface, and carrier molecules that after binding are internalized and intracellularly processed. This will be discussed further in Section 7.7.

7.4.1 Target Epitopes on Inflammatory Endothelium

A rational approach in the development of drug targeting carriers for endothelial cells in inflamed tissue is to identify disease-induced target epitopes in these cells [54,55]. As discussed in Sections 7.2.1 and 7.3, in this respect E- and P-selectin, VCAM-1 and ICAM-1 are considered candidate target epitopes.

Over the past decade, many target epitopes have been described to be upregulated on angiogenenic blood vessels. Recently, several targets that are under investigation for the clinical development of anti-tumour agents, have also been studied under conditions of chronic inflammation. One of them is the $\alpha v\beta 3$ integrin receptor, which is upregulated on synovial blood vessels in antigen-induced arthritis (AIA) in rabbits and in human RA [36]. Furthermore, the VEGF-receptors VEGF-R1 (Flt-1/flt-1) and VEGF-R2 (KDR/flk-1) are strongly expressed by microvascular endothelial cells in RA, psoriatic epidermis and atherosclerotic lesions [42,56,57].

In addition to the above-mentioned target epitopes, other endothelial molecules can be considered as targets. For instance, chemokine receptors have been identified to be present on activated endothelial cells [58]. A chemokine that may be useful in endothelial cell targeting is Fractalkine, a chemokine with a CX_3C motif on a membrane-bound mucin-like stalk. This molecule has chemoattractant activity and promotes strong adhesion of T cells and monocytes through the Fractalkine receptor CX_3CR1. It can, however, also exist in a soluble form and thereby provide an undesirable sink for Fractalkine-targeted compounds. Furthermore, it is also expressed to various extents by monocytes and microglial cells [59–61]. Besides the specificity of cellular expression of the target epitopes, other important considerations should be taken into account when selecting a target epitope. These considerations will be discussed in more detail in Section 7.7.

At present, many questions regarding the use of the target epitopes for endothelium-directed drug targeting strategies still remain to be answered. The efforts that are being put into the development of drug targeting strategies directed at these epitopes should, in the near future, eventually lead to a better understanding of the potential of this area of drug development.

7.4.2 Targeting Devices

7.4.2.1 Monoclonal Antibodies

Various antibody-based targeting moieties have been described, either as directly acting compounds or as targeting devices. Antibodies against adhesion molecules have been widely used as blocking agents, presumably because of their high specificity and their relative ease of production. For instance, an anti-$\alpha 4$ integrin monoclonal antibody significantly attenuated colitis in the cotton-top tamarin by intervention in leucocyte adhesion and possibly in other integrin-mediated events such as T cell aggregation, T cell-stromal interactions and lymphocyte homing [62]. Furthermore, Jamar *et al.* reported the use of an anti-E-selectin antibody for imaging rheumatoid arthritis in humans. This antibody was ^{111}In-labelled and its localization was compared with a non-specific ^{99}Tc^{m}-labelled immunoglobulin. The anti-E-selectin antibody was found to have a more specific distribution and higher sensitivity of detection than the control antibody. As the ^{99}Tc^{m}-label is preferred for its physical properties, further investigations now comprise the production of a ^{99}Tc^{m}-labelled F(ab′)$_2$-fragment of the anti-E-selectin antibody [63].

In studies of tumour-induced angiogenesis, the monoclonal antibody LM609 against the αvβ3 integrin receptor was used for imaging purposes [64]. Its humanized counterpart Vitaxin, which has now entered clinical anti-cancer trials, could in theory also be used for targeting to endothelium in chronically inflamed sites where αvβ3 is upregulated.

There are a few examples of the use of antibodies as targeting devices to vascular endothelium. Kiely *et al.* used the monoclonal antibody H18/7 to target hirudin, an anti-thrombin agent, to the surface of activated human vascular endothelial cells *in vitro* [65]. The same antibody was also incorporated in liposomes to target doxorubucin to activated endothelium [66]. A bispecific antibody recognizing both E-selectin and an adenovirus (AdZ.FLAG) was used to specifically transduce activated endothelium with a gene encoding β-galactosidase, resulting in the production of this protein [67]. However, all of these studies were carried out *in vitro* only and at present, no definite conclusions on their pharmacological potential *in vivo* can be drawn.

In another approach, antibodies directed against ICAM-1, angiotensin converting enzyme (ACE) or CD31 conjugated to catalase were successfully used to protect perfused rat lungs against oxidative stress [68,69]. The lungs receive the entire cardiac blood output and contain 30% of the endothelial cell population in the body. In addition, ICAM-1 and ACE are constitutively expressed on the pulmonary vasculature. Therefore, this approach may result in significant accumulation in the lungs of ICAM-1- or ACE-directed conjugates, even when inflammatory processes in other organs are the desired target. Consequently, these conjugates are likely to be useful only for the treatment of lung disorders.

7.4.2.2 Peptides

Peptides, like antibodies, have until now mostly been exploited as direct-acting moieties in imaging or inhibition studies. An E-selectin binding peptide has been used for imaging in arthritis models. This peptide has the advantage of recognizing and binding to murine, rat and human E-selectin, in contrast to antibodies which lack cross reactivity with E-selectins from other species [70,71]. Peptides recognizing and blocking the function of other selectins, ICAM-1, VCAM-1 and chemokines have also been documented [1,72].

Peptides containing an RGD motif can be used to target the αvβ3 integrin receptor on angiogenic blood vessels. Specifically the dicyclic peptide RGD-4C, an 11mer with two disulfide bridges, exhibits a high affinity for the αvβ3 integrin receptor due to its constrained conformation. Arap *et al.* showed that this peptide can selectively deliver chemotherapeutics at the αvβ3 integrin receptor on angiogenic blood vessels in solid tumours [73]. Furthermore, an apoptosis-inducing peptide was specifically delivered to angiogenic endothelium by a chemically derived peptide dimer [74]. Recently, Storgard and colleagues demonstrated that monocyclic RGDfV peptides target the synovial blood vessels in antigen-induced arthritis in rabbits [36]. A general consideration concerning peptides is the relatively short half-life of these molecules compared to larger proteins [75]. Whether this is an advantage or disadvantage depends on the application. For nuclear imaging purposes a short half-life is favourable, whereas a longer circulation time would be useful for therapeutic purposes.

Peptides as well as oligosaccharides (discussed below) that specifically bind to molecules on the endothelium may be used as homing ligands in larger constructs. Covalent attachment to e.g. a protein backbone or liposomes may lead to carriers with multivalent binding sites. Furthermore, drug loading of the peptide/oligosaccharide-modified carrier can be increased, in contrast to the 1 : 1 ratio in the case of direct coupling of the drug to the peptide or oligosaccharide.

7.4.2.3 Oligosaccharides

Selectins mediate contact by binding to carbohydrate-containing receptors on leucocytes through their N-terminal lectin domain. Sialyl-Lewis X (Neu5Acα2-3-Galβ1-4(Fucα1-3)Glc-NAc) and derivatives thereof were shown to bind to the selectins and subsequently inhibit ischaemia-induced leucocyte infiltration in the liver [76–78]. A similar compound prevented antigen-induced late bronchial responses and airway hyper-responsiveness in allergic sheep [79].

The binding potency of the native sialyl-Lewis X can be increased with three orders of magnitudes by conjugating this saccharide to BSA. This is probably due to the clustering of saccharides that may favour binding to E-selectin [80].

It should be borne in mind, that targeting the selectins with Sialyl-Lewis X-derived homing devices will not result in selective targeting to the endothelium. As well as the activated endothelium, naïve T-lymphocytes bearing L-selectin will also be a target for such a preparation. Whether this poses a problem by creating a site of non-target cell binding and hence loss of the drug targeting preparation, or whether it is beneficial from a therapeutic point of view because it may inhibit T cell activation, remains to be investigated.

7.4.3 Drugs Inhibiting Endothelial Activation

7.4.3.1 Inhibitors of NFκB and Other Intracellular Signalling Pathways

Based on its central role in pro-inflammatory reactions, the transcription factor NFκB has attracted much of attention as a target for therapeutic intervention. Drugs like pyrrolidine dithiocarbamate (PDTC), N-acetylcysteine or α-tocopheryl succinate prevent IκB degradation, thereby stabilizing the NFκB complex and inhibiting cytokine production, adhesion molecule expression and preventing vascular injury both *in vitro* and *in vivo* [81–86]. However, these drugs work in a rather non-specific fashion since they influence general cellular mechanisms such as the redox state and proteasome function.

A more selective inhibition of NFκB can be achieved by transfecting cells with DNA coding for the natural inhibitor IκBα or a mutant IκB protein that lacks 36 N-terminal amino acids, and consequently becomes proteolysis resistant. In this way expression of adhesion molecules and monocyte adhesion and transmigration can be inhibited [87,88]. The potentials and limitations of these latter types of therapy are however not fully understood as yet. Different transfection systems (adenoviral, retroviral, non-viral) are available for gene delivery purposes, all with their own potentials and restrictions.

Inhibitors of IκBα phosphorylation have been described which irreversibly inhibit cytokine-induced phosphorylation without affecting constitutive phosphorylation. One such compound (Bay 11-7083 ((E)3-[4-*t*-butylphenyl)-sulfonyl]-2-propenenitrile)) was found to be effective in two animal models of inflammation after intraperitoneal administration [89]. In addition to the effect it has on the expression of adhesion molecules in pro-inflammatory responses, inhibition of the transcription factor NFκB will also have an effect on angiogenesis. Endothelial cells can produce growth factors and cytokines which have pro-angiogenic effects. Some of these factors, e.g. IL-8, TNFα and MCP-1 are known to be produced via NFκB-mediated endothelial cell activation [90,91]. The importance of NFκB-mediated responses in pro-angiogenic endothelium was reflected in studies in which the NFκB inhibitor PDTC decreased retinal neovascularization in the eye of mice [92].

Taking their central role in cytokine signalling into account, it seems evident that molecules that inhibit JAK/STAT signalling are also interesting candidates for targeting to endothelial cells in chronic inflammatory lesions. Endogenous factors such as suppressors of cytokine signalling (SOCS), IL-4, STAT-induced STAT-inhibitor-1 (SSI-1) and JAK binding protein (JAB), have been reported to intervene effectively at this level of cytokine signalling in a variety of cell types [93–95]. In addition, a number of new chemical entities are under development to fulfil this function [96–99]. Although at present most of these compounds have been tested in oncology research, it seems likely that they can also affect JAK/STAT signalling in other cells such as pro-inflammatory endothelium.

It is anticipated that elucidation of the fine tuning of the regulatory processes in endothelial cells under pro-inflammatory conditions, will lead to the identification of additional novel pharmacological targets in the near future.

7.4.3.2 Glucocorticoids, NSAIDs and Others

Glucocorticoids are commonly used in inflammatory disorders owing to their broad anti-inflammatory and immune suppressive effects in a wide variety of diseases. A major drawback of these compounds is the serious side-effects associated with their use. They have at least a partial inhibitory effect on the expression of adhesion molecules by the endothelium and inhibit cytokine production [2]. For instance, the glucocorticoid dexamethasone impaired the increase of E-selectin and ICAM-1 expression on HUVEC, at least partially by an inhibitory effect on NFκB (Figure 7.4) [100–102]. In IBD patients, prednisolone, another glucocorticoid, blocked NFκB activation thereby inducing healing of colonic inflammation [103]. Barnes and Karin also underlined the beneficial effect of glucocorticoids in chronic inflammatory diseases via NFκB-mediated mechanisms [90]. The anti-angiogenic effects of glucocorticoids were demonstrated by the ability of dexamethasone to inhibit capillary

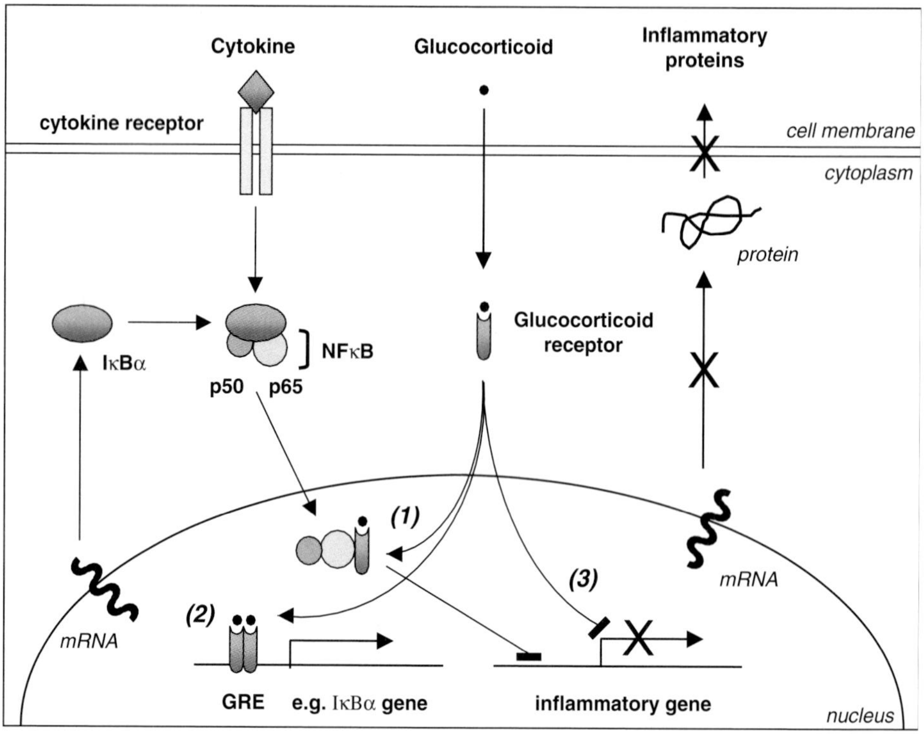

Figure 7.4. Glucocorticoids affect intracellular signal transduction pathways and gene transcription in (at least) three ways. Glucocorticoid–receptor complexes can prevent NFκB-mediated activation of pro-inflammatory genes by binding to the p65 subunit of NFκB (1). Glucocorticoid–receptor complexes can furthermore bind as dimers to the glucocorticoid-responsive element (GRE) in the promoter region of the IκBα gene, leading to IκBα expression and subsequent complexation with, and hence inactivation of NFκB (2). Lastly, the monomeric glucocorticoid–receptor complex can intervene in transcriptional activation of various genes which are activated when NFκB binds to κB sites in their promotor region (3). Adapted from reference [90].

tube formation *in vitro* and to decrease the vascular density of brain tumours *in vivo* [104,105].

Non-steroidal anti-inflammatory drugs (NSAIDs) exert their effect by inhibiting the enzymes cyclooxygenase-1 and -2 (COX-1 and -2), which are instrumental in the synthesis of prostaglandins. COX-2 is thought to be responsible for the enhanced production of prostaglandins in inflammation, whereas COX-1 is associated with the production of protective prostaglandins in e.g. the gastrointestinal tract. The well-known gastrointestinal side-effects of NSAIDs are likely to be mediated by the non-selective inhibition of both COX enzymes, resulting in e.g. a decreased protection of the gut by COX-1-related prostaglandins. NSAIDs are thought to predominantly mediate downregulation of leucocyte adhesion molecules such as L-selectin [106]. Furthermore, they inhibit expression of endothelial cell adhesion molecules. For example, diclofenac inhibited the adhesion of HL60 cells to HUVEC, *in vitro* [107]. NSAIDs such as ibuprofen and acetylsalicylic acid did not exert such an effect in this experimental set-up. In a study by Hofbauer *et al.*, ibuprofen was able to inhibit leucocyte migration through endothelial monolayers [108]. Sodium salicylate decreased E-selectin, ICAM-1 and VCAM-1 expression, probably via inhibition of IκB-phosphorylation [109], and P-selectin expression in an NFκB-independent fashion [110]. COX enzymes are unlikely to be involved in these effects, since indomethacin, a non-salicylate COX-inhibitor had no effect on surface expression of adhesion molecules. Recently, it was demonstrated that both COX selective and non-selective NSAIDs inhibited angiogenesis through a direct interaction with molecular processes such as inhibition of MAP kinase (ERK2) activity and intervention in ERK nuclear translocation in endothelial cells [111].

Finally there has been a growing interest in the development of selective leukotriene inhibitors. Bay Y 1015 (R-(-)-2-cycloheptyl-N-methylsulfonyl-(4-(2-quinolinyl-methoxy) phenyl)-acetamide) is a quinoline-type 5-lipoxygenase-activating protein inhibitor which was effective in inhibiting inflammation in a dextran sulfate model of mouse colitis [112]. Whether these compounds can also exert their anti-inflammatory action through inhibition of endothelial cell activation needs to be established.

7.4.3.3 Antisense Oligonucleotides

A selective method of preventing the expression of adhesion molecules or cytokines is the use of antisense oligonucleotides. These oligonucleotides are short sequences of nucleic acids complementary to mRNA sequences of specific proteins of interest. If delivered to the cytoplasmic compartment of cells these oligonucleotides are able to form a complex with their target mRNA. In this way the translation of mRNA into protein by ribosomes is inhibited. The subsequent mRNA degradation by RNAse H results in reduced expression of the protein (see also Chapter 5 for a description of antisense oligonucleotides as therapeutic modalities).

ICAM-1, VCAM-1 and E-selectin synthesis was successfully blocked *in vitro* using these types of molecules [113]. *In vivo*, the systemic administration of ICAM-1 antisense oligonucleotide prevented and reversed murine colitis without serious side-effects [114]. In a placebo-controlled trial of the human analogue, the antisense oligonucleotide was effective and well tolerated [115].

Local and systemic administration of an NFκB p65 subunit antisense phophorothioate oligonucleotide effectively inhibited experimental colitis in mice [116].

It is important to realize however that these antisense molecules were not specifically targeted to the endothelium. Consequently, the contribution of the endothelial cells to the effects observed is unknown. Furthermore, in these studies adequate control experiments with mismatched oligonucleotides are essential, since polyanionic agents such as antisense oligonucleotides can exert a broad range of non-specific antisense effects due to non-specific binding to proteins [117].

This non-antisense-based feature of oligonucleotides is used in Systematic Evolution of Ligands by Exponential Enrichment (SELEX) technology, a process based on oligonucleotide combinatorial chemistry. SELEX can lead to the development of high affinity binding antagonists for various molecules, including adhesion molecules, growth factors, nucleic acid binding proteins, enzymes, and peptides. This is exemplified by the development of so-called aptamers that bind to P-selectin with subnanomolar affinities, and, in theory, may serve as candidates for homing devices in drug targeting preparations [118].

7.4.3.4 Drugs that Inhibit Angiogenesis-associated Events

Inhibition of angiogenesis is a relatively new approach in the treatment of chronic inflammatory diseases. Angiogenesis is a complex, multi-step process that can be inhibited at many levels. For example, antibodies or soluble receptors can block the action of angiogenic factors such as VEGF and bFGF. Matrix degradation can be inhibited by MMP inhibitors such as Marimastat or by plasminogen activator inhibitors, whereas endothelial cell proliferation and migration can be affected by the fumagillin analogue AGM-1470 or by an anti-$\alpha v\beta 3$ monoclonal antibody (see Chapter 9 for a more detailed description of the application of these anti-angiogenic drugs in cancer therapy).

That angiogenesis inhibition positively affects chronic inflammatory disorders was demonstrated in several animal studies. The fumagillin analogue AGM-1470 (also called TNP-470) prevented pannus formation in rat arthritis models [119,120] and also reduced artherosclerosis in aortas of apoE –/– mice [121]. Intra-articular administration of cyclic RGD peptides in bFGF-augmented antigen-induced arthritis (AIA) in rabbits resulted in increased vascular apoptosis leading to inhibition of synovial angiogenesis. This therapy also reduced the symptoms of arthritis such as joint swelling, synovial infiltrate, and pannus formation [36]. Thalidomide, a drug with many pharmacological activities, may also be a suitable drug for use in anti-angiogenic strategies in inflammatory disorders [122,123]. Besides having an immunomodulatory effect, it is believed to inhibit the upregulation of endothelial integrin expression [124].

It should be noted, however, that mechanisms of action of most anti-angiogenic compounds are not well understood at present. For example, trombospondin-1 (TSP-1) is able to inhibit tumour-associated angiogenesis, but when TSP-1 pellets were implanted into the ankles of AIA rats, it enhanced joint swelling and body weight loss in a dose- and time-dependent manner. These, possibly indirect, effects may be due to the involvement of TSP-1 in cell adhesion, as well as to its interactions with other adhesion molecules and inflammatory mediators [125].

Although in animal studies little to no toxicity of the various angiogenesis inhibitors was reported, the recent withdrawal of two MMP inhibitors from clinical trials indicates that these drugs may have serious, deleterious side-effects [126]. A novel approach that is therefore worthwhile investigating, is the selective delivery of anti-angiogenic drugs to the pro-angiogenic endothelium. It can be envisioned that increased availability of these agents at the site of the angiogenic endothelium may improve therapeutic outcome and diminish toxicity. In general, many drugs with various mechanisms of action were shown to be potent inhibitors of inflammatory responses in animals. Lack of effectiveness and/or severe toxicity in early clinical testing justifies studying the added value of selective targeting of these compounds to the endothelial compartment at the diseased site.

7.5 *In Vitro* Techniques for Studying Endothelial Cell Activation

7.5.1 Cell Cultures

Endothelial cells from various origins in the body can be used in endothelial cell research (see also Chapter 9). The most widely used cells are human umbilical vein endothelial cells (HUVEC), primary cells that can be isolated from the umbilical cord. Because of limited availability of umbilical cords and the limitations associated with the number of cell passages that can be used, several immortalized endothelial cell lines have been developed [127,128]. Although these cell lines are simple to use, most lines have lost particular characteristics. For example, the spontaneously transformed HUVEC cell line ECV304 and the cell line EaHy926 do not express VCAM-1 and E-selectin in response to TNFα stimulation [129]. Furthermore, chemokine receptor expression and responses to pharmacological stimuli are differentially regulated in HUVEC and ECV304 [58]. The endothelium differs to a large extent between larger and smaller vessels with respect to responses to stimuli and their regulatory processes, thereby e.g. influencing the relative expression of adhesion molecules [130]. Therefore, the use of microvascular cells, for instance human dermal microvascular endothelial cells (HDMEC), may be an alternative option. In this cell type E-selectin was shown to be degraded and internalized more slowly than in HUVEC [3]. This seems to be a general characteristic of microvasculature endothelium: human intestinal microvascular endothelial cells also displayed prolonged expression of E-selectin after stimulation [131].

Murine endothelial cell lines can be used to more explicitly follow up on observations in *in vivo* mouse models or for rapid screening purposes of murine-directed drug targeting preparations. Examples of these cell lines are H5V, 1G11 and SIEC, which, to a variable extent, express the adhesion molecules E-selectin, ICAM-1 and VCAM-1 on stimulation with e.g. TNFα [127]. The binding and processing of carrier molecules, in addition to the pharmacological effects of intracellularly delivered drugs can in principle be studied in the cell types mentioned.

7.5.2 Read-out Systems

As stated earlier, activation of endothelial cells by pro-inflammatory stimuli leads to the expression of cell adhesion molecules and cytokines such as IL-6 and IL-8. The expression and hence modulation of surface expressed adhesion molecules by e.g. targeted delivery of inhibitors of NFκB, can be measured using flow cytometric analysis or whole cell ELISA techniques. Cytokine production can be measured in the supernatant of cultured cells or in biological fluids. Furthermore, competitive or quantitative RT-PCR analysis of mRNA levels of cell adhesion molecules or cytokines, allows the transcriptional activity of the genes of interest to be estimated.

To investigate the effects of drugs on NFκB activation at the molecular level, the Electric Mobility Shift Assay (EMSA) is a useful read-out system. With this technique the nuclear localization of this transcription factor following activation and subsequent translocation can be semi-quantitatively assessed. Drugs intervening in this process can thus be tested. Some drugs, however, inhibit NFκB-mediated gene expression at levels downstream of NFκB binding to its DNA consensus sequence. For these drugs the EMSA technique is not suitable. Furthermore, antibodies specifically recognizing the NFκB nuclear localization sequence can be applied in immunohistochemical analysis to determine the activation and nuclear localization of this transcription factor [132].

EMSA assays can also be exploited to measure STAT nuclear localization, which is, similar to NFκB localization, a measure of STAT activity. Determination of JAK phosphorylation is carried out by immunoprecipitation of the JAK proteins from cell lysates, followed by SDS-PAGE electrophoresis, immunoblotting with antiphosphotyrosine antibody and JAK-specific antibody re-probing [99].

Another molecular biological approach to the measurement of (inhibition of) NFκB activity exploits transfection of endothelial cells with artificial reporter genes. These genes contain a κB-promotor region preceding a gene encoding a reporter protein such as Green Fluorescent Protein (GFP) or luciferase. If NFκB becomes activated it will activate the promotor of the gene construct, resulting in the production of the reporter protein. The expression level of this protein can subsequently be determined by flow cytometry or enzymatically and is a direct measure of NFκB activity. The same principle can be applied using, for instance, a reporter construct with a promotor containing a Glucocorticoid Responsive Element (GRE) to study the cellular delivery of targeted glucocorticoids such dexamethasone. These transfection systems can be useful tools in the investigation of the basic principles and characteristics of drug targeting to endothelial cells.

The general aim in inhibiting endothelial cell function by the targeted delivery of anti-inflammatory drugs is to inhibit local leucocyte recruitment. *In vitro*, interactions between leucocytes and endothelium, and pharmacological intervention can be analysed using quantitative flow cytometry. In such an assay, endothelium is cultured on collagen gels. After incubation of leucocytes with endothelium under different experimental conditions, adherent leucocytes are released by trypsin treatment. Migrated leucocytes are retrieved by digestion of the collagen using collagenase. For quantitation purposes, known amounts of fluorescent beads are added to the cell samples prior to flow cytometry analysis. In combination with cell identification based on forward scatter/side scatter characteristics, accurate determination of the numbers of leucocytes which adhered to or migrated through the endothelium can be obtained [133].

The various steps of angiogenesis can be investigated *in vitro* in assays studying endothelial cell migration, proliferation or tube formation. These types of assays will be described in detail in Chapter 9, as most of them have been applied in cancer research. Because angiogenesis *in vivo* does not occur with only one cell type present, cell co-culture models were developed. In these models endothelial cells were grown in the presence of other cells, e.g. tumour cells, keratinocytes, astroglia cells or fibroblasts, which can induce angiogenesis by producing soluble factors or by inducing cell–cell contact. Villaschi and Nicosia have examined the communication between endothelial cells and fibroblasts in a rat aorta explant model. In these cultures, fibroblasts stabilized microvessel sprouts and hence allowed the study of processes such as proliferation, migration, tube formation, and later events such as pericyte migration and extracellular matrix deposition [35,134].

7.6 *In Vivo* Animal Models for Studying Inflammation

It is difficult to mimic *in vitro* the complex environment of the endothelium that prevails at sites of inflammation. The extracellular matrix (ECM) is subjected to a continuous remodelling during endothelial cell activation and leucocyte recruitment. Changes in blood flow velocities may alter the cellular behaviour of endothelial and non-endothelial support cells. Therefore, drug targeting strategies aimed at the endothelium need to be investigated *in vivo* to establish their effects on pathological processes.

Many animal models have been described in which inflammation is induced and activation of the endothelium occurs. Here, some disease-specific models, in addition to some general inflammatory experimental models will be summarized.

7.6.1 Rheumatoid Arthritis

There are several spontaneous and induced animal models of arthritis available. In MRL/*lpr* mice that have a defective Fas gene, arthritis develops spontaneously, but the immune mechanisms underlying joint disease in these mice are not known [135]. T cell-mediated arthritis can be induced in susceptible strains of mice and rats by immunization with type II collagen, the collagen type found in cartilage. This model has long been used to study mechanisms and effects of anti-arthritic drugs, for instance liposome-encapsulated drugs [136]. However, in the human situation there is no evidence for collagen-specific immunity. Arthritis can also be induced using various bacterial antigens. However, in these models the resemblance to human disease is, at best, questionable [137].

7.6.2 Inflammatory Bowel Disease

There is a wide variety of animal models that mimic aspects of inflammatory bowel disease. The choice of an appropriate model depends on the question being addressed in a particular study. There are multiple factors contributing to IBD, including for example, environmental or genetic susceptibility. Common experimental models differ largely with regard to the factors that contribute most prominently to the pathogenesis of IBD. Therefore not only should the model be chosen carefully, but the intrinsic limitations in the interpretation of results should also be recognized. For instance, to test new anti-inflammatory drugs a simple and reproducible model involving non-specific inflammation might be selected. Examples are dextran sodium sulphate (DSS)-induced colitis in which the cellular toxicity of DSS induces an inflammatory reaction, or trinitrobenzene sulfonic acid (TNBS)-induced colitis, which is a delayed-type hypersensitivity response to this contact allergen. In addition to these 'non-specific' models, several murine knock-out models have become available for IBD research. These include IL-2 and IL-10 knock-out mice, as well as mice with disrupted T-cell receptor α-chain genes, which all develop severe colitis. The intestinal lesions in these animal models resemble at least to some extent those that can be found in the human situation, including for instance, cellular infiltrates or ulcerations. In a detailed review Elson *et al.* discussed these different models and their most useful applications [138].

7.6.3 Atherosclerosis

Various animal species have been exploited in experimental atherosclerosis research, but nowadays most of the research is performed in mice [139]. This species has the advantage of developing atherosclerotic lesions in a relatively short period of time. The emergence of a broad variety of knock-out and transgenic mouse strains has led to a huge increase in the atherosclerosis research performed in this species. Mice are resistant to atherosclerosis when fed a normal low-fat chow diet, but they can develop atherosclerotic lesions after hypercholesterolaemia has been induced. The three most widely used models in research on atherosclerosis are diet-induced, apoE deficiency-induced or LDL receptor deficiency-induced [139].

Rabbits are the second most used species in this type of research, in particular the LDL receptor deficient Watanabe heritable hyperlipidaemic rabbits. These animals show expression of VCAM-1 in atherosclerotic lesions [140], but resemblance to human lesions is generally low [139,141].

7.6.4 Angiogenesis

One of the animal models of angiogenesis that is driven by inflammatory stimuli is ocular neovascularization. Ocular angiogenesis is induced by implantation of pellets containing pro-angiogenic factors in the avascular cornea in mice. Angiogenic effects were observed with pellets containing VEGF or bFGF, and also with IL-1β or TNFα. Implantation of inflammatory cells such as macrophages or fibroblasts also induces angiogenesis in this model [35]. A vast number of animal models used for angiogenesis research in general have been developed over recent years. Although it is believed that in many conditions angiogenesis and inflammation are directly related, the variations in inflammatory responses in the various models warrants great care in extrapolating data from one model to another.

7.6.5 General Inflammation Models

In addition to disease-specific models, several models of general inflammation, involving leucocyte adherence and transmigration into inflamed sites have been developed. These can be used to study general kinetics, homing characteristics and effects of anti-inflammatory drug targeting conjugates. An example of a rapid method of general induction of adhesion molecule expression *in vivo*, is the systemic administration of bacterial lipopolysaccharide (LPS). This leads to expression of E-selectin, P-selectin, ICAM-1 and VCAM-1 within different vascular beds [142]. In a similar manner, systemic IL-1 administration in pigs leads to generalized E-selectin expression as determined by using a radiolabelled monoclonal antibody [143]. Intra-dermal administration of inflammatory mediators such as IL-1, LPS or TNFα results in a local inflammatory reaction and adhesion molecule expression in the endothelium of the skin [143].

Several immune-regulated models have been described, of which the delayed-type hypersensitivity reaction in the skin is an example. In several species application of allergic or contact-sensitizing substances results in a local inflammatory reaction involving adhesion molecules and leucocyte recruitment [144,145]. These models of pathological mechanisms and effects of anti-inflammatory drugs have been studied in mice in particular [146].

The models described herein are just a few examples of general experimental models used to study inflammatory reactions. A wide variety of other models exist which can be used in drug targeting research. It is however beyond the scope of this chapter to list all the models which have been used.

7.7 General Considerations and Practical Directions for Endothelial Cell Targeting Research

7.7.1 The Choice of a Target Epitope

Identifying an appropriate epitope for the desired drug targeting strategies is a complex process in which several considerations have to be taken into account. First, an important feature of a target epitope is its cellular processing, i.e. its internalization characteristics and its route into the cell. For instance, E-selectin is directed to the lysosomes and subsequently degraded, whereas P-selectin is re-routed through the Golgi-apparatus to the Weibel Palade bodies in which it is stored [147–150]. In contrast, antibodies directed against ICAM-1 or VCAM-1 are not internalized, but remain surface bound to the endothelial cell [147,148]. Second, the tissue-specific distribution pattern of constitutively expressed target epitopes may largely determine the selectivity of accumulation of the chosen targeting device in the diseased tissue. For instance, under normal physiological circumstances ICAM-1 and VCAM-1 are expressed on non-inflammatory endothelium, dendritic cells and leucocytes. Another example is the constitutive expression of P-selectin on platelets, in addition to being expressed on activated endothelial cells. When aiming at target epitopes that are not specifically induced in diseased tissue, effective targeting to chronically activated endothelium does not seem to be feasible. Third, the level of expression of the target epitope has to be taken into account. Hypothetically, differences in target epitope density may allow discrimination between diseased and non-diseased tissues, as has been reported for tumour-associated antigens. In this case the mechanism of action and the therapeutic window of the drug will determine whether delivery of small amounts to the non-diseased endothelium will lead to undesirable toxicity. Fourth, cleavage of target molecules from the target cell membrane, resulting in soluble adhesion molecules, may frustrate the process of targeting to the endothelial cell as a result of undesirable systemic complex formation.

Bearing these considerations in mind, of the presently identified endothelial adhesion molecules, E-selectin in particular seems to be a suitable epitope for targeting chronic inflammatory endothelial cells.

7.7.2 Disease Stage

In tumour models different stages of angiogenesis are thought to prevail, leading to varying degrees of responsiveness to anti-angiogenic treatment. Therefore anti-angiogenic drugs may prove most efficacious when targeted at distinct stages of the angiogenic process [151].

The same may hold true for the different stages in endothelial cell activation during flare-ups in chronic inflammatory disorders. Therefore, the use of a combination of different drug targeting preparations aimed at disease stage-specific epitopes is likely to be a prerequisite for targeting endothelial cells at various stages of activation.

7.7.3 Drugs of Choice

Similar to drug targeting strategies aimed at multiple target epitopes, a drug that intervenes at various stages of cell activation may be exploited for effective blockade of endothelial cell involvement in chronic inflammation.

In angiogenic processes in cancer the induction of endothelial apoptosis is a promising strategy in the battle against tumour growth. Not only does this approach inhibit the formation of new blood vessels, but the change in phenotype of apoptotic endothelial cells resulting in their becoming pro-adhesive to non-activated platelets, contributes to the anti-tumour effects, since this leads to prothrombotic effects and coagulation [152]. Unexpectedly, the induction of endothelial apoptosis by blocking the $\alpha v\beta 3$ integrin receptor with either antibodies or peptides also resulted in an improved therapeutic outcome in an experimental arthritis model [36]. It needs to be established whether under inflammatory conditions the pro-adhesiveness of endothelial cells contributes to the effects observed. These studies exemplify the complexity of the immune system in inflammatory and angiogenic processes, and stress the need for further investigation in this regard. Finally, it should be noted that information concerning angiogenic processes in inflammatory disorders is still limited, and this is particularly true with regard to attempts to inhibit angiogenesis in the treatment of chronic inflammatory disorders. It is the remit of future research to further clarify the potentials and limitations of these approaches.

Some of the new drugs described, for instance inhibitors of NFκB and JAK/STAT signal transduction pathways, have been selected because of their potency in *in vitro* screening procedures. In most cases no data are available on *in vivo* effectiveness and toxicity, either in animals or in human patients. Taking into account the high potency but lack of cell selectivity of these compounds, considerable toxicity can be expected. In these cases, drug targeting technology should be considered as a valuable tool to improve the chances of these compounds becoming therapeutically useful in the future.

Apart from the primary delivery process, the rate of intracellular release of the drug from the carrier and the potential back-flux of the released agent from the target cell into the system have to be considered. In fact, the targeting efficiency (the cellular levels obtained with a targeted drug in comparison with the levels obtained with a non-targeted drug) will be determined by the half-life of elimination from the target cell after cell selective delivery and the half-life following parent drug administration. A slow (rate-limiting) release from the carrier will obviously be favourable since this will determine the length of time that the drug remains in the target cell (see also Chapter 13 on pharmacokinetic considerations in drug targeting).

A large amount of clinical data is available on some of the common anti-inflammatory drugs such as glucocorticoids and NSAIDs. Coupling of these drugs to drug carriers may favourably affect kinetics and metabolism, thereby improving effectiveness and safety. So, despite the available knowledge concerning these classical drugs, new evaluation of their use as a part of drug targeting conjugates in a clinical setting seems obligatory.

7.8 Conclusions

The importance of the endothelium in the pathology of chronic inflammatory diseases is nowadays appreciated. Therefore, the endothelium is an attractive target for therapeutic intervention. Drug targeting strategies can be helpful in the treatment of these disorders. Endothelial cells have the advantage of being easily accessible for drug targeting conjugates. Preliminary results from *in vitro* studies demonstrated the potential of endothelium-directed targeting strategies. This has stimulated current efforts to determine the therapeutic potential of delivery modalities in *in vivo* models of chronic inflammation. Future studies will elucidate whether intervention in endothelial activation is able to abrogate the vicious circle of leucocyte recruitment and tissue damage, thereby improving the diseased state.

References

[1] Osborn L, In: *Adhesion Molecules and Chemokines in Lymphocyte Trafficking* (Ed. Hamann A), pp. 217–236. Harwood Academic Publ. GmbH, Chur, **1997.**
[2] Bell CJ, Wallace JL, *BioDrugs* **1997,** *7*, 273–284.
[3] Kluger MS, Johnson DR, Pober JS, *J. Immunol.* **1997,** *158*, 887–896.
[4] Panes J, Perry M, Granger DN, *Br. J. Pharmacol.* **1999,** *126*, 537–550.
[5] Malik AB, Lo SK, *Pharmacol. Rev.* **1996,** *48*, 213–229.
[6] Wuyts A, Struyf S, Proost P, Van Damme J, In: *The Cytokine Network and Immune Function* (Ed. Theze J), pp. 125–145. Oxford University Press, Oxford, **1999.**
[7] Newman PJ, *J. Clin. Invest.* **1999,** *103*, 5–9.
[8] Mantovani A, Sozzani S, Vecchi A, Introna M, Allavena P, *Thromb. Haemost.* **1997,** *78*, 406–414.
[9] Carlos TM, Harlan JM, *Blood* **1994,** *84*, 2068–2101.
[10] Bianchi E, Bender JR, Blasi F, Pardi R, *Immunol. Today* **1997,** *18*, 586–591.
[11] Gonzalez-Amaro R, Diaz-Gonzalez F, Sanchez-Madrid F, *Drugs* **1998,** *56*, 977–988.
[12] Orteu CH, Poulter LW, Rustin MH, Sabin CA, Salmon M, Akbar AN, *J. Immunol.* **1998,** *161*, 1619–1629.
[13] Daemen MA, van't Veer C, Denecker G, Heemskerk VH, Wolfs TG, Clauss M, Vandenabeele P, Buurman WA, *J. Clin. Invest.* **1999,** *104*, 541–549.
[14] Brostoff J, Scadding GK, Male D, Roitt IM, *Clinical Immunology*. Mosby, London, **1994.**
[15] Koch AE, Burrows JC, Haines GK, Carlos TM, Harlan JM, Leibovich SJ, *Lab. Invest.* **1991,** *64*, 313–320.
[16] Kriegsmann J, Keyszer GM, Geiler T, Lagoo AS, Lagoo-Deenadayalan S, Gay RE, Gay S, *Arthritis Rheum.* **1995,** *38*, 750–754.
[17] Oppenheimer-Marks N, Brezinschek RI, Mohamadzadeh M, Vita R, Lipsky PE, *J. Clin. Invest.* **1998,** *101*, 1261–1272.
[18] Fox DA, *Arch. Intern. Med.* **2000,** *160*, 437–444.
[19] Amin AR, Attur M, Abramson SB, *Curr. Opin. Rheumatol.* **1999,** *11*, 202–209.
[20] Brand K, Page S, Rogler G, Bartsch A, Brandl R, Knuechel R, Page M, Kaltschmidt C, Baeuerle PA, Neumeier D, *J. Clin. Invest.* **1996,** *97*, 1715–1722.
[21] Karmann K, Min W, Fanslow WC, Pober JS, *J. Exp. Med.* **1996,** *184*, 173–182.
[22] Mach F, Schonbeck U, Libby P, *Atherosclerosis* **1998,** *137 (Suppl.)*, S89–S95.
[23] Mach F, Schonbeck U, Fabunmi RP, Murphy C, Atkinson E, Bonnefoy JY, Graber P, Libby P, *Am. J. Pathol.* **1999,** *154,* 229–238.
[24] Stallmach A, Strober W, Macdonald TT, Lochs H, Zeitz M, *Immunol. Today* **1998,** 438–441.

[25] Powrie F, Mauze S, Coffman RL, *Res. Immunol.* **1997,** *148*, 576–581.
[26] Groux H, Powrie F, *Immunol. Today* **1999,** *20*, 442–445.
[27] Ina K, Kusugami K, Hosokawa T, Imada A, Shimizu T, Yamaguchi T, Ohsuga M, Kyokane K, Sakai T, Nishio Y, Yokoyama Y, Ando T, *J. Gastroenterol. Hepatol.* **1999,** *14*, 46–53.
[28] Nikolaus S, Bauditz J, Gionchetti P, Witt C, Lochs H, Schreiber S, *Gut* **1998,** *42*, 470–476.
[29] Murch SH, Braegger CP, Walker Smith JA, Macdonald TT, *Gut* **1993,** *34*, 1705–1709.
[30] Braegger CP, Nicholls S, Murch SH, Stephens S, Macdonald TT, *Lancet* **1992,** *339*, 89–91.
[31] Uguccioni M, Gionchetti P, Robbiani DF, Rizzello F, Peruzzo S, Campieri M, Baggiolini M, *Am. J. Pathol.* **1999,** *155*, 331–336.
[32] Nakamura S, Ohtani H, Watanabe Y, Fukushima K, Matsumoto T, Kitano A, Kobayashi K, Nagura H, *Lab. Invest.* **1993,** *69*, 77–85.
[33] McGeer PL, McGeer EG, *J. Leukoc. Biol.* **1999,** *65*, 409–415.
[34] Folkman J, Angiogenesis in cancer, vascular, rheumatoid and other disease. *Nature Med.* **1995,** *1*, 27–31.
[35] Jackson JR, Seed MP, Kircher CH, Willoughby DA, Winkler JD, *FASEB J.* **1997,** *11*, 457–465.
[36] Storgard CM, Stupack DG, Jonczyk A, Goodman SL, Fox RI, Cheresh DA, *J Clin. Invest.* **1999,** *103*, 47–54.
[37] Firestein GS, *J. Clin. Invest.* **1999,** *103*, 3–4.
[38] Koch AE, *Arthritis Rheum.* **1998,** *41,* 951–962.
[39] Nickoloff BJ, Mitra RS, Varani J, Dixit VM, Polverini PJ, *Am. J. Pathol.* **1994,** *144*, 820–828.
[40] Chen YX, Nakashima Y, Tanaka K, Shiraishi S, Nakagawa K, Sueishi K, *Arterioscler. Thromb. Vasc. Biol.* **1999,** *19*, 131–139.
[41] Couffinhal T, Kearney M, Witzenbichler B, Chen D, Murohara T, Losordo DW, Symes J, Isner JM, *Am. J. Pathol.* **1997,** *150*, 1673–1685.
[42] Inoue M, Itoh H, Ueda M, Naruko T, Kojima A, Komatsu R, Doi K, Ogawa Y, Tamura N, Takaya K, Igaki T, Yamashita J, Chun TH, Masatsugu K, Becker AE, Nakao K, *Circulation* **1998,** *98*, 2108–2116.
[43] Bousvaros A, Leichtner A, Zurakowski D, Kwon J, Law T, Keough K, Fishman S, *Dig. Dis. Sci.* **1999,** *44*, 424–430.
[44] Hunt BJ, Jurd KM, *Br. Med. J.* **1998,** *316,* 1328–1329.
[45] Verma IM, Stevenson J, *Proc. Natl Acad. Sci. USA* **1997,** *94*, 11758–11760.
[46] Read MA, Neish AS, Gerritsen ME, Collins T, *J. Immunol.* **1996,** *157*, 3472–3479.
[47] Barnes PJ, *Int. J. Biochem. Cell. Biol.* **1997,** *29*, 867–870.
[48] Umansky V, Hehner SP, Dumont A, Hofmann TG, Schirrmacher V, Droge W, Schmitz ML, *Eur. J. Immunol.* **1998,** *28*, 2276–2282.
[49] Bach FH, Hancock WW, Ferran C, *Immunol. Today* **1997,** *18,* 483–486.
[50] Bach FH, Ferran C, Hechenleitner P, Mark W, Koyamada N, Miyatake T, Winkler H, Badrichani A, Candinas D, Hancock WW, *Nature Med.* **1997,** *3*, 196–204.
[51] Itoh T, Arai KI, In: *The Cytokine Network and Immune Function* (Ed. Theze J), pp. 169–190. Oxford University Press, Oxford, **1999.**
[52] Ip YT, Davis RJ, *Curr. Opin. Cell Biol.* **1998,** *10*, 205–219.
[53] Jain RK, *Annu. Rev. Biomed. Eng.* **1999,** *01*, 241–263.
[54] Jacobson BS, Stolz DB, Schnitzer JE, *Nature Med.* **1996,** *2*, 482–484.
[55] Beljaars L, Molema G, Schuppan D, Geerts A, De Bleser PJ, Weert B, Meijer DKF, Poelstra K, *J. Biol. Chem.* **2000,** *275*, 12743–12751.
[56] Fava RA, Olsen NJ, Spencer-Green G, Yeo KT, Yeo TK, Berse B, Jackman RW, Senger DR, Dvorak HF, Brown LF, *J. Exp. Med.* **1994,** *180,* 341–346.
[57] Detmar M, Brown LF, Claffey KP, Yeo KT, Kocher O, Jackman RW, Berse B, Dvorak HF, *J. Exp. Med.* **1994,** *180*, 1141–1146.
[58] Murdoch C, Monk PN, Finn A, *Cytokine* **1999,** *11*, 704–712.
[59] Bazan JF, Bacon KB, Hardiman G, Wang W, Soo K, Rossi D, Greaves DR, Zlotnik A, Schall TJ, *Nature* **1997,** *385*, 640–644.
[60] Imai T, Hieshima K, Haskell C, Baba M, Nagira M, Nishimura M, Kakizaki M, Takagi S, Nomiyama H, Schall TJ, Yoshie O, *Cell* **1997,** *91*, 521–530.
[61] Wang JM, Shen W, Su S, *Trends Cardiovasc. Med.* **1998,** *8*, 169–174.
[62] Podolsky DK, Lobb R, King N, Benjamin CD, Pepinsky B, Sehgal P, deBeaumont M, *J. Clin. Invest.* **1993,** *92*, 372–380.
[63] Jamar F, Chapman PT, Manicourt DH, Glass DM, Haskard DO, Peters AM, *Br. J. Radiol.* **1997,** *70*, 473–481.

[64] Sipkins DA, Cheresh DA, Kazemi MR, Nevin LM, Bednarski MD, Li KCP, *Nature Med.* **1998,** *4*, 623–626.
[65] Kiely JM, Cybulsky MI, Luscinskas FW, Gimbrone Jr MA, *Arterioscler. Thromb. Vasc. Biol.* **1995,** *15*, 1211–1218.
[66] Spragg DD, Alford DR, Greferath R, Larsen CE, Lee KD, Gurtner GC, Cybulsky MI, Tosi PF, Nicolau C, Gimbrone MA, *Proc. Natl Acad. Sci. USA* **1997,** *94*, 8795–8800.
[67] Harari OA, Wickham TJ, Stocker CJ, Kovesdi I, Segal DM, Huehns TY, Sarraf C, Haskard DO, *Gene Ther.* **1999,** *6*, 801–807.
[68] Atochina EN, Balyasnikova IV, Danilov SM, Granger DN, Fisher AB, Muzykantov VR, *Am. J. Physiol.* **1998,** *275*, L806–L817.
[69] Muzykantov VR, Christofidou Solomidou M, Balyasnikova I, Harshaw DW, Schultz L, Fisher AB, Albelda SM, *Proc. Natl Acad. Sci. USA* **1999,** *96*, 2379–2384.
[70] Martens CL, Cwirla SE, Lee RY, Whitehorn E, Chen EY, Bakker A, Martin EL, Wagstrom C, Gopalan P, Smith CW, Tate E, Koller KJ, Schatz PJ, Dower WJ, Barrett RW, *J. Biol. Chem.* **1995,** *270*, 21129–21136.
[71] Zinn KR, Chaudhuri TR, Smyth CA, Wu Q, Liu HG, Fleck M, Mountz JD, Mountz JM, *Arthritis Rheum.* **1999,** *42*, 641–649.
[72] Reckless J, Grainger DJ, *Biochem. J.* **1999,** *340 (Pt 3)*, 803–811.
[73] Arap W, Pasqualini R, Ruoslahti E, *Science* **1998,** *279*, 377–380.
[74] Ellerby HM, Arap W, Ellerby LM, Kain R, Andrusiak R, Rio GD, Krajewski S, Lombardo CR, Rao R, Ruoslahti E, Bredesen DE, Pasqualini R, *Nature Med.***1999,** *5*, 1032–1038.
[75] Haubner R, Wester HJ, Reuning U, Senekowitsch-Schmidtke R, Diefenbach B, Kessler H, Stocklin G, Schwaiger M, *J. Nucl. Med.* **1999,** *40*, 1061–1071.
[76] Palma-Vargas JM, Toledo-Pereyra L, Dean RE, Harkema JM, Dixon RA, Kogan TP, *J. Am. Coll. Surg.* **1997,** *185*, 365–372.
[77] Kogan TP, Dupre B, Bui H, McAbee KL, Kassir JM, Scott IL, Hu X, Vanderslice P, Beck PJ, Dixon RA, *J. Med. Chem.* **1998,** *41*, 1099–1111.
[78] Simanek EE, McGarvey GJ, Jablonowski JA, Wong CH, *Chem. Rev.* **1998,** *98*, 833–862.
[79] Abraham WM, Ahmed A, Sabater JR, Lauredo IT, Botvinnikova Y, Bjercke RJ, Hu X, Revelle BM, Kogan TP, Scott IL, Dixon RA, Yeh ET, Beck PJ, *Am. J. Respir. Crit. Care. Med.* **1999,** *159*, 1205–1214.
[80] Welply JK, Abbas SZ, Scudder P, Keene JL, Broschat K, Casnocha S, Gorka C, Steininger C, Howard SC, Schmuke JJ, Graneto M, Rotsaert JM, Manger ID, Jacob GS, *Glycobiology* **1994,** *4*, 259–265.
[81] Liu SF, Ye X, Malik AB, *Mol. Pharmacol.* **1999,** *55*, 658–667.
[82] Munoz C, Pascual Salcedo D, Castellanos MC, Alfranca A, Aragones J, Vara A, Redondo MJ, de Landazuri MO, *Blood* **1996,** *88*, 3482–3490.
[83] Ferran C, Millan MT, Csizmadia V, Cooper JT, Brostjan C, Bach FH, Winkler H, *Biochem. Biophys. Res. Commun.* **1995,** *214*, 212–223.
[84] Faruqi RM, Poptic EJ, Faruqi TR, De La Motte C, DiCorleto PE, *Am. J. Physiol.* **1997,** *273*, H817–H826.
[85] Erl W, Weber C, Wardemann C, Weber PC, *Am. J. Physiol.* **1997,** *273*, H634–H640.
[86] Boyle Jr EM, Kovacich JC, Canty Jr TG, Morgan EN, Chi E, Verrier ED, Pohlman TH, *Circulation* **1998,** *98*, II282–II288.
[87] Lockyer JM, Colladay JS, AlperinLea WL, Hammond T, Buda AJ, *Circ. Res.* **1998,** *82*, 314–320.
[88] Weber KS, Draude G, Erl W, de Martin R, Weber C, *Blood* **1999,** *93*, 3685–3693.
[89] Pierce JW, Schoenleber R, Jesmok G, Best J, Moore SA, Collins T, Gerritsen ME, *J. Biol. Chem.* **1997,** *272*, 21096–21103.
[90] Barnes PJ, Karin M, *N. Engl. J. Med.* **1997,** *336*, 1066–1071.
[91] Marumo T, Schini-Kerth VB, Busse R, *Diabetes* **1999,** *48*, 1131–1137.
[92] Yoshida A, Yoshida S, Ishibashi T, Kuwano M, Inomata H, *Ophthalmol. Vis. Sci.* **1999,** *40*, 1624–1629.
[93] Castro A, Sengupta TK, Ruiz DC, Yang E, Ivashkiv LB, *J. Immunol.* **1999,** *162*, 1261–1269.
[94] Naka T, Narazaki M, Hirata M, Matsumoto T, Minamoto S, Aono A, Nishimoto N, Kajita T, Taga T, Yoshizaki K, Akira S, Kishimoto T, *Nature* **1997,** *387*, 924–929.
[95] Sakamoto H, Yasukawa H, Masuhara M, Tanimura S, Sasaki A, Yuge K, Ohtsubo M, Ohtsuka A, Fujita T, Ohta T, Furukawa Y, Iwase S, Yamada H, Yoshimura A, *Blood* **1998,** *92*, 1668–1676.
[96] Levitzki A, *Pharmacol. Ther.* **1999,** *82*, 231–239.

[97] Sudbeck EA, Liu XP, Narla RK, Mahajan S, Ghosh S, Mao C, Uckun FM, *Clin. Cancer Res.* **1999,** *5*, 1569–1582.
[98] Wasik MA, Nowak I, Zhang Q, Shaw LM, *Leuk. Lymph.* **1998,** *28*, 551–560.
[99] Wang LH, Kirken RA, Erwin RA, Yu CR, *J. Immunol.* **1999,** *162*, 3897–3904.
[100] Cronstein BN, Kimmel SC, Levin RI, Martiniuk F, Weissmann G, *Proc. Natl. Acad. Sci. USA* **1992,** *89*, 9991–9995.
[101] Didonato JA, Saatcioglu F, Karin M, *Am. J. Respir. Crit. Care Med.* **1996,** *154*, S11–S15.
[102] Brostjan C, Anrather J, Csizmadia V, Natarajan G, Winkler H, *J. Immunol.* **1997,** *158*, 3836–3844.
[103] Ardite E, Panes J, Miranda M, Salas A, Elizalde JI, Sans M, Arce Y, Bordas JM, Fernandezcheca JC, Pique JM, *Br. J. Pharmacol.* **1998,** *124*, 431–433.
[104] Lansink M, Koolwijk P, Van Hinsberg V, Kooistra T, *Blood* **1998,** *92*, 927–938.
[105] Wolff JE, Molenkamp G, Hotfilder M, Laterra J, *Klin. Padiatr.* **1997,** *209*, 275–277.
[106] Diaz Gonzalez F, Sanchez Madrid F, *Immunol. Today* **1998,** *19*, 169–172.
[107] Sakai A, *Life Sci.* **1996,** *58*, 2377–2387.
[108] Hofbauer R, Speiser W, Kapiotis S, *Life Sci.* **1998,** *62*, 1775–1781.
[109] Pierce JW, Read MA, Ding H, Luscinskas FW, Collins T, *J. Immunol.* **1996,** *156*, 3961–3969.
[110] Xia L, Pan J, Yao L, McEver RP, *Blood* **1998,** *91*, 1625–1632.
[111] Jones MK, Wang H, Peskar BM, Levin E, Itani RM, Sarfeh IJ, Tarnawski AS, *Nature Med.* **1999,** *5*, 1418–1423.
[112] Murthy S, Murthy NS, Coppola D, Wood DL, *Inflamm. Res.* **1997,** *46*, 224–233.
[113] Bennett CF, Condon TP, Grimm S, Chan H, Chiang MY, *J. Immunol.* **1994,** *152*, 3530–3540.
[114] Bennett CF, Kornbrust D, Henry S, Stecker K, Howard R, Cooper S, Dutson S, Hall W, Jacoby HI, *J. Pharmacol. Exp. Ther.* **1997**, *280*, 988–1000.
[115] Yacyshyn BR, BowenYacyshyn MB, Jewell L, Tami JA, Bennett CF, Kisner DL, Shanahan WR, *Gastroenterology* **1998,** *114*, 1133–1142.
[116] Neurath MF, Pettersson S, Meyer zum Büschenfelde K-H, Strober W, *Nature Med.* **1996,** *2*, 998–1004.
[117] Akhtar S and Agrawal S, *Trends Pharmacol. Sci.* **1997,** *18*, 12–18.
[118] Jenison RD, Jennings SD, Walker DW, Bargatze RF, Parma D, *Antisense Nucleic Acid Drug Dev.* **1998,** *8*, 265–279.
[119] Peacock DJ, Banquerigo ML, Brahn E, *J. Exp. Med.* **1992,** *175*, 1135–1138.
[120] Oliver SJ, Banquerigo ML, Brahn E, *Cell Immunol.* **1994,** *157*, 291–299.
[121] Moulton KS, Heller E, Konerding MA, Flynn E, Palinski W, Folkman J, *Circulation* **1999,** *99*, 1726–1732.
[122] D'Amato RJ, Loughnan MS, Flynn E, Folkman J, *Proc. Natl Acad. Sci.USA* **1994,** *91*, 4082–4085.
[123] Moreira AL, Sampaio EP, Zmuidzinas A, Frindt P, Smith KA, Kaplan G, *J. Exp. Med.* **1993**, *177*, 1675–1680.
[124] Marriott JB, Muller G, Dalgleish AG, *Immunol. Today* **1999,** *20*, 538–540.
[125] Koch AE, Szekanecz Z, Friedman J, Haines GK, Langman CB, Bouck NP, *Clin. Immunol. Immunopathol.* **1998,** *86*, 199–208.
[126] Anonymous. *SCRIP* **1999,** *October (2477)*, 20.
[127] Dong QG, Bernasconi S, Lostaglio S, De Calmanovici RW, Martin Padura I, Breviario F, Garlanda C, Ramponi S, Mantovani A, Vecchi A, *Arterioscler. Thromb. Vasc. Biol.* **1997,** *17*, 1599–1604.
[128] van Leeuwen EBM, Veenstra R, van Wijk R, Molema G, Hoekstra A, Ruiters MHJ, van der Meer J, *Blood Coagul. Fibrinol.* **2000,** *11*, 15–25.
[129] Lidington EA, Moyes DL, McCormack AM, Rose ML, Transpl. Immunol. **1999**, 7, 239–246.
[130] Tamaru M, Tomura K, Sakamoto S, Tezuka K, Tamatani T, Narumi S, *Arterioscler. Thromb. Vasc. Biol.* **1998,** *18*, 1292–1303.
[131] Haraldsen G, Kvale D, Lien B, Farstad IN, Brandtzaeg P, *J. Immunol.* **1996,** *156*, 2558–2565.
[132] Wilson SJ, Leone BA, Anderson D, Manning A, Holgate ST, *J. Pathol.* **1999,** *189*, 265–272.
[133] Molema G, Mesander G, Kroesen BJ, Helfrich W, Meijer DK, de Leij LF, *Cytometry* **1998,** *32*, 37–43.
[134] Villaschi S, Nicosia RF, *Lab. Invest.* **1994,** *71*, 291–299.
[135] Horizoe T, Nagakura N, Chiba K, Shirota H, Shinoda M, Numata H, Kobayashi S, Abe C, *Inflamm. Res.* **1999,** *48*, 432–436.
[136] Konigsberg PJ, Debrick JE, Pawlowski TJ, Staerz UD, *Biochim. Biophys. Acta* **1999,** *1421*, 149–162.

[137] Abbas AK, Lichtman AH, Pober JS, *Cellular and Molecular Immunology*, pp. 424–438. W. B. Saunders Company, Philadelphia, **1997.**
[138] Elson CO, Sartor RB, Tennyson GS, Riddell RH, *Gastroenterology* **1995,** *109*, 1344–1367.
[139] Smith JD, Breslow JL, *J. Intern. Med.* **1997,** *242*, 99–109.
[140] Cybulsky MI, Gimbrone Jr MA, *Science* **1991,** *251*, 788–791.
[141] Carew TE, Schwenke DC, Steinberg D, *Proc. Natl Acad. Sci. USA* **1987,** *84*, 7725–7729.
[142] Mori N, Horie Y, Gerritsen ME, Anderson DC, Granger DN, *Gut* **1999,** *44*, 186–195.
[143] Keelan ETM, Licence ST, Peters AM, Binns RM, Haskard DO, *Am. J. Physiol.* **1994,** *266*, H278–H290.
[144] Silber A, Newman W, Reimann KA, Hendricks E, Walsh D, Ringler DJ, *Lab. Invest.* **1994,** *70*, 163–175.
[145] Grabbe S, Schwarz T, *Immunol. Today* **1998,** *19*, 37–44.
[146] Austrup F, Vestweber D, Borges E, Lohning M, Brauer R, Herz U, Renz H, Hallmann R, Scheffold A, Radbruch A, Hamann A, *Nature* **1997,** *385*, 81–83.
[147] von Asmuth EJU, Smeets EF, Ginsel LA, Onderwater JJM, Leeuwenberg JFM, Buurman WA, *Eur. J. Immunol.* **1992,** *22*, 2519–2526.
[148] Kuijpers TW, Raleigh M, Kavanagh T, Janssen H, Calafat J, Roos D, Harlan JM, *J. Immunol.* **1994,** *152*, 5060–5069.
[149] Chuang PI, Young BA, Thiagarajan RR, Cornejo C, Winn RK, Harlan JM, *J. Biol. Chem.* **1997,** *272*, 24813–24818.
[150] Yoshida M, Szente BE, Kiely JM, Rosenzweig A, Gimbrone Jr MA, *J. Immunol.* **1998,** *161*, 933–941.
[151] Bergers G, Javaherian K, Lo KM, Folkman J, Hanahan D, *Science* **1999,** *284*, 808–812.
[152] Bombeli T, Schwartz BR, Harlan JM, *Blood* **1999,** *93*, 3831–3838.
[153] Gastroenterology, Electronic reference and review, Lippincott-Raven.

8 Strategies for Specific Drug Targeting to Tumour Cells

Jos G. W. Kosterink, Wijnand Helfrich, Lou F. M. H. de Leij

8.1 Introduction

Cancer is the second most common cause of death among adults in most Western countries. Great progress has been made in the treatment of selected malignancies and approximately 50% of all malignancies can be cured by current treatment strategies. The majority of these cures are achieved by surgery, that is if the disease has not spread throughout the whole body. Radiotherapy and chemotherapy used alone or in combination have greatly improved the management of patients with a variety of solid and haematologic malignancies. Chemotherapy has curative potentials in patients with various haematologic malignancies, testicular cancer and germ cell tumours. Despite improvements in the treatment of most metastatic solid tumours, these remain largely incurable. Reasons for this are insufficient tumour selectivity of anti-cancer agents and poor penetration within the tumour mass [1,2]. Another problem is that after surgical removal of the solid tumour, metastatic cells that are resistant to conventional chemotherapy often remain. The same holds for tumours with high metastatic capacity and high proliferation rates, even though these might be sensitive initially to chemo- or radiotherapy. Relapse may occur with therapy-resistant recurrences. New therapeutic approaches are under investigation to address these obstacles.

The purpose of this chapter is to describe briefly the pathology of cancer, the cell types involved in cancer disease, the currently available therapeutics and problems/hurdles to tumour-directed drug delivery/targeting. The following will be discussed in more detail: surface epitopes for targeting, molecular targets within the cells for therapeutic intervention, suitable targeting devices, drugs of choice for targeting and *in-vitro* and *in-vivo* techniques which are available to study the various approaches. Finally the current clinical experience with targeted anti-cancer drugs will be discussed.

8.2 Cancer Pathology

8.2.1 Cell Biology of Cancer

In the cell cycle, dividing cells undergo one mitosis (M) after another, passing through G_1, S (DNA synthesis phase), and G_2 phases. Some cells leave the cycle temporarily, entering a G_0 state from which they can be rescued by appropriate mitogenic stimuli. Other cells leave the cycle permanently, entering terminal differentiation.

Any population of cells can grow in number by any one of three mechanisms: shortening the length of the cell cycle, decreasing the rate of cell death, and moving G_0 cells into the cell cycle. All three mechanisms operate in normal and abnormal growth. In most tumours, all three mechanisms are important in determining the growth of the tumour, which is best characterized by its doubling time. Doubling time of tumours range from as little as 17 days for Ewing sarcoma to more than 600 days for certain adenocarcinomas of the colon and rectum. However, the fastest growing tumour is probably Burkitt's lymphoma, with a mean doubling time of less than 3 days.

Cancer is a multi-step process in which multiple genetic alterations must occur, usually over a span of years, to have a cumulative effect on the control of cell differentiation, cell division, and growth [3].

As in cancer predisposing syndromes, these genetic alterations are sometimes carried in the germline. Among human tumours, heritable mutations are an exception. Most alterations are acquired in somatic life in the form of chromosomal translocations, deletions, inversions, amplifications or point mutations. Certain oncogenic viruses play important roles in a few human tumours. Examples are human papilloma-virus in cervical cancer and skin tumours, Epstein-Barr virus in nasopharyngeal carcinoma and Burkitt's lymphoma, and human T-cell leukaemia viruses (e.g. HTLV-I, HTLV-II) in T-cell leukaemia.

In recent past decades there has been an extraordinary progress in the understanding of the mechanisms of oncogenesis. The application of molecular biological techniques in the field of tumour virology, cytogenetics, and cell biology led to the discovery of the transforming genes of tumour viruses, the genes activated at the breakpoints of non-random chromosomal translocations of lymphomas and leukaemias, the correlation between growth factors or growth factor receptors and certain transforming genes, and the existence of transforming genes that are activated *in vivo* and *in vitro* by direct-acting chemical carcinogens [4–6]. The transforming genes are collectively called oncogenes. Oncogene products are positive effectors of transformation. They impose their activity on the cell to elicit the transformed phenotype and can be considered positive regulators of growth. To the transformed cell, they represent a gain in function. Tumour suppressor gene products are negative growth regulators and their loss of function results in expression of the transformed phenotype.

The normally functioning cellular counterparts of the oncogenes, called protooncogenes are also important regulators of biological processes. They are localized in different cell compartments, are expressed at different stages of the cell cycle, and appear to be involved in the cascade of events that maintain the ordered procession through the cell cycle.

The cell cycle is regulated by external mitogens (e.g. growth factors, peptide and steroid hormones, lymphokines), which activate a process called signal transduction by which specific signals are transmitted within the cell to the nucleus. The process is also mediated by non-integral-membrane-associated proteins belonging to the tyrosine kinase, RAS gene families, and members of the MAPK family. Signals generated by mitogenic stimulation can lead to the expression of specific genes coding for proteins localized in the nucleus. Certain members of the nuclear oncogene protein family have been shown to be transactivators of specific RNA transcripts.

8.2.2 Histogenesis

The traditional principle of tumour histogenesis is that neoplasms characterized by a certain phenotype arise from normal cells of similar phenotype. Considerable evidence has accumulated in recent years which indicates that this histogenetic assumption is incorrect. Most, if not all, neoplasms arise from immature cells, which in the course of neoplastic transformation acquire phenotypic features equivalent to those of one or more normal cell types. Often this differentiation develops along lines analogous to those expected under normal conditions for that particular cell.

The characterization and genesis of neoplasms on the basis of morphological (shape-related) features is needed for the evaluation of the common traits and differences among the many types of neoplasms that can affect the human body. Such identification and classification confirms that many different cell types can be involved in cancer.

The traditional classification of neoplasms on the basis of their behaviour postulates benign (non-metastatic) and malignant (metastatic) types. These designations are determined by the expected behaviour of the tumour rather than its microscopic appearance. This division of tumours into benign and malignant represents a gross oversimplification of the wide behavioural range exhibited by tumour cells in terms of local aggressiveness and metastatic potential.

Determination of the microscopic type of a malignancy does not always provide all the information needed to predict the clinical course of the disease or to choose the appropriate therapy. Microscopic grading is an attempt to determine the degree of malignancy independently from cell type and is based on the evaluation of several parameters, which vary depending on the system being studied. They include cellularity, pleomorphism, mitotic activity, type of margins, amount of matrix formation, and presence of haemorrhage, necrosis and inflammation.

The number of grades varies from system to system, but in general the three-grade system (well differentiated, moderately differentiated, and poorly differentiated, or undifferentiated; or grades I, II and III, respectively) has proved to be the most reproducible and the best suited to prediction of survival.

For more detailed information about cancer pathology readers are referred to De Vita *et al.* [7].

8.3 Currently Available Therapeutics

Non-surgical methods of cancer treatment, primarily radiation therapy and chemotherapy, rely almost exclusively on procedures that kill cells. The main problem with these treatments is that they do not provide specificity for cancer cells. In the case of radiation therapy, a degree of specificity is achieved by localizing the radiation to the tumour and its immediate surrounding normal tissue. For anti-cancer drugs, it is primarily the rapid proliferation of many of the cancer cells that makes them more sensitive to cell killing than their normal counterparts. However, both modalities are limited by their cytotoxic effects on normal cells. In the case of radiotherapy, normal tissue surrounding the tumour limits the radiation dose, whereas for anti-cancer drugs, it is usually the killing of rapidly dividing normal cells such as those in the bone marrow, hair follicles, and epithelial cells lining the gastrointestinal tract, that limits the dose that can be given.

8.4 Barriers in Tumour-directed Therapies/Strategies

The success of treating tumours, especially solid tumours, by systemic therapy depends on various characteristics of the tumour. Besides the importance of intrinsic drug activity and the potential targets within the tumour cells, drug pharmacokinetics and whole body distribution, site of delivery and the ability of site-specific targeting (affinity) are important features.

In the following sections tumour cell-directed targeting and intracellular delivery of drugs will be discussed. This includes crucial factors such as tumour structure and physiology as well as physiological, cellular, molecular, biochemical and pharmacokinetic barriers.

8.4.1 Tumour Structure and Physiology

At the simplest level, the successful delivery of cytotoxic agents, either as small molecules or associated with polymers or liposomes, to a solid tumour depends on the relationship between the tumour cells and the blood vessels supporting their growth. Therefore the first requirement for effective delivery is a fully functional vasculature with respect to perfusion function. For those strategies of treatment where blockade of the tumour blood flow or inhibition of tumour growth is associated with angiogenesis, other considerations need to be taken into account as will be discussed in Chapter 9.

In solid tumours the criterion of adequate perfusion is rarely met. Solid tumours comprise of sheets or nests of neoplastic cells interspersed within a supporting stroma. The stromal component of the tumour is composed of fibroblasts, inflammatory cells, and blood vessels, and may represent as much as 90% of the mass of a tumour, depending on the tumour type [8]. The supporting stroma plays a critical role, in particular in the formation of new blood vessels in the growth of solid tumours [9]. It is not possible for a tumour to grow in excess of 1–2 mm in diameter without evoking a new blood supply. Neovascularization is necessary for growth of the tumour in order to maintain the supply of nutrients and to remove the resultant catabolites. This process of new vessel formation, or angiogenesis, is the result of a complex programme of proteolytic and migratory events involving the endothelial cell [10] (see also Chapter 9). There is much evidence to support the observation that this process is mediated by growth factors produced by tumour cells or by immune competent effector cells infiltrating the tumour parenchyma, or both [11,12]. As a result of the intense local angiogenic pressures, the vasculature of many tumours appears abnormal [13]. This abnormality occurs at the level of the vessel wall itself which is often characterized by an interrupted endothelium and/or an incomplete basement membrane.

Abnormalities of vessel architecture on a macroscopic scale are also frequently observed. Pre-existing arterioles and venules inevitably incorporated into the growing tumour mass may become obstructed and compressed, while other arterioles appear to be maximally dilated, displaying loss of vasomotion. Similarly, the neovasculature arising from pre-existing venules often displays a range of abnormalities, including increased blood vessel tortuosity and elongation, as well as abnormal and heterogeneous capillary density. The overall picture will depend on the nature and developmental stage of the tumour.

8.4.2 Physiological Barriers

Overall, the pattern of perfusion in human tumours is non-uniform, and human tumours contain well-perfused, rapidly growing regions, as well as poorly-perfused, often necrotic, regions. So the first obstacle to effective systemic treatments is the heterogenicity of the distribution of areas of growth within the tumour.

The next barrier to appropriate delivery of cytotoxic agents is the transport of agents across the blood vessel wall into the interstitium. In normal tissues an intact endothelium acts as a selective barrier to all but the smallest molecules and ions. Larger molecules may penetrate by para- or trans-cellular pathways and in some cases by active transport. Barrier function in tumours is often inadequate due to compromised endothelial integrity. Because of this reduced integrity, access for drugs and macromolecules such as antibodies and liposomes can be increased. However, hydrodynamics and solute behaviour influence the movement of such agents and the net effect of diffusive and convective forces may differ considerably from that predicted from observations on normal tissues [14].

Diffusion, particularly of macromolecules, plays a minor role in transport across this barrier. Convection due to leaky blood vessels, on the other hand, should enhance delivery; yet the movement of drugs and macromolecules into the interstitium is often surprisingly limited. This is generally attributed to a diminished hydrostatic pressure gradient between the vascular compartment and the interstitium, which is explained by decreased vascular pressure or increased interstitial pressure, or both.

There are several consequences of these anomalies in pressure gradients for the delivery and distribution of drugs and macromolecules within the tumour interstitium. First, high interstitial pressures mean that the central regions of the tumour, already poorly perfused, demonstrate low or non-existent convective flow into the interstitium. Furthermore, interstitial convective flow will tend to radiate outward from the centre, towards the periphery and regions of lower interstitial pressure. Therefore, only small amounts of drugs or macromolecules will reach cells in the centre of the tumour. At the tumour periphery, where convective transfer across the blood vessel wall might take place, further movement towards the centre of the tumour will be impeded by bulk flow in the opposite direction.

In summary, in solid tumours the laws of hydrodynamics and transport of solutes mitigate against the successful delivery of drugs and macromolecules to tumour cells.

8.4.3 Cellular and Biochemical Barriers, Multi-drug Resistance

The first barrier at the level of the single cell is the cell membrane. Although the majority of drugs gains access into cells by passive diffusion, a number of anti-metabolites is actively transported. Also, there are certain membrane proteins which act as energy-dependent efflux pumps for a number of commonly used chemotherapy drugs. Examples of these proteins are P-glycoprotein (P-gp), first described by Juliano and Ling [15], as well as the multi-drug resistance related protein (MRP), [16] and the lung resistance related protein (LRP) [17,18]. These proteins are either alone or in concert operative in the phenomenon known as multidrug resistance (MDR).

Frequently used cytostatic agents which are involved in MDR are the anthracyclines (doxorubicin, daunorubicin), vinca-alkaloids (vincristine, vinblastine), epipodohyllotoxins (etoposide), and taxanes (paclitaxel). The most extensively studied mechanism is the overexpression of P-gp, which is a 170-kDa transmembrane drug efflux pump encoded by the MDR_1 gene in humans. Another mechanism is the over-expression of MRP, a 190-kDa drug efflux pump, encoded by the MRP_1 gene. A third mechanism which is involved in MDR is the heterotopic expression of LRP. This protein is extensively expressed in a variety of normal tissues, especially in the bronchus, renal proximal tubulus, canalicular domain of the hepatocyte [19], macrophages and adrenal cortex. *In vitro* studies also suggest that LRP has a role in the compartmentalization and transport of chemotherapeutic drugs out of the tumour cells.

Once a drug has entered the cell, detoxification mechanisms within the cytoplasm can potentially inactivate cytotoxic drugs. These include the activity of glutathione and the glutathione-S-transferase enzyme. At the nuclear level there is a wide variety of proteins available to protect the cell against chemotherapy-induced damage. The topoisomerase enzymes [20] are common targets for cytotoxic drugs. Topoisomerases are nuclear enzymes, which are involved in DNA replication. Inhibitors of the topoisomerase-1 include agents based on the camptothecin structure, topotecan and irinotecan. They stabilize the covalent complex between DNA and topoisomerase-1 resulting in DNA breakdown and finally cell death. Inhibitors of topoisomerase-2 include etoposide, teniposide and doxorubicin.

The malignant cell, similar to the normal cell, has a complex array of enzymes involved in recognizing and repairing DNA damage. Increased levels of DNA repair enzymes have been identified in models of resistance to cytotoxic drugs, in particular to methylating agents, with elevations in O-methyltransferase, and in resistance to platinum-based drugs. So, in addition to the tumour structure and physiological barriers, there is a variety of ways by which an individual tumour cell can escape adequate targeting of drugs and/or their cytotoxic effects.

8.4.4 Pharmacokinetic Barriers

Before reaching the site of action (tumour cells), basic pharmacokinetic tolerance and whole body distribution patterns of cytotoxic drugs play an important role in the final outcome of drug treatment. As a result of unfavourable pharmacokinetics, patients are often unable to tolerate effective doses due to unacceptable toxicity. This holds true especially for the more conventional cytotoxic drugs. There is also a variability between patients in pharmacokinetic tolerance of cytotoxic drugs, e.g. in parameters such as oral bio-availability of drugs, differences in excretion rate (partially P-gp mediated), and altered metabolism through variations in cytochrome P-450 iso-enzyme activities, particularly in the elderly. The vast majority of cytotoxic drugs are metabolized via cytochrome P-450-dependent mechanisms, and many of these drugs are excreted through the kidneys and liver at least partially by the P-gp systems [19].

The use of so-called reversal agents to block P-gp in order to decrease multi-drug resistance, will therefore also affect the elimination rate of those anti-cancer agents that are substrates for this transport system [19].

The pharmacokinetic processing of macromolecules used as targeting devices or drug carrier systems is different from that of conventional cytotoxic drugs and plays an important role in e.g. the targeting efficiency of these cytotoxic agents coupled to the macromolecules.

8.5 Strategies to Deliver Drugs to Targets within the Tumour (Cells)

As discussed above there are several hurdles to overcome in attempting to enhance the delivery of the drug to the tumour cell. In addition to the use of high dose chemotherapy with concomitant protection of normal tissues, a number of other approaches have been developed. Local perfusion is used with significant benefit in some cancers. This technique is however limited to cancers localized to a single site, e.g. to one of the extremities. This approach will not be discussed here.

Other approaches have been exploited in attempting to increase the therapeutic index by improving the specificity and efficacy of the drug and reducing the toxicity. One example of this is to target the cytotoxic agent to the tumour cells. To increase specificity and reduce toxicity, trigger mechanisms have been designed to activate cytotoxic agents synthesized in their pro-drug/inactive forms, in a site selective manner. Triggering signals can be either exogenous factors such as light or chemicals or endogenous (cellular) factors such as enzymes. The inherent features of cancer cells can also be used in the development of targeting agents for tumour cells. Cancer cells often over-express specific (tumour) antigens, carbohydrate structures, or growth factor receptors on their cell surface. In addition to tumour cell membrane-specific antigens, some cells also express unique proteases. Based on the above concepts, various strategies for targeting cytotoxic agents are under development and are currently being tested in pre-clinical and/or clinical settings. These include:

(1) Monoclonal antibodies (MAb) against tumour-associated antigens or growth factors using their intrinsic activity or used as carriers to target cytotoxic drugs, radionuclides and toxins (Section 8.5.1).
(2) Bispecific monoclonal antibodies (BsMAb) which combine the specificity of two different antibodies within one molecule and cross-link an effector cell or a toxic molecule with the target cell (Section 8.5.2).
(3) Pro-drugs in conjunction with enzymes or enzyme–MAb conjugates (Section 8.5.3).
(4) Synthetic copolymers as drug carriers (Section 8.5.4);
(5) Liposomes as carriers for drug delivery (Section 8.5.5).

The following sections will discuss these different approaches in more detail. Only those approaches which are of interest for potential development into clinical strategies will be discussed.

8.5.1 Monoclonal Antibody-mediated Therapeutics

The ground-breaking development of monoclonal antibodies by Köhler and Milstein [21] initiated the development of antibody-mediated therapeutics for cancer. Because of their unique specificity, MAb were predicted to become the magic bullets in the battle against cancer. Over the last two and a half decades MAbs have moved from clone to clinic for the treatment of various malignancies. Several MAbs are currently entering clinical trials and should appear on the market in the next few years. The first MAb for cancer therapy was approved in the US in 1997.

MAbs have been used in a natural, fragmented, chemically modified, or recombinant form in a variety of settings (Figure 8.1a–d). They have been coupled to drugs, toxins, enzymes, radionuclides, cytokines, superantigens and drug-filled liposomes (Figure 8.1e–f). The development of each construct, their advantages and disadvantages will be discussed as well as their applications in animal models and patient populations. For a more detailed review readers are also referred to Farah *et al.* [22].

As the specificity and availability of the target epitope expressed by the tumour cells are important determinants for therapeutic outcome, the most interesting antigenic targets will be discussed below.

8.5.1.1 Antigenic Targets

Many different tumour-associated antigens (TAAs) have been described for targeted immunotherapy. General considerations that rationalize the choice of a target antigen are:

(1) The expression of the antigen on the tumour cells should be homogenous throughout the tumour and high enough to ensure the effective binding of the antibody of choice.

▶

Figure 8.1. MAb Constructs. (**a**) A murine immunglobulin IgG molecule contains two heavy chains (H) and two light chains (L). The heavy and light chains are linked by an interchain disulfide bond indicated by a horizontal line between the CH1 and CL domains. The heavy chains are also linked to each other by one or more interchain disulfide bonds (indicated by horizontal lines in the CH2 domain). (**b**) Fragments of IgG generated by enzymatic digestion. (1) $F(ab)'_2$ fragments are generated by pepsin digestion of IgG. (2) F(ab)′ fragments are generated by reduction of $F(ab)'_2$ fragments. (3) Fab fragments are generated by papain digestion. (**c**) Recombinant MAbs. (1) Chimeric MAbs have mouse $F(ab)'_2$ portions (white) and human Fc portions (black). (2) Humanized MAbs contain mouse hypervariable regions (CDRs) (white) and human framework regions (black). (3) Recombinant fragments: (I) Fv fragments are heterodimers composed of noncovalenty associated variable domains of the heavy and light chains, (II) single chain Fvs (scFvs) are covalently linked through a polypeptide linker that can be introduced to stabilize interchain association, (III) a diabody consists of two scFv molecules, (IV) a triabody consists of three Fvs. These are a few of more common constructs that have been described to date. (**d**) Bispecific MAbs. (1) Tetravalent, chemically cross-linked molecules. Two MAbs are held together covalently by a chemical cross-linker. This construct has four binding sites: two for antigen A, and two for antigen B. (2) Divalent quadromas are obtained from the secretions of hybrid hybridomas. The quadroma receives one set of heavy and light chains from each parent hybridoma, creating one binding site for antigen A and one for antigen B. (3) Divalent recombinant molecules are single-chain fusion proteins with one binding site for antigen A and for antigen B. (**e**) Immunoconjugates are MAbs linked to toxic agents or effector molecules. The linkage can be chemical or the immunconjugate can be generated by genetic engineering. (**f**) Immunoliposomes consists of a toxic agent encapsulated within a lipid vesicle with multiple MAbs attached to the vesicle as targeting moieties.

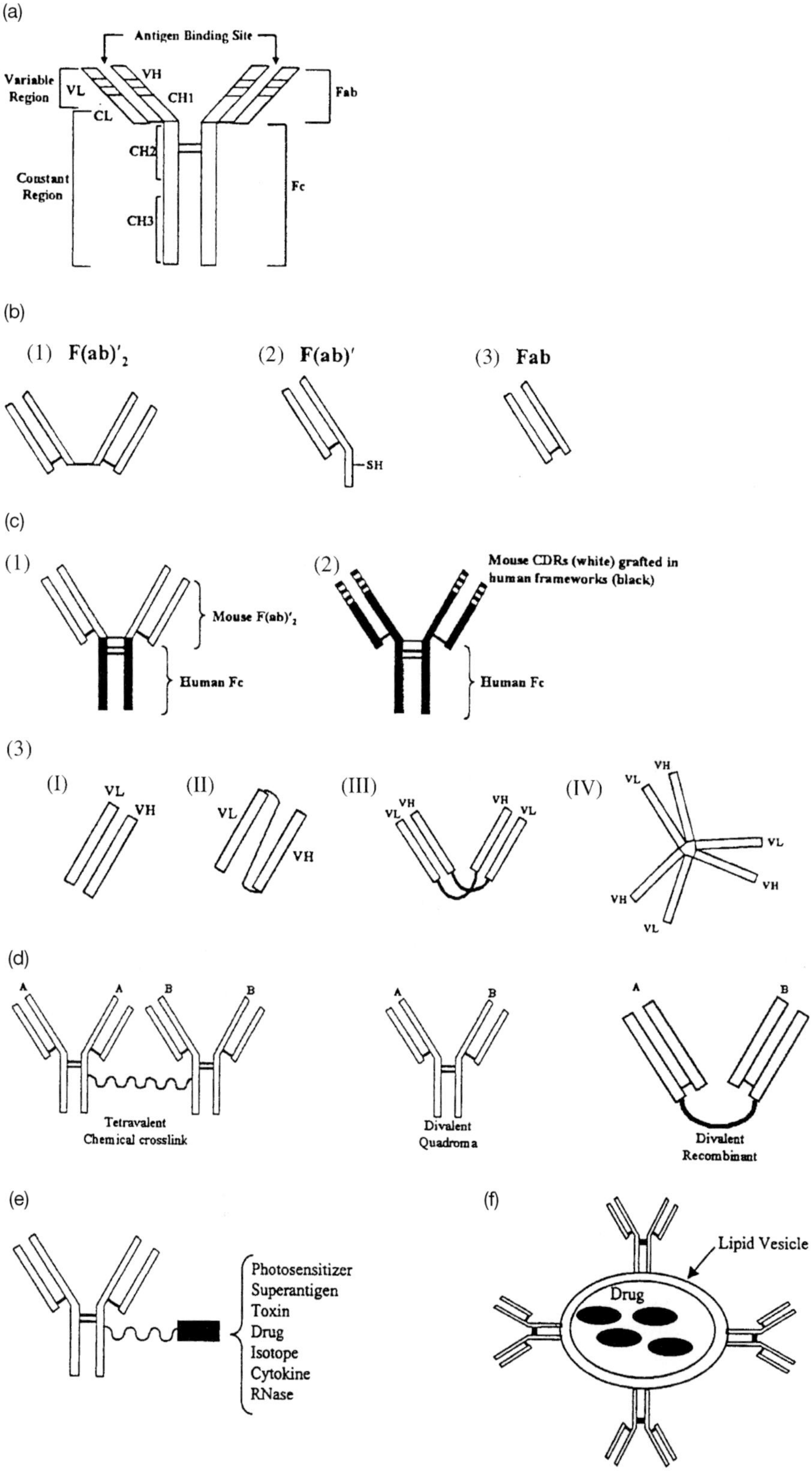
(a)
Antigen Binding Site
Variable Region
VL
VH
CH1
CL
Fab
CH2
Constant Region
Fc
CH3
(b)
(1) F(ab)'2
(2) F(ab)'
(3) Fab
SH
(c)
(1)
Mouse F(ab)'2
Human Fc
(2)
Mouse CDRs (white) grafted in human frameworks (black)
Human Fc
(3)
(I) VL VH
(II) VL VH
(III) VH VL VH VL
(IV) VH VL VL VH VH VL
(d)
A A B B
Tetravalent
Chemical crosslink
A B
Divalent
Quadroma
A B
Divalent
Recombinant
(e)
Photosensitizer
Superantigen
Toxin
Drug
Isotope
Cytokine
RNase
(f)
Lipid Vesicle
Drug

(2) Expression of the antigen by normal tissues should be limited or, if the antigen is expressed on normal tissue, it should be inaccessible to antibodies in these tissues.
(3) The antigen should be membrane bound and not shed from the cell surface. One of the positive exceptions to this rule is carcinoembryonic antigen (CEA) which is also present in the serum of patients in significant concentrations.

The heterogeneity of tumours as well as the fact that their antigenic make-up resembles that of the equivalent normal tissues, has made it difficult to identify suitable target molecules. In the following, a number of potential target antigens for such an approach are discussed [23]. The surface Ig idiotype sequences present in B-cell malignancies are close to ideal with respect to specificity as they truly represent a tumour specific antigen. However, anti-idiotype targeting has several drawbacks that are difficult to overcome. First, the unique intrinsic specificity of the surface Ig implies that new antibodies have to be generated for every distinct B-cell clone. Second, soluble malignant B-cell-produced antibody present in the serum may act as a scavenger for the therapeutic anti-idiotypic antibodies thereby preventing them from binding to their membrane bound target [24]. Other B-cell-specific target antigens include the normal B-cell markers such as CD19 or CD20, which are present on a wide range of B-cell-derived malignancies. Immunotherapy directed against normal B-cell-specific markers holds the risk of compromising the natural immune response by eradication of the complete B-cell repertoire. However it may be anticipated that this immune 'gap' can be restored by new, bone-marrow-derived B-cells.

Carcinomas are frequently occurring solid tumours. Examples of carcinoma-associated antigens that have been exploited in therapeutic protocols are c-erbB-1 or epidermal growth factor (EGF) receptor, c-erbB-2 or HERs/*neu* antigen, the folate receptor or folate-binding protein (FBP) and the epithelial glycoprotein-2 (EGP-2) [25]. Over-expression of the c-erbB-1 proto-oncogene product was reported in squamous cell carcinomas of the lung [26,27], adenocarcinomas and large cell carcinomas [28]. The proto-oncogene product c-erbB-2 is amplified in a variety of adenocarcinomas and squamous cell carcinomas, including lung, breast, gastric and colon cancer [29,30]. The antigen is also expressed in normal lung tissue [28].

A number of both pre-clinical and clinical studies have used the folate receptor or FBP as a target for immunotherapy of ovarian carcinoma [31,32]. Expression of this tumour-associated antigen by normal tissues is restricted [33]. The carcinoma-associated antigen, EGP-2, also called EpCAM, is a 38-kDA transmembrane glycoprotein, present on the majority of simple, stratified and transitional epithelia [34]. The biological function of EGP-2 has not yet been established.

Another approach in solid tumour therapy is to target antibodies to antigens expressed on the tumour vasculature, rather than to tumour-associated antigens of solid tumours. This has shown impressive activity in pre-clinical models [35,36]. Directing therapy to the accessible vascular compartment reduces the impact of the physical barriers of solid tumours, such as heterogeneous blood flow and elevated interstitial pressure [14]. Identification of appropriate target antigens that are expressed on the tumour vasculature, but not on cells of normal vessels, is an area of ongoing interest (see also Chapter 9).

Monoclonal antibodies against tumour-associated antigens or growth factors have been used to target the delivery of cytotoxic drugs, radionuclides and (bacterial) toxins [22]. Simi-

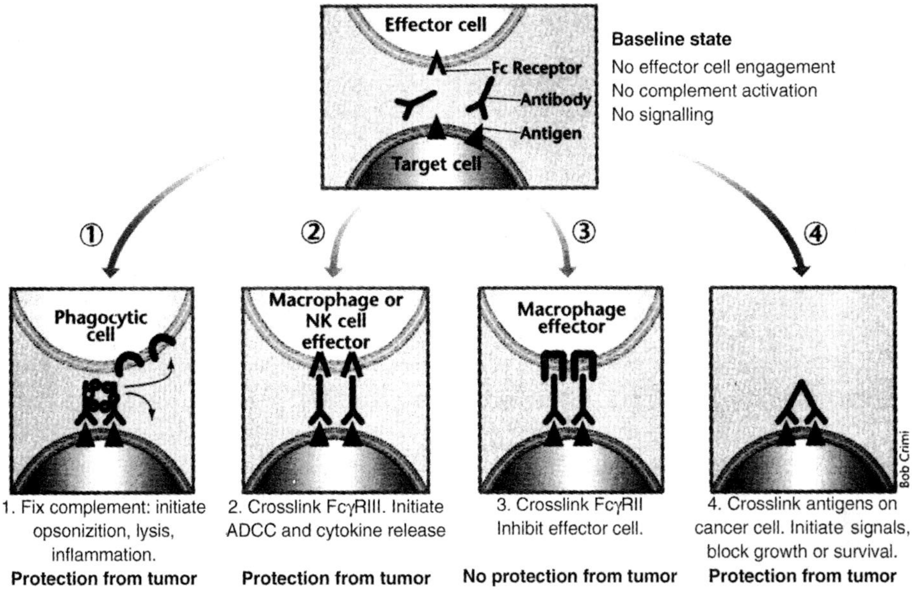

Figure 8.2. Monoclonal antibodies can block tumor growth using many mechanisms. Top, monoclonal antibodies recognize antigens on the target cell, in this case a cancer cell. 1. Monoclonal antibody bound to antigen activates complement components (small ring between the two antibody molecules), leading to opsonization of cancer cells by phagocytic cells expressing complement receptors (half-circles), direct lysis of tumor cells and inflammation with recruitment of inflammatory cells. 2. Monoclonal antibody binds to activating Fc receptors on the effector cells, leading to anti-body.dependent cellular cytotoxicity (ADCC) or release of cytokines. 3. Monoclonal antibody binds to inhibitory Fc receptors (or to both activation and inhibitory Fc receptors), inhibiting effector cell activation. 4. Monoclonal antibody binds directly to growth factor receptors or other signaling molecules on the cancer cell, leading to cell death reprinted with permission from [155].

larly cytotoxic immune effector cells have been redirected to kill tumour cells using bispecific antibodies [37]. These approaches will be discussed below.

8.5.1.2 Unconjugated Antibodies

Some unconjugated or 'naked' MAbs can induce anti-tumour effects by mechanisms that include the activation of the effector cells of the immune system, or the fixation of complement (C) (Figure 8.2). The former, called antibody dependent cellular cytotoxicity (ADCC), depends on the ability of lymphocytes, macrophages, and granulocytes to recognize the Fc region (see Figure 8.1a) of the tumour cell-bound antibody. The latter involves activation of the C cascade that eventually punches holes in the plasma membrane of the target cell. Unfortunately, one of the inherent weaknesses of using mouse MAbs to treat humans is their inability to effectively activate human ADCC or human C because of structural differences between the Fc portions of mouse and human Igs [38]. Of the different subclasses of mouse

IgGs, IgG2a is the one which is most efficient in mediating human ADCC, whereas IgG3 can mediate potent C-dependent cytolysis [39].

Some MAbs have the ability to signal target cells to undergo cell cycle arrest (CCA) or apoptosis. The prototypic example of such a MAb is anti-Fas that signals apoptosis in all Fas-positive cells [40,41]. However, because of the ubiquitous expression of Fas, administration of anti-Fas is lethal. Other MAbs, particularly when used as homodimers [42] which hyper-crosslink their antigenic targets, can induce CCA or apoptosis. Both anti-CD19 and anti-CD22 induce CCA in several Burkitt's lymphoma cell lines both *in vitro* and in mice xenografted with human tumours [43]. Anti-Id MAbs are also thought to be of therapeutic value because of their ability to direct negative signals to tumour cells [44].

More recently, our knowledge of cellular signalling pathways has led to the development of MAbs which target molecules involved in the regulation of tumour cell growth. Cytostatic or cytotoxic effects can result from the binding of a MAb to growth factors or cellular growth factor receptors which are required for tumour cell survival [45]. For example, many adult carcinomas depend, in part, on the autocrine or paracrine effects of epidermal growth factor (EGF) or transforming growth factor-α (TGF-α). As a result, some anti-EGF receptor MAbs have anti-tumour activity in tumours of the breast, vulva, cervix, and in squamous cell carcinomas [45]. Other MAbs targeting various cell surface growth factor receptors have also effectively induced CCA or apoptosis in tumour cells [40,46,47].

To potentiate the cytotoxic effects of MAbs with low endogenous activity, cytokines and activated effector cells have been co-administered [48]. Cytokines can increase extravasation of MAbs into the tumour and, by inducing local inflammatory responses, enhance the influx of effector cells. For example, the addition of interleukin-2 (IL-2) or the concomitant adoptive transfer of lymphokine-activated killer cells (LAKs) can enhance the activity of MAbs. Other cytokines, such as interferon-gamma (IFNγ) and IFNα can augment the delivery of MAbs to tumour targets by upregulating antigen expression [48,49]. The use of activated effector cells (peripheral blood mononuclear cells or granulocytes) in combination with MAbs has also resulted in their increased cytotoxicity to various tumours [48].

8.5.1.2.1 Potential Disadvantages and Limitations of the MAb Approach

Unfortunately, the clinical efficacy of MAb-directed therapy is often limited. One important factor in this respect is that the target antigen is expressed on normal as well as malignant cells. With the exception of MAbs to idiotypic domains of B-lymphocytes, MAbs which are exclusively tumour-specific have not been identified. Rather, most currently used MAbs recognize tumour-associated antigens expressed at higher density on malignant cells relative to normal cells. Furthermore, MAbs are murine in origin, particularly those used in past research. As a consequence, human anti-mouse antibody (HAMA) responses developed in patients treated with murine MAbs, led to accelerated clearance of the administered MAb and blocking of the therapeutic effect.

As mentioned in Section 8.4.2, elevated interstitial pressure, heterogeneous and reduced functional vasculature, and the relatively large distances that Mabs have to travel in the tumour interstitium, are hurdles which need to be overcome in the pursuit of efficient drug targeting. The relatively large molecular weight of Mabs (approximately 150 kDa) [2,14], also contributes to limited tumour penetration and minimal efficacy especially if MAb-directed therapies are used as single agents in patients with advanced disease.

Several modifications have been explored to improve efficacy. The problem of relatively large molecular weight can be partially resolved by using fragments of IgG generated by enzymatic digestion (Figure 8.1b). With advances in protein engineering, efforts are being aimed at reducing the size of the MAb, as well as reducing immunogenicity by using chimeric or humanized MAbs [39] (see Section 8.5.1.3). Despite these concerns, adverse effects of naked MAb, even after repetitive administration, are uncommon, and when they occur they are usually readily reversible.

8.5.1.3 Recombinant Antibodies

8.5.1.3.1 Recombinant DNA Technology

Recombinant DNA technology can be exploited to deal with the above-mentioned problems and has been used not only to manipulate the size, but also the shape, affinity, and immunogenicity of the MAb molecule. Chimeric versions of murine MAbs can be constructed through combination of variable chains of the original murine MAb with the constant domains of human Ig. This serves to enhance effector functions and reduce the chances of a HAMA response occurring (Figure 8.1c). Alternatively, the six hypervariable loops (complementarity determining regions, CDRs) forming the antigen binding site of a murine antibody can be transplanted into a human framework resulting in a CDR-grafted or humanized antibody (Figure 8.1c).

8.5.1.3.2 Single Chain Fv Antibody Fragments

In addition to modified complete Ig molecules, recombinant DNA technology has been used to construct small antibody-like molecules called single chain Fv fragments (scFv) [50,51] (Figure 8.1c). Briefly scFvs are recombinant antibody fragments consisting of only the variable light chain (VL) and variable heavy chain (VH) domains covalently connected by a flexible polypeptide linker typically composed of 15 amino acid residues ((Glycine4Serine)3). Due to their relatively small size (approximately 26 kDa), scFvs are rapidly distributed and a significant improvement in penetration into solid tumours has been shown *in vivo*. Murine scFv fragments can be produced by PCR-based gene assembly using mRNA templates isolated from the corresponding hybridoma cell line [52].

Functional expression (display) of scFv proteins on the surface of bacteriophage has been widely exploited in the selection of scFvs which have retained the binding properties characteristic of the MAb from which they were derived (see also Chapter 10 on the application of phage display for target antigen-specific scFv identification). The inherent advantage of phage display technology is its direct link between the DNA sequence and the protein function [53]. Large numbers of clones can be rapidly screened for antigen binding, making it the method of choice for hybridoma Ig cloning. It is clear that molecular cloning and sequencing of scFv forms the basis for further antibody engineering and modelling.

8.5.1.3.3 Phage Display Library

As an alternative to immunization and hybridoma construction procedures, it is possible now to construct large (synthetic) human antibody gene repertoires entirely *in vitro* (see also Chapter 10). This procedure can generate a huge library of recombinant filamentous bacteriophages that express hundreds of millions of different human scFvs on their tips fused to the phage minor coat protein III [38,54]. The scFvs displayed by these phages can show antigen binding activity and phages with the desired binding characteristics and specificity can be selected by panning on the antigen. The selected phage (including the genetic information of the displayed scFv inside) can be rescued and grown after each round of panning after which the 'enriched' phage library is again subjected to selection so that even rare phages ($< 1/10^8$) can be isolated. Using this strategy human antibodies and/or their fragments have been isolated with specificities against foreign and self antigens [55].

8.5.1.3.4 Transgenic 'Human' Animals

A further advance in antibody technology is the development of transgenic mouse 'human' strains. XenoMouse animals have been engineered in such a way that they now produce exclusively human antibodies rather than murine antibodies when immunized. The use of XenoMouse animals to produce MAbs avoids the need for any engineering of the antibody genes, since the products are already 100% human protein. XenoMouse animals are fully compatible with standard hybridoma technology and can be readily adopted by laboratories experienced in monoclonal antibody production [56].

8.5.1.3.5 Considerations for Recombinant Antibody Production

When the biodistribution of scFv (Figure 8.1c), Fab′, $(Fab)_2'$ (Figure 8.1b), and IgG (Figure 8.1a) were compared, most of the intact IgG delivered to tumours was concentrated in the region immediately adjacent to the blood vessels. The Fab′ and $F(ab)_2'$ fragments demonstrated intermediate degrees of tumour penetration, while the scFv was distributed more evenly throughout the tumour [57].

In mice with human breast carcinoma xenografts, a humanized IgG anti-HER-2 MAb eradicated well-established tumours [58]. In addition, a humanized version of an IgG anti-CD33 MAb (HuM 195) mediated ADCC *in vitro* [59] and had an 8.6-fold higher avidity than the parent murine Mab. Recombinant antibody fragments may have valuable properties as discussed above, but their biophysical behaviour, production yield and low thermostability leaves much to be desired and thereby limits their usefulness for *in-vivo* applications so far [60]. One possibility to improve these characteristics of scFv fragments with suboptimal stability and/or folding yield, is the grafting of their CDRs onto the framework of a different, more stable scFv [61,62].

Another valuable tool for the development of scFv-based therapeutics consists of a versatile expression vector for the rapid construction and evaluation of scFv-based fusion proteins and bispecific scFv [63]. The vector was used for grafting a number of biological effector principles onto anti-EGP-2 scFv. Biologically-active fusion proteins were produced by directing them through the endoplasmic reticulum-based protein folding machinery of eukaryotic cells. This procedure may help to identify those fusion proteins that which desirable characteristics such stability and biological activity in the presence of serum and at low protein concentrations.

Biophysical properties such as high thermal stability are thus of paramount importance in the decision as to whether or not these molecules are useful *in vivo*. The above described approaches may provide a strategy to meet these requirements and may eventually result in attractive modalities for the targeting of solid tumours in patients.

8.5.1.4 Immunotoxins (ITs)

The conjugates referred to as ITs are hybrid molecules consisting of MAbs linked to powerful toxins (or toxin subunits) purified from plants, fungi, or bacteria [64] (Figure 8.1e and Table 8.1). These toxins inhibit protein synthesis after internalization, leading to death of the targeted cell. Small quantities of ITs when compared with unconjugated MAbs, are needed for effective target cell killing. In fact, a single toxin molecule in the cytosol can kill a target cell, and, unlike chemotherapeutic agents, ITs will kill both resting and dividing cells.

Limitations to IT therapy include their immunogenicity and toxicity. Dose-limiting side-effects of IT therapy include hepatotoxicity and vascular leak syndrome.

Table 8.1. Toxins ued for the preparation of ITs.

Source	Toxin	Enzymatic activity
Plant	Type I RIPs* (single chain)	N-glycosidase for 28s rRNA
	Pokeweed anti-viral protein (PAP)	
	Saporin (SAP)	
	Gelonin	
	Momordin	
	Trichosanthin	
	Barley toxin	
	Type II RIPs (two chains)	
	Abrin	
	Ricin	
	Viscumin	
Bacteria	Diphtheria toxin (DT)	ADP Ribosylation of EF2
	Pseudomonas exotoxin (PE)	
Fungi	α-Sarcin	Ribonuclease for 28 S RNA
	Restrictocin	

* RIP: Ribosome-Inactivating Proteins

8.5.1.5 Monoclonal Antibody–Drug Conjugates

MAb–drug conjugates offer the advantages of improving the therapeutic index by increasing drug uptake by tumour cells, reducing drug toxicity to normal cells, and prolonging bioavailability of the drug and thus more extensive exposure to tumour cells.

Conventional cytotoxic drugs such as doxorubicin, idarubicin, bleomycin, methotrexate, cytosine arabinoside, chlorambucil, cisplatin, vinca alkaloids, and mitomycin C have all been conjugated to tumour-binding MAbs [65–69] (Figure 8.1e). Drug conjugates can be prepared by covalently coupling drugs directly to a MAb or indirectly through an intermediate spacer molecule such as dextran, human serum albumin, poly-glutamic acid, carboxymethyl dextran, or amino-dextran. An indirect linkage facilitates the attachment of more drug molecules to one MAb molecule, in theory resulting in an increased delivery of drug molecules to the tumour (see Chapter 11 on drug–carrier conjugate synthesis strategies).

The members of the enediyne family of antibiotics are highly potent drugs that are good candidates for attachment to MAbs [70]. Calicheamicin conjugates of e.g. the MAb CT-M-01, exerted strong cell specific activity against s.c. breast-tumour xenografts in athymic mice [71]. Similar impressive anti-tumour activity was shown with a calicheamicin conjugate of anti-ganglioside-GD2 MAb used to treat experimental liver metastases in immunocompetent mice [72].

Table 8.2. MAb-drug conjugates that have been developed for cancer therapy.

Class of anti-neoplastic drug	Drug	Disease
Antimetabolites	Methotrexate	Lung cancer, Colon cancer, Teratocarcinoma, T cell lymphoma
	5-Fluorouracil	B leukaemia
	Cytosine arabinoside	B leukaemia
	Aminopterin	Murine thymoma
	5-Fluoro-2'-deoxyuridine	Colon carcinoma
Alkylating agents	Chlorambucil	Murine thymoma, Murine lymphoma, Melanoma
	Melphalan	Colon Cancer, Murine thymoma
	Mitomycin C	Lung cancer, Various disseminated refractory malignancies, Gastric cancer
	Cisplatinum	Ovarian carcinoma
	Trenimon	Hepatoma
Anthracyclines	Doxorubicin/adriamycin	Melanoma, Ovarian carcinoma, T-cell lymphoma, Colon carcinoma, B-cell lymphoma, Various disseminated refractory malignancies, Breast cancer, Lung cancer, Pancreatic cancer, Liver cancer, Neuroblastoma
	Daunomycin	Soft-tissue sarcomas, Mammalian carcinoma, Hepatoma
Antimitotic agents	Vinca alkaloids	Lung adenocarcinoma
Miscellaneous agents	Bleomycin	Leukaemia
	Idarubicin	Murine thymoma
	Maytansine	Colon cancer
	Calicheamicins	Human breast carcinoma xenografts

In the case of drug–monoclonal antibody conjugates, the entire conjugate may be internalized after which the drug can be released intracellularly. The drug may also be cleaved extracellularly and subsequently taken up into tumour cells by diffusion or active transport. Success with drug–MAb conjugates has been limited thus far because of poor uptake of the MAb–conjugates especially in solid tumours. Drug delivery is also limited by the number of drug molecules that can be efficiently carried by each antibody molecule. Furthermore chemical conjugation is usually a complex procedure that can damage both the MAb and the drug and MAb–conjugates require intra- or extracellularly active biochemicals and/or enzymes to cleave the active drug from the antibody. Table 8.2 gives an overview of the various antibody–drug conjugates that have been developed for cancer therapy to date.

8.5.1.6 Radioimmunoconjugates

Another way of using MAbs as therapeutic agents is to couple them to radionuclides (Figure 8.1e and Table 8.3). Radioimmunoconjugates offer many advantages in the treatment of cancer. Cell killing does not rely on the host's immune system and occurs by the ionizing effects of emitted radioactive particles [73]. These radioactive cytotoxic particles are effective over a distance of several cell diameters, allowing eradication of antigen-negative cells by the radioimmunoconjugate bound to the adjacent antigen-positive tumour cells. This is useful considering the heterogeneity of antigen expression in some tumours. Finally, the amount of radioactive MAb delivered to a tumour can be measured non-invasively by imaging [67]. The most important factors for therapeutic efficacy of radioimmunoconjugates are good penetration, favourable pharmacokinetics, and a prolonged time of retention in the tumour [74].

Table 8.3. Isotopes used for radioimmunotherapy in cancer

	Radioisotope
Beta-emitters	Iodine-131
	Yttrium-90
	Rhenium-188
	Rhenium-186
	Copper-67
Alpha-emitters	Bismuth-212
	Astatine-211
Electron capture	Iodine-125

8.5.2 Bispecific Monoclonal Antibodies

Another approach to selectively inducing tumour cell killing is by the use of bispecific monoclonal antibodies (BsMAb). They combine the specificity of two separate antibodies within one molecule and cross-link an effector killer cell or a toxic molecule with the target cell to be destroyed [75]. There are three major approaches for creating BsMAbs. They can be obtained by chemical cross-linking of two MAbs, by fusing two hybridomas [76], or by genetic engineering [77]. Each method has its advantages and disadvantages. Chemical conjugates have a well-defined linkage and can be produced in high yield. However, there is lot-to-lot variability in purity and activity. Quadromas, produced by fusing two hybridomas, can also produce large quantities of BsMAbs. However, in addition to the desired BsMAb, the parental MAb and every possible combination of heavy and light chain matches and mismatches are also produced. Furthermore, quadromas are often genetically unstable and require frequent subcloning. With recombinant fusion proteins, it is possible to make new combinations of Fab or Fv segments or to combine human and mouse gene segments. Yields and correct folding of the purified fusion protein can present problems as discussed earlier.

Cytotoxic drugs including toxins such as saporin, ricin A chain, vinca alkaloids, and radioisotopes have been delivered to tumour cells with BsMAbs that bind to the drug/toxin with one arm and to a surface molecule on the targeted cell with the other arm. This approach has proven successful in animals as e.g. shown by Schmidt *et al.* [78].

Cytotoxic effector cells have also been cross-linked to tumour cells via BsMAb (Figure 8.3). The BsMAb activates the cytotoxic activity of the effector cell on bridging it to the target cell. Several effector cells, including phagocytic cells, natural killer (NK) cells and T lymphocytes, can mediate cellular cytotoxity [37,75,79,80]. Adequate pre-activation of the effector cells is an important requirement in these methods of drug delivery. In the case of T

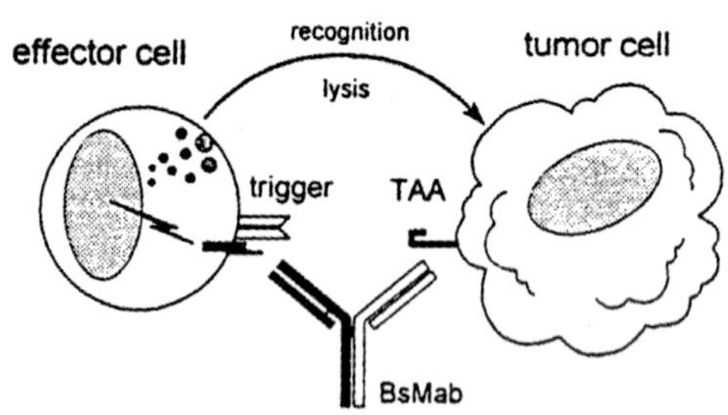

Effector cell	**Trigger epitope**	**Tumor associated antigen**
T cells	TcR / CD3	*Carcinomas:*
	CD2	CEA
		EGP-2
NK cells	CD16	EGF-R
	CD44	Her-2 / neu
	CD69	
	CD38	*B cell lymphomas:*
	CD56	CD19
		CD20
Neutrophils	CD64 (activatedPMNs)	CD22
	CD89	
	CR3 (CD11b/CD18)	*T cell leukemia:*
		CD25
Monocytes / macrophages	CD64	CD52
	CD89	*a.o.*

Figure 8.3. Schematic representation of bispecific antobody mediated tumor cell recognition by an immune effector cell. Summarised are effector cell types, trigger molecules and tumor associated antigens used as a targed as reported in the literature. From reference [37].

cells, the presence of co-stimulatory molecules such as CD28 and cytokines is a prerequisite to achieve this, whereas granulocytes and macrophages can be activated with granulocyte macrophage colony stimulating factor (GM-CSF).

Heteroconjugates made by cross-linking two different IgGs are twice as large as MAbs and thus are limited in their ability to penetrate tumours, although this problem can be solved by combining two different scFvs resulting in BsMAb formats with minimal molecular mass. As is the case for all mouse Mab-based therapies, HAMA is generated, but with the help of humanizing and chimerizing technologies this should become less of a problem in the future.

8.5.3 Pro-drug Strategy

8.5.3.1 Antibody-directed Enzyme Pro-drug Therapy (ADEPT)

Pro-drugs in combination with enzyme–MAb conjugates can also be used to target tumour cells [81,82]. The so-called antibody-directed enzyme pro-drug therapy (ADEPT) approach involves the use of antibody–enzyme conjugates directed against tumour-associated antigens that achieve *in situ* activation of subsequently administered pro-drugs. Pro-drugs are inactive drug precursors that are not readily taken up by cells and hence are less toxic to healthy cells. The pro-drug can be converted locally in the tumour into the active drug by a specific enzyme which is covalently linked to tumour-specific antigen-targeted MAbs. When the active form of the drug is released, it will then distribute to the nearby tumour cells, resulting in cell death. A number of such pro-drug/MAb–enzyme conjugates have been developed and tested *in vitro* and *in vivo* [83,84]. One of the significant advantages of this approach is that the targeted enzyme can be effective without being endocytosed. Another beneficial aspect is that a large amount of the drug can be enzymatically generated at the tumour site. Table 8.4 shows some ADEPT strategies developed in recent years. Limitations of ADEPT include suboptimal tumour uptake due to heterogeneity in antigen expression, development of immune responses against the enzyme component, the risk of diffusion of the active drug away from the tumour site and the complexity of dosing schedules.

8.5.3.2 Pro-drug Monotherapy

Another pro-drug strategy under development is the concept of 'monotherapy'. An attractive feature of pro-drug monotherapy, unlike ADEPT, is that antibody–enzyme conjugates are not required. In the case of pro-drug monotherapy, local production of elevated levels of enzymes by the tumour is exploited to release the active drug. Pro-drug monotherapy works well with anthracycline pro-drugs that are activated by β-glucuronidase [85,86] which can be found in elevated concentrations in necrotic areas of tumour tissue [87]. De Groot *et al.* [88] also developed anthracycline pro-drugs that can be activated by the tumour-associated protease plasmin. The plasmin system plays a key role in tumour invasion and metastasis by its matrix degrading activity and its involvement in tumour growth, most likely by its participation in growth factor activation and angiogenesis.

Table 8.4. ADEPT strategies developed for cancer therapy.

Enzyme	Pro-drug	Active drug
Carboxypeptidase G2	Benzoic acid Mustard glutamates	Benzoic acid mustards
Carboxypeptidase A	Methotrexate-alanine	Methotrexate
Alkaline phosphatase	Etoposide phosphate	Etoposide
	Mitomycin phosphate	Mitomycin
	Doxorubicin phosphate	Doxorubicin
	Phenolmustard phosphate	Phenolmustard
	Mitomycin phosphate	Mitomycin
Beta-glucuronidase	Phenolmustard-glucuronide	Phenolmustard
	Epirubicin-glucuronide	Epirubicin
	Doxorubicin-glucuronide	Doxorubicin
Penicillin amidase	Palytoxin-4-hydroxyphenylacetamide	Palytoxin
	Doxorubicin-phenoxyacetamide	Doxorubicin
	Melphalan-phenoxyacetamide	Melphalan
Beta-lactamase	Cephalosporin vinca alkaloid	Desacetylvinblastine hydrazide
	Cephalosporin mustard	Phenylenediamine mustard
	Cephalosporin mitomycin C	Mitomycin C
	Cephalosporin doxorubicin	Doxorubicin
Carboxylesterase	Paclitaxel carbonate	Paclitaxel
	Carbonyloxycamptothecin	Camptothecin
Cytosine deaminase	5-Fluorocytosine	Fluorouracil
Plasmin	Doxorubicin tripartate	Doxorubicin
	Daunorubicin tripartate	Daunorubicin

One of the limitations of pro-drug monotherapy may be the risk of diffusion of the active drug away from the tumour site.

8.5.4 (Synthetic) (co)Polymers

Polymers or synthetic copolymers are believed to accumulate in solid tumours due to enhanced vascular permeability of tumour blood vessels combined with a lack of lymphatic drainage in the tumour tissue [89,90]. Polymer-based targeting strategies can be divided into two main categories, i.e. polymer–protein conjugates (so far the most widely studied) and polymer–drug conjugates, particularly those containing conventional anti-tumour agents. Polymer–drug conjugation can be used to alter the biodistribution, elimination rate and rate of metabolism of covalently bound drugs. In the case of protein adducts, polymer conjugation can prolong the protein plasma elimination half-life, reduce proteolytic degradation and may have the added benefit of reducing immunogenicity. Polyethylene glycol (PEG) is the most widely used polymer for protein conjugation (Figure 8.4).

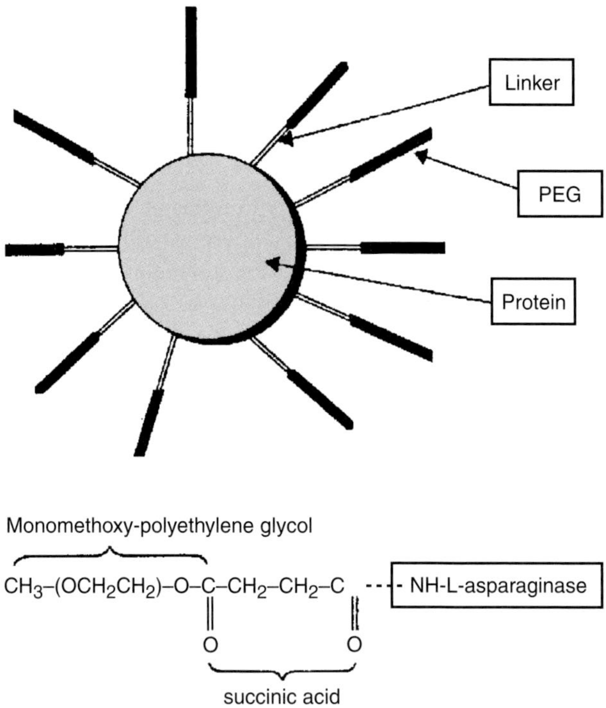

Figure 8.4. Schematic diagram showing the structure of a typical polyethylene glycol (PEG) conjugate and the chemical structure of PEG-asparaginase.

Soluble polymer conjugates have also been proposed as macromolecular pro-drugs for controlled release and targeting of various low molecular weight, (non-protein) chemicals [91,92]. In this case, polymer conjugation not only serves to alter drug biodistribution by restricting cellular capture to the lysosomotropic route, but the polymer–drug linkage can also be designed to allow site-specific enzymatic or hydrolytic cleavage. Thus, both the rate and the site of drug delivery can in principle be controlled. Enhanced permeability of the microvasculature at certain sites, particularly within solid tumours, can be exploited to facilitate site-specific accumulation of polymer–drug conjugates [93]. Other (co)polymers of interest besides PEG are SMANCS (styrene-co-maleic anhydride neocarzinostatin; zinostatin stimalar) and HPMA (N-(2-hydroxypropyl) methylacrylamide).

Targeting moieties such as sugars (galactose, mannose), proteins and antibodies have been incorporated into the conjugates to promote receptor-mediated recognition. Thus, cell- or organ-specific localization of therapy may be achieved [98]. It should be noted however, that various cell types in the liver and spleen are important target cells for sugar-derivatized proteins (see Chapter 4) and that hepatic clearance will always compete with extrahepatic distribution.

Polymer conjugates are most useful in the context of immuno-conjugates. Other protein constructs such as fusion proteins can assist their future development. Soluble polymer conjugates have now also been introduced into clinical practice [90,94] as will be described in Section 8.6.3.

8.5.5 Liposomes

Selective targeting of drugs using liposomes is expected to optimize the pharmacological effect and toxicities of encapsulated drugs with the advantage that liposomal components are non-toxic, non-immunogenic and biodegradable [95,96]. Through encapsulation of drugs in a macromolecular carrier such as a liposome, the volume of distribution is significantly reduced and the concentration of drug in the tumour is increased. Under optimal conditions, the drug is carried through the body associated with the lipid and/or aqueous moiety of the liposome. On arrival at the tumour the system should leak at a sufficient rate for the encapsulated drug to become bioavailable. The liposome protects the drug from metabolism and inactivation in the plasma. Due to size limitations in the transport of large molecules across healthy endothelium, the drug will accumulate to a reduced extent in healthy tissues. Discontinuities in the endothelium of the tumour vasculature, on the other hand, may result in an increased extravasation of large molecules and increased accumulation of liposomal drug in the tumour. However, this increased penetration phenomenon may be highly dependent on the type of tumour and the stage of tumour development.

Initially, liposomes seemed to be optimal drug carrier systems, but further research in general showed disappointing results [97]. The clinical utility of what are now called 'conventional' liposomes was limited by their rapid uptake by phagocytic cells of the immune system, predominantly in the liver and spleen, resulting in their largely uncontrollable properties upon administration *in vivo*.

Interest in liposomes as drug carriers was rejuvenated by the introduction of new ideas from membrane biophysics. Liposomes can now be designed as non-reactive (sterically stabilized) particles, as well as cationic or fusogenic liposomes. The non-reactive liposomes can also be designed to induce tumour-specific targeting, while cationic or fusogenic liposomes can exhibit high specificity for nucleic acid and cell membrane interactions. Because of their reduced recognition and uptake by the phagocytic system, these liposomes are referred to as 'stealth' liposomes [98]. These may prove to be useful in cancer therapy, although it should be noted that even if distribution to macrophages is slowed down, they will eventually find their way into these cells.

In sterically-stabilized liposomes, the lipid bilayer contains hydrophilic polymers or hydrophilic glycolipids, PEG and the ganglioside GM being the subjects of the most detailed studies. These liposomes remain in the blood for up to 100 times longer than conventional liposomes and thus can increase the pharmacological efficacy of encapsulated agents. Consequently, on chronic administration side-effects related to macrophage function are certainly not excluded and can become dose limiting [99].

The choice of the drug for delivery via liposomes is essential. To be effective as a carrier, a liposome must be able to efficiently balance stability in the circulation with the ability to make the drug bioavailable at the tumour. Furthermore the drug must have adequate activity against the chosen tumour. A drug such as doxorubicin with a relatively broad activity against a variety of tumour types is an ideal choice in this regard [100]. The drug also has to be efficiently loaded into the liposomal carrier [101,102]. Liposomes bearing attached antibodies or other ligands accumulate much more readily in target cells than conventional liposomes (Figure 8.1f). The encapsulation of drugs in MAb-targeted liposomes can be used to selectively increase the concentration of drug delivered to antigen-expressing cells [103–105]. Immunoliposomes utilizing internalizing MAbs, such as anti-HER-2 [106] or anti-CD19 [107], can be used to selectively deliver high concentrations of drug into the cytoplasm of antigen-expressing cells.

Encapsulation of immunomodulators, e.g. muramyl tripeptide analogues, into liposomes has been designed to stimulate host immunity [108] and can be used in combination with other treatment modalities. The systemic activation of macrophages provides an additional therapeutic modality for the eradication of cancer and cancer metastases.

Liposomes have also been tested as carriers in gene therapy. Cationic lipids in particular, can condense DNA and increase transfection yields *in vitro* by several orders of magnitude. Reports on transfections *in vivo* stimulated intense interest in the use of liposomes for gene therapy, but data so far have been quite disappointing [109]. With the generation of more sophisticated multifunctional liposomal systems containing steric stabilization, homing devices and/or fusogenic/controlled released properties, studies on liposomal gene delivery systems are now focused on the development of small, circulation-stable lipid–DNA complexes. These complexes can be administered systemically and, once accumulated at the tumour site, be specifically taken up by tumour cells via endocytosis or direct fusion with the tumour plasma cell membrane [110,111].

Perhaps the most interesting and potentially most powerful therapeutic application of liposome technology for cancer therapy, may be in combining therapeutics aimed at various (newly identified) molecular targets with conventional cytotoxic drugs. Furthermore, optimal tumour specificity and therapeutic activity may be achieved by synergistically combining the selectivity benefits of tumour cell molecular targets with pharmacokinetic targeting. In fact, both therapeutic modalities can be delivered in liposomal form [112].

For a review on optimizing liposomes for delivery of chemotherapeutic agents to solid tumours, readers are referred to Drummond *et al.* [113].

8.6 Clinical Studies

8.6.1 MAb and MAb-based Constructs

MAb-based constructs represent, as described, a heterogeneous class of anti-tumour agents with remarkable efficacy in the treatment of experimental cancers in animals. Several MAb and immunoconjugates have been evaluated further in cancer patients, and the results have indicated that some have activity at safe doses.

Clinical trials with MAbs began in the late 1970s, primarily in patients with haematological malignancies. The first successful results were obtained using anti-Id MAb, where impressive, long-lasting responses were induced in patients with Non-Hodgkin's lymphoma (NHL) [114]. Unfortunately, anti-Ids were expensive to generate and were useful in only a limited number of patients. Therefore, commercial development was considered risky. Other successful results have been observed using anti-IL-2 receptor MAb in T-cell acute lymphoblastic leukaemia (ALL) [115,116]. An early trial using a MAb against CD10 showed dramatic reductions in peripheral blood leukaemia cells in three patients [115]. A MAb against the Lewis Y antigen which is abundantly expressed on the surface of cells from several human carcinomas, induced four minor responses in 12 patients with advanced breast cancer [58]. Finally, MAbs against the asialoganglioside antigens, GD2 and GD3, which are present on tumours of neuroectodermal origin including melanoma and neuroblastoma, were successfully applied in a number of studies [117–121].

An important demonstration of the efficacy of a MAb in minimal residual disease was achieved using MAb 17-1A (directed against the EGP-2 or EpCAM antigen as described previously) in patients with stage III colorectal cancer. Following surgical resection, MAb therapy reduced the overall death rate by 32% and the rate of recurrence by 23% [122].

In general, experience with unlabelled MAbs has clearly demonstrated that therapy is rarely associated with toxicity even when HAMA responses are evoked, although circulating immune complexes can lead to serum sickness and organ damage in rare instances. Clinical trials with murine and human MAbs have clearly led to an appreciation of the complexity of treating patients with these reagents. Hence, issues such as (1) the pharmacokinetics of MAb *vis-à-vis* the presence of circulating target antigens; (2) the number of available target antigens and binding avidity of the MAb; and (3) the importance of dosing schedules *vis-à-vis* the generation of HAMA, have been addressed in numerous trials. In general, it has been concluded that MAb therapy will probably be most effective in the treatment of minimal residual disease and/or of micrometastasis, following routine chemotherapy, radiotherapy, or surgery.

Humanized and chimerized MAbs have been developed for the treatment of non-Hodgkin lymphoma, renal cell carcinoma, ovarian carcinoma, breast cancer, melanoma, and neuroblastoma [117,119,120,123,124]. Patients with relapsed or refractory myeloid leukaemias that have been treated with HuM95, did not develop significant HAMA responses [59].

Multicentre studies have demonstrated the efficacy of rituximab, a chimeric IgG-1 directed against CD20, in the treatment of relapsed low-grade and follicular non-Hodgkin lymphoma. In 166 patients, receiving 375 mg • m^{-2} rituximab in four weekly doses, the overall response rate was 50% in 161 evaluable patients who had previously received chemotherapy [125]. In addition, combination therapy trials of rituximab with CHOP (cyclophosphamide, doxorubicin, vincristine, prednisone) in refractory and newly diagnosed patients suggests that rituximab may also have a role in the eradication of residual disease. This combination appears to be a viable treatment option for relapsed low-grade non-Hodgkin lymphoma [126].

CD52 MAb (Campath-1H) has also been extensively evaluated for its capacity to lyse malignant lymphopoietic cells [127]. CD52 MAb proved to be effective for chronic leukaemias of T-cell or B-cell origin that may be resistant to conventional chemotherapy. Patients with T-cell polylymphocytic leukaemia, including those with a large tumour burden and high peripheral-blood-cell counts, showed complete remission after using the Campath-1H MAb. Trastuzumab, the recombinant humanized anti-p185 HER2/neu monoclonal antibody, has been investigated in the treatment of breast cancer. In a single-arm clinical study of 222 pa-

Table 8.5. Unconjugated monoclonal antibodies that have been developed to be used therapeutically to treat cancer.

Disease[a]	**Target antigen**
NHL	Ig (Id), CD20, CD52, CD22, CD19
ALL	CD10, CD25
CLL	CD5
AML	CD33, CD45
T-cell ALL	CD25, CD5, CD7, CD25
T-cell Lymphoma	CD4, CD5
Multiple myeloma	IL-6
Colorectal cancer	CEA, CO17-1A
Melanoma	p240, p97, GD2, GD3
Breast cancer	LG6, HER-2/neu, LewisY
Prostate cancer	PSA, EGFR
Ovary	LG6
Lung cancer	SCLC, EGFR, BLP
Neuroblastoma	GD2
Renal cell cancer	G250
Glioma	EGFR
Head and neck cancer	EGFR

[a] NHL, Non-Hodgkin's lymphoma; ALL/CLL, acute/chronic lymphoblastic leukaemia; AML, acute myeloid leukaemia.

tients, treatment with a loading dose of traztuzumab 4 mg • kg^{-1} i.v., followed by weekly doses of 2 mg • kg^{-1} produced an overall response rate of 14% [128]. In other clinical studies patients with metastatic breast carcinoma were randomized to a cytostatic regimen with or without trastuzumab [129,130]. Table 8.5 gives an overview of 'naked' MAbs that have been used therapeutically to treat cancer.

Thus far, studies with MAbs conjugated to toxins, drugs and isotopes are mostly in an early stage of investigation and therefore limited data are available for the evaluation of the efficacy of these agents. ITs used in clinical trials for cancer (see Farah *et al.* [22] for a concise summary) are MAb-conjugated to saporin, Pseudomonas exotoxin, and ricin. They have shown spectacular results in haematological cancers, but poor results in large solid carcinomas [64].

Encouraging results with MAb–drug conjugates were seen in a phase I study with a conjugate of calicheamicin γI_1 and a humanized anti-CD33 MAb in patients with refractory or relapsed AML [131]. Results support further evaluation in a setting of newly diagnosed or minimal-residual disease.

The immunoconjugate of doxorubicin with a chimeric anti-Le^y-related, tumour-associated antigen expressed on most human carcinomas, was evaluated in phase I [132] and phase II [133] clinical trials. The phase II trial performed in patients with metastatic breast carcinoma showed low clinical response rates. These data together with tumour biopsy analysis suggest that the dose that could be safely administered was insufficient to maintain the intra-tumoural concentration of doxorubicin at a level required to achieve regression. It is likely therefore that immunoconjugates of doxorubicin will be only effective in minimal-disease settings.

In general, trials with radioimmunoconjugates reported higher response rates in patients with haematological malignancies when compared with patients with solid tumours. The most impressive studies with non-myeloablative regimens reported objective responses in 70–80% of patients with chemotherapy-refractory B-cell lymphomas, a median response duration of 12 months and minimal toxicity using 131-I-labelled anti-CD20 [134]. In trials with myeloablative regimens, performed in conjunction with autologous haematopoeitic bone marrow or stem cell transplantation, responses have been seen in 95% of patients, complete responses in 85%, with a progression-free survival of 62% and overall survival of 93% with a median follow-up of 2 years [135,136]. The estimated 4-year overall and progression-free survival rates were reported to be 68 and 42%, respectively [137].

Anti-CD45 antibody BC8 labelled with 131-I may reduce the rate of tumour relapse and has acceptable toxicity in patients with AML, ALL or myelodysplastic syndrome who underwent stem-cell rescue [138]. Of 25 patients treated with advanced AML and myelodysplastic syndrome, seven were disease-free at 15–89 months post-transplantation. Of nine patients with advanced ALL, three were disease-free at 23, 58 and 70 months post-transplantation, respectively.

As with the monoclonal antibody therapies described above, B-cell malignancies would be the most attractive targets for anti-CD3 bispecific monoclonal antibody-based immunotherapy [139]. Several phase I/II clinical trials have been described using BsMAb for non B-cell malignancies. BIS-1, a BsMAb directed against the TAA EGP-2 and the CD3/T-cell receptor complex on T lymphocytes was studied in renal [140] and lung cancer [141]. In carcinoma patients with EGP-2 positive malignant ascites or pleural exudates, local administration of *ex-vivo* IL-2-activated autologous lymphocytes and BIS-1, resulted in both anti-tumour effects and a strong local inflammatory reaction [141]. In patients with advanced breast or ovarian cancer administration of BsMAb directed against FcγRI or FcγRIII and HER-2/neu resulted in elevated plasma levels of cytokines [142,143].

The biological effects observed at tumour sites indicate that BsMAbs effectively penetrate tissue. However, trials are limited by the toxicity caused by induction of a 'cytokine storm' or by the complex pharmacokinetics. Furthermore, T-cell directed BsMAb approaches are hindered by difficulties in mobilizing and activating T and NK effector cells. Recent attention, therefore, has focused on BsMAbs which target myeloid effectors [79].

8.6.2 Pro-drugs

ADEPT strategies have been described but only the carboxypeptidase G2 approach has been tested in patients so far. In a phase I clinical trial, patients with non-resectable metastatic or locally recurrent colorectal carcinoma were treated with ADEPT. Carboxypeptidase G2 activity was found in metastatic tumour biopsies. The pro-drug was converted into the active drug but leakage into the bloodstream also occurred [22,144].

Clinical trials with 'monotherapy' pro-drug strategies are in progress, but have not yet been evaluated.

8.6.3 (Synthetic) (co)Polymers

Soluble polymer conjugates have been introduced into clinical practice in the last decade [90,94]. Several PEG conjugates have been evaluated clinically [145] for cancer therapy including a PEG conjugate of asparaginase in the treatment of ALL in patients hypersensitive to the native enzyme [146], and a PEG conjugate of IL-2 [147].

A phase I clinical and pharmacokinetic study of PK1 comprising doxorubicin covalently bound to N-(2-hydroxypropyl)-methacrylamide copolymer by a peptidyl linker, was carried out in 36 patients with refractory or resistant cancers [94]. PK1 demonstrated anti-tumour activity, and that polymer–drug conjugation decreased doxorubicin dose-limiting toxicities. Phase II studies are in progress.

8.6.4 Liposomes

There are three liposomal forms of doxorubicin or daunorubicin on the market (Table 8.6). Doxil® and DaunoXome® have been approved for the treatment of AIDS-related Kaposi's sarcoma and are being evaluated in clinical trials for the treatment of a variety of cancers [148–151]. Evacet® (liposomal doxorubicin) has recently been tested in large phase II and III clinical trials for the treatment of metastatic breast cancer and is awaiting approval by the FDA [151]. Data obtained from trials thus far suggest that all three liposomal drugs offer significant therapeutic benefit compared with the free drug [113].

Table 8.6. Monoclonal antibody based products and liposome formulations registered for cancer therapy in the USA and/or Europe.

Generic name	Trade name	Company	Indication
Rituximab[a]	Rituxan (Mabthera)[c]	Roche	Relapsed or refractory low-grade/follicular non-Hodgkin's lymphoma
Trastuzumab[a]	Herceptin	Genentech-Roche	Metastatic breast cancer
Edrecolomab[a]	Panorex	Glaxo-Wellcome	Post-operative adjuvant therapy Dukes C colorectal carcinoma
Doxorubicin[b]	Doxil (Caelyx)[d]	Alza-corporation	AIDS-related Kaposi's sarcoma
Daunorubicin[b]	DaunoXome	Nexstar Pharmaceuticals	AIDS-related Kaposi's sarcoma
Doxorubicin[b]	Evacet (Myocet)[e]	The Liposome Company, Inc.	Metastatic breast cancer

[a] MAb-based.
[b] Liposome formulation.
[c] Rituxan is known as Mabthera in Europe.
[d] Doxil is known as Caelyx in Europe.
[e] Evacet is known as Myocet in Europe.

Liposomal preparations can be therapeutically beneficial based on their ability to decrease non-specific toxicities associated with the drug, or by being more efficacious against a specific type of cancer, increasing the response frequency, and/or the average time to relapse or response duration.

Liposome-encapsulated immunomodulators are currently under investigation in different patient groups although this development has certainly not advanced as far as that with the liposomal anthracyclines. MLV-MTP-PE (multilamellar vesicles-muramyl tripeptide-phosphatidylethanolamine) was studied in several clinical trials in osteosarcoma patients who developed pulmonary metastases during adjuvant chemotherapy [108]. The intravenous administration of MLV-MTP-PE induced tumouricidal properties in monocytes as well as increase in serum IL-1 shortly after intravenous infusion. Furthermore elevations in C-reactive protein, β2-microglobulin and ceruloplasmin were frequently observed. Even higher anti-tumour activity was observed in combination with ifosfamide. These preliminary results suggests that liposome-encapsulated immunomodulators in combination with chemotherapy may be an appropriate treatment for recurrent disease.

In Table 8.6 a summary is given of MAb and liposome-based formulations registered for cancer therapy.

8.7 Animal Models: their Predictive Value

Studies in pre-clinical models with human tumours are often carried out in (immuno)deficient mice. However, particularly in the case of monoclonal antibody-directed therapy, it is important to recognize that these models, while useful, frequently over-predict activity and under-predict toxicity because the target antigen is tumour-specific in the animal but only tumour-associated in man.

Besides mouse models, rat models are used for experimental *in vivo* studies of MAbs for immunotherapy of cancers. Kroesen *et al.* [80] described a rat model to investigate the potential of targeted anti-tumour treatment against EGP-2-positive, rat syngenic tumours. However, possible adverse side-effects on healthy tissues could not be studied in this model as the target epitope was not expressed on tissues other than tumour cells. To overcome this problem, a transgenetic rat model expressing the EGP-2 protein in various organs was developed for evaluation of anti-carcinoma-associated-antigen-EGP-2-directed immunotherapy strategies [152].

Other limitations of mouse/rat models are that tumour growth is different in the animal model compared to the situation in man. Tumour growth is more rapid in the rat/mouse model which has an effect on vascularization and intra-tumoural pressure for example. These factors can, as discussed in Section 8.4, have great impact on tumour penetration and uptake of the MAb-based drug-targeting constructs.

8.8 Conclusions and Future Perspectives

The studies presented above represent the most advanced pre-clinical and clinical therapy programmes for cancer management including drug targeting strategies. Patience is required to evaluate the large variety of results, to understand the reasons for limited success, and the great number of concepts to improve the treatment strategies. It should be emphasized that after more than 15 years of experience with monoclonal antibody therapy of cancer, we have no definite idea of essential mechanisms required for clinical activity. The findings of e.g. Clynes *et al.* [153,154] support the idea that the anti-tumour effects of MAbs depend on many mechanisms. For instance, myeloid cells, probably macrophages and monocytes that express both activation and inhibitory Fc receptors, are essential immune effector cells for MAb therapy. Blocking inhibitory receptors or MAbs that selectively trigger activating Fc receptors without affecting inhibitory Fc receptors could increase potency. The underlying mechanisms of the therapeutic effects of MAbs may also vary from one MAb to another. These findings should be considered in the improvement of drug-targeting strategies based on MAbs, including Mab derivatives such as those used in ADEPT. The same holds true for approaches such as modulating growth factor receptor functions or inducing immune responses using BsMAbs or the combination of tumour cell-directed strategies combined with BsMAbs delivering agents which inhibit the supply of blood to the tumour.

Liposomal drugs have been suggested to be the 'magic bullet' of cancer therapy due to their ability to accumulate selectively in the tumour. The problem remains that not all cancers and patients respond in the same way. The drug actually delivered to the required site of action plays an important role in the response achieved, while the potential toxicity of the surface modified liposomes in the macrophage system have to be taken into account on chronic administration.

Multi-drug resistance represents a significant obstacle in the use of standard chemotherapy regimens to cure cancer. It is unlikely that tumours resistant to free drug therapy will be eradicated by liposome-encapsulated or any other targeted form of drug. Additional studies that attempt to encapsulate drugs which exhibit non-overlapping modes of drug resistance and significant activity against a particular form of cancer, or combine free drugs with non-overlapping modes of drug resistance with currently available liposomal drugs, will address the potential and limitations of these approaches. There is also a need to develop liposomal formulations of other drugs and therapeutic regimens designed to stimulate host immunity alone or in combination with other treatment modalities.

In conclusion, research needs to be aimed at improving the potency and reducing the immunogenicity and toxicity of drug formulations in order to increase the therapeutic index. Methods to increase the overall therapeutic index of a drug, e.g. by using tumour-specific ligands, increasing extravasation of liposomes into tumours and increasing the bioavailability of the drug selectively at the site of the tumour, should be explored further.

In addition to this, research must focus on optimizing the biophysical properties of the carriers and drug-conjugates, as these properties are of major importance to the decision of whether or not to use these molecules *in vivo.* A wealth of innovative strategies involving recombinant molecules with novel effector functions are in the early stages of clinical evaluation and hold great promise for the future. These developments will initiate other studies which should focus on long-term safety and cost–benefit analyses. Only adequately designed case–control studies supported by international collaboration can show the long-term safety of these new therapies. The same holds true for studies evaluating the costs of these therapeutic regimens and the resultant improvement in the quality of life, in relation to the cumulative social and health-care costs of current therapies. These multidisciplinary collaborative efforts will be the basis for the introduction of new therapeutics with optimum efficacy and cost, into clinical practice.

References

[1] Jain RK, *Sci. Am.* **1994,** *271*, 58–65.
[2] Jain RK, *Science* **1996,** *271*, 1079–1080.
[3] Foulds L, *J. Chron. Dis.* **1958,** *8*, 2–37.
[4] Bishop JM, *Annu. Rev. Biochem.* **1983,** *52*, 301–354.
[5] Hunter T, Cooper JA, In: *Enzyme Control by Protein Phosphorylation.* (Eds Boyer PD, Krebs EG), p. 191. Academic Press, New York, **1986.**
[6] Marshall CJ, *Cell* **1991,** *64*, 313–326.
[7] De Vita Jr VT, Hellman S, Rosenberg SA (Eds) *Cancer, Principles and Practice of Oncology*, 5th edn. J. B. Lippincott Company, Philadelphia, PA, **1997.**
[8] Nagy JA, Brown LF, Senger DR, Lanir N, van de Water L, Dvorak AM, Dvorak HF, *Biochim. Biophys. Acta* **1988,** *948*, 305–326.
[9] Folkman J, *J. Natl Cancer. Inst.* **1990,** *82*, 4–6.
[10] Griffioen AW, Molema G, *Pharmacological Rev.* **2000,** *52*, 237–268.
[11] Bicknell R, Harris AL, *Eur. J. Cancer* **1991,** *27*, 781–785.
[12] Polverini PJ, Leibovich SJ, *Lab. Invest.* **1984,** *51*, 635–642.
[13] Konerding MA, Steinberg F, Streffer C, *Acta Anat.* **1989,** *136*, 21–26.
[14] Jain RK, *J. Control. Rel.* **1998,** *53*, 49–67.
[15] Juliano RL, Ling V, *Biochim. Biophys. Acta* **1976,** *455*, 152–162.
[16] Cole SPC, Bhardwaj G, Gerlach JH, Machie JE, Grant CE, Almquist KC, Stewart AJ, Kurz EU, Duncan AMV, Deeley RG, *Science* **1992,** *258*, 1650–1654.
[17] Scheffer GL, Wijngaard PLJ, Flens JM, Isquierdo MA, Slovak ML, Pinedo HM, Meijer CJLM, Clevers HC, Scheper RJ, *Nature Med.* **1995,** *1*, 578–582.
[18] Izquierdo MA, Scheffer GL, Flens MJ, Giaccone G, Boxterman HJ, Meijer CJLM, van der Valk P, Scheper RJ, *Am. J. Pathol.* **1996,** *148*, 877–887.
[19] Meijer DK, Smit JW, Hooiveld GJ, Van Montfoort JE, Jansen PL, Muller M, *Pharm. Biotechnol.* **1999,** *12*, 89–157.
[20] Wang JC, *Biochim. Biophys. Acta* **1987,** *909*, 1–9.
[21] Köhler G, Milstein C, *Nature* **1997,** *256*, 495–497.
[22] Farah RA, Clinchy B, Herrera L, Vitetta ES, *Crit. Rev. Euk. Gene Expr.* **1998,** *8*, 321–356.
[23] De Leij L, *Encycloped. Cancer* **1997,** *1*, 1818–1839.
[24] Demanet C, Brissincle J, De Jonge, Thielemans K, *Blood* **1996,** *81*, 4390–4398.
[25] Mellstedt H, Frodin JE, Masucci G, Ragnhammer P, Fagerberg J, Hjelm AL, Shetye J, Wersall P, Osterborg A, *Sem. Oncol.* **1991,** *18*, 462–477.
[26] Hendler FJ, Ozanne BW, *J. Clin. Invest.* **1984,** *74*, 647–651.
[27] Berger MS, Gullick WJ, Greenfield C, *J. Pathol.* **1987,** *52*, 297–307.
[28] Tateishi M, Ishida T, Mitsudomi T, Sugimachi K, *Cancer Res.* **1990,** *50*, 7077–7080.
[29] Schneider PM, Hung MC, Chiocca SM, *Cancer Res.* **1989,** *49*, 4968–4971.
[30] Weiner DB, Nordberg J, Robinson R, *Cancer Res.* **1990,** *50*, 5184–5191.
[31] Mezzanzanica D, Canevari S, Colnaghi MI, *Int. J. Clin. Lab. Res.* **1991,** *21*, 159–164.
[32] Bolhuis RLH, Lamers CHJ, Goey SH, *Int. J. Cancer* **1992,** *S7*, 78–81.
[33] Miotti S, Canevari S, Menard S, Mezzanzanica D, Porro G, *Int. J. Cancer* **1987,** *39*, 297–303.
[34] Bumol TF, Marder P, van de Herdt S, Borowitz MJ, Apelgren LD, *Hybridoma* **1988,** *7*, 407–415.
[35] Huang X, Molema G, King S, Watkins L, Edgington TS, Thorpe PE, *Science* **1997,** *275*, 547–550.
[36] Ran S, Gao B, Duffy S, Watkins L, Rote N, Thorpe PE, *Cancer Res.* **1998,** *58*, 4646–4653.
[37] Molema G, Kroesen BJ, Helfrich W, Meijer DKF, de Leij LFMH, *J. Control. Rel.* **2000,** *64*, 229–239.
[38] Clackson T, Hoogenboom M, Griffiths AD, Winter G, *Nature* **1991,** *352*, 624–28.
[39] Lobuglio AF, Saleh MN, *Am. J. Med. Sci.* **1992,** *304*, 214–224.
[40] Debatin K-M, Krammer PH, *Leukaemia* **1995,** *9*, 815–820.
[41] Trauth BC, Klas C, Peters AMJ, Matzhu S, Moller P, Falk W, Debatin K-M, Krammer PH, *Science* **1989,** *245*, 301–305.
[42] Gethie M-A, Podar EM, Ilgen A, Gordon BE, Uhr JW, Vitetta ES, *Proc. Natl Acad. Sci. USA* **1997,** *94*, 7509–7514.

[43] Chaouchi NA, Vazquez, Galanaud P, Leprince C, *J. Immunol.* **1995,** *154*, 3096–3104.
[44] Vuist W.MJ, Levy R, Maloney DG, *Blood* **1994,** *83,* 899–906.
[45] Baselga J, Mendelsohn J, *Pharmacol. Ther.* **1994,** *64*, 127–154.
[46] Shepard HM, Lewis GD, Sarup JC, Fendly BM, Maneval D, Mordenti J, Figari I, Kotts CE, Palladino MA, Ullrich A, Slamon D, *J. Clin. Immunol.* **1991,** *11*, 117–127.
[47] Wu X, Fan Z, Masui H, Rosen S, Mendelsohn J, *J. Clin. Invest.* **1995,** *95*, 1897–1905.
[48] Schneider-Gadicke E, Riethmuller G, *Eur. J. Cancer* **1995,** *31*, 1326–1330.
[49] Greiner JW, Guadagni F, Roselli M, Nieroda CA, *Anti-cancer Res.* **1996,** *16*, 2129–2133.
[50] Krebber A, Bornhauser S, Burmester J, Honegger A, Willuda J, Bosshard HR, Plückthun A, *J. Immonol. Methods* **1997,** *201*, 35–55.
[51] Reiter Y, Brinkmann U, Lee B, Pastan I, *Nature Biotechnol.* **1996,** *14*, 1239–1245.
[52] Orlandi R, Gussow DH, Jones PT, Winter G, *Proc. Natl Acad. Sci. USA* **1989,** *86*, 3833–3837.
[53] Hoogenboom M, Griffiths AD, Johnson KS, Chiswell DJ, Hudson P, Winter G, *Nucleic Acids Res.* **1991,** *19*, 4133–4137.
[54] Winter G, Griffiths AD, Hawkins RE, Hoogenboom HR, *Annu. Rev. Immunol.* **1994,** *12*, 433–455.
[55] Huls GA, Heijnen IA, Cuomo ME, Koningsberger JC, Wiegman L, Boel E, Van der Vuurst de Vries AR, Loyson SA, Helfrich W, Van Berge Henegouwen GP, Van Meijer M, de Kruif J, Logtenberg T, *Nature Biotechnol.* **1999,** *17*, 276–281.
[56] Davis CG, Gallo ML, Corvalon, *Cancer Metast. Rev.* **1999,** *18*, 421–425.
[57] Yokota T, Milenic DE, Whitlow M, Schlom J, *Cancer Res.* **1992,** *52*, 3402–3408.
[58] Jurcic JG, Scheinberg DA, Houghton AN, In: *Cancer Chemotherapy and Biological Response Modifiers Annual 16* (Eds Pinedo HM, Longo DL, Chabner BA), pp. 168–188. Elsevier Science, Amsterdam, **1996.**
[59] Caron PC, Scheinberg DA, *Curr. Opin. Oncol.* **1994,** *6*, 14–22.
[60] Dall'Acqua W, Carter P, *Curr. Opin. Struct. Biol.* **1998,** *8*, 443–450.
[61] Jung S, Plückthun A, *Protein Eng.* **1997,** *10*, 959–966.
[62] Willuda J, Honegger A, Waibel R, Schubiger PA, Stahel R, Zangemeister- Wittke U, Plückthun A, *Cancer Res.* **1999,** *59*, 5758–5767.
[63] Helfrich W, Haisma HJ, Magdolen V, Luther T, Blom VJJ, Westra J, van der Hoeven R, Kroesen BJ, Molema G, de Leij L, *J. Immunol. Methods* **2000,** *237*, 131–145.
[64] Thrush GR, Lark LR, Clinchy BC, Vitetta ES, *Annu. Rev. Immunol.* **1996,** *14*, 49–71.
[65] Dillman RO, Johnson DE, Ogden J, Beidler D, *Mol. Biotherapy.* **1989,** *1*, 250–255.
[66] Ghose T, Norvell ST, Guclu A, Cameron D, Bodurtha A, MacDonald AS, *Br. Med. J.* **1972,** *3*, 495–499.
[67] Goldenberg DM, *Am. J. Med.* **1993,** *94*, 297–312.
[68] Trail PA, Willner D, Hellstrom KE, *Drug Deliv. Res.* **1995,** *34*, 196–209.
[69] King HD, Yurgaitis D, Willner D, Firestone RA, Yang MB, Lasch SJ, Hellstrom KE, Trail PA, *Bioconjug. Chem.* **1999,** *10*, 279–288.
[70] Smith AL, Nicolaou KC, *J. Med. Chem.* **1996,** *39*, 2103–2117.
[71] Hinman LM, Hamann PR, Wallace R, Menendez AT, Durr FE, Upeslacis J, *Cancer Res.* **1993,** *53*, 3336–3342.
[72] Lode HN, Reisfeld RA, Handgretinger R, Nicolaou KC, Gaedicke Gand Wrasidlo W, *Cancer Res.* **1998,** *58*, 2925–2928.
[73] Wilder RB, DeNardo GL, DeNardo SJ, *J. Clin. Oncol.* **1996,** *14*, 1383–1400.
[74] Stigbrand T, Ullen A, Sandstrom P, Mirzaie-Joniani H, Sundstrom B, Nilsson B, Arlesting L, Norrlund RR, Ahlstrom KR, Hietala S, *Acta Oncol.* **1996,** *35*, 259–265.
[75] Kroesen BJ, Helfrich W, Molema G and de Leij L, *Adv. Drug. Del. Rev.* **1998,** *31*, 105–129.
[76] Staerz UD, Kanagawa O, Bevan MJ, *Nature* **1985,** *314*, 628–631.
[77] Pluckthun A, Pack P, *Immunotechnology* **1997,** *3*, 83–105.
[78] Schmidt M, Wels W, *Br. J. Cancer* **1996,** *74*, 853–862.
[79] Van Spriel AB, Ojik HH, van de Winkel JGJ, *Immunol. Today* **2000,** *21*, 391–397.
[80] Kroesen BJ, Helfrich W, Bakker A, Wubbema AS, Bakker H, Kal HB, The TH, de Leij L, *Int. J. Cancer* **1995,** *61*, 812–818.
[81] Huennekens FM, *Trend. Biotechnol.* **1994,** *12*, 234–239.
[82] Bagshawe KD, *Clin. Pharmacokinet.* **1994,** *27*, 368–376.
[83] Kerr DE, Schreiber GJ, Vrudhula VM, Svensson HP, Hellström I, Hellström KE, Senter PD, *Cancer Res.* **1995,** *55*, 3558–3563.
[84] Haisma HJ, Boven E, van Muijen M, de Jong J, van der Vijgh WJ, Pinedo HM, *Br. J. Cancer* **1992,** *66*, 474–478.

[85] Leenders RGG, Damen EWP, Bijsterveld EJA, Scheeren HW, Houba PHJ, van der Meulen-Muileman IH, Boven E, Haisma HJ, *Bioorg. Med. Chem. Lett.* **1999,** *718*, 1597–1610.
[86] Bosslet K, Straub R, Blumrich M, Czech J, Gerken M, Sperker B, Kroemer HH, Gesson JP, Koch M, Monneret C, *Cancer Res.* **1998,** *58*, 1195–1219.
[87] Bosslet K, Czech J, Hoffman D, *Tumour Target.* **1995,** *1*, 45–50.
[88] de Groot FMH, de Bart ACW, Verheijen JH, Scharen HW, *J. Med. Chem.* **1999,** *42*, 5277–5283.
[89] Takakura Y, Hashida M, *Crit. Rev. Oncol. Hematol.* **1995,** *18*, 207–231.
[90] Duncan R, Spreafico F, *Clin. Pharmacokinet.* **1994,** *27*, 290–306,.
[91] Duncan R, *Anti-Cancer Drugs* **1992,** *3*, 175–210.
[92] Sezaki H, Takakura Y, Hashida M, *Adv. Drug Delivery Rev.* **1989,** *3*, 247–266.
[93] Seymour LW, *CRC Crit. Rev. Ther. Drug Carrier. Syst.* **1992,** *9*, 135–187.
[94] Vasey PA, Kaye SB, Morrison R, Twelves C, Wilson P, Duncan R, Thomson AH, Murray LS, Hilditech TE, Murray T, Burtles S, Fraier D, Frigerio E, Cassidy J, *Clin. Cancer Res.* **1999,** *5*, 83–94.
[95] Lasic DD, Papahadjopoulos D, *Science* **1995,** *267*, 1275–1276.
[96] Gregoriadis G, *Trends Biotechnol.* **1995,** *13*, 527–537.
[97] Lasic DD, *From Physics to Applications.* Elsevier, Amsterdam, **1993.**
[98] Lasic DD, Martin FJ, (Eds) *Stealth Liposomes.* CRC, Boca Raton, FL, **1995.**
[99] Daemen T, Regts J, Meesters M, Ten Kate MT, Bakker-Woudenberg IAJM, Scherphof GL, *J. Control. Rel.* **1997,** *44*, 1–9.
[100] Doroshaw JH, *Cancer Chemotherapy and Biotherapy: Principles and Practice.* (eds Chabner BA and Longo DL) pp. 409–433. Lippincott-Raven, Philadelphia, PA. **1996.**
[101] Harrigan PR, Wong KF, Redelmeier TE, Wheeler JJ, Cullis PR, *Biochim. Biophys. Acta* **1993,** *1149*, 329–338.
[102] Cullis PR, Hope MJ, Bally MB, Madden TD, Mayer LD, Fenske DB, *Biochem. Biophys. Acta* **1997,** *1331*, 187–211.
[103] Park JW, Hong K, Kirpotin DB, Meyer O, Papahadjopoulos D, Benz CC, *Cancer Lett.* **1997,** *118*, 153–160.
[104] Crosasso P, Brusa P, Dosio F, Arpicco S, Pacchioni D, Schuber F, Cattel L, *J Pharm. Sci.* **1997,** *86*, 832–839.
[105] Tseng YL, Hong RL, Tao MH, Chang FH, *Int. J. Cancer* **1999,** *80*, 723–730.
[106] Kirpotin D, Park JW, Hong K, Zalipsky S, Li WL, Carter P, Benz CC, Papahadjopoulos D, *Biochemistry* **1997,** *36*, 66–75.
[107] Lopes de Menezes DE, Pilarski LM, Allen TM, *Cancer Res.* **1998,** *58*, 3320–3330.
[108] Killion JJ, Fidler IJ, *Pharmacol. Ther.* **1998,** *78*, 141–154.
[109] Dass CR, Walker TL, DeCruz EE, Burton MA, *Drug Deliv.* **1997,** *4*, 151–165.
[110] Li S, Huang L, *Gene Therapy* **1997,** *4*, 891–900.
[111] Kaneda Y, Saeki Y, Morischita R, *Mol. Med. Today* **1999,** *5*, 298–303.
[112] Mayer LD, *Cancer Metast. Rev.* **1998,** *17*, 211–218.
[113] Drummond DC, Meijer O, Hong K, Kirpotin DB, Papahadjopoulos D, *Pharmacol. Rev.* **1999,** *51*, 691–743.
[114] Brown SL, Miller RA, Levy R, *Semin. Oncol.* **1989,** *16*, 199–210.
[115] Dillman RO, *J. Clin. Oncol.* **1994,** *12*, 1497–1515.
[116] Renner C, Trümper L, Pfreundschuh M, *Leukaemia* **1997,** *11*, S55–S59.
[117] Irie RF, Morton DL, *Proc. Natl Acad. Sci. USA* **1986,** *83*, 8694–8698.
[118] Cheung N-KV, Lazarus H, Miraldi FD, Abramowsky CR, Kallick S, Saarinen UM, Spitzer T, Strandjord SE, Coccia PF, Berger NA, *J. Clin. Oncol.* **1987,** *51*, 430–1440.
[119] Vadhan-Raj S, Cordon-Cardo C, Carswell E, Mintzer D, Dantis L, Duteau C, Templeton MA, Oettgen HF, Old LJ, Houghton AN, *J. Clin. Oncol.* **1988,** *6*, 1636–1648.
[120] Handgretinger R, Baader P, Dopfer R, Klingebiel T, Reuland P, Treuner J, Reisfeld RA, Niethammer D, *Cancer Immunol. Immunother.* **1992,** *35*, 199–204.
[121] Saleh MN, Khazaeli MB, Wheeler RH, Dropcho E, Liu T, Urist M, Miller DM, Lawson S, Dixon P, Russell CH, LoBuglio A, *Cancer Res.* **1992,** *52*, 4342–4347.
[122] Riethmuller G, Holz e, Schlimok G, Schmiegel W, Raab R, Hoffken K, Gruber R, Funke I, Pichlmaier H, Hirche H, *J. Clin. Oncol.* **1998,** *16*, 1788–1794.
[123] Oosterwijk E, Debruyne FMJ, Schalken JA, *Semin. Oncol.* **1995,** *22*, 34–41.
[124] Weiner LM, *Semin. Oncol.* **1999,** *26*, 41–50.
[125] McLaughlin P, Grillo-Lopez AJ, Link BK, Levy R, Czuczman MS, Williams ME, Heyman MR, Bence-Bruchler I, White CA, Canabillas F, Jain V, Ho AD, Frister J, Wey K, Shen D, Dallaize BK, *J. Clin. Oncol.* **1998,** *16*, 2824–2833.

[126] Czuczman MS, Grillo-Lopez AJ, White CA, Saleh M, Gordon L, LoBuglio AF, Jonas C, Klippenstein D, Dallaize B, Warms C, *J. Clin. Oncol.* **1999,** *26*, 268–276.
[127] Dyer MJS, *Semin. Oncol.* **1999,** *26*, 52–57.
[128] Baselga J, Tripathy D, Mendelsohn J, Baughman S, Benz CC, Dantis L, Shlarin NT, Seidman AD, Hudis CA, Moore J, Rosen PP, Twaddell T, Henderson IC, Norton L, *J. Clin. Oncol.* **1996,** *14*, 737–744.
[129] Pegram MD, Lipton A, Hayes DF, Weber BL, Baselga MM, Tripathy D, Baly D, Baighman SA, Twaddell T, Glaspy JA, Slalom DJ, *J. Clin. Oncol.* **1998,** *16*, 2659–2663.
[130] Baselga J, Tripathy D, Mendelsohn J, *et al., Semin. Oncol.* **1999,** *26*, 78–83.
[131] Sievers EL, Appelbaum FR, Spielberger RT, Forman SJ, Flowers D, Smith FO, Shannon-Dorcy K, Berger MS, Berstein ID, *Blood*, **1999,** *93*, 3678–3684.
[132] Saleh MN, LoBuglio AF, Trail PA, In: *Monoclonal Antibody-Based Therapy of Cancer* (Ed. Grossband M), pp. 397–416. Marcel Dekker, New York, **1998.**
[133] Tolcher AW, Sugarman S, Gelmon KA, Cohen R, Saleh M, Isaacs C, Young L, Healey D, Onetto N Slichenmyer W, *J. Clin. Oncol.* **1999,** *17*, 478–484.
[134] Kaminski MS, Zasadny KR, Francis IR, Fenner MC, Ross CW, Milik AW, Estes J, Tude M, Regan D, Fisher S, Glenn SD, Wahl RL, *J. Clin. Oncol.* **1996,** *14*, 1974–1981.
[135] Press OW, Eary JF, Appelbaum FR, Martin PJ, Badger CC, Nelp WB, Glenn S, Butchko G, Fisher D, Porter B, *N. Engl. J. Med.* **1993,** *329*, 1219–1224.
[136] Press OW, Eary JF, Appelbaum FR, Martin PJ, Nelp WB, Glenn S, Fisher DR, Porter B, Matthews DC, Gooley T and Bernstein ID, *Lancet* **1995,** *346*, 336–340.
[137] Liu SY, Eary JF, Petersdorf SH, Martin PJ, Maloney DG, Appelbaum FR, Matthews DC, Bush SA Duracke LD, *J. Clin. Oncol.* **1998,** *16*, 3270–3278.
[138] Matthews DC, Appelbaum FR, Eary JF, Fischer DR, Durack LD, Hui TE, Martin PJ, Mitchell D, Press OW, Storb R, Bernstein ID, *Blood* **1999,** *94*, 1237–1247.
[139] De Gast GC, Haagen IA, van Houten AA, Klein SC, Duits AJ, de Weyer RA, Vroom TM, Clark MR, Phillips J, Van Dijk AJ, *Cancer Immunol. Immunother.* **1995,** *40*, 390–396.
[140] Janssen RAJ, Kroesen BJ, Buter J, Mesander G, Sleijfer DT, The TH, Mulder NH, de Leij L, *Br. J. Cancer* **1995,** *72*, 795–799.
[141] Kroesen BJ, Nieken J, Sleijfer DT, Molema G, de Vries EGE, Groen HJM, Helfrich W, The TH, Mulder NH, de Leij L, *Cancer Immunol. Immunother.* **1997,** *45*, 203–206.
[142] Valone FH, Kaufman PA, Guyre PM, Lewis LD, Memoli V, Deo Y, Graziano R, Fisher JL, Meyer L, Mrozek–Orlowski M, Wardwell K, Guyre V, Morley TL, Arvinu C, Fanger MW, *J. Clin. Oncol.* **1995,** *13*, 2281–2292.
[143] Weiner LM, Clark JI, Davey M, Li WS, Garcia de Palazzo I, Ring DB, Alpaugh RK, *Cancer Res.* **1995,** *55*, 4586–4593.
[144] Martin J, Stribbling SM, Poon GK, Begent RH, Napier M, Sharma SK, Springer CJ, *Cancer Chemother. Pharmacol.* **1997,** *40*, 189–201.
[145] Fuertges F, Abuchowski A, *J. Control. Rel.* **1990,** *11*, 139–148.
[146] Ho DH, Brown NS, Yen A, Holmes R, Keating M, Abuchowski A, Newman RA, Kozahoff IH, *Drug Metab. Dispos.* **1986,** *14*, 349–352.
[147] Meyers FJ, Paradise C, Scudder SA, Goodman G, Konrad M, *Clin. Pharmacol. Ther.* **1991,** *49*, 307–313.
[148] Martin FJ, *Medical Applications of Liposomes* (Eds frasic DD and Papahadjopoulos D) pp. 635–688, Elsevier, Amsterdam, **1998.**
[149] Eckhardt JR, Campbell E, Burris HA, Wiess GR, Rodriguez GI, Fields SM, Thurman AM, Peacock NW, Cobb P, Rothenberg ML, Ross ME, Von Hoff DD, *Am. J. Clin. Oncol.* **1994,** *17*, 498–501.
[150] Gill PS, Espina BM, Muggia F, Cabriales S, Tulpule A, Esplin JA, *J. Clin. Oncol.* **1995,** *13*, 996–1003.
[151] Valero V, Buzdar AU, Theriault RL, Azarnia N, Fonseca GA, Willey J, Ewer M, Walter RS, Mackay B, Podoloff D, Booser D, Lee LW, Hortobagyi GN, *J. Clin. Oncol.* **1999,** *17*, 1425–1434.
[152] McLaughlin PMJ, Kroesen BJ, Dokter HA, van der Molen H, de Groot M, Brinker MGL, Kok K, Ruiters MHJ, Buys CHCM, de Leij LFMH, *Cancer Immunol. Immunother.* **1999,** *48*, 303–311.
[153] Clynes RA, Takechi Y, MoroiY, Houghton A, Ravetch JV, *Proc. Natl Acad. Sci. USA* **1998,** *95*, 652–656.
[154] Clynes RA, Towers TL, Persta LG, Ravetch JV, *Nature Med.* **2000,** *6*, 443–446.
[155] Houghton AN, Scheinberg DA, *Nature Med.* **2000,** *6*, 373–375.

9 Tumour Vasculature Targeting

Daisy W. J. van der Schaft, S. Ramakrishnan, Grietje Molema, Arjan W. Griffioen

9.1 Introduction

Various stages in solid tumour growth can be discriminated. After multiple genetic changes, cellular growth progresses via hyperplasia and dysplasia towards an *in situ* tumour. Only after the so-called 'angiogenic switch', an as yet, ill-defined set of molecular changes leading to the formation of new blood vessels, does a tumour progress towards a mass extending more than several millimetres in diameter [1]. By recruitment of new blood vessels, the tumour provides itself with a continuous supply of nutrients and paves the way for metastasis formation.

The architecture of a solid tumour is such, that numerous layers of tumour cells are fed by one blood vessel. This poses a significant barrier for macromolecular drug targeting preparations aimed at the tumour cells that need to extravasate from the blood into the tumour tissue to reach the target cells: the larger the chosen carrier, the less accessible the tumour tissue will be. In contrast, endothelial cells lining the tumour vasculature are easily accessible for these macromolecular preparations. Tumour endothelial cell specific delivery of generally toxic anti-neoplastic or blood coagulation-inducing agents presents an attractive option for increasing therapeutic efficacy and reducing toxic side-effects elsewhere in the body. Furthermore, the pivotal role of endothelial cells in the maintenance of tumour cell survival and growth also makes them an interesting target for therapeutic intervention.

Tumour growth does not expand more than several millimetres in diameter in the absence of new blood vessel formation. Many tumours diagnosed in the clinic have a size beyond 0.5 cm in diameter and are considered pro-angiogenic. Endothelial cells are important executioner cells in the angiogenic cascade. Disruption of these pro-angiogenic characteristics by inhibiting pro-angiogenic signal transduction in endothelial cells therefore provides a powerful tool with which to intervene in tumour growth. The majority of research on inhibiting the function of tumour vasculature focuses at the development of drugs that may have an explicit action on the endothelial cells in the tumour. Based on their mechanism(s) of action and the fact that angiogenesis is a normal physiological process, these drugs in theory can also interrupt the functioning of healthy endothelial cells in the body. Recently, several angiogenesis inhibitors were withdrawn from clinical studies, most likely as a result of lack of selectivity for tumour endothelium. This emphasizes the need for systems that can deliver potent angiogenesis inhibitors at/in the tumour endothelial cells only.

At present, most target epitopes capable of discriminating tumour endothelium from normal endothelium are molecules that are expressed during angiogenesis. In this chapter a summary of current knowledge regarding tumour growth-related angiogenic endothelial processes in angiogenesis will be given. Furthermore, target epitopes and drug-targeting approaches exploited to date, with potential for future therapeutic application will be discussed.

9.1.1 Functions of Vascular Endothelial Cells in the Body

The vasculature can be considered to be one of the crucial organs in the body, extending more than 900 m^2 and playing a major role in maintaining the body's integrity in various ways. Blood vessels consist of endothelial cells that are directly in contact with the blood, pericytes located beneath the endothelium, smooth muscle cells, fibroblasts, basement membrane, and extracellular matrix (ECM). Depending on the location in the body and the organ microenvironment, the cellular constituents, basement membrane and ECM differ in phenotype, composition, and function [2]. The endothelial cells form a monolayer in every blood vessel in the circulation. They are actively involved in several regulatory processes in the body. Besides being metabolically active and selectively permeable for small solutes and peptides/proteins, endothelial cells are actively involved in regulating haemostasis [3]. Furthermore, they are able to recruit cells of the immune system to specific sites of e.g. infection or inflammation by virtue of the regulated expression of cell adhesion molecules and production of cytokines and chemokines [4] (see also Chapter 7). In addition, endothelial cells are actively involved in vascular remodelling in for example ovulation, wound healing, tumour growth and diabetic retinopathy [5,6].

9.1.2 Molecular Control of Tumour Growth-related Angiogenesis

Physiological stimuli during wound healing and during the reproductive cycle in women lead to controlled angiogenesis. However, pathologic conditions such as tumour growth, rheumatoid arthritis, and diabetic retinopathy are also characterized by abundant angiogenesis.
Angiogenesis is rapidly initiated in response to hypoxic or ischaemic conditions. In tumour growth, this active vascular remodelling is reflected by enhanced tumour endothelial cell

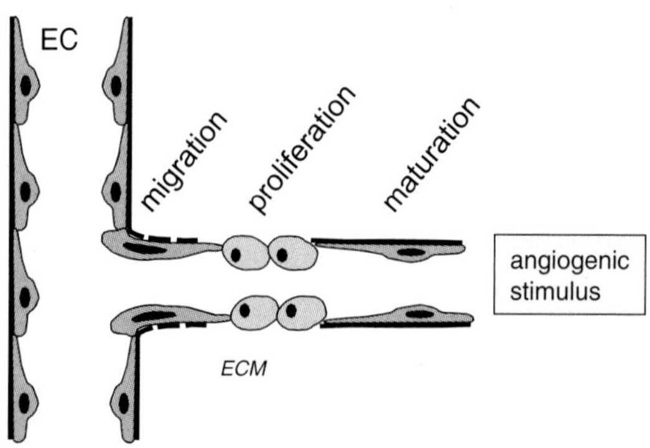

Figure 9.1. In tumours that have undergone the angiogenic switch, the expression of a variety of growth factors and other soluble molecules, and the changes in the extracellular matrix (ECM), among others, lead to angiogenesis, the formation of new blood vessels from pre-existing ones. In this process, endothelial cells (EC) covering the blood vessel wall become activated, migrate into the tissue, proliferate and eventually differentiate into mature blood vessels. These vessels serve as a supply of nutrients for the ongoing demands of the growing tumour cells.

proliferation to up to 20–2000 times faster than in healthy adult endothelium [7]. In all types of angiogenesis, either under physiological or pathologicaal conditions, endothelial cell activation is followed by matrix degradation, cellular migration, proliferation, and ultimately neovasculature maturation (Figure 9.1).

9.1.2.1 Role of Growth Factors VEGF and FGF-2

More than 20 cytokines from various sources have now been reported to be involved in the processes taking place during angiogenesis [8]. Vascular endothelial growth factor (VEGF) and basic Fibroblast Growth Factor (bFGF or FGF-2) are the two growth factors whose roles in angiogenesis have been most extensively studied to date. VEGF (also known as VEGF-A) isoforms VEGF-121, 145, 165, 183, 189 and 205 are produced through alternative splicing from a single VEGF gene located on chromosome 6 [9]. The isoforms differ in their molecular composition and weight, and biological properties. The larger forms of VEGF differ from VEGF-121 by the presence of a heparin-binding domain, which is encoded by the exon-7. VEGF is abundantly produced by hypoxic tumour cells, macrophages and other cells of the immune system [10]. It induces vasodilatation via endothelial nitric oxide production and increases endothelial cell permeability [11]. This allows plasma proteins to enter the tissue to form a fibrin-rich provisional network [12]. VEGF also induces the expression of various proteases and receptors important in cellular invasion and tissue remodelling, it activates cellular proliferation and prevents endothelial cell apoptosis [13,14]. The two VEGF-specific tyrosine kinase receptors, VEGFR-1 (Flt-1) and VEGFR-2 (KDR/flk-1), are expressed on vascular endothelium, and to a lesser extent on monocytes/macrophages and certain tumour cell types. Interaction of VEGF with VEGFR-2 is a critical requirement to induce the full spectrum of VEGF biological responses. In addition to the two VEGF receptors, VEGF-165 has been found to bind neuropilin-1, which is also expressed on endothelial cells [15]. A recent study using genetic deletion methods has determined that neuropilin-1 is important for embryonic vessel formation [16].

Endothelial cells exploit various proteases such as matrix metalloproteinases to penetrate into new areas of the body by degrading the basement membrane. Furthermore, urokinase-plasminogen activator and tissue type-plasminogen activator convert the ubiquitous plasma protein plasminogen to plasmin. Plasmin is believed to be the most important protease for the mobilization of FGF-2 from the ECM pool. FGF-2 induces endothelial cell motility, proliferation and proteinase activity, and modulates integrin levels [17,18]. The cellular effects of FGFs are mediated via specific binding to high-affinity tyrosine kinase receptors [17]. In addition, low affinity FGF receptors consist of polysaccharide components of heparan sulfate proteoglycans on cell surfaces and ECM. Binding to these components present in the ECM has been proposed as a mechanism to stabilize and protect FGF from inactivation. Heparan sulfate on cell surfaces, on the other hand, plays a more active role in displacing ECM-bound FGF-2 and its subsequent presentation to the high affinity signal transducing receptors [19]. Angiogenesis seems exquisitely sensitive to small changes in factors such as VEGF and FGF-2, which may have important therapeutic implications in treatment of angiogenesis-driven disorders [20,21].

9.1.2.2 Role of Integrins

Integrins are transmembrane proteins composed of an α and β subunit in over 20 different $\alpha\beta$ heterodimeric combinations. They bind to ECM proteins or cell surface ligands through short peptide sequences and are implicated in angiogenesis control. Combinations of different integrins on (endothelial) cell surfaces allow cells to recognize and respond to a variety of different ECM proteins [22]. They are able to transduce signals from within the cells to the outside as well as from the outside into the cell [23]. Integrin-mediated cell adhesion has impact on two key aspects of growth regulation. First, it can influence the activity of the basal cell cycle machinery consisting of cyclin-dependent kinase complexes. Second, integrins play a vital role in anchorage-dependent cell death or anoikis [24,25]. For example, integrin $\alpha_V\beta_3$ mediates endothelial cell adhesion to vitronectin, fibrinogen, laminin, collagen, von Willebrand Factor or osteopontin through their exposed tripeptide Arg-Gly-Asp (RGD) moiety [26]. Since $\alpha_V\beta_3$ is minimally expressed on normal resting endothelium, but significantly upregulated on tumour and other activated endothelium, it is believed to play a critical role in the process of angiogenesis. Both peptide and antibody inhibitors of $\alpha_V\beta_3$ induced endothelial cell apoptosis, suggesting a role for this integrin in endothelial cell survival during angiogenesis [27]. Another α_V integrin associated with angiogenesis is $\alpha_V\beta_5$. Whereas *in vivo* FGF-2 or tumour necrosis factor α (TNFα) induced $\alpha_V\beta_3$-dependent angiogenesis, VEGF or transforming growth factor β (TGF-β) initiated an angiogenesis pathway dependent only on $\alpha_V\beta_5$ [28].

9.1.2.3 Role of the Extracellular Matrix

Components of the ECM play an important role in the regulation of endothelial cell morphology and function. Thrombospondin (TSP), for example, can affect endothelial cell proliferation negatively as well as positively, depending on the endothelial microenvironment. Furthermore, through binding to and activation of TGF-β and affecting protease activity, TSP may be able to influence cell growth, migration and differentiation [29]. Laminin also plays a role in cell attachment, growth promotion, protease secretion and interactions with other ECM components. It can bind to cell surface binding proteins including integrins which leads to integrin signalling [30]. SPARC (Secreted Protein Acidic and Rich in Cysteine), also known as BM40 or osteonectin, is a protein whose expression is elevated under stress conditions. Transient expression of SPARC during endothelial cell injury and cellular activation indicate a role in tissue repair, remodelling and angiogenesis [31]. Exogenously added SPARC or SPARC-derived peptides were able to modify endothelial cell behaviour via the induction of proteases and inhibitors of plasmin generation [32,33].

9.1.2.4 Role of Subendothelial Support Cells

Endothelial cell interaction with ECM and mesenchymal cells is a prerequisite to form a stable vasculature. Therefore, after endothelial cell proliferation and maturation, and the formation of endothelial tube structures, surrounding vessel layers composed of mural cells (pericytes in small vessels and smooth muscle cells in large vessels) need to be recruited. Endothelial cells accomplish this via the synthesis and secretion of platelet-derived growth factor (PDGF), a mitogen and chemoattractant for a variety of mesenchymal cells. Subsequent differentiation of the mural precursor cells into pericytes and smooth muscle cells is believed to be a cell–cell contact-dependent process. Upon endothelial cell–mural cell contact, a latent form of TGF-β, produced by both endothelium and mural cells, is activated in a plasmin-mediated process. Activated TGF-β can induce changes in myofibroblasts and pericytes which contributes to the formation of a mature vessel, ECM production and maintenance of growth control [34]. The coincident investment of growing capillaries by pericytes with the deposition of basement membrane and cessation of vessel growth during wound healing, also indicates vessel growth regulation by pericytes [35]. The FGF-1 receptor is also implicated in endothelial cell differentiation leading to vascular tube formation. In addition to inducing plasminogen activator, and endothelial cell proliferation and migration, FGF-1 receptor signalling resulted in endothelial tube formation in collagen [36].

9.2 Angiogenesis Assays and Models

The specificity of blood vessel-targeting to eradicate solid tumours depends on altered physiological processes in the tumour vasculature relative to normal vasculature in healthy tissues. It is therefore necessary to investigate the fundamental properties of the vascular biology and cell biological events in vessel formation. A number of different assay systems and angiogenesis models can be used for the research underlying development of vascular targeting techniques for the treatment of cancer.

9.2.1 Endothelial Cell Sources

The availability of viable endothelial cells is crucial for research on vessel formation, angiogenesis and angiogenic endothelial cell targeting. Endothelial cells can be obtained from various tissues, purified and cultured *in vitro*. The best available source of human endothelial cells is the large vein in the umbilical cord. Because this vein is not branched, it can be filled with collagenase which will enzymatically detach the endothelial cells from the vessel wall. Endothelial cells in culture begin cell cycling and form a confluent monolayer on the tissue-culture plastic. It should always be borne in mind that cultured endothelial cells are never in a quiescent state and can therefore not serve as a model for resting vasculature. For reasons of simple isolation most laboratories make use of the human umbilical vein endothelial cells (HUVEC). The major drawback of these cells is their macrovascular origin, which makes them less suitable for studies on the microvascular processes occurring during angiogenesis. Although more laborious, human microvascular endothelial cells can be isolated from other organs such as the foreskin or adipose tissue.

Endothelial cells in culture need a constant supply of growth factors such as FGFs and VEGF in order to continue cell cycling. In most cultures the addition of serum to the culture medium is sufficient to maintain a low level of endothelial cell proliferation. Cells cultured this way can be subcultured at a split ratio of 1 : 3 for four to five passages without significant loss of growth potential. Addition of low concentrations of recombinant or purified growth factors such as FGF-2 will increase the number of possible passages to 10–12. These limited subculture possibilities make repeated isolations necessary, thereby introducing significant variation in the endothelial cell source. To circumvent these drawbacks, it is possible to immortalize endothelial cells with viral oncogenes such as simian virus-40 large T antigen. Transfection of endothelial cells with a DNA construct containing the gene for this molecule can result in cell lines that can be subcultured for over 60 passages. Examples of such cell lines are HMEC-1 [37], EA.hy926 [38], ECL4n [39], and EvL [40]. The major problem with these immortalized endothelial cells is, however, that they are not genetically stable, resulting in loss of phenotypic and functional characteristics and functional resemblance to their *in vivo* counterparts. Of note, many endothelial cell lines can be cultured on plastic tissue-culture material, without being dependent on integrin signalling via ECM molecules for growth and survival. Careful interpretation of results obtained with these cell lines is necessary.

Endothelial cells can also be prepared from tissue from other species. Capillary endothelial cells of bovine origin are used quite frequently, because these cells are rather sensitive to treatment with angiogenesis inhibitors such as angiostatin and endostatin (Griffioen *et al.*, unpublished results). It should be noted however, that species-dependent responses to drugs can occur.

9.2.2 Functional Assays with Endothelial Cells

Angiogenesis is a multi-step process, depending on activation, migration, proliferation and differentiation of endothelial cells, and all of these stages in the cascade can be studied independently.

9.2.2.1 Cell Growth Assays

Proliferation of endothelial cells can be studied using various assays. Since the cell cycle potential of endothelial cells is low, with cell doubling times of sometimes over 35 h, cell counting is a time-consuming and insensitive method of assessing cell growth. A much better way to determine proliferation is by the measurement of [^{3}H]-thymidine incorporation [41,42]. Alternative assays for the study of endothelial cell proliferation and other mechanisms of cell activation that do not involve the S-phase of the cell cycle, are based on colorimetric systems which measure mitochondrial activity. Furthermore, proliferation of endothelial cells can be analysed by DNA profiling, for example by flow cytometric analysis of cells in G0/G1 phase (2n DNA), cells in G2/M phase (4n DNA) and cells in S phase ($2 < n < 4$) after permeabilization and staining with propidium iodide. In addition, proliferation can be quantified by determination of cell cycle-dependent expression of molecules such as proliferating cell nuclear antigen (PCNA) [43] or Ki-67 [44].

The number of cells present depends on the level of cell growth and cell death. Therefore, detection of cell death is a commonly used approach to average cell growth. Apoptosis induction can be studied most easily by detection of subdiploid cells or analysis of DNA degradation profiles on the flow cytometer after DNA extraction and propidium iodide staining.

Apoptosis can also be assessed by quantification of cells with fragmented nuclei by staining with e.g. propidium iodide or acridyl orange. Nick-end labelling in terminal transferase-mediated UTP nick-end labelling (TUNEL)-analysis by immunohistochemistry or flow cytometry can be regarded as a method specific for the detection of apoptosis. Qualitative indications of apoptosis induction can be detected by analysing DNA ladder formation in an agarose gel. Measurement of annexin-V binding to cells [45], which is based on changes in membrane asymmetry, can be employed to detect early signs of apoptosis.

9.2.2.2 Adhesion and Migration Assays

In order to form new blood vessels, endothelial cells need to migrate through the extracellular matrix. Two sequential steps in the angiogenesis cascade are fundamental to this process. The first step is the production of matrix metalloproteinases (MMPs) which dissolve the extracellular matrix to facilitate migration of endothelial cells. Measurement of MMP1 (collagenase-1), MMP2 (gelatinase A), MMP3 (stromelysin-1) and MMP9 (gelatinase B), whose expression is correlated to angiogenesis and tumour growth, can be carried out with ELISA, histochemistry or Western blot analysis.

In the second step, endothelial cells use their adhesion molecule make-up to initiate the actual migration towards the angiogenic stimulus. Expression of adhesion molecules involved in matrix binding such as $\alpha_V\beta_3$-, $\alpha_V\beta_5$- and β_1-containing integrins, and functional studies investigating adhesion molecule binding to matrix components is an important part of angiogenesis research [46]. For the development of carrier molecules for drug targeting strategies aimed at $\alpha_V\beta_3$ integrin expressed by tumour endothelium, these adhesion assays are also suitable systems in which to test carrier specificity [47].

More functional studies that address the process of migration use the so-called Boyden chamber [48]. In this assay the migratory capacity of endothelial cells after activation with chemoattractants or pro-angiogenic stimuli is studied. Velocity of migration and the percentage of cells that are capable of migrating through an ECM-coated membrane into another compartment can be determined.

The wound assay [49] is another method of measuring endothelial cell migration. This assay is based on damaging or wounding a confluent monolayer of endothelial cells and the subsequent repair or closing of the wound by migration of endothelial cells. This assay can be carried out using different matrix components.

The technologies described above can be used to pinpoint the mechanism(s) of action of angiogenic or angiostatic agents in specific steps in the angiogenic cascade. For instance, application of these systems revealed that IFNα and angiostatin inhibit cell migration whereas endostatin and platelet factor-4 function primarily as inhibitors of endothelial cell proliferation.

9.2.3 *In Vitro* Angiogenesis Assays

The advantage of the assays described above is the control over a limited number of parameters involved. A more complex experimental set-up which studies consecutive steps in the angiogenic cascade is represented by three-dimensional *in vitro* models. In these models, endothelial cells are cultured on top of a matrix gel (i.e. collagen [50], fibrin [51], or matrigel [52]) and are induced to form sprouts into the matrix by stimulation with either growth factors, tumour biopsies or tumour cell lines grown as spheroids. An essential feature in all of these assays is that a lumen is formed in these sprouts and that not merely migration and cell rearrangement has taken place.

Endothelial cells grown on gelatin-coated beads embedded in a matrix can be induced to form sprouts into the matrix [53]. The sprouting can be quantified either by measuring the distance over which vessels were formed or by computer-aided measurement of total vessel length. Other methods reflect the *in vivo* situation even more closely. These methods are based on endothelial cell sprouting from freshly isolated tissues embedded in matrix gels, including the rat aortic ring [54] and human placenta tissue [55]. This procedure is not applicable for all tissues. Tumour biopsies, for example, often produce an excess of proteases. As a result, the matrix is digested and endothelial cell sprouting prevented [50]. A recently published *in vitro* angiogenesis assay that mimics the *in vivo* situation quite well, exploits the use of embryoid bodies [56]. *In vitro* cultured mouse blastocyst cells [57] recapitulate several steps of murine embryogenesis, including vasculogenesis and angiogenesis [58]. There is complete blood vessel development in these embryoid bodies [59] which makes this system suitable for studying the effects of a wide spectrum of angiogenesis modulators.

The advantages of *in vitro* assays are (1) the ability to control the assay variables, (2) the potential opportunity to study the various steps within the complete process, (3) cellular and molecular events can be more carefully monitored and (4) the costs and the duration of the experiments are often lower than those of *in vivo* assays. The disadvantages of *in vitro* assays are that the cells, reagents and conditions used in different laboratories are not standardized and that the *in vitro* effects seen, do not always match the activities observed *in vivo*. This has been demonstrated for e.g. TNFα, which inhibits angiogenesis *in vitro*, but induces angiogenesis *in vivo* [6]. Particularly in the light of the possible influence of various cells of the immune system on angiogenesis, extrapolation of data from *in vitro* to *in vivo* needs to be carefully addressed.

9.2.4 *In Vivo* Assays to Study Angiogenesis and Targeting of Angiogenic Blood Vessels

It is well recognized that *in vitro* angiogenesis assays can have clear advantages. However, the major drawback of all of these assays is that they require the endothelial cells to be removed from their natural microenvironment, which alters their physiological properties. To study angiogenesis *in vivo*, the most frequently used assay systems exploit chicken chorio-allantoic membrane (CAM) [28,60], the corneal pocket [61], transparent chamber preparations such as the dorsal skin fold chamber [62,63], the cheek pouch window [64] and polymer matrix implants [65,66].

The CAM assay is based on the developmental angiogenesis occurring in the CAM during chick embryo development. The developing vasculature can easily be observed and regulators of angiogenesis can be tested in this model by intra-vessel injection or by addition of soluble compounds, either by release from within a silicone ring placed onto the membrane or by release from a methylcellulose or alginate pellet. This assay is relatively inexpensive and is not considered an animal experiment by law. This can be an advantage in countries with strict animal experimentation legislation.

The corneal pocket assay and the window preparations are designed to measure vessel formation after addition of stimulators. These assays can for instance be used for the study of angiogenic potential of human tumours. These models are also suitable for pre-clinical testing of angiogenesis inhibitors.

Implantation of polymer matrices that contain angiogenic factors requires quantification of the extent of vessel ingrowth. This can either be analysed immunohistochemically or by haemoglobin/red blood cell count in the tissue. These models generally do not allow analysis of the time course of vascularization since this would require the sacrifice of animals. Application in a dorsal skin fold chamber circumvents this experimental problem, as it provides the opportunity to monitor vessel formation at various time points during the experiment.

In vivo assays however, also have a number of disadvantages. For example, the pharmacokinetics, necessary for correct interpretation of results, are often unknown, and in addition the host might respond nonspecifically to the implantation. For a review on *in vivo* angiogenesis models and their potentials and problems, the reader is referred to reference [67].

9.3 Tumour Vasculature Targeting and Pre-clinical Experience

Endothelial cells are structurally and functionally different depending on the tissue of origin. Cell surface markers expressed selectively on tumour vascular endothelium offer, in theory, a unique opportunity to target cytotoxic and otherwise bioactive molecules. The majority of currently known endothelial markers have been identified by either reactivity to a monoclonal antibody or by histochemical methods. Yet, recent advances in molecular biological techniques are making significant impact in identifying new markers. Using suppression subtractive PCR and differential display libraries, many new (endothelial) cell-specific markers are being discovered [2,68]. Similarly, targeted gene deletion approaches have provided valuable information about the role of many, until now uncharacterized, proteins in angiogenesis. Some of these components are responsible for intracellular transcriptional regulation of endothelial-specific gene expression (e.g. Id1 and Id3 [69]), while others are expressed on the cell surface (e.g ephrin-B4 and ephrin-B2). The Id proteins may inhibit the DNA binding of transcription factors. Id1 and Id3 knockout mice display vascular malformations in the forebrain and an absence of branching and sprouting blood vessels into the neuroectoderm. Furthermore, in Id knockout mice, tumour growth is either completely blocked or impaired, resulting in poorly vascularized and necrotic tumours [69]. Ephrin-B2 and ephrin-B4 define the boundary between arteries and veins [70].

In general, endothelial-specific markers can be grouped into three major categories:

(1) Pan-specific endothelial cell markers. This class of markers includes antigens, which are generically expressed on endothelial cells (e.g. CD31).
(2) Tissue and organ specific markers expressed on endothelial cells.
(3) Disease associated markers (e.g. endoglin and endosialin) that are selectively expressed (tumour endothelium specific antigens) or over-expressed (tumour endothelium associated antigens) by the tumour vasculature.

Table 9.1. Epitopes on pro-angiogenic vascular endothelium that may differentiate between healthy and diseased vasculature and therefore be suitable for drug targeting or diagnostic purposes. Target epitopes presented by molecules specific for tumour-associated basement membrane, extracellular matrix or non-endothelial cell components have not been included.

Target epitope	**Reference**
30.5-kDa antigen	Hagemeier *et al.* [108]
CD34	Schlingemann *et al.* [109]
VEGF/VEGF receptor complex	Ramakrishnan *et al.* [76]
VEGF receptor	Dvorak *et al.* [110]
Endosialin	Rettig *et al.* [96]
E-selectin	Nguyen *et al.* [111]
α_V integrins	Brooks *et al.* [112]
Endoglin	Burrows *et al.* [113]
Tie-2	Sato *et al.* [114]
TNFα receptor	Eggermont *et al.* [115]
CD44	Griffioen *et al.* [97]
Angiostatin receptor	Moser *et al.* [102]
Endostatin receptor	Chang *et al.* [104]
CM101 binding protein	Not identified at present
MMP-2/MMP-9	Koivunen *et al.* [116]

A number of molecules in groups 2 and 3 have been identified by the differential homing capacity of phage display libraries and combination peptide libraries [71]. Biochemical strategies such as the application of 2D gel electrophoresis on protein extracts from endothelial cell surfaces have also proven useful in this respect [72].

For tumour vasculature targeting purposes the latter group of markers is the most relevant. Numerous putative target epitopes have been put forward in the last two decades (see Table 9.1 and for a review of this subject see Griffioen and Molema [73]), of which some have been studied extensively for drug targeting purposes. Most of the target epitopes are associated with the pro-angiogenic character of the tumour vasculature. It should be noted, however, that targeting strategies based on these target epitope characteristics will not be able to differentiate between tumour growth-associated angiogenesis and physiological angiogenesis that occurs e.g. in wound healing.

Most of the targeting strategies focus on delivering a cytotoxic molecule into an intracellular compartment of endothelial cells. This approach necessitates three important attributes to the carrier molecules, (1) selective binding with high affinity, (2) internalization, and (3) intracellular routing favouring translocation of the targeted agent to the cytoplasm. Other methods useful for anti-angiogenic therapy include surface localization of a bioactive molecule capable of inducing blood coagulation or endothelial apoptosis, or delivering radiation. It is noteworthy that these latter approaches are independent of internalization of cell surface-bound targeting moieties. All of the strategies however have the same purpose, the inhibition of tumour blood flow and thereby tumour growth.

9.3.1 Growth Factor Receptor Targeting

Endothelial cells in normal vasculature are quiescent and divide approximately once every 6 months [74] and about 0.01% of endothelial cells is in S-phase at any given time. In contrast, at areas of active angiogenesis, e.g. in tumour growth, wound healing and in reproductive tissues undergoing remodelling, endothelial cells divide rapidly. Increased proliferation in these areas is accompanied by over-expression of growth factor receptors involved in angiogenesis. Some of the well-characterized receptor systems involved in angiogenic response are VEGF receptors, angiopoietin receptor (Tie-2), FGF receptor and endoglin. With the exception of Tie-2 receptor, these receptor systems have all been studied for their application as targets in order to selectively inhibit tumour endothelial cell proliferation and function.

9.3.1.1 VEGF Receptor Targeting

VEGFR-1 and VEGFR-2 are over-expressed on tumour vasculature, while being present at low density in the surrounding normal tissues [75]. The upregulation of VEGFR expression is mediated by hypoxia and autocrine stimulation. Since growth factor receptors undergo endocytosis upon ligand binding, VEGF was initially studied for its ability to deliver toxin polypeptides. In these studies, VEGF-165 was chemically linked to a truncated form of diphtheria toxin (DT385) by a disulfide bond. The toxin molecule used is truncated at position 385 by genetic deletion to eliminate direct binding of diphtheria toxin to endothelial cells. The resultant molecule has the catalytic domain (A-chain) of diphtheria toxin and the translocation domain of the B-chain. The A-chain of diphtheria toxin possesses ADP-ribosylase activity and ribosylates elongation factor-2 (EF-2) at a specific, post-translationally modified histidine residue called diphthamide. Consequently, EF-2 is irreversibly inactivated leading to precipitous inhibition of protein synthesis in the target cells. The VEGF–toxin conjugate was found to be quite effective in inhibiting endothelial cell proliferation *in vitro* and experimental angiogenesis *in vivo* [76]. Cytotoxicity to endothelial cells was specific and dependent on VEGFR expression. Free toxin molecules did not show any effect on endothelial cell viability. VEGF–toxin conjugate treatment of tumour-bearing mice resulted in selective vascular damage in the tumour tissue and inhibited tumour growth [77]. Histological studies demonstrated that conjugate treatment spared the blood vessels of normal tissues such as liver, lung and kidney from being damaged. The differential effects of VEGFR targeting on tumour vasculature can possibly be attributed to three factors, (1) over-expression of VEGFR in tumour vessels leading to increased homing of the VEGF–toxin conjugate, (2) proliferation-dependent sensitivity to the effector moiety in the conjugate, and (3) polarized distribution of VEGFR. Over-expression of VEGFR on tumour vessels has been well documented. Recent *in vitro* studies suggest that only proliferating endothelial cells are sensitive to the VEGF–toxin conjugate. Quiescent, confluent endothelial cells were found to be totally resistant to the cytotoxic activity of VEGF–toxin. Endothelial cells in healthy tissues are quiescent and therefore may have escaped the cytotoxicity of the VEGF–toxin conjugate. Definite evidence for the third possibility is forthcoming. It is likely that the VEGFR is abluminally distributed and therefore inaccessible to systemically circulating VEGF–toxin conjugate. At the tumour site however, increased permeability and vascular leakage may facilitate

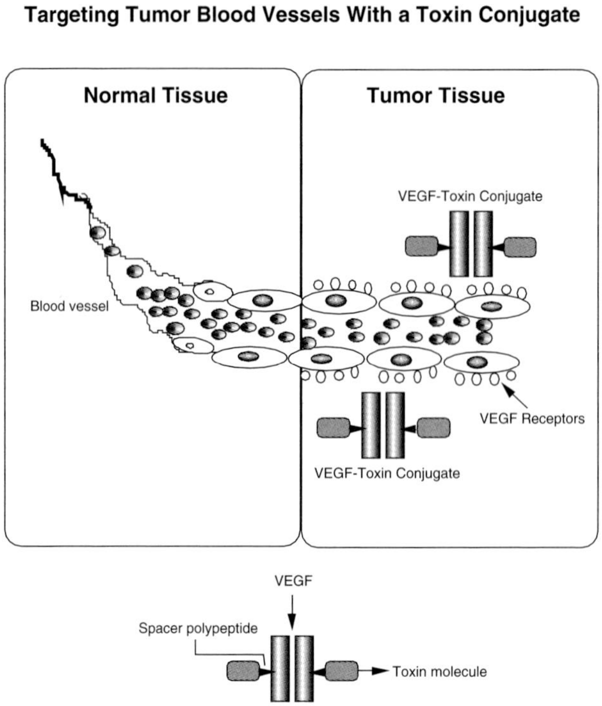

Figure 9.2. Inhibition of tumour neovascularization with VEGF–toxin conjugate. Schematic diagram showing higher levels of VEGF-receptor expression in the vascular endothelial cells of the tumour tissue compared to normal tissue vasculature. Note that the receptors for VEGF are located at the abluminal side of the endothelium. Both the location and density of VEGF receptors provide a unique opportunity for targeting toxin molecules selectively to the tumour blood vessels. VEGF–toxins administered into the systemic circulation will not affect the normal vessels due to the lack of sufficient numbers of receptor molecules and their abluminal distribution. When the conjugates reach tumour tissues, increased vascular leakage allows the conjugates to extravasate and home onto the abluminally-located VEGF-receptors. VEGF–toxin conjugates are internalized by receptor-mediated endocytosis and selectively inhibit endothelial cell proliferation and tumour angiogenesis.

extravasation of the VEGF–toxin conjugate. As a result, extravascular VEGF–toxin can easily bind to the abluminally distributed receptors (Figure 9.2). Among the different splice variants, both VEGF-165 and VEGF-121 were found to be equally efficient in delivering toxin polypeptides to endothelial cells. Conjugate treatment not only decreased vessel density in tumour tissue but also decreased the number of branch points and nodes.

Chemical conjugates used in earlier studies were prepared by random derivatization of VEGF with bifunctional reagents. Such methods often result in a heterogenous mixture of different VEGF/toxin stoichiometry. To avoid batch-to-batch variations and to obtain structurally well defined toxin conjugates, the coding region of VEGF and toxin polypeptides were fused at the DNA level. Unlike the chemical conjugates (dimeric VEGF linked to a toxin moiety), genetically fused proteins were expressed as monomeric VEGF fused to a toxin moiety. Interestingly, the monomeric constructs were found to be biologically active and inhibited tumour growth in mice [78]. Based on these observations, it was concluded that it is possible to re-engineer VEGF–toxin conjugates to optimize their anti-angiogenic effect. For example, VEGF could be separated from toxin polypeptides by introducing a spacer polypeptide, which can reduce steric hindrance. The spacer is 15 amino acids in size and its sequence is (Gly-Gly-Gly-Gly-Ser)$_3$. The spacer molecule provides adequate flexibility and allows unhindered interaction between VEGF and its receptor present on the cell surface. Pharmacokinetic properties of the construct can be improved, e.g. by fusion with human Fc fragments. Such strategies will improve the anti-angiogenic and anti-tumour activity of vascular targeting reagents.

9.3.1.2 Other Growth Factor Receptors Used for Targeting Strategies

Other growth factor receptors over-expressed in tumour vessels are FGF receptors and Tie-2 receptors. There have been few studies thus far to evaluate the relative merits of targeting via FGF to inhibit tumour blood vessels. Davol *et al.* prepared an endothelial cell-specific cytotoxic conjugate [79] by chemically linking the plant-derived ribosomal inhibitory protein saporin to FGF. The FGF–saporin conjugate inhibited proliferation of endothelial cells effectively *in vitro*. Similarly, a fusion protein containing placenta growth factor and saporin was found to exert anti-angiogenic activity [80]. Since proliferating endothelial cells show up-regulation of a variety of proliferation-associated cell markers such as transferrin receptors, it is also possible to target transferrin receptors to inhibit angiogenesis or directly attack tumour blood flow. Monoclonal antibodies reactive to human transferrin receptors have been used to prepare cytotoxic conjugates containing recombinant ricin A chain. These constructs were found to inhibit human corneal endothelial cells in a proliferation-dependent manner [81]. Targeting transferrin receptors will affect both proliferating endothelium and tumour cells and may therefore be useful at least for local application (e.g. for excessive proliferation of blood vessels following eye injury), if not for systemic therapy. Although extensive experience with transferrin-mediated targeting has been obtained in brain targeting research (as discussed in Chapter 2), this application has not been widely studied in tumour vasculature targeting strategies.

9.3.2 Endoglin Targeting

Endoglin (CD105) is a transmembrane glycoprotein, which is expressed on the surface of vascular endothelial cells (chicken, rodent and human). Endoglin is intricately associated with TGF-β receptor complex and is considered to be an ancillary, non-signalling receptor for TGF-β [82]. Genetic studies and gene deletion experiments have shown that endoglin plays a critical role in the development of the vascular system [83]. Specifically, endoglin modulates the communication between vascular endothelial cells and vascular smooth muscle cells, an important step in the maturation of blood vessels. Furthermore, endoglin binds to TGF-β1 and TGF-β3 in conjunction with TGF-β type II receptor. TGF-β1 mediated signalling is necessary for the differentiation of newly recruited mesenchymal (vascular smooth muscle) cells by endothelial cells. Genetic deletion of TGF-β1 in mice leads to embryonic lethality with a significant defect in developmental angiogenesis [84].

Endoglin is over-expressed in the vasculature of tumours and other tissues undergoing vascular remodelling [85]. Differential upregulation of endoglin presents an opportunity to target cytotoxic molecules to endothelial cells. Several monoclonal antibodies specific for human endoglin have been produced [86]. Some of the monoclonal antibodies generated against human endoglin were also found to cross react with mouse endothelial cells. In spite of low binding, monoclonal antibodies (SN6f and SN6j) were readily internalized by mouse endothelial cells. Using such a cross-reactive antibody, Seon *et al.* chemically linked ricin A chain or deglycosylated ricin A chain (the catalytic subunit of the plant toxin ricin) [86]. Ricin A chain is a N-glycosidase which specifically cleaves a single adenine residue from the 28-S ribosomal RNA. Depurination irreversibly impairs the function of ribosomes and thereby inhibits protein synthesis. Conjugates of endoglin-specific antibody and ricin A chain showed specific cytotoxicity against endothelial cells *in vitro* and inhibited experimental angiogenesis *in vivo*. The anti-angiogenic properties of anti-endoglin–ricin A chain conjugate were demonstrated in a dorsal air sac model system. Most importantly, endoglin-specific immunotoxin showed strong anti-tumour activity in a SCID mouse tumour model. In these studies, MCF-7, a human breast cancer cell line was transplanted into mice and then treated with endoglin-specific immunotoxin. These studies showed complete inhibition of tumour growth in all of the treated mice without any apparent toxicity to normal tissues [87]. Apart from intracellular targeting, endoglin-specific antibodies were also successfully used to localize radionuclide on the cell surface. Radioiodinated monoclonal antibodies (10 μCi) given to tumour-bearing animals significantly inhibited tumour growth, indicating the clinical potential of targeting endoglin [88].

9.3.3 Targeting Integrins to Tumour Vasculature

Interaction between cell surface-anchored integrins and extracellular matrix components constitutes an additional pathway necessary for angiogenesis control. In fact, studies have identified two cytokine-mediated, integrin-dependent angiogenic pathways. One of these pathways is associated with $\alpha_V\beta_3$ integrin, which selectively influences FGF-2 mediated angiogenic signals [28]. A second, non-overlapping pathway is represented by cross-talk between $\alpha_V\beta_5$ integrin and PKC-dependent, VEGF- or TNFα-induced, signalling [89]. Tumour angiogenesis can therefore be inhibited by blocking the interaction between integrins and the RGD motif-containing extracellular matrix proteins. Furthermore, the integrins present on tumour endothelium can serve as target epitopes via which toxic compounds can be delivered to the endothelial cells of the tumour.

Erkki Ruoslahti and colleagues [90] developed a novel targeting strategy by using polypeptides capable of delivering cytotoxic drugs to integrins. An *in vivo* selection of phage display libraries identified peptides that specifically home to components of tumour blood vessels (see Chapter 10 for details relating to this technology of ligand/target finding). Ruoslahti's research group identified two major classes of peptides, one containing the RGD motif and the other containing an NGR motif. These polypeptides were then chemically linked to the anti-cancer drug doxorubicin. Treatment of breast carcinoma-bearing mice with the conjugated doxorubicin caused vascular damage in the tumours and a strong anti-tumour effect at a 10–40 x lower concentration than that of free doxorubicin, while liver and heart toxicity was reduced compared to that observed with free doxorubicin [90]. Whether this effect was caused by the selective delivery of the chemotherapeutic drug to the tumour endothelial cells and/or tumour cells, direct caspase-3 activation [91], or a combination, has not as yet been established. Their results illustrate the potential of targeting therapeutic agents to integrins expressed on the vasculature of tumours as an effective means of cancer treatment.

9.3.4 Tumour Vasculature-specific Blood Coagulation Induction

Toxins as effector molecules have been widely studied *in vitro* and applied *in vivo* in pre-clinical and clinical studies. A frequent observation with immunotoxins in the clinic is the occurrence of vascular leak syndrome. This toxicity is associated with the toxin moiety of the immunotoxin and sometimes demands cessation of therapy or administration of sub-optimal dosages [92]. Another approach with high potential is to selectively inhibit tumour blood flow by selectively targeting the blood coagulation-inducing activity of the tumour endothelium.

In one study on coagulation-inducing capacity, a mouse model for tumour vasculature was exploited in which the expression of MHC Class II was artificially induced on tumour endothelial cells [93]. Bispecific antibodies (BsAb) against MHC Class II and a truncated form of the activator of the extrinsic coagulation pathway, truncated Tissue Factor (tTF), were chemically prepared. Intravenous administration of a mixture consisting of BsAb and tTF (BsAb * tTF), the so-called 'coaguligand' formulation, to mice with clinically relevant tumour burden, resulted in dramatic tumour reduction without concurrent toxicity in other organs (Figure 9.3). Site-specific blood coagulation in the tumour blood vessels caused an almost instantaneous and persisting blockade of tumour blood flow. Treatment of mice bearing

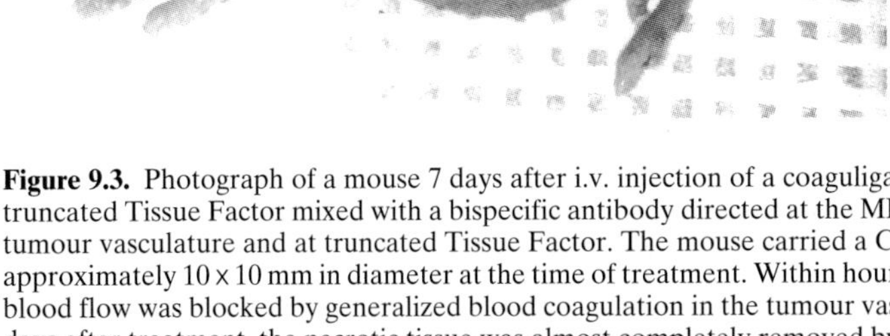

Figure 9.3. Photograph of a mouse 7 days after i.v. injection of a coaguligand formulation consisting of truncated Tissue Factor mixed with a bispecific antibody directed at the MHC Class II molecules on the tumour vasculature and at truncated Tissue Factor. The mouse carried a C1300 muγ tumour measuring approximately 10 x 10 mm in diameter at the time of treatment. Within hours after treatment the tumour blood flow was blocked by generalized blood coagulation in the tumour vasculature (not shown). Seven days after treatment, the necrotic tissue was almost completely removed by the host immune cells.

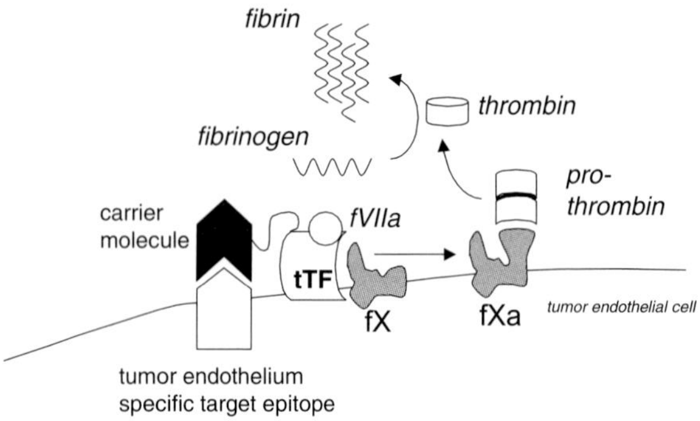

Figure 9.4. Schematic representation of the mechanism of action of the coaguligand approach. Cross linking of truncated Tissue Factor to tumour endothelial cells leads to local blood coagulation via the tTF/fVIIa complex. tTF, truncated Tissue Factor; fVIIa, factor VIIa; fX (A), factor X (A).

subcutaneous tumours, twice with BsAb * tTF coaguligand led to 38% complete tumour regressions and 24% partial responses [94]. The attractiveness of the coaguligand approach is the use of a truncated form of TF which is devoid of coagulation-inducing activity as long as it is prevented from complexing with the lipophilic factor X on cell membranes. Upon cross linking of the hydrophilic tTF with the target cell membranes by the BsAb, tTF becomes complexed with factor X. In the presence of factor VII/VIIa, this leads to the induction of blood coagulation (Figure 9.4). It is thought that there is a threshold in the number of tTF cross linked to cell membranes, above which the coagulation cascade is initiated. In theory, this allows tumour endothelium-associated target epitopes to be utilized which are highly but not exclusively, expressed on tumour endothelium. The level of expression on other vascular beds is then too low to trigger coagulation after cross linking of the coaguligand.

Using a similar approach of tumour infarction, mouse solid Hodgkin's tumours spontaneously expressing endothelial VCAM-1 were significantly retarded in outgrowth [95]. The anti-tumour effect was not as dramatic as seen in the MHC Class II model. Possibly, the number of tTF molecules delivered at the site of the tumour endothelium was not sufficient to create a rapid and more or less generalized pro-coagulant situation throughout the tumour vasculature. Only if coagulation is induced in the majority of vessels, will the number of tumour cells deprived of nutrients be sufficient enough to show strong anti-tumour effects. Furthermore, anti-coagulation activities may be strong enough to counteract the coaguligand effects when the kinetics of coagulation induction are insufficient to imbalance local pro- and anti-coagulation activities.

The coagulation induction potency of coaguligand formulations are likely to be determined by the following factors: (i) the number of target epitopes on the tumour endothelium that allow BsAb-mediated interaction between tTF and factor X on the target cell membrane; (ii) local anti-coagulation activity which may be regulated in a species-specific manner; and (iii) the kinetics of cross linking of the BsAb and the target epitopes in relation to the kinetics of coagulation induction capacity. The number of MHC Class II and VCAM-1 molecules expressed on the tumour vasculature of the animal models discussed, were high, as were the affinities of the antibodies used. This enabled a significant number of tTF to be rapidly cross linked to the target cell membrane. For clinically relevant target epitopes and targeting devices, the importance of these characteristics needs to be established.

In addition to the targeting of toxins and coagulation-inducing effector moieties to tumour vasculature, inhibitors of angiogenesis-related signal transduction pathways are candidates for selective targeting to tumour endothelium. Although quite effective in various animal models, recent observations of severe toxicity in clinical studies justifies more selective delivery of these molecules into the pro-angiogenic endothelium. At present, however, no data are available on such strategies.

9.3.5 Other Potential Targets

Of the target epitopes suggested for use in tumour vasculature directed drug targeting strategies (Table 9.1), those discussed above seem to be the most promising for development for clinical application. A few have not been extensively studied for this purpose but may also be interesting candidates, and are therefore discussed below.

Endosialin is a cell surface glycoprotein that was identified in various human tumours including sarcomas, carcinomas and neuroectodermal tumours. It comprises a core polypeptide of about 95 kDa and is highly glycosylated (O-linked oligosaccharides). Its biological function and the importance of its expression on tumour vascular endothelium is not yet understood. This antigen is thought to be located on the luminal surface of tumour endothelial cells which represents an obvious advantage for targeting [96]. Apparently, monoclonal antibody (FB5) reactive to endosialin did not show any detectable binding to the vasculature of normal tissues. Although it was suggested that radiolabelled FB5 was rapidly internalized into endosialin-expressing cells, no follow-up on this was reported [96].

Griffioen *et al.* [97] investigated the potential of targeting the activation antigen CD44. Their studies showed that endothelial cells from tumour vasculature displayed an increased expression of CD44 as compared to endothelial cells from normal tissue. CD44-targeted immunotoxin produced efficient inhibition of CD44-positive endothelial cells with high specificity. Further pre-clinical studies are currently in progress.

Targeted radioimmunotherapy of pulmonary micrometastases was feasible in mice with an antibody directed against thrombomodulin, expressed selectively and in large amounts on the luminal surfaces of capillaries and small blood vessels in the lungs. The short-lived ($t^1/2$ = 45 min) a-particle emitter ^{213}Bi, conjugated to the antibody was delivered to healthy lung and tumour capillaries, resulting in significant tumour growth reduction and an extended life-span of animals treated at low doses. At higher doses, tumours almost completely regressed. However, animals died of lung fibrosis as a result of concurrent damage to healthy tissue [98].

A breakthrough in the search for novel anti-angiogenic compounds occurred when the hypothesis that a primary tumour, while capable of stimulating angiogenesis for its own blood supply, can produce angiogenesis inhibitors which suppress the outgrowth of distant metastases, was proven to hold true. This hypothesis came from the observation that the removal of primary tumours could lead to the accelerated growth of metastases [99]. To test this hypothesis the Lewis lung carcinoma mouse model was used, in which the primary tumour completely suppressed the growth of its metastases. From the urine of these mice a cleavage fragment of plasminogen, called angiostatin, was purified and found to replace the inhibitory activity of the primary tumour completely [100]. Treatment of tumour-bearing mice with angiostatin almost completely prevented metastasis formation in the lung. In theory, these inhibitor proteins could serve as carrier molecules for drug targeting, provided they specifically bind to tumour vasculature. They could then form the basis for dual targeting strategies, in which the carrier itself exerts an effect in addition to the effect of the attached drug. Depending on the mechanisms of action of both active components, synergistic effects might be expected [101]. The target for angiostatin on endothelial cells has recently been discovered to be ATP synthase [102]. Whether this binding site is expressed in tumour vasculature and can be exploited as a target epitope with angiostatin as a carrier molecule, needs to be investigated.

Using a similar strategy endostatin was discovered [103]. Although the exact identity of the binding site for endostatin is not known, Chang *et al.* demonstrated that endostatin can bind to blood vessels of different calibre in various organs. In breast carcinoma binding of endostatin co-localized with FGF-2, but FGF-2 and heparin did not compete for endostatin binding [104]. The lack of selectivity for tumour vasculature probably excludes this molecule from being used as a carrier molecule in drug targeting strategies.

To summarize, some major steps forward have been made in the development of novel drug targeting approaches aimed at selectively killing tumour endothelial cells. The extensive 'from the bench to the bed' experience with tumour cell-targeted immunotoxins [105] has paved the way for further development of these tumour endothelial cell-targeted strategies. In this context it is of primary importance that the handling of clinically relevant target epitopes and their drug targeting ligands by endothelial cells, be established under pathological conditions.

9.4 Tumour Vasculature Targeting Potentials: Extrapolation of Animal Studies to the Human Situation

From the above, it is clear that tumour vasculature-directed drug targeting approaches to blocking tumour blood flow can be potent strategies for the therapy of large solid tumours. At present, however, only pre-clinical data are available in this area of research, and no sensible extrapolation from pre-clinical experiments with human or animal tumours can be made from the animal model to the clinical setting. One important difference between human tumours and tumours grown in animals is the level of vascular permeability. Although this parameter can vary significantly between the various animal tumours [106], it is believed that the vasculature of animal tumours is in general more permeable. This may be a result of the fact that the majority of animal tumours grow more rapidly than those developing in humans. Another consequence of this rapid tumour growth, is that the majority of blood vessels in animal tumours are in a pro-angiogenic state. As a result, anti-angiogenic therapy or angiogenesis-related epitope-targeted therapy will affect a greater proportion of the blood vessels in an animal tumour. In human tumours the vasculature is more heterogeneous. Therefore, the selective targeting of drugs to different epitopes covering a broad range of angiogenesis-related markers seems most appropriate strategy to gain access to the majority of tumour blood vessels.

Many of the drugs that inhibit endothelial cell proliferation, migration and maturation in the angiogenic process act at the level of cell death induction. The thrombo-embolisms observed in the clinic with several anti-angiogenic compounds may indicate that enhanced endothelial cell apoptosis in humans can lead to enhanced micro-thrombus formation and severe toxicity. This observation is in line with the description of the enhanced coagulation-inducing capacity of endothelial cells *in vitro*, when endothelial apoptosis is triggered [107]. Whether the delivery of anti-angiogenic drugs or coagulation-inducing effector molecules into/at tumour endothelial cells via drug targeting will have a similar detrimental effect in humans needs to be carefully addressed.

9.5 Summary and Future Perspectives

Targeting active agents to tumour vasculature to selectively induce tumour blood flow blockade was shown to be highly effective in inhibiting clinically significant tumour burdens in animals. One of the potential advantages of such a treatment may be the absence of drug resistance, as tumour endothelial cells are considered genetically stable. The next step will be to develop similar strategies for use in the clinic. For this purpose, target epitopes on human tumour endothelium need to be identified and studied for their suitability for such strategies. Although some interesting target candidates have been put forward, proof of concept in the human situation needs to be validated. Furthermore, the choice of the drugs to be selectively delivered at or into the pro-angiogenic endothelium will require extensive research in the coming years.

Some 20 different anti-angiogenic agents are currently in clinical trials. Examples of these are marimastat, AG3340, neovastat, TNP-470, thalidomide, CAI, SU5416 , anti-VEGF antibody, and angiostatin (see NCI homepage). It needs to be established whether these drugs can be considered as candidates for use in future drug targeting strategies. Since tumour therapy aims at the complete eradication of tumour cells, a combination of tumour vasculature-directed strategies (anti-angiogenic drugs as such, or targeted drugs as discussed in this chapter) together with tumour cell-directed chemo- and/or immunotherapy, may provide the way forward in the search for optimal treatment for future clinical application.

References

[1] Hanahan D, Folkman J, *Cell* **1996,** *86*, 353–364.
[2] Rajotte D, Arap W, Hagedorn M, Koivunen E, Pasqualini R, Ruoslahti E, *J. Clin. Invest.* **1998,** *102*, 430–437.
[3] Verstraete M, In: *The Endothelial Cell in Health and Disease* (Eds Vane JR, Born GVR, Welzel D), pp. 147–164. Schattauer, Stuttgart, New York, **1995.**
[4] Carlos TM, Harlan JM, *Blood* **1994,** *84*, 2068–2101.
[5] Risau W, *Nature* **1997,** *386*, 671–674.
[6] Frater-Schroder M, Risau W, Hallmann R, Gautschi P, Bohlen P, *Proc. Natl Acad. Sci. USA* **1987,** *84*, 5277–5281.
[7] Denekamp J, *Acta Radiol. Oncol.* **1984,** *23*, 217–225.
[8] Klagsbrun M, Moses MA, *Chem. Biol.* **1999,** *6*, R217–R224.
[9] Mattei MG, Borg JP, Rosnet O, Marme D, Birnbaum D, *Genomics* **1996,** *32*, 168–169.

[10] Brown LF, Detmar M, Claffey K, Nagy JA, Feng D, Dvorak AM, Dvorak HF, *EXS* **1997,** *79*, 233–269.
[11] Ziche M, Morbidelli L, Choudhuri R, Zhang HT, Donnini S, Granger HJ, Bicknell R, *J. Clin. Invest.* **1997,** *99*, 2625–2634.
[12] Dvorak HF, Galli SJ, Dvorak AM, *Hum. Pathol.* **1986,** *17*, 122–137.
[13] Gupta K, Kshirsagar S, Li W, Gui L, Ramakrishnan S, Gupta P, Law PY, Hebbel RP, *Exp. Cell. Res.* **1999,** *247*, 495–504.
[14] Ferrara N, Keyt B, *EXS* **1997,** *79*, 209–232.
[15] Soker S, Takashima S, Miao HQ, Neufeld G, Klagsbrun M, *Cell* **1998,** *92*, 735—745.
[16] Kawasaki T, Kitsukawa T, Bekku Y, Matsuda Y, Sanbo M, Yagi T, Fujisawa H, *Development* **1999,** *126*, 4895–4902.
[17] Klein S, Roghani M, Rifkin DB, *EXS* **1997,** *79*, 159–192.
[18] Gleizes PE, Rifkin DB, *Pathol. Biol.* **1999,** *47*, 322–329.
[19] Miao HQ, Ishai Michaeli R, Atzmon R, Peretz T, Vlodavsky I, *J. Biol. Chem.* **1996,** *271*, 4879–4886.
[20] Czubayko F, Liaudet-Coopman ED, Aigner A, Tuveson AT, Berchem GJ, Wellstein A, *Nature Med.* **1997,** *3*, 1137–1140.
[21] Rak J, Kerbel RS, *Nature Med.* **1997,** *3*, 1083–1084.
[22] Varner JA, *EXS* **1997,** *79*, 361–390.
[23] Aplin AE, Howe A, Alahari SK, Juliano RL, *Pharmacol. Rev* **1998,** *50*, 197–263.
[24] Frisch SM, Ruoslahti E, *Curr. Opin. Cell Biol.* **1997,** *9*, 701–706.
[25] Howe A, Aplin AE, Alahari SK, Juliano RL, *Curr. Opin. Cell Biol.* **1998,** *10*, 220–231.
[26] Cheresh DA, *Adv. Molec. Cell Biol.* **1993,** *6*, 225–252.
[27] Brooks PC, Clark RA, Cheresh DA, *Science* **1994,** *264*, 569–571.
[28] Friedlander M, Brooks PC, Shaffer RW, Kincaid CM, Varner JA, Cheresh DA, *Science* **1995,** *270*, 1500–1502.
[29] DiPietro LA, *EXS* **1997,** *79*, 295–314.
[30] Grant DS, Kleinman HK, *EXS* **1997,** *79*, 317–333.
[31] Jendraschak E, Sage EH, *Semin. Cancer Biol.* **1996,** *7*, 139–146.
[32] Tremble PM, Lane TF, Sage EH, Werb Z, *J. Cell Biol.* **1993,** *121*, 1433–1444.
[33] Hasselaar P, Loskutoff DJ, Sawdey M, Sage EH, *J. Biol. Chem.* **1991,** *266*, 13178–13184.
[34] Hirschi KK, D'Amore PA, *EXS* **1997,** *79*, 419–428.
[35] Crocker DJ, Murad TM, Geer JC, *Exp. Mol. Pathol.* **1970,** *13*, 51–65.
[36] Kanda S, Landgren E, Ljungstrom M, Claesson-Welsh L *Cell Growth Differ.* **1996,** *7*, 383–395.
[37] Ades EW, Candal FJ, Swerlick RA, George VG, Summers S, Bosse DC, Lawley TJ, *J. Invest. Dermatol.* **1992,** *99*, 683–690.
[38] Edgell CJ, McDonald CC, Graham JB, *Proc. Natl Acad. Sci. USA* **1983,** *80*, 3734–3737.
[39] Griffioen AW, Damen CA, Martinotti S, Blijham GH, Groenewegen G, *Cancer Res.* **1996,** *56*, 1111–1117.
[40] van Leeuwen EBM, Veenstra R, van Wijk R, Molema G, Hoekstra A, Ruiters MHJ, van der Meer J, *Blood Coagul. Fibrinol.* **2000,** *11*, 15–25.
[41] Gupta SK, Singh JP, *J. Cell Biol.* **1994,** *127*, 1121–1127.
[42] Relou IAM, Damen CA, van der Schaft DWJ, Groenewegen GG, Griffioen AW, *Tissue Cell* **1998,** *30*, 525–530.
[43] Tsukada T, Kubota A, Ueda M, Amano J, Shimokado K, Numano F, *Jpn. Circ. J.* **1991,** *55*, 996–1002.
[44] Barrett TL, Smith KJ, Williams J, Corner SW, Hodge JJ, Skelton HG, *Mod. Pathol.* **1999,** *12*, 450–455.
[45] Allen RT, Hunter WJ, III, Agrawal DK, *J. Pharmacol. Toxicol. Methods* **1997,** *37*, 215–228.
[46] Ruegg C, Yilmaz A, Bieler G, Bamat J, Chaubert P, Lejeune FJ, *Nature Med.* **1998,** *4*, 408–414.
[47] Schraa AJ, Kok RJ, van Veen-Hof IH, Moorlag H, Meijer DKF, de Leij LFMH, Molema G, *International Meeting on Angiogenesis*, Amsterdam April 27–29 **2000.**
[48] Glaser BM, D'Amore PA, Seppa H, Seppa S, Schiffmann E, *Nature* **1980,** *288*, 483–484.
[49] Sholley MM, Gimbrone Jr MA, Cotran RS, *Lab. Invest.* **1977,** *36*, 18–25.
[50] Barendsz-Janson AF, Griffioen AW, Muller AD, van Dam-Mieras MCE, Hillen HFP, *J. Vasc. Res.* **1998,** *35*, 109–114.
[51] Vailhe B, Ronot X, Tracqui P, Usson Y, Tranqui L, *In Vitro Cell Dev. Biol. Anim.* **1997,** *33*, 763–773.
[52] Grant DS, Tashiro K, Segui Real B, Yamada Y, Martin GR, Kleinman HK, *Cell* **1989,** *58*, 933–943.

[53] Nehls V, Drenckhahn D, *Microvasc. Res.* **1995,** *50*, 311–322.
[54] Nicosia RF, Bonanno E, Smith M, *J. Cell Physiol.* **1993,** *154*, 654–661.
[55] Brown KJ, Maynes SF, Bezos A, Maguire DJ, Ford MD, Parish CR, *Lab. Invest.* **1996,** *75*, 539–555.
[56] Wartenberg M, Gunther J, Hescheler J, Sauer H, *Lab. Invest.* **1998,** *78*, 1301–1314.
[57] Evans MJ, Kaufman MH, *Nature* **1981,** *292*, 154–156.
[58] Risau W, Sariola H, Zerwes HG, Sasse J, Ekblom P, Kemler R, Doetschman T, *Development* **1988,** *102*, 471–478.
[59] Vittet D, Prandini MH, Berthier R, Schweitzer A, Martin-Sisteron H, Uzan G, Dejana E, *Blood* **1996,** *88*, 3424–3431.
[60] Nguyen M, Shing Y, Folkman J, *Microvasc. Res.* **1994,** *47*, 31–40.
[61] Conrad TJ, Chandler DB, Corless JM, Klintworth GK, *Lab. Invest.* **1994,** *70*, 426–434.
[62] Algire GH, *J. Natl Cancer Inst.* **1943,** *4*, 1–11.
[63] Lichtenbeld HC, Barendsz-Janson AF, van Essen H, Struijker Boudier H, Griffioen AW, Hillen HFP, *Int. J. Cancer* **1998,** *77*, 455–459.
[64] Shubik P, Feldman R, Garcia H, Warren BA, *J. Natl Cancer Inst.* **1976,** *57*, 769–774.
[65] Mahadevan V, Hart IR, Lewis GP, *Cancer Res.* **1989,** *49*, 415–419.
[66] Plunkett ML, Hailey JA, *Lab. Invest.* **1990,** *62*, 510–517.
[67] Jain RK, Schlenger K, Hockel M, Yuan F, *Nature Med.* **1997,** *3*, 1203–1208.
[68] Girard JP, Baekkevold ES, Yamanaka T, Haraldsen G, Brandtzaeg P, Amalric F, *Am. J. Pathol.* **1999,** *155*, 2043–2055.
[69] Lyden D, Young AZ, Zagzag D, Yan W, Gerald W, O'Reilly R, Bader BL, Hynes RO, Zhuang Y, Manova K, Benezra R, *Nature* **1999,** *401*, 670–677.
[70] Wang HV, Chen ZF, Anderson DJ, *Cell* **1998,** *93*, 741–753.
[71] Pasqualini R, Ruoslahti E, *Mol. Psychiatry* **1996,** *1*, 423–423.
[72] Jacobson BS, Stolz DB, Schnitzer JE, *Nature Med.* **1996,** *4*, 408–414.
[73] Griffioen AW, Molema G, *Pharmacol. Rev.* **2000,** *52*, 237–268.
[74] Folkman J, *Eur. J. Cancer* **1996,** *32A*, 2534–2539.
[75] Plate KH, Breier G, Millauer B, Ullrich A, Risau W, *Cancer Res.* **1993,** *53*, 5822–5827.
[76] Ramakrishnan S, Olson TA, Bautch VL, Mohanraj D, *Cancer Res.* **1996,** *56*, 1324–1330.
[77] Olson TA, Mohanraj D, Roy S, Ramakrishnan S, *Int. J. Cancer* **1997,** *73*, 865–870.
[78] Arora N, Masood R, Zheng T, Cai J, Smith DL, Gill PS, *Cancer Res.* **1999,** *59*, 183–188.
[79] Davol P, Beitz JG, Mohler M, Ying W, Cook J, Lappi DA, Frackelton Jr AR, *Cancer* **1995,** *76*, 79–85.
[80] Chiaramonte R, Polizzi D, Bartolini E, Petroni D, Comi P, *Anticancer Drug Des.* **1997,** *12*, 649–657.
[81] Fulcher S, Lui GM, Houston LL, Ramakrishnan S, Burris T, Polansky J, Alvarado J, *Invest. Ophthalmol. Vis. Sci.* **1988,** *29*, 755–759.
[82] Arthur HM, Ure J, Smith AJ, Renforth G, Wilson DI, Tornsey E, Charlton R, Parums DV, Jowett T, Marchuk DA, Burn J, Diamond AG, *Dev. Biol.* **2000,** *217*, 42-53.
[83] Li DY, Sorensen LK, Brooke BS, Urness LD, Davis EC, Taylor DG, Boak BB, Wendel DP, *Science* **1999,** *284*, 1534–1537.
[84] Oshima M, Oshima H, Taketo MM, *Dev. Biol.* **1996,** *179*, 297–302.
[85] Miller DW, Grulich W, Karges B, Stahl S, Ernst M, Ramaswamy A, Sedlacek HH, Muller R, Adamkiewicz J, *Int. J. Cancer* **1999,** *81*, 568–572.
[86] Seon BK, Matsuno F, Haruta Y, Kondo M, Barcos M, *Clin. Cancer Res.* **1997,** *3*, 1031–1044.
[87] Matsuno F, Haruta Y, Kondo M, Tsai H, Barcos M, Seon BK, *Clin. Cancer Res.* **1999,** *5*, 371–382.
[88] Tabata M, Konda M, Haruta Y, Seon BK, *Int. J. Cancer.* **1999,** *82*, 737–742.
[89] Eliceiri BP, Cheresh DA, *J. Clin. Invest.* **1999,** *103*, 1227–1230.
[90] Arap W, Pasqualini R, Ruoslahti E, *Science* **1998,** *279*, 377–380.
[91] Buckley CD, Pilling D, Henriquez NV, Parsonage G, Threlfall K, Scheel-Toellner D, Simmons DL, Akbar AN, Lord JM, Salmon M, *Nature* **1999,** *397*, 534–539.
[92] Baluna R, Rizo J, Gordon BE, Ghetie V, Vitetta ES, *Proc. Natl Acad. Sci. USA* **1999,** *96*, 3957–3962.
[93] Burrows FJ, Thorpe PE, *Proc. Natl Acad. Sci. USA* **1993,** *90*, 8996–9000.
[94] Huang X, Molema G, King S, Watkins L, Edgington TS, Thorpe PE, *Science* **1997,** *275*, 547–550.
[95] Ran S, Gao B, Duffy S, Watkins L, Rote N, Thorpe PE, *Cancer Res.* **1998,** *58*, 4646–4653.
[96] Rettig WJ, Garin-Chesa P, Healy JH, Su SL, Jaffe EA, Old LJ, *Proc. Natl Acad. Sci. USA* **1992,** *89*, 10832–10836.

[97] Griffioen AW, Coenen MJ, Damen CA, Hellwig SM, van Weering DH, Vooys W, Blijham GH, Groenewegen G, *Blood* **1997,** *90*, 1150–1159.
[98] Kennel SJ, Boll R, Stabin M, Schuller HM, Mirzadeh S, *Br. J. Cancer* **1999,** *80*, 175–184.
[99] Sugarbaker EV, Thornthwaite J, Ketcham AS, In: *Progress in Cancer Research and Therapy* (Eds Day B, Meyers WPL), pp. 227–240. Raven Press, New York, **1977.**
[100] O'Reilly MS, Holmgren L, Shing Y, Chen C, Rosenthal RA, Moses M, Lane WS, Cao Y, Sage EH, Folkman J, *Cell* **1994,** *79*, 315–328.
[101] Meijer DKF, Molema G, Moolenaar F, De Zeeuw D, Swart PJ, *J. Control. Rel.* **1996,** *39*, 163–172.
[102] Moser TL, Stacks MS, Asplin I, Enghild JJ, Hojrup P, Everitt L, Hubchak S, Schnaper HW, Pizzo SV, *Proc. Natl Acad. Sci. USA* **1999,** *96*, 2811–2816.
[103] O'Reilly MS, Boehm T, Shing Y, Fukai N, Vasios G, Lane WS, Flynn E, Birkhead JR, Olsen BR, Folkman J, *Cell* **1997,** *88*, 277–285.
[104] Chang Z, Choon A, Friedl A, *Am. J. Pathol.* **1999,** *155*, 71–76.
[105] Frankel A-E, Fitzgerald D, Siegall C, Press O-W, *Cancer Res.* **1996,** *56*, 926–932.
[106] Yuan F, Dellian M, Fukumura D, Leunig M, Berk DA, Torchilin VP, Jain RK, *Cancer Res.* **1995,** *55*, 3752–3756.
[107] Bombeli T, Schwartz BR, Harlan JM, *Blood* **1999,** *93*, 3831–3838.
[108] Hagemeier HH, Vollmer E, Goerdt S, Schulze-Osthoff K, Sorg C, *Int. J. Cancer* **1986,** *38*, 481–488.
[109] Schlingemann RO, Rietveld FJ, de Waal RM, Bradley NJ, Skene AI, Davies AJ, Greaves MF, Denekamp J, Ruiter DJ, *Lab. Invest.* **1990,** *62*, 690–696.
[110] Dvorak HF, Brown LF, Detmar M, Dvorak AM, *Am. J. Pathol.* **1995,** *146*, 1029–1039.
[111] Nguyen M, Corless CL, Kraling BM, Tran C, Atha T, Bischoff J, Barsky SH, *Am. J. Pathol.* **1997,** *150*, 1307–1314.
[112] Brooks PC, Montgomery AM, Rosenfeld M, Reisfeld RA, Hu T, Klier G, Cheresh DA, *Cell* **1994,** *79*, 1157–1164.
[113] Burrows FJ, Derbyshire EJ, Tazzari PL, Amlot P, Gazdar AF, King SW, Letarte M, Vitetta ES, Thorpe PE, *Clin. Cancer Res.* **1995,** *1*, 1623–1634.
[114] Sato TN, Qin Y, Kozak CA, Audus KL, *Proc. Natl Acad. Sci. USA* **1993,** *90*, 9355–9358.
[115] Eggermont AM, *Anticancer Res.* **1998,** *18*, 3899–3905.
[116] Koivunen E, Arap W, Valtanen H, Rainisalo A, Medina OP, Heikkila P, Kantor C, Gahmberg CG, Salo T, Konttinen YT, Sorsa T, Ruoslahti E, and Pasqualini R, *Nature Biotechnol.* **1999,** *17*, 768–774.

10 Phage Display Technology for Target Discovery in Drug Delivery Research

Ricardo Mutuberria, Jan-Willem Arends, Arjan W. Griffioen, Hennie R. Hoogenboom

10.1 Introduction

Phage display technology has revolutionized the search for proteins, peptides, and antibodies that bind to molecular targets. The creation of ligand-displaying libraries in combination with powerful selection methods has opened up a wide range of possibilities not only for the search and generation of new ligands amenable for drug targeting, but also in the field of drug discovery and drug design. Libraries containing billions of such ligands can be displayed on phage and enriched for target-binding clones under rationally designed selection techniques, producing molecules with the desired specificity for a given target. The linkage of genotype and phenotype present within the phage particle allows further manipulation of the ligand encoding genes to achieve the desired targeting entity. Selected ligands can be provided with desired effector functions and employed for therapeutic purposes.

For drug targeting applications, where high affinity ligands are needed which specifically recognize cell- and/or disease-induced surface molecules and are routed into the desired cellular compartments, this technology may be of great importance. This chapter will address how this technology is being used for drug targeting research and target discovery, through the selection of ligands to known and novel targets, and for the engineering and optimization of phage display selected ligands for therapeutic applications. Finally, an update on the current applications of the technology in drug targeting is presented.

10.2 Phage Display Technology

10.2.1 Introduction to the Technology

Phage display libraries have been used extensively for the selection of peptides, antibody fragments or protein variants binding to structures such as proteins, peptides, carbohydrates, nucleic acids or small molecular weight compounds. The ability to rationally design and construct libraries with large molecular diversity makes possible the identification of novel ligands or variants of known ligands with desired binding specificity and characteristics. The display of ligands on the surface of bacteriophage (Figure 10.1) is accomplished by the cloning of protein- or peptide-encoding DNA into the phage genome by fusion to one of the phage coat proteins, pVI, pIII or pVIII. The coat protein fusion is therefore incorporated into the mature phage resulting in the ligand being displayed on the phage surface, while its genetic

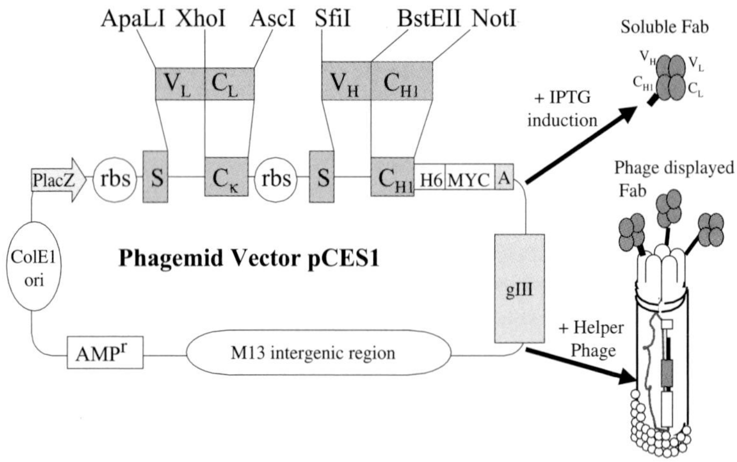

Figure 10.1. Display of a Fab fragment on filamentous phage. Fab fragments can be displayed on phage using phagemid vector pCES1 which expresses the heavy chain fragment containing the variable domain and the first constant domain fused to the coat protein gene III, in combination with separate expression of the partner (light) chain. Bacteria harbouring this vector are infected with helper phage to drive the production of phage particles carrying the Fab fragment as a fusion product with the phage coat protein pIII on the surface, while the immunoglobulin encoding genes reside within the phage genome. Alternatively IPTG induction drives the generation of the soluble Fab fragments on the bacterial periplasm. AMPr, ampicillin resistance; H6, histidine tag for purification purposes; MYC, myc tag for detection purposes; A, amber stop codon (TAG) which allows expression of the soluble antibody fragment in non-suppressor strains; gIII, phage gene III; rbs, ribosomal binding site; S, signal sequence directing the expressed protein to the bacterial periplasm; ColE1 ori, *E. coli* origin of replication; PlacZ, LacZ promoter.

material resides within the phage genome. When the cloned DNA encodes variants of a certain ligand, a phage display library is created. Phage display was first achieved in 1985 by the expression of a peptide on the surface of bacteriophage M13 [1]. Five years later the first random peptide libraries [2–4], and antibody fragment libraries [5] were constructed. Today a large number of moieties have been successfully displayed on the surface of filamentous phages. These include peptides (reviewed by Cesarini *et al.*) [6], antibody fragments (reviewed by Hoogenboom) [7], enzymes (reviewed by Soumillon *et al.*) [8], protein scaffolds (reviewed by Nygren and Uhlen) [9], cDNA libraries, (reviewed by Hufton *et al.*) [10], protease inhibitors [11], transcription factors [12], cytokines [13], and extracellular domains of receptors [14].

In order to retrieve ligands with the desired specificity, phage display libraries are enriched for target binding clones by subjecting the phage libraries to repetitive rounds of selection. This includes incubation with antigen or 'biopanning', washing of non-bound phage, elution and re-infection of selected phages into bacteria (Figure 10.2). Antigen binding phage is generally eluted by low or high pH treatment, which drives the dissociation of the ligand from its target without substantially altering the infectivity of the phage for bacteria. A selected filamentous phage is propagated in bacteria, which secrete multiple copies of the phage displaying a particular insert. Selection is repeated until a population of binding clones is enriched

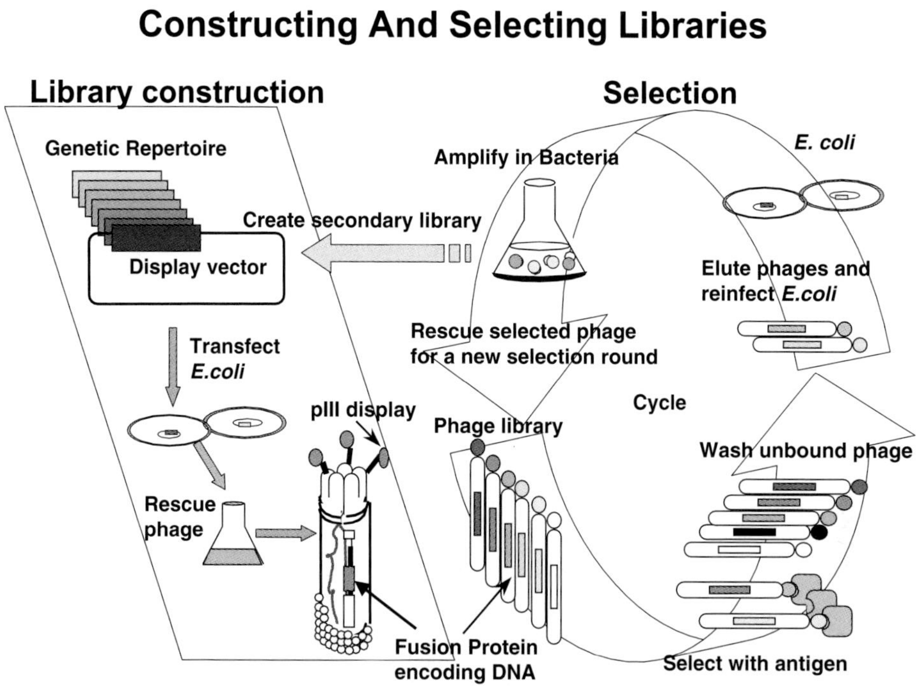

Figure 10.2. The phage display cycle, DNA encoding for millions of variants of a certain ligand (e.g. peptides, proteins or protein fragments) is batch cloned into the phage genome as part of the phage coat protein pIII (coat proteins pVI and pVIII can also be used for display). From this repertoire, phage carrying specific binding ligands can be isolated by a series of cycles of selections on the antigen, each of which involves binding to antigen, washing unbound phage, elution of bound phage and re-amplification in the bacterial cell.

and eventually individual clones are screened for binding to the target. Any procedure that efficiently separates binding clones from those that do not bind, can be used as a selection approach. This has given rise to a large variety of selection methods (Figure 10.3).

Some phage display vectors allow ligands with desired specificity to be produced both as soluble or phage displayed molecules. Alternatively, selected ligands can be recloned for production of the soluble form, or synthesized *in vitro* as is commonly done for peptide ligands. Phage displayed moieties often exhibit the same or similar functional characteristics as their native counterparts in solution. Soluble forms of the ligand are commonly used to determine relative specificity or affinity to the target. Genetic manipulation of the selected clones allows further optimization of the selected ligand to meet the requirements necessary for drug targeting purposes.

10.2.2 Phage Display Libraries

10.2.2.1 Introduction

The Phage Display process generally starts by creating a large collection of phage, known as a 'library', containing up to hundreds of millions of related proteins (such as antibody fragments or peptides) that are displayed on the surface of filamentous phage. The fusion protein is generally cloned in frame with gene III or gene VIII and downstream of a signal sequence that directs proteins to the periplasm, where many recombinant proteins will fold correctly. Phagemids, small plasmid vectors with higher transformation efficiencies, are the commonly used vectors for library creation. Phagemids carry the fusion protein gene with appropriate cloning sites for exogenous DNA insertion and with a phage packaging signal [15,16]. Phagemids make large amounts of recombinant protein, but are unable to make phage unless the bacteria is also infected by helper phage (fd phage with disabled packaging signal and antibiotic resistance gene), which supplies all the other proteins required to make functional phage. Phagemids can also be formatted for direct secretion of soluble, non-fused product, for the production of the soluble form of the displayed ligand [17].

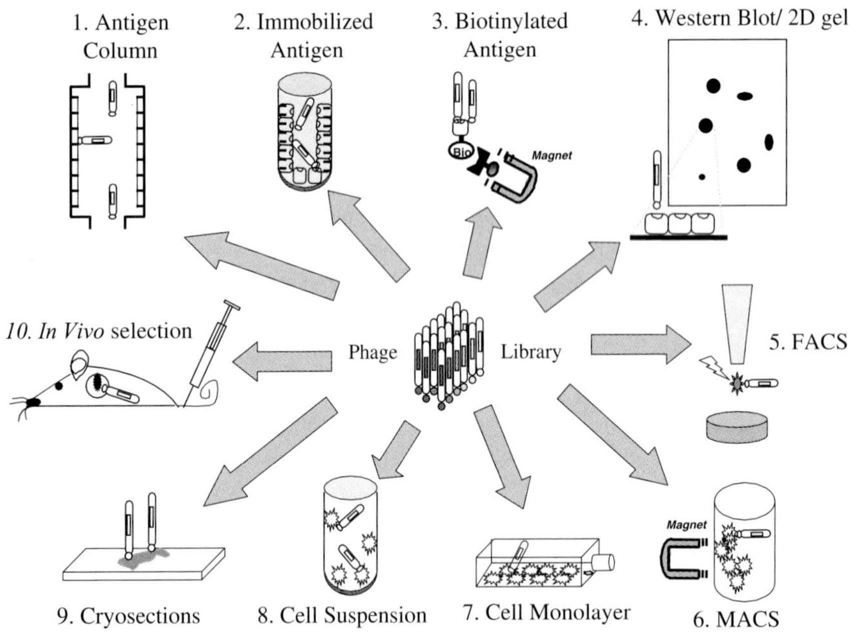

Figure 10.3. Selection strategies for obtaining specific phage ligands.(1) panning on an antigen column; (2) panning on an antigen absorbed onto a solid support; (3) to avoid conformational changes of the antigen during coating, selection on biotinylated antigen in solution is preferable, bound and unbound phage antibodies are separated using streptavidin-coated magnetic beads; (4) selection on proteins isolated by electrophoresis; (5) selection and subtraction via FACS; (6) selection and subtraction via MACS; (7) selection on cell monolayers; (8) selection on cells in suspension; (9) selection on tissues on cryosections; (10) *In vivo* selections.

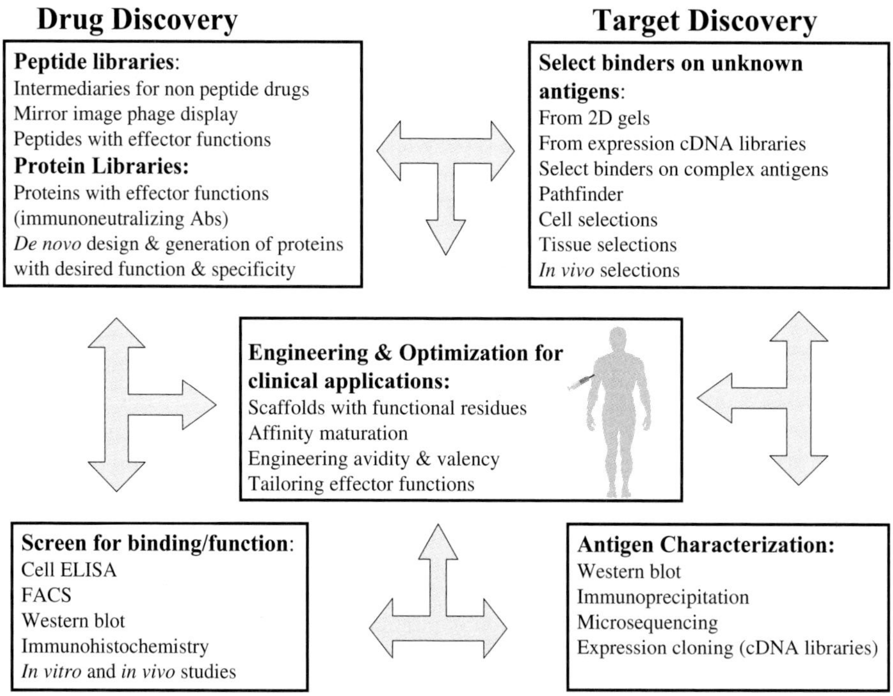

Figure 10.4. Application of phage display in drug and target discovery.

10.2.2.2 Peptide Display

Since many proteins exert their biological activity through relatively small regions of their folded surfaces, their interactions may in principle be reproduced by much smaller peptides that retain these localized bioactive surfaces. Peptides can provide information about the molecular interactions of protein binding sites [18], can be used as ligands for targeting purposes [19], are useful intermediates in the development and design of non-peptide drugs [20], and can act as small molecule drugs with biological activity [21]. Peptides are readily synthesized and are characterized by rapid tissue penetration having potentially improved properties for certain *in vivo* applications [22] (Figure 10.4).

The first reported random peptide libraries were constructed in 1990, and fused to pIII [2–4]. These where followed by random libraries fused to pVIII [23]. The phage peptide libraries are enriched on antigen, and clones analysed by sequencing to identify possible consensus residues. Once consensus amino acid sequences are found among selected peptides, the encoded peptides may be synthesized to confirm their specificity and determine affinity for the target. The peptides themselves can be used as specific binding ligands in research or diagnosis, for drug development, or they can form the starting point for the synthesis of smaller bioactive molecules.

Phage peptide libraries may contain either constrained [24] or linear peptides. For the former, the peptide is presented in a structurally constrained, thus more rigid form, for example using cysteine bridging. Linear peptides can adopt more conformations in solution; they are particularly useful for determining the consensus for the recognition and modification sites of proteases, kinases etc. [25]. Constrained peptides often produce higher affinity ligands [26,27].

In general, relatively large peptides can be displayed by fusion to pIII, whereas for pVIII display, the maximum size of the peptide is usually limited to around 15 amino acid residues. This size restriction is less pronounced when using phagemids, in which wild-type coat protein competes with the fusion product for insertion into the phage coat. In the case of pVIII fusion, polyvalent display is achieved, since over 2000 copies of the coat protein are incorporated into the viral particle. With phagemid systems lower numbers of fusion proteins per phage particle are obtained, but the phage coat protein may be engineered to allow more efficient display [28]. Generally polyvalent display leads to selection of low (micromolar) affinity peptide binders [3]. On the other hand 'monovalent' display, obtained by using pIII as the fusion partner in a phagemid system, allows the selection of higher affinity (nanomolar) variants [29].

Libraries can also be made from naturally occurring peptides (e.g. melanocortin [30] and somatostatin [31]) and their variants, or be derived from protein-fragments. The former have been employed mainly for the identification of peptide variants with improved characteristics, and the latter for the identification of antigenic determinants present on targets, for example by selecting for binders to monoclonal antibodies specific for an antigen-fragment peptide library.

Naturally occurring peptides can be altered in size and structure, and randomization of residues can be introduced into libraries for the identification of peptides with improved characteristics or improved binding specificity. Such an approach has been used to select receptor-specific variants of atrial natriuretic peptide (ANP) with improved expression in *E. coli* and specificity for just one form of the ANP receptors [29].

Phage displayed peptide libraries have been successfully employed for the identification of peptide ligands with a variety of applications. Recent reviews describe the developments in the generation and screening of peptide libraries [6], their application in the identification of receptor ligands [32] and in drug development [33].

10.2.2.3 Antibody Display

One of the widest and most powerful applications of phage display technology has been in the generation of recombinant antibodies. The first protein to be successfully displayed on phage was single-chain antibody variable fragment (scFv) [5]. This was achieved by fusing the coding sequence of the antibody variable regions to the amino terminus of the phage coat protein pIII (Figure 10.1).

Large and highly diverse combinatorial antibody repertoires can be constructed on phage, displaying antibodies with diverse antigen combining sites, generated by the PCR amplification of antibody variable (V) genes [7,34]. Antibodies to any chosen antigen can be selected from universal libraries made from non-immune sources of antibody genes. Human antibodies may be generated from such non-immune 'naïve' libraries built with the V-genes of B cells from human donors [35], or from synthetic libraries built with *in vitro* rearranged variable germline genes [36]. Alternatively, for some applications, antigen-biased antibody repertoires can be created using V-genes from immune sources (reviewed by Hoogenboom *et al.*) [7].

Today several large single pot libraries with over 10^{10} independent clones have been generated. From these libraries high affinity antibodies against virtually any antigen can be isolated [7]. The size of the library significantly improves the likelihood of identifying antibody fragments with high specificity and high affinity due to the sampling of a larger diversity.

Phage displayed antibody libraries have been successfully used to isolate antibodies against a variety of antigens such as self-antigens [37], haptens [38], carbohydrates [39], DNA [40] and RNA [41]. Selected antibodies may also have remarkable specificity for the antigen, for example, antibodies which are able to distinguish between two variant forms of a native antigen which differ in a single amino acid, have been selected [42].

The selected antibody fragments have been used in many applications, varying from antibody-based biochemical assays to *in vivo* imaging of tumours. The relative ease with which they form multimers and fusion proteins, and the possibility of high levels of expression of these molecules in a variety of different hosts, makes them ideal protein-based diagnostic and possibly future therapeutic reagents (reviewed by Hudson and Kortt) [43] (Figure 10.4).

10.2.2.4 Protein Scaffolds

Until the advent of the phage display era, antibodies were traditionally the sole source of antigen binding molecules. By variation of surface residues of a protein, it is in principle possible to build a library of proteins each with a different topological surface. From these libraries, variants with a particular binding specificity can be selected using phage display. Indeed, protein display libraries based on proteins with very different folding patterns and thus varying structural frameworks, have been constructed. Ideally protein scaffolds to be used in drug targeting should: (1) be small single domains, (2) be of human origin, (3) have predictable pharmacokinetics (for human therapeutics), (4) have available sites or surfaces for the introduction or transfer of functional sites, and, (5) have a sufficiently large antigen-binding surface for affinity or specificity maturation. They should also be (6) suitable for engineering into multivalent, multi-specific molecules, or molecules with effector functions, and (7) yield high level expression in multiple hosts with the capacity to fold *in vitro* and *in vivo.* Although antibodies fulfil most of these requirements, there are applications where other proteins may be more suitable.

Scaffold libraries have been created for several applications in affinity chromatography, or for the generation of ligands to disease-related targets. The most developed scaffold is a two alpha helix-containing variant of protein A from *Staphylococcus*, protein Z. so called 'Affibodies' [44], showing selective binding to respiratory syncytial virus (RSV) G protein, have been selected [45]. Some libraries have been used for the transfer of bioactive peptides or functional residues from one protein to a new protein scaffold. In our laboratory, human cytotoxic T lymphocyte associated protein-4 (CTLA-4) has been used as a protein scaffold to display the 14mer somatostatin hormone, or RGD sequence-containing peptides. This has produced a series of ligands to the somatostatin receptor and $\alpha v\beta 3$ integrin respectively, the latter being important in angiogenesis (see Chapters 9 and 7), with possible application as human therapeutics [46,46a].

10.2.2.5 Engineering Proteins with Phage Libraries

Phage display technology offers a valuable tool for the *in vitro* evolution of molecules. By targeted or random mutagenesis, libraries containing variants of a protein can be constructed, and selection of these libraries under controlled conditions can result in the generation of variants with improved characteristics. A large variety of proteins have been successfully displayed on phage with the purpose of selecting desired variants. Hormone variants with higher affinity for their receptors [47], with similar biopotency and reduced size [48], or with improved specificity for one of its receptors [49] have been selected from phage libraries. Proteins with higher affinity or specificity for their receptor [50] or enzymes with higher or new catalytic activities [51,52] have also been generated.

Minimizing proteins into significantly smaller polypeptides has also been achieved via both rational design processes and selection from vast combinatorial phage display libraries. Such 'mini-proteins' represent a potential intermediate step toward the development of drugs targeted to protein–protein interfaces [53]. Finally, binding ligands with desired function and specificity can also be generated using a combination of phage display technology and semi-rational protein design. In this way peptides that bind to the extracellular domain of the fibroblast growth factor (FGF) receptor, were selected from a peptide library. Guided by the knowledge that agonist activity of FGF is conferred by the ability to cause receptor dimerization, the selected peptide was expressed as a fused protein coupled to domains that undergo spontaneous dimer formation, enhancing binding affinity to the receptor.

Furthermore, one of these fusion proteins also possessed agonist activity *in vitro*. This has generated a small protein with no homology to FGF which binds to the FGF receptor reproducing the biological activity of FGF [54].

10.2.2.6 cDNA Expression Libraries

The display of cDNA libraries was first achieved on pIII by an indirect display method, based on the interaction between the Jun and Fos leucine zippers [55,56]. This approach was successfully applied for the selection of allergenic proteins from *Aspergillus fumigatus* using serum from affected patients [57]. An alternative approach was later described where cDNA fusion occurred at the C-teminal domain of the minor coat protein pVI [58]. cDNA libraries are being used for the screening and selective isolation of genes by specific gene-product/ligand interaction. For the identification of disease-related targets a cDNA fragment display library can be selected against homogeneous ligands [59], such as natural ligands or ligands selected from antibody or peptide libraries against novel targets, or heterogeneous ligands such as patient sera or polyclonal antibodies. The latter application has been validated by our group for the selection of products from a colorectal carcinoma cDNA library using both homogeneous and heterogeneous ligands [10]. Display of cDNA or cDNA fragment libraries on lambda phage is complementary to the filamentous phage display libraries. With M13 phage there will be a bias to display naturally secreted cDNAs or extracellular domains (intracellularly expressed cDNAs can also be expressed on phage, e.g. zinc finger proteins). Phage lambda display has the advantage of cytoplasmic expression of the cDNAs and higher display levels [60]. Selection of cDNA or cDNA fragments using cells or tissues, and the rapid cloning of cDNAs expressing proteins specific for these ligands should facilitate the search for novel targets and biologically important antigens (Figure 10.4).

10.3 Generation of Ligands Amenable for Targeting

In the following sections, procedures and approaches for retrieving ligands specific for well-defined targets will be discussed, followed by discussions relating to the application of refined selection methods for the identification of novel targets.

10.3.1 Selection of Ligands to Defined Targets

Most phage selection strategies rely on the availability of purified or recombinant antigen. This allows common selection strategies such as biopanning on immobilized antigen coated onto solid supports [35], or on columns [38]. However, coating/immobilization of the antigen may alter the conformational integrity of the antigen and as a consequence, antibodies that do not recognize the native form may be selected. To circumvent this problem, indirect coating methods can be applied. Antibodies may be used to capture the antigen, and non-specific Ig of the same species and isotype may be added during the selection procedure to avoid selection of antibodies to the capturing agent [61]. However, the most frequently used method today, is based on selection with labelled soluble antigen, such as biotinylated antigen, followed by capture on a biotin-binding surface [62]. After incubation of phage with the biotinylated antigen, unbound phage can be removed and the antigen-bound phage can be retrieved with streptavidin-coated paramagnetic beads and a magnet. This selection method allows for the accurate control of selection parameters such as antigen concentration, which may result in selection according to the affinity for the antigen, e.g. decreasing the antigen concentration to favour the selection of high affinity antibodies [63]. Phage libraries can also be incubated with biotinylated antigen and diluted into excess unlabelled antigen for variable times prior to capture on streptavidin-coated magnetic beads. This allows for selection on the basis of the kinetics of dissociation (off-rate) from the antigen; e.g. longer incubation times with unlabelled antigen will favour selection of slow off-rate antibodies [62].

Phage display allows further variations to favour the selection of ligands directed towards specific epitopes present in the target antigen. Similar to the presence of immunodominant epitopes *in vivo*, selection-dominant epitopes exist *in vitro* [64]. Ligands directed to immunodominant epitopes can be included during the panning procedure. By such 'epitope blocking panning', a broader range of specific antibodies from combination libraries can be rescued. This approach has been used to isolate neutralizing human antibodies to weakly immunogenic epitopes of human immunodeficiency virus 1 (HIV-1) gp120 [65] and RSV [66].

When the target molecules are not easily purified, or are cell surface antigens that require the presence of a lipid bilayer for maintaining their native form, selection on complex antigen mixtures is necessary. To avoid selection of non-relevant phage and to select antigen or, in the case of cell selections, to obtain cell type specific phage ligands, depletion and/or subtraction methods may be employed. Depletion is achieved by the incubation of the phage library with non-target cell population(s), which do not display the target antigen, previous to the incubation with the target cell population. Subtraction, on the other hand, is performed when the phage library is incubated with both target and non-target cell populations simultaneously, and the target cell population with the bound phage is subsequently isolated. Competitive elution with an antibody or the antigen itself [67] can be used to elute only those antibodies homing to the desired antigen. Selection with alternating different sources of antigen to select only those ligands that bind to antigen(s) which are common to all sources [68], may also be employed. Alternatively, a method named 'pathfinder' has been devised for guided selection on complex antigen sources, using ligands directed towards the target molecule. These ligands, conjugated to horseradish peroxidase (HRP) can be used as pathfinders during the panning procedure. In the presence of biotin tyramine these molecules catalyse biotinylation of phage binding in close proximity to the target antigen, allowing specific recovery of 'tagged' phage from the total population using streptavidin. In this way, phage binding to the target itself, or in its immediate proximity, are selectively recovered [69]. This technique has been applied for the selection of phage antibodies against antigens including carcinoembryonic antigen (CEA), E- and P-selectins, and for the selection of novel antibodies which recognize immobilized purified antigen [70]. This technique could also be applied in the future for the selection of ligands to novel targets such as unknown receptors of known ligands or any molecules involved in other protein–protein interactions (Figure 10.4).

10.3.2 Phage Display for Target Identification

10.3.2.1 *In Vitro* Selections on Complex Antigens

We have seen the versatility of phage display technology for the selection of ligands directed to known target molecules. However, one of the major advantages of this technology is based on its applicability for the selection of ligands directed to novel targets present on certain tissues, cell types, or cells in a specific stage of differentiation or in disease-induced states. The selection methods developed for defined targets may now be used to detect, by virtue of the phage, novel ligands or epitopes in the antigenic mixture. In addition, specific protocols have been developed to direct the selection process towards the isolation of tissue- or cell type-specific ligands (Figure 10.2).

In principle, whole live cells or crude preparations derived from cells or tissues can be used as an antigen source to identify novel targets. For example, measles virus-infected cell lysates have been used to select antibodies which reacted specifically with measles virus-infected cells [71]. Protein mixtures from cells or tissues may first be separated to direct the selection to specific cell compartments or antigens. By SDS-PAGE, the antigen(s) of interest can be identified and isolated, e.g. antigens detected in 2-D gels of cell lysates of target cells but not in lysates of non-relevant cells. The target antigen(s) can then be blotted onto a membrane, or eluted from the gel for later use as an antigen source for selection. This principle has been proven by the selection of antibodies recognizing the native form of the ED-B domain of fibronectin, a marker of angiogenesis [72].

Selection techniques are often adapted to conditions where antigenic expression is as close as possible to the *in vivo* situation. Antigen expression on cells is highly dependent on the cellular environment and on cell–cell and cell–matrix interactions. In order to maintain some cell–cell and partial cell–matrix interactions, cell panning can be carried out on *in vitro* cultured cell monolayers, where culture conditions can also be further modified in order to mimic the antigenic expression *in vivo*. Cell panning can also be carried out with cells in solution. Unbound phage can be removed after the cells and bound phage are separated from the supernatant by centrifugation or by magnetic retrieval with magnetically labelled beads which bind to the cells (reviewed by Mutuberria *et al.*) [73].

A highly pure cell population, with its specific bound phages, can be isolated from complex cell mixtures by the use of magnetic activated cell sorting (MACS) [74] or by fluorescent activated cell sorting (FACS) [75,76]. Cells bearing the antigens of interest are magnetically or fluorescently labelled via cell-specific antibodies. Following incubation of the cell mixture with the phage library, the antigen-positive cell population and bound phage is retrieved from the antigen-negative cells using MACS or FACS. When the phage library is incubated with the cell mixture prior to target cell isolation (subtraction), selection of target cell-specific antibodies may be favoured, since negative non-target cells will compete for antibodies that bind to common antigens. MACS selections on a model system has shown clear advantages over other cell selection methods such as selections on monolayers [73]. This may be attributed to the improved interaction between cells and phage, the efficient elimination of non-relevant phage by washing the cells immobilized on a magnetic column, and the high cell viability and integrity during this mild selection procedure. We have applied this selection method using a large naive antibody library on mildly fixed endothelial cells immobilized on a magnetic separation column. Angiogenic factors and tumour-conditioned media were used to induce tumour angiogenesis-associated antigen expression on cultured human umbilical vein endothelial cells. A large number of endothelial cell binders have been selected by this approach, most of which bind preferentially to angiogenic vasculature, and therefore tumour vasculature. A further advantage of the MACS selection approach is that it is based on the isolation of positive target cells on magnetic columns, from which irrelevant phage can be simultaneously washed off, using simple, efficient and gentle methods.

As an alternative to selections on cells, other antigen sources, which better maintain the *in vivo* antigen expression profile and that allow appropriate *in vitro* selection procedures, are available. Selection on tissue cryosections may result in antibodies directed to intra- and extracellular antigens on any cell type present in the tissue section, as well as antibodies binding to matrix components [77].

Model selections, in which phage antibodies with defined specificity are mixed with non-specific phage and the enrichment and yield for a selection procedure is measured, provides a rapid experimental approach for studying such complex selection procedures. In order to compare different selection approaches on complex preparations, and to determine the best selection parameters for each approach, an extensive study of various *in vitro* and *in vivo* model selections was recently carried out by our group [73].

10.3.2.2 *In Vivo* Selections and Selections for Functional Activity

In an original approach named '*in vivo* selection' [19], phage capable of selective homing to different tissues, such as the vasculature of lung, skin, and pancreas [78], were recovered from a phage display peptide library following intravenous administration of the library to a living mouse. *In vivo* selections have also been carried out in animals with a human tumour xenograft. Phage specifically bound to the murine tumour vasculature could then be recovered from the tumour tissue and amplified to yield tumour-specific endothelial cell binding peptides [79]. So far *in vivo* selections strategies have been limited to the selection of peptides directed to murine endothelial cell markers. This technique is producing ligands that may have extensive application in vasculature targeting. This was demonstrated with *in vivo* selected anti-integrin peptides coupled to the anti-cancer drug doxorubicin. The targeted drug–peptide complex enhanced the efficacy of the drug against human breast cancer xenografts in nude mice with reduced toxicity [79].

All the selection methods described are applicable for the generation of ligands with biological activity. Often ligands selected by conventional techniques can be screened for a fortuitous biological function: immunoneutralizing antibodies, receptor agonists or antagonists can be identified from a pool of selected ligands when screened for biological function. Alternatively, selections may be targeted towards biologically active sites of the antigens.

Some selection methods have been specifically designed for the selection of ligands with a particular biological effector function. Such 'selection for function' has been used for the retrieval of catalytic antibodies [80] and cell-internalizing phage antibodies [81], the latter being highly useful molecules for the delivery of drugs, toxins, or DNA into the cytoplasm of mammalian cells. Selection of antibody or peptide ligands for their function may, in the future, be directed towards cell survival or killing upon ligand binding, cell transfection, inhibition of cell surface molecules such as drug transport molecules and inhibition of viral entry and receptor cross linking or triggering [7].

10.4 Engineering and Optimization for Targeting

Once the specific ligand has been selected, large arrays of possibilities are available in order to reshape the ligand to obtain the best targeting results. In drug targeting applications, pharmacokinetics, biodistribution, penetration, and bioactivity are strictly governed by the characteristics of the ligand.

Certain therapeutic applications require high affinity ligands. Ligands such as peptides with intrinsic low affinity for target antigens can be affinity matured. Secondary libraries of the selected peptide can be created by selectively incorporating mutations, and variants with higher affinities can be selected. Often the display format will change from multi- to monovalent display to aid a genuine affinity selection [82].

Affinity maturation of antibodies has been achieved by the introduction of diversity into the V-genes, which then creates diversity within the antigen binding sites. This secondary library is then subjected to a selection that will enrich high affinity variants. More or less random diversity may be introduced by altering variable domain pairings in a process called chain shuffling [38], by error prone PCR [62], by using bacterial mutating strains [83], or by DNA shuffling [84]. Alternatively, the antigen binding region may be mutated using oligonucleotide-directed mutagenesis or by the introduction, with limited frequency, of random mutations using PCR. The highest affinity increases in antibodies have been achieved when directing the mutations to the complementary determining region 3 (CDR3), yielding in some cases picomolar affinity antibodies [85,86]. Similar methodologies can be used with different ligands. In addition, it is not only library-derived ligands which can be affinity matured. Hybridoma-derived monoclonal antibodies, as well as other proteins or protein fragments may be affinity matured by some of the methods mentioned above. Affinity maturation of the ligand can also result in high affinity molecules with improved *in vivo* biodistribution. This was demonstrated by the anti-human CEA specific scFv, isolated and affinity matured by phage display technology. Both the original and affinity-matured antibody showed targeting to tumour xenografts. However, although no difference was detected in tumour uptake, the affinity-matured antibody with improved off-rates, was retained in the tumour for a longer period. [87]. Occasionally, maximizing the affinity *in vitro* may result in modification of the antibody specificity which could complicate the use of the resulting ligands for therapeutic applications [88].

The possibility of specifically tailoring antigen-binding properties can also be directed towards the engineering of avidity and valency. Antibody fragments are known to have more rapid tissue penetration than full antibody molecules, as demonstrated in tumour targeting [89]. Phage display has made easier the use of recombinant DNA technology for the development of multi-specific and multivalent molecules, previously generated using non-recombinant methods. Dimeric antibody fragments, or 'diabodies,' can be designed for bivalent or bispecific interactions [90]. Phage libraries displaying bivalent bispecific antibody fragments have also been constructed. Diabody libraries enable the selection of the most appropriate bispecific molecules, with the highest affinity for binding, epitope recognition and stability [91]. Multi-specific ligands can be used to direct targeted drugs to one or more cellular antigens which may be present in either a particular cell type or diverse cell types, or to stromal or secreted proteins found in the target tissue. Multivalent ligands have been exploited mainly for immunotherapy, in the stimulation of cytotoxic pathways *in vivo* to treat cancer [92]. The use of phage display in immunotherapy has been recently reviewed [93].

Ligands can be tailored to improve their *in vivo* stability. To improve the *in vivo* thermal stability of the tumour targeting monoclonal anti-epithelial glycoprotein-2 (EGP-2) antibody, the antigen binding residues were grafted onto the framework of a highly stable scFv resulting in increased serum stability at 37°C [94]. The stability of peptides for *in vivo* use has been approached in a different manner, using selection of peptide libraries on the D-form of the antigen [95]. This technique named 'mirror image phage display' has recently been employed in the generation of D-peptide inhibitors of HIV-1. Peptides directed against the mirror image (chemically synthesized with D-amino acids) of a pocket of a viral protein involved in viral entry, were first selected. The D-peptide mirror images of the selected consensus sequences, binding to the natural 'L' form of the target, were then chemically synthesized. Because these D-peptides are not subject to degradation by naturally occurring enzymes, they can be used as the starting point for the development of new drugs or as effective orally administered pharmaceuticals [96].

A phage display-selected scFv directed against CEA has been designed to have effector functions and tested for its potential in antibody-directed enzyme pro-drug therapy (ADEPT). A biologically active recombinant fusion protein containing anti-CEA scFv and the enzyme pseudomonas carboxypeptidase (CPG2) has been produced as a recombinant protein in *E. coli* and was shown to localize in human colon tumour xenografts. The tumour targeting properties combined with the biological properties of the enzyme can be exploited to induce tumour-specific pro-drug activation, in cases where a non-toxic pro-drug is converted by the action of the targeted enzyme into a highly cytotoxic drug [97].

Finally, phage display may help to address the issue of immunogenicity, particularly of non-human antibodies. Antibody humanization is often achieved by grafting CDR-loops into human antibody fragments. However, humanization often requires the replacement of key residues involved in antigen binding in the framework regions, with corresponding residues from the parent non-human antibody. A phage display method that allows selection of framework mutations which increase the binding of humanized antibodies has been described for the humanization of the anti-vascular endothelial growth factor (VEGF) murine antibody A4.6.1 [98]. The humanized version of this antibody is currently in phase II clinical trials to evaluate the inhibitory effect of this antibody-drug on tumour growth and neovascularization. Furthermore, by guided selection [99], a rodent antibody may be rebuilt into a fully human antibody. In two consecutive rounds of chain shuffling, the rodent antibody genes are replaced by human genes, which will mediate binding to a highly similar if not identical epitope on the antigen, differing only in their chemistry of interaction [100].

10.5 Discovery of Novel Therapeutics Using Phage Display Technology

Although phage display technology is becoming a widespread research tool, few ligands generated by this technology have reached clinical trials. From the large array of ligands generated by phage display with possible therapeutic applications, there are five antibodies selected from phage libraries currently undergoing clinical trials as drug candidates. Preliminary data with a fully human anti-tumour necrosis factor alpha (TNFα)-neutralizing monoclonal antibody for the treatment of rheumatoid arthritis, demonstrated that the antibody was safe and effective in early clinical trials [101]. Other phage library-derived antibodies that are undergoing clinical trials include antibodies to interleukin 2 (IL-2) for the treatment of autoimmune and inflammatory disorders, an anti-transforming growth factor beta-1 (TGFβ1) antibody as an anti-fibrotic drug, and an anti-TGFβ2 antibody for the treatment of proliferative vitreoretinopathy and prevention of scarring in the eye following glaucoma therapy. A large series of ligands which may have possible therapeutic application are undergoing further characterization and optimization.

Cancer has become a major target in the application of this technology, and human antibodies against tumour antigens such as CEA, EGP-2 and Mucin-1 (MUC-1) are already available. Also a series of peptides and antibodies directed against angiogenesis-related markers such as basic fibroblast growth factor (bFGF), VEGFs and their receptors, tumour-associated fibronectin, tenascin isoforms, integrins or metalloproteases, have been selected for their potential application in tumour targeting, inhibition of tumour growth and in other angiogenesis-related diseases (see Chapter 9 for drug targeting strategies aimed at angiogenesis-related molecules).

Identification of cell receptor ligands has generated a series of molecules with excellent receptor specificity and occasionally with desired effector functions. This is the case for peptide antagonists of the human estrogen receptor, or cyclic peptides capable of activating the erythropoietin receptors (EPOR) isolated from phage display libraries. Other ligands with effector functions include thrombin inhibitors, viral inhibitors and even intracellular targets such as a reverse transcriptase inhibitory antibody. Table 10.1 lists a number of ligands, which may in the future have widespread application as therapeutic agents.

Table 10.1. Example of phage display ligands with therapeutic applications.

Name of ligand	Type of ligand	Target and applicability	Reference
D2E7	scFv	TNFα/Rheumatoid arthritis	[101]
CAT-152	scFv	TGFβ2/Glaucoma surgery, proliferative vitreo retinopathy	http://www.catplc.co.uk
CAT-192	scFv	TGFβ1/Antifibrogenic treatment	http://www.catplc.co.uk
J695	scFv	IL-2/Autoimmunity, Inflammation	http://www.catplc.co.uk
Anti CEA antibodies		CEA/Tumour targeting	Reviewed in reference [102]
UBS-54	scFv	Ep-CAM (EGP-2)/Colorectal carcinoma	[103,104]
L19	scFv	Fibronectin ED-B/Angiogenesis (ocular neovasculature and tumour targeting)	[105,106,107]
TN11	scFv	Tenascin C isoform/High grade astrocytoma	[108]
10A PH1	scFv Fab	Mucin-1/Adenocarcinoma	[109] [110]
Human Fab Y0317	Fab	VEGF/Treatment of macular degeneration, angiogenesis inhibition and tumour growth.	[111] http://www.gene.com/Pipeline/index.html
Anti GP 41 D-pep	Peptide	HIV GP41/Inhibition of HIV-1 infection	[96]
RGD-Doxorubicin	Peptide	αvβ3 integrin/Targeting angiogenic vasculature	[79]
PLAEIDGIELTY	Peptide	α9β1 integrin/Targeting lung epithelia (cystic fibrosis)	[112]
REA18 (rNAP)	Peptide	ANP-A receptor / Induction of natriuresis and diuresis. Treatment of acute renal or heart failure	[29,49]

10.6 Conclusions

Phage display has become the most efficient and effective method developed to date for rapidly identifying peptides, antibodies and other proteins that bind to molecular targets. Some of these ligands have biological effector functions or are candidates for ligand-based drug targeting. Identification of the antigens targeted by phage-displayed ligands has been simplified by this technology, by the use of DNA display libraries and by coupling gene isolation and clone identification with affinity selection, facilitating the search for novel targets and biologically important antigens. Selection for function allows the retrieval of ligands with effector functions for use in the generation of drugs. Selection for cellular internalization has opened a new door for the delivery of macromolecular constructs and coupled drugs into the cytoplasm of mammalian cells.

The physical bond between the selected ligand and its encoding gene allows further manipulation of the ligands to obtain optimal affinity and avidity, size, and valency. Furthermore, the coupling of a drug to the targeting ligands can be readily engineered.

The design features required for an optimally-effective drug and/or targeting agent can be readily obtained by phage display technology. Many features give phage display technology clear advantages over conventional approaches for the generation of reagents for drug targeting purposes. These include diversity in protein type and sequence space of combinatorial phage libraries, the power of filamentous phage-based selection, the possibilities of genetic manipulation to generate more effective ligands and specific effector functions, and the adaptability of the system for the production of therapeutic ligands derived from the libraries.

In the near future phage display technology may have further application in the rapid analysis and comparison of protein profiles of large proteomes, such as human cells and tissue. Arrays of phage-displayed proteins such as antibodies or peptides, which selectively bind to proteins from a complex mixture can be generated. Proteins absorbed by the antibody arrays can then be analysed by mass spectrometry. Such technologies will be especially useful for identifying differences between cell or tissue samples such as healthy versus diseased states, and may lead to the identification of drug targets. Once a protein of interest has been identified, its corresponding antibody or peptide ligand can be retrieved and used to monitor protein expression or modification in a range of cell or tissue samples, and can also be used for cloning the target antigen.

Phage display has become a powerful method for the generation of protein-based binding and biologically active reagents. In the forthcoming years an extensive application of this technique is predicted for the development of drugs and drug targeting entities.

References

[1] Smith GP, *Science* **1985,** *228,* 1315–1317.
[2] Devlin JJ, Panganiban LC, Devlin PE, *Science* **1990,** *249*, 404–406.
[3] Cwirla SE, Peters EA, Barrett RW, Dower WJ, *Proc. Natl Acad. Sci. USA* **1990**, *87*, 6378–6382.
[4] Scott JK, Smith GP, *Science* **1990,** *249*, 386–390.
[5] McCafferty J, Griffiths AD, Winter G, Chiswell DJ, *Nature* **1990,** *348*, 552–554.
[6] Cesareni G, Castagnoli L, Cestra G, *Comb. Chem. High Throughput Screen* **1999,** *2*, 1–17.

[7] Hoogenboom HR, de Bruine AP, Hufton SE, Hoet RM, Arends JW, Roovers RC, *Immunotechnology* **1998,** *4*, 1–20.
[8] Soumillion P, Jespers L, Bouchet M, Marchand-Brynaert J, Sartiaux P, Fastrez J, *Appl. Biochem. Biotechnol.* **1994,** *47*, 175–190.
[9] Nygren PA, Uhlen M, *Curr. Opin. Struct. Biol.* **1997,** *7*, 463–469.
[10] Hufton SE, Moerkerk PT, Meulemans EV, de Bruine A, Arends J, Hoogenboom HR, *J. Immunol. Methods* **1999,** *231*, 39–51.
[11] Dennis MS, Lazarus RA, *J. Biol. Chem.* **1994,** *269*, 22129–22136,
[12] Wu H, Yang WP, Barbas CF, III, *Proc. Natl Acad. Sci. USA* **1995,** *92*, 344–348.
[13] Gram H, Strittmatter U, Lorenz M, Gluck D, Zenke G, *J. Immunol. Methods* **1993,** *161*, 169–176.
[14] Scarselli E, Esposito G, Traboni C, *FEBS Lett.* **1993,** *329*, 223–226.
[15] Garrard LJ, Yang M, O'Connell MP, Kelley RF, Henner DJ, *BioTechnology* **1991,** *9*, 1373–1377.
[16] Bass S, Greene R, Wells JA, *Proteins* **1990,** *8*, 309–314.
[17] Hoogenboom HR, Griffiths AD, Johnson KS, Chiswell DJ, Hudson P, Winter G, *Nucleic Acids Res.* **1991,** *19*, 4133–4137.
[18] Geysen HM, Rodda SJ, Mason TJ, Tribbick G, Schoofs PG, *J. Immunol. Methods* **1987,** *102*, 259–274.
[19] Pasqualini R, Ruoslahti E, *Nature* **1996,** *380*, 364–366.
[20] Blackburn BK, Lee A, Baier M, Kohl B, Olivero AG, Matamoros R, Robarge KD, McDowell RS, *J. Med. Chem.* **1997,** *40*, 717–729.
[21] Buckley CD, Pilling D, Henriquez NV, Parsonage G, Threlfall K, Scheel-Toellner D, Simmons DL, Akbar AN, Lord JM, Salmon M, *Nature* **1999,** *397*, 534–539.
[22] Lister-James J, Moyer BR, Dean T, *Q. J. Nucl. Med.* **1996,** *40*, 221–233.
[23] Felici F, Castagnoli L, Musacchio A, Jappelli R, Cesareni G, *J. Mol. Biol.* **1991,** *222*, 301–310.
[24] Luzzago A, Felici F, Tramontano A, Pessi A, Cortese R, *Gene* **1993,** *128*, 51–57.
[25] Matthews DJ, Wells JA, *Science* **1993,** *260*, 1113–1117.
[26] Ladner RC, *Trends Biotechnol.* **1995,** *13*, 426–430.
[27] Zang X, Yu Z, Chu YH, *Bioorg. Med. Chem. Lett.* **1998,** *8*, 2327–2332.
[28] Sidhu SS, Weiss GA, Wells JA, *J. Mol. Biol.* **2000,** *296*, 487–495.
[29] Cunningham BC, Lowe DG, Li B, Bennett BD, Wells JA, *EMBO J.* **1994,** *13*, 2508–2515.
[30] Szardenings M, Tornroth S, Mutulis F, Muceniece R, Keinanen K, Kuusinen A, Wikberg JE, *J. Biol. Chem.* **1997,** *272*, 27943–27948.
[31] Rousch M, Lutgerink JT, Coote J, de Bruine A, Arends JW, Hoogenboom HR, *Br. J. Pharmacol.* **1998,** *125*, 5–16.
[32] Koivunen E, Arap W, Rajotte D, Lahdenranta J, Pasqualini R, *J. Nucl. Med.* **1999,** *40*, 883–888.
[33] Lowman HB, *Annu. Rev. Biophys. Biomol. Struct.* **1997,** *26*, 401–424.
[34] Winter G, Griffiths AD, Hawkins RE, Hoogenboom HR, *Ann. Rev. Immunol.* **1994,** *12*, 433–455.
[35] Marks JD, Hoogenboom HR, Bonnert TP, McCafferty J, Griffiths AD, Winter G, *J. Mol. Biol.* **1991,** *222*, 581–597.
[36] Hoogenboom HR, Winter G, *J. Mol. Biol.* **1992,** *227*, 381–388.
[37] Griffiths AD, Malmqvist M, Marks JD, Bye JM, Embleton MJ, McCafferty J, Baier M, Holliger KP, Gorick BD, Hughes JN, Hoogenboom HR, Winter G, *EMBO J.* **1993,** *12*, 725–734.
[38] Clackson T, Hoogenboom HR, Griffiths AD, Winter G, *Nature* **1991,** *352*, 624–628.
[39] Deng SJ, MacKenzie CR, Sadowska J, Michniewicz J, Young NM, Bundle DR, Narang SA, *J. Biol. Chem.* **1994,** *269*, 9533–9538.
[40] Barbas SM, Ditzel HJ, Salonen EM, Yang WP, Silverman GJ, Burton DR, *Proc. Natl Acad. Sci. USA* **1995,** *92*, 2529–2533.
[41] Powers JE, Marchbank MT, Deutscher SL, *Nucleic Acids Symp. Ser.* **1995,** *33*, 240–243.
[42] Griffin HM, Ouwehand WH, *Blood* **1995,** *86*, 4430–4436.
[43] Hudson PJ, Kortt AA, *J. Immunol. Methods* **1999,** *231*, 177–189.
[44] Nord K, Gunneriusson E, Ringdahl J, Stahl S, Uhlen M, Nygren PA, *Nature Biotechnol.* **1997,** *15*, 772-777.
[45] Hansson M, Ringdahl J, Robert A, Power U, Goetsch L, Nguyen TN, Uhlen M, Stahl S, Nygren PA, *Immunotechnology* **1999,** *4*, 237–252.
[46] Nuttall SD, Rousch MJ, Irving RA, Hufton SE, Hoogenboom HR, Hudson PJ, *Proteins* **1999,** *36*, 217–227.
[46a] Hufton SE, van Neer N, van den Beuken T, Desmet J, Sablon E, Hoogenboom HR, *FEBS Lett.*, **2000**, *475*, 225–231.
[47] Lowman HB, Wells JA, *J. Mol. Biol.* **1993,** *234*, 564578.

[48] Li B, Tom JY, Oare D, Yen R, Fairbrother WJ, Wells JA, Cunningham BC, *Science* **1995,** *270*, 1657–1660.
[49] Jin H, Li B, Cunningham B, Tom J, Yang R, Sehl P, Thomas GR, Ko A, Oare D, Lowe DG, *J. Clin. Invest.* **1996,** *98*, 969–976.
[50] Ballinger MD, Jones JT, Lofgren JA, Fairbrother WJ, Akita RW, Sliwkowski MX, Wells JA, *J. Biol. Chem.* **1998,** *273*, 11675–11684.
[51] Hansson LO, Widersten M , Mannervik B, *Biochem. J.* **1999,** *344*, 93–100.
[52] Demartis S, Huber A, Viti F, Lozzi L, Giovannoni L, Neri P, Winter G, Neri D, *J. Mol. Biol.* **1999,** *286*, 617–633.
[53] Cunningham BC, Wells JA, *Curr. Opin. Struct. Biol.* **1997,** *7*, 457–462.
[54] Ballinger MD, Shyamala V, Forrest LD, Deuter-Reinhard M, Doyle LV, Wang J, Panganiban-Lustan L, Stratton JR, Apell G, Winter JA, Doyle MV, Rosenberg S, Kavanaugh WM, *Nature Biotechnol.* **1999,** *17*, 1199–1204.
[55] Crameri R, Suter M, *Gene* **1993,** *137*, 69–75.
[56] Crameri R, Jaussi R, Menz G, Blaser K, *Eur. J. Biochem.* **1994,** *226*, 53–58.
[57] Crameri R, Faith A, Hemmann S, Jaussi R, Ismail C, Menz G, Blaser K, *J. Exp. Med.* **1996,** *184*, 265–270.
[58] Jespers L, Messens JH, De Keyser A, Eeckhout D, Van den Brande J, Gandemans Y, Lauwereys MJ, Vlasuk GP, Stanssens PE, *Bio/Technol.* **1995,** *13*, 378–382.
[59] Sche PP, McKenzie KM, White JD, Austin DJ, *Chem. Biol.* **1999,** *6*, 707–716.
[60] Santi E, Capone S, Mennuni C, Lahm A, Tramontano A, Luzzago A, Nicosia A, *J. Mol. Biol.* **2000,** *296*, 497–508.
[61] Sanna PP, Williamson RA, De LA, Bloom FE, Burton DR, *Proc. Natl Acad. Sci. USA* **1995,** *92*, 6439–6443.
[62] Hawkins RE, Russell SJ, Winter G, *J. Mol. Biol.* **1992b,** *226*, 889–896.
[63] Hawkins RE, Russell SJ, Baier M, Winter G, *J. Mol. Biol.* **1993,** *234*, 958–964.
[64] Hoogenboom HR, Lutgerink JT, Pelsers MM, Rousch MJ, Coote J, Van Neer N, De Bruine A, Van Nieuwenhoven FA, Glatz JF, Arends JW, *Eur. J. Biochem.* **1999,** *260*, 774–784.
[65] Ditzel HJ, Binley JM, Moore JP, Sodroski J, Sullivan N, Sawyer LS, Hendry RM, Yang WP, Barbas CF, Burton DR, *J. Immunol.* **1995,** *154*, 893–906.
[66] Ping T, Tornetta MA, Ames RS, Bankosky BC, Griego S, Silverman C, Porter T, Moore G, Sweet RW, *J. Immunol.* **1996,** *157*, 772–780.
[67] Meulemans EV, Slobbe R, Wasterval P, Ramaekers FC, van EG, *J. Mol. Biol.* **1994,** *244*, 353–360.
[68] Andersen PS, Stryhn A, Hansen BE, Fugger L, Engberg J, Buus S, *Proc. Natl Acad. Sci. USA* **1996,** *93*, 1820–1824.
[69] Osbourn JK, Earnshaw JC, Johnson KS, Parmentier M, Timmermans V, McCafferty J, *Nature Biotechnol.* **1998,** *16*, 778–781.
[70] Osbourn JK, Derbyshire EJ, Vaughan TJ, Field AW, Johnson KS, *Immunotechnology* **1998,** *3*, 293–302.
[71] Burgoon MP, Williamson RA, Owens GP, Ghausi O, Bastidas RB, Burton DR, Gilden DH, *J. Neuroimmunol.* **1999,** *94*, 204–211.
[72] Pini A, Viti F, Santucci A, Carnemolla B, Zardi L, Neri P, Neri D, *J. Biol. Chem.* **1998,** *273*, 21769–21776.
[73] Mutuberria R, Hoogenboom HR, van der Linden E, de Bruine AP, Roovers RC, *J. Immunol. Methods* **1999,** *231*, 65–81.
[74] Siegel DL, Chang TY, Russell SL, Bunya VY, *J. Immunol. Methods* **1997,** *206*, 73–85.
[75] de Kruif J, Boel E, Logtenberg T, *J. Mol. Biol.* **1995,** *248*, 97–105.
[76] Palmer DB, George AJ, Ritter MA, *Immunology* **1997,** *91*, 473–478.
[77] Tordsson J, Abrahmsen L, Kalland T, Ljung C, Ingvar C, Brodin T, *J. Immunol. Methods* **1997,** *210*, 11–23.
[78] Rajotte D, Arap W, Hagedorn M, Koivunen E, Pasqualini R, Ruoslahti E, *J. Clin. Invest.* **1998,** *102*, 430–437.
[79] Arap W, Pasqualini R, Ruoslahti E, *Science* **1998,** *279*, 377–380.
[80] Janda KD, Lo CH, Li T, Barbas CF, Wirsching P, Lerner RA, *Proc. Natl Acad. Sci. USA* **1994,** *91*, 2532–2536.
[81] Becerril B, Poul MA, Marks JD, *Biochem. Biophys. Res. Commun.* **1999,** *255*, 386–393.
[82] Wrighton NC, Farrell FX, Chang R, Kashyap AK, Barbone FP, Mulcahy LS, Johnson DL, Barrett RW, Jolliffe LK, Dower WJ, *Science* **1996,** *273*, 458–464.
[83] Low NM, Holliger PH, Winter G, *J. Mol. Biol.* **1996,** *260*, 359–368.

[84] Crameri A, Cwirla S, Stemmer WP, *Nature Med.* **1996,** *2*, 100–102.
[85] Yang WP, Green K, Pinz SS, Briones AT, Burton DR, Barbas Cr, *J. Mol. Biol.* **1995,** *254*, 392–403.
[86] Schier R, McCall A, Adams GP, Marshall KW, Merritt H, Yim M, Crawford RS, Weiner LM, Marks C, Marks JD, *J. Mol. Biol.* **1996,** *263*, 551–567.
[87] Jackson H, Bacon L, Pedley RB, Derbyshire E, Field A, Osbourn J, Allen D, *Br. J. Cancer.* **1998,** *78*, 181–188.
[88] Casson LP, Manser T, *J. Exp. Med.* **1995,** *182*, 743–750.
[89] Chester KA, Begent RH, Robson L, Keep P, Pedley RB, Boden JA, Boxer G, Green A, Winter G, Cochet O, *Lancet* **1994,** *343*, 455–456.
[90] Holliger P, Prospero T, Winter G, *Proc. Natl Acad. Sci. USA* **1993,** *90*, 6444–6448.
[91] McGuinness BT, Walter G, FitzGerald K, Schuler P, Mahoney W, Duncan AR, Hoogenboom HR, *Nature Biotechnol.* **1996,** *14*, 1149–1154.
[92] Bauer S, Renner C, Juwana JP, Held G, Ohnesorge S, Gerlach K, Pfreundschuh M, *Cancer Res.* **1999,** *59*, 1961–1965.
[93] de Haard H, Henderikx P, Hoogenboom HR, *Adv. Drug Del. Rev.* **1998,** *31*, 5–31.
[94] Willuda J, Honegger A, Waibel R, Schubiger PA, Stahel R, Zangemeister-Wittke U, Pluckthun A, *Cancer Res.* **1999,** *59*, 5758–5767.
[95] Schumacher TN, Mayr LM, Minor Jr DL, Milhollen MA, Burgess MW, Kim PS, *Science* **1996,** *271*, 1854–1857.
[96] Eckert DM, Malashkevich VN, Hong LH, Carr PA, Kim PS, *Cell* **1999,** *99*, 103–115.
[97] Michael NP, Chester KA, Melton RG, Robson L, Nicholas W, Boden JA, Pedley RB, Begent RH, Sherwood RF, Minton NP, *Immunotechnology* **1996,** *2*, 47–57.
[98] Baca M, Presta LG, SJ OC, Wells JA, *J. Biol. Chem.* **1997,** *272*, 10678–10684.
[99] Jespers LS, Roberts A, Mahler SM, Winter G, Hoogenboom HR, *Biotechnology* **1994,** *12*, 899–903.
[100] Beiboer SH, Reurs A, Roovers RC, Arends JW, Whitelegg NR, Rees AR, Hoogenboom HR, ***J. Mol. Biol.*** **2000,** *296*, 833–849.
[101] Kempeni J, *Ann. Rheum. Dis.* **1999,** *58 (Suppl. 1)*, I70–I72.
[102] Mayer A, Chester KA, Flynn AA, Begent RH, *J. Immunol. Methods* **1999,** *231*, 261–273.
[103] Huls G, Heijnen IA, Cuomo E, van der Linden J, Boel E, van de Winkel JG, Logtenberg T, *Cancer Res.* **1999,** *59*, 5778–5784.
[104] Huls GA, Heijnen IA, Cuomo ME, Koningsberger JC, Wiegman L, Boel E, van der Vuurst de Vries AR, Loyson SA, Helfrich W, van Berge Henegouwen GP, van Meijer M, de Kruif J, Logtenberg T, *Nature Biotechnol.* **1999,** *17*, 276–281.
[105] Carnemolla B, Neri N, Castellani P, Veirana N, Neri G, Pini A, Winter G ,Zardi L, *Int. J. Cancer* **1996,** *68*, 397–405.
[106] Neri D, Carnemolla B, Nissim A, Leprini A, Querze G, Balza E, Pini A, Tarli L, Halin C, Neri P, Zardi L, Winter G, *Nature Biotechnol.* **1997,** *15*, 1271–1275.
[107] Birchler M, Viti F, Zardi L, Spiess B, Neri D, *Nature Biotechnol.* **1999,** *17*, 984–988.
[108] Carnemolla B, Castellani P, Ponassi M, Borsi L, Urbini S, Nicolo G, Dorcaratto A, Viale G, Winter G, Neri D, Zardi L, *Am. J. Pathol.* **1999,** *154*, 1345–1352.
[109] Henderikx P, Kandilogiannaki M, Petrarca C, von Mensdorff-Pouilly S, Hilgers JH, Krambovitis E, Arends JW, Hoogenboom HR, *Cancer Res.* **1998,** *58*, 4324–4332.
[110] de Haard HJ, van Neer N, Reurs A, Hufton SE, Roovers RC, Henderikx P, de Bruine AP, Arends JW, Hoogenboom HR, *J. Biol. Chem.* **1999,** *274*, 18218–18230.
[111] Chen Y, Wiesmann C, Fuh G, Li B, Christinger HW, McKay P, de Vos AM, Lowman HB, *J. Mol. Biol.* **1999,** *293*, 865–881.
[112] Schneider H, Harbottle RP, Yokosaki Y, Jost P, Coutelle C, *FEBS Lett.* **1999,** *458*, 329–332.

11 Development of Proteinaceous Drug Targeting Constructs Using Chemical and Recombinant DNA Approaches

Robbert J. Kok, Sigridur A. Ásgeirsdóttir, Willem R. Verweij

11.1 Introduction

Several techniques have been developed to selectively increase the accumulation of drugs in specific organs and tissues. One of these drug targeting techniques is the covalent conjugation of the drug to a macromolecule that accumulates at the target site. For this purpose, proteins as well as various other types of (polymeric) macromolecules can be used as drug carriers. This chapter will focus on the design and preparation of proteinaceous drug targeting structures.

Figure 11.1 shows the functional domains that are present in such a drug targeting structure. The core of the construct is the *carrier* backbone which, in the case of a proteinaceous construct, consists of a protein. Among other factors, the size of the carrier protein has a major influence on the distribution of the drug–carrier construct within the body. More specific targeting of the construct can be achieved by incorporation of site-directing ligands *(homing devices)* into the protein. By binding to specific receptors, the homing device is instrumental in delivering the construct to its target site. The homing device can be either a conjugated antennary group, or simply an intrinsic domain of the protein. Finally, an essential part of a drug

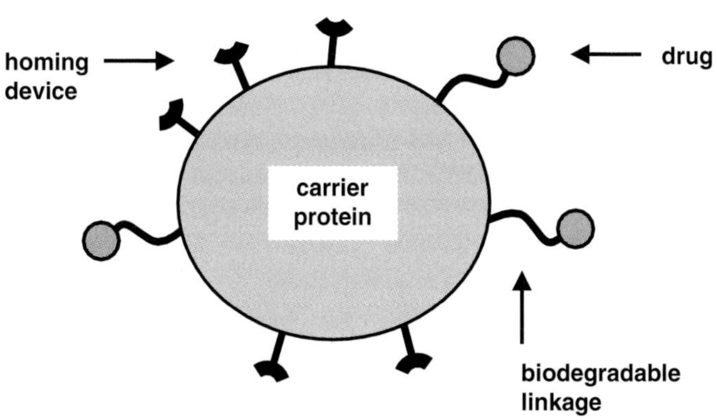

Figure 11.1. General structure of a proteinaceous drug targeting construct. Functional domains present in a drug targeting construct are (1) the carrier protein, (2) homing devices or site-directing ligands and (3) the drug moiety. In order to ensure its activation, the drug is often attached via a biodegradable linkage to the carrier.

delivery structure is the *active drug moiety*. Several types of bioactive compounds can be included in a targeting construct. Small drug molecules and other more specifically toxic molecules (toxins) can be linked covalently to the carrier protein. Since the release of the drug from the carrier can be decisive in its pharmacological activity, various types of linkages have been developed which are degraded specifically at the target site. Some of the more generally applicable approaches will be presented here.

As an alternative to linking drug molecules chemically to a core protein, therapeutically active proteins can be used for the preparation of drug targeting constructs. Classically, the latter type of constructs are prepared by chemical modification of the core protein. More recently, proteinaceous drug targeting constructs have also been prepared by recombinant DNA technology. By reconstructing the characteristics of therapeutic proteins at the DNA level and by genetic fusion of functional domains of different proteins into one construct, several interesting new drug delivery approaches have been initiated. These will be addressed at the end of this chapter.

11.2 The Carrier

Since the scope of this chapter is limited to proteinaceous drug targeting constructs, we will not discuss other types of macromolecular drug carriers, such as liposomes and polymeric drug carriers [1–3]. Often, polymeric drug carriers are preferred over proteinaceous drug carriers for their chemical stability, versatility in coupling reactions, high drug loading capacity and lower immunogenicity [2]. However, some general characteristics of proteins have advantages over those of other macromolecular carriers. Proteins are often biodegradable and, generally, biocompatible. Another important feature of a proteinaceous carrier is that the

Table 11.1. Proteins used in drug targeting constructs.

Protein	Remarks
Serum albumin	Used as core protein for various types of homing devices
Lysozyme (LZM)	Accumulates in the kidney
Transferrin	Passage through the blood–brain barrier; uptake in tumour cells and other proliferating cells
Monoclonal antibodies and antibody fragments	Targeting via specific binding to cell surface receptors; fragments have been prepared chemically and by recombinant expression
Catalase (CAT)	Therapeutically active protein (detoxification of reactive oxygen species)
Superoxide dismutase (SOD)	Therapeutically active protein (detoxification of reactive oxygen species)
Bacterial and plant toxins; toxin fragments	Toxic after entry into target cell; fragments produced by recombinant technology have been used as targeting moiety and as effector moiety
Cytokines, interleukines and growth factors; interleukin fragments	Therapeutic proteins produced by recombinant technology; receptor binding fragments have been used as targeting moiety

carrier is an homogenous product. In contrast, polymeric carriers are non-homogeneous by nature. Finally, some proteins have unique characteristics that relate to their complex tertiary or quaternary structures. For instance, the binding affinity of the protein to its natural receptor can be the driving force for the selective targeting of the construct. Other interesting carrier proteins are those that are pharmacologically active themselves. Such intrinsically active proteins can be used as active drug substances, or as carriers for small drug molecules in so-called dual active conjugates [4]. Table 11.1 lists examples of carrier proteins that will be discussed in the following sections of this chapter.

11.2.1 Albumin

An important feature of the carrier protein is the size of the macromolecule. Small proteins with a molecular weight lower than about 60 kDa, are rapidly cleared from the bloodstream by glomerular filtration in the kidney [5]. By choosing a carrier protein with an adequately high molecular weight, renal filtration can be prevented. Being sufficiently large to prevent renal filtration, but at the same time small enough for efficient tissue penetration, the albumin molecule has an ideal molecular weight (70 kDa). For this reason, serum albumins from different origins, such as human (HSA), bovine (BSA) or the albumin type autologous to the species in which the conjugate is being tested, have been the carrier of choice for many drug targeting preparations. Two types of drug–albumin conjugates have been reported. First, simple drug–albumin conjugates which accumulate in the target tissue by passive extravasation have been described for the delivery of various anti-cancer drugs to solid tumours [6]. Due to several factors, such as elevated levels of vascular permeability factors and an impaired lymphatic drainage, tumour blood vessels show an enhanced permeability and retention of macromolecules [7]. Preferably, a low number of drug molecules should be conjugated per albumin molecule, since otherwise the construct may undergo enhanced uptake by scavenger receptors in the liver and spleen [8]. On the other hand, specific uptake by scavenger receptors has been exploited for the delivery of anti-inflammatory drugs to the liver [9–11].

The second type of drug–albumin conjugate comprises those in which the albumin protein functions as a backbone for both conjugated drug molecules and homing devices. This type of structure will selectively accumulate in the target tissue by binding to cell surface receptors, a process called active targeting. The binding of the construct to the target cells is primarily driven by the qualities of the homing device, such as type and spatial orientation of the targeting moiety, rather than by the characteristics of the original carrier backbone. Therefore, similar structures can be prepared using carrier proteins other than albumin.

11.2.2 Low Molecular Weight Proteins

As stated earlier, proteins with a molecular weight lower than that of albumin are able to pass through the glomerular membrane in the kidney. Consequently, such low molecular weight proteins (LMWPs) are rapidly removed from the bloodstream. Following glomerular filtration, LMWPs are reabsorbed in the proximal tubular cells of the kidney. Since this process makes the kidney the major catabolic site for these proteins, they can be used as carriers for renal drug delivery [12,13]. A typical example of a protein that has been used for this targeting purpose is lysozyme (LZM, 14 kDa) [14,15] (see Chapter 5 for a more detailed discussion on the development of renal targeting preparations).

11.2.3 Monoclonal Antibodies

Monoclonal antibodies have been extensively reported on as carriers for targeted drug delivery. Starting in the late 1970s, the production of monoclonal antibodies has now evolved into a routine technique that has yielded many potential carrier molecules. Particularly in the field of cancer therapy, monoclonal antibodies are being used for the delivery of diagnostic and therapeutic agents [16] (see also Chapter 8). The original antibodies were of mouse origin, evoking a human anti-mouse antibody (HAMA) immunological response when administered in humans. The use of new recombinant techniques has enabled the preparation of humanized antibodies, in which the mouse recognition domain has been grafted onto a human antibody structure [17].

Whole IgG antibodies with a molecular weight of 150 kDa, are often unable to penetrate tumour tissue as efficiently as smaller molecules [18]. Therefore, smaller antibody fragments and genetically-engineered antibody derivatives have been investigated as drug carriers (see Figure 11.5). These carrier molecules will be discussed in Section 11.8.1.

11.2.4 Transferrin

Some proteins are excellent carriers for drug targeting since they bind to more or less specific receptors on the target cells. In addition to monoclonal antibodies, which in theory can bind to any kind of receptor, several natural ligands for cell-surface receptors have been explored as carrier proteins. This approach is exemplified by constructs that have been developed for targeting via the transferrin receptor (TfR). The transferrin receptor is expressed on most proliferating cells, as well as in a few non-proliferating tissues, among which is the endothelium of brain capillaries. This distribution pattern has inspired the development of transferrin-based constructs and anti-TfR antibodies as carriers for cancer therapy, as well as for the delivery of compounds across the blood–brain barrier [19,20] (see also Chapter 2). With respect to the latter, the ability of the TfR to undergo transcytosis results in the release of the carrier complex in the brain, rather than in endocytosis by the endothelium of the blood–brain barrier. Once inside the central nervous system, the drug delivery construct can bind to TfR-positive cells, such as brain tumours, which can be regarded as a second step in the delivery process.

As an alternative to targeting brain tumours which express the TfR, the transferrin approach can be used for the delivery of fusion proteins which bind to pharmacological receptors inside the central nervous system. An example of this is the construct consisting of nerve growth factor (NGF) and transferrin described in Section 11.8.2.3. The transferrin moiety in this type of construct will enable it to enter the brain, upon which the drug moiety will act by binding to its receptor. This approach seems especially suitable for compounds that cannot pass the blood–brain barrier, such as peptides and other hydrophilic substances.

11.3 The Homing Device

The binding of a protein to cell surface receptors can greatly enhance its selective accumulation and retention in the target tissue. However, most of the natural protein ligands are only available in limited amounts, insufficient for their application as carriers in drug delivery studies. Therefore, many researchers have tried to develop non-natural receptor ligands. Figure 11.2 shows some general approaches that have been followed to introduce receptor-binding domains into a carrier protein. First, recognition domains, such as carbohydrates, peptides or other molecules, can be covalently attached to side-chain residues of the core protein (Figure 11.2a). Second, receptor binding domains can be introduced by recombinant DNA engineering of the protein backbone (Figure 11.2b). Third, the receptor binding properties of a protein can be altered by the removal or modification of residues on the surface of the molecule. Examples of such strategies are the partial deglycosylation of a protein (Figure 11.2c) or modifications in the surface-charge of the protein (Figure 11.2d).

A major advantage of coupling receptor binding ligands to a carrier protein is the multivalent character of the construct obtained, which generally results in a drastic increase in the

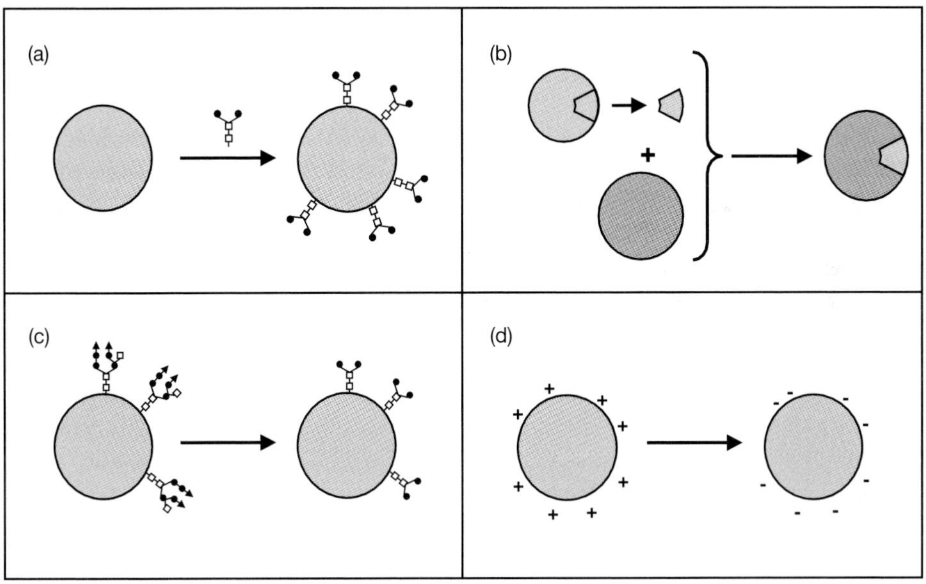

Figure 11.2. Different approaches for the introduction of homing devices into proteinaceous drug targeting constructs. (a) Site-directing ligands such as carbohydrates or short peptide sequences can be reacted to side-chain residues of the core protein. Typically, such homing devices are reacted to primary amino groups of the protein. (b) Using recombinant DNA techniques, the receptor binding domain of a protein can be grafted onto another protein. (c) Receptor binding domains can become available in a carrier protein by chemical or enzymatic (partial) deglycosylation. The removal of terminal sugar residues results in the de-masking of the targeting groups. (d) Modification of charged groups on the protein surface can result in increased affinity for specific receptors, e.g. the scavenger receptor when negative charge is introduced into the protein.

overall binding constant compared to the single recognition domain. Particularly in the case of low affinity ligands, such as sugar molecules, multivalent interactions may be essential for sufficiently strong binding to the target receptor to take place. In addition, multivalent binding can contribute to receptor dimerization, a process often observed in the binding of natural ligands to their receptors [21]. Dimerization and the subsequent intracellular signalling between the receptor molecules may be an essential step for the internalization of the receptor and its bound ligand, and thus for receptor-mediated endocytosis of the targeting construct. Apart from receptor internalization, intracellular signalling by the carrier may result in a pharmacological response in the target cells. Such carrier-mediated pharmacological activity may be of therapeutic value or, inversely, may be counter-productive to the therapeutic actions of the coupled drug moiety.

11.3.1 Carbohydrate Ligands

Historically, the first proteinaceous carriers containing man-made site-directing ligands were prepared by partial deglycosylation of natural glycoproteins [22]. For instance, desialylation of bovine fetuin produces asialofetuin, which contains clustered galactose residues at the end of its carbohydrate antennas. Asialofetuin, and other similarly prepared asiologlycoproteins, proved efficient carriers for hepatic delivery since the galactose residues are recognized by the asialoglycoprotein receptor on hepatocytes [23]. Nevertheless, large scale preparation of such modified proteins is cumbersome and may lead to heterogeneous and enzyme-contaminated preparations [24].

Another approach to the preparation of glycoproteins is the derivatization of a non-glycosylated protein, such as serum albumin, with simple sugar residues such as galactose or mannose. The synthetic preparation of such neoglycoproteins offers the advantages of a more predictable composition of the carrier and the possibility of large-scale production [25]. Many methods are available for the attachment of sugar groups to side-chain residues of the protein, either at aromatic groups of tyrosine and phenylalanine side-chains, or at primary amino groups of lysine residues [26].

Although neoglycoproteins show high affinity for carbohydrate-recognizing receptors (lectins), they demonstrate only moderate specificity for a specific receptor. This moderate specificity relates to the relative simple orientation of the oligosaccharide residues compared to those of the natural glycoprotein ligands which often display multibranched arrays of different sugar molecules [27]. Consequently, these glycoprotein ligands show a much higher receptor specificity.

Many approaches have been undertaken to obtain complex carbohydrate ligands with a high receptor specificity, either by *de novo* synthesis or by stripping the endogenous oligosaccharides from bulk amounts of natural glycoproteins [28]. In addition, carbohydrate mimetics are being developed in which part of the glycosyl antenna is substituted by other functional groups. For example, the modification of hydroxyl residues in fucose resulted in an increased specificity for either Kuppfer cells or for a tumour cell line, depending on the type and position of the modified hydroxyl function [29].

As an alternative to chemical synthesis, complex glycoproteins can be prepared by molecular biology techniques [30,31]. Typically, the recombinant protein is expressed in mammalian cell lines that have acquired glycosyltransferases by genetic engineering. Biosynthesis of glycoproteins results in the natural linkage of the carbohydrate antenna to asparagine residues of the protein backbone, rather than to lysine residues, which is the case with chemical conjugation of carbohydrates.

11.3.2 Folate

Apart from carbohydrate ligands, several other molecules can function as homing devices by binding to cell-surface receptors. A well-known example is the folate vitamin, which has been used as a targeting moiety in proteinaceous constructs, liposomes and low molecular weight prodrugs [32]. The folate receptor is expressed on a wide range of tumours, but also in the proximal tubule of the kidney. As a result of the latter, extensive renal accumulation of low molecular weight folate complexes may occur (see Chapter 5). This distribution can be prevented by using a carrier protein that is not filtered into urine, since the folate receptor is only expressed on the luminal brush border, i.e. on the urinary side of the cells.

11.3.3 Peptide Ligands

Many natural protein ligands bind to their receptors via interactions of a specific area of the protein backbone. The receptor binding domain of such a protein can be transferred into another protein, for instance a therapeutically active one. This technique is commonly applied in the preparation of recombinant targeting constructs, and will be discussed in Section 11.8.2.

When the receptor binding domain is encoded in a small peptide sequence, the peptide ligand can also be synthesized and conjugated chemically to the carrier protein. This approach was followed in our laboratory by Beljaars *et al.* for the development of carriers aimed at the hepatic stellate cell, a cell type involved in liver fibrosis [33] (see also Chapter 4). A peptide sequence derived from the receptor binding domains of collagen VI was incorporated into a cyclic peptide homing device, and subsequently conjugated to lysine residues of HSA. This carrier bound selectively to activated hepatic stellate cells and rapidly accumulated in the livers of fibrotic rats.

A new technique in the field of drug delivery is the screening of phage display libraries for peptide ligands that can be used as homing devices [1,34,35]. Briefly, these ligands are obtained by selective enrichment of a large library of different peptide sequences onto isolated receptors, whole cells or even in whole organs after *in vivo* administration (see also Chapter 10). After several rounds of enrichment, a pool of ligands is obtained with increased affinity for the target of interest. Table 11.2 lists some interesting studies on the identification of peptide homing devices and their corresponding (molecular) targets. Of special interest are the peptide ligands in which the RGD (Arg-Gly-Asp) sequence is incorporated, since this motif is present in many receptor binding domains of protein ligands [43]. Specially constrained conformations of the RGD motif are recognized by the $\alpha v\beta 3$ and $\alpha v\beta 5$ integrins, matrix-binding receptors upregulated in tumour blood vessels [34] (see Chapter 9 for a detailed description of the use of these peptides in drug targeting strategies).

Table 11.2. Targets for peptide homing devices identified by phage display.

Target tissue	Molecular target	References
Tumour cells; angiogenic endothelial cells	Melanoma- associated antigen	[36]
Lung vasculature	Membrane dipeptidase	[37]
Endothelial cells in inflammatory lesions	E-selectin	[38]
Skeletal and cardiac muscle	Unknown	[35]
Tumour vasculature	avβ3 and αvβ5 integrin	[34] [39, 40]
Tumour vasculature	Aaminopeptidase N	[41]
Tumour cells	Sialyl Lewis x antigen	[42]

11.3.4 Modifications of the Physicochemical Properties of the Protein

Apart from the incorporation of small functional groups in a protein that recognize specific receptors, a more general modification of the physicochemical properties of the protein can also lead to selective drug targeting. For instance, the coupling of lipophilic drug molecules to a carrier protein like serum albumin can result in cell-specific uptake of the conjugate by the non-parenchymal cells of the liver, i.e. the Kupffer and endothelial cells [8,11]. Competition experiments with poly-anionic substrates for the type B scavenger receptor, demonstrated that this receptor was involved in the clearance of such conjugates [8]. Two changes in the physicochemical properties of the protein in particular conjugates could be responsible for the affinity for the scavenger receptor: either the increase in net negative charge, due to the removal of primary amino groups in the protein, or the increased hydrophobicity of the proteinaceous structure [11].

Another approach that results in an altered surface-charge of the protein, is the derivatization of lysine residues with negatively charged substituents. Several carboxylic acids, such as succinic acid (Suc), aconitic acid (Aco) or maleic acid (Mal), have been used for this purpose. Anionized proteins are preferentially accumulated in the liver endothelial cells, but uptake by macrophages in the liver and spleen has also been observed [9,44]. Interestingly, while this approach resulted in an increased scavenger receptor-mediated uptake in the liver of large proteins like albumin or catalase, the uptake of anionized LMWPs like superoxide dismutase (SOD) and LZM was only slightly affected [44]. Instead, Suc-LZM might be an interesting carrier for delivery of drug substances to the bladder, since its excretion into the urine was greater than that of drug–LZM conjugates [45]. Although liver targeting of SOD could not be achieved via succinylation of the LMWP, the attachment of negatively charged polymeric substituents like DIVEMA (DIVEMA: copolymer of divinyl ether and maleic anhydride) did result in an increased liver accumulation, as did the attachment of galactose or mannose sugar residues [46,47].

Another target defined for anionized albumins are cells of the immune system that have been infected with the human immunodeficiency virus (HIV). Suc-HSA and Aco-HSA are potent inhibitors of HIV-1 replication *in vitro* [48]. Thus, anionized albumins can be regarded as pharmacologically active proteins, that can be used either as such, or as dual-active conjugates for the delivery of other anti-HIV drugs such as azidothymidine-monophosphate (AZTMP) [49].

When high doses of anionized albumins are administered, the uptake by the scavenger receptors in the liver and spleen becomes saturated [50]. Under these conditions, the negatively charged albumins were shown to distribute rapidly into the lymph. Furthermore, high concentrations of the protein were sustained in the lymph, which may be advantageous in relation to the antiviral effects of the drug, since the virus also resides in the lymphoid tissue [51].

11.4 The Active Drug

While being the main reason from a therapeutic point of view, for developing drug targeting structures, the choice of active drug substance is often determined by rather opportunistic reasons such as the availability of the substance or its suitability for common coupling reactions. Many proof-of-principle studies have been carried out with the same model drug compounds, an example of this being doxorubicin. Whether those studies can be extrapolated to other drug compounds depends on factors related to the drug, target cell and targeting construct. The following considerations should offer some guidelines that may be of use in the design of new targeting constructs.

The first step in the selection of a drug substance is to determine whether the drug actually exerts its action in the target cell or tissue. This question might seem obvious for drugs that act by killing their target cells, such as anti-neoplastic drugs, which are toxic for proliferating cells. Such drug molecules have frequently been incorporated into drug delivery preparations. Since tumour therapy is associated with a high incidence of severe side-effects, these drug molecules make excellent candidates for targeted delivery. However, other categories of drugs exert their therapeutic effects by influencing multiple cell types or organs, which makes it difficult to define the therapeutic target site. Examples of this latter class include for instance, cardiovascular-active compounds, which can be active in blood vessels, the heart, the kidney and even the central nervous system. Thus, it might well be that their beneficial effects on a specific organ relate to multiple effects somewhere else in the body. Selective drug delivery of such drug substances might even result in an impaired therapeutic response if an essential target is missed by the 'delivered' drug. Although such drug delivery constructs will prove to have poor therapeutic properties, they can however be important in elucidating the potential therapeutic actions of a drug molecule *in vivo*. Many claims regarding the therapeutic potential of drugs under investigation are based on *in vitro* experiments, often performed on non-relevant cell types or using therapeutically unattainable doses. The selective delivery of such compounds may corroborate such studies or, conversely, prove their irrelevancy.

After a therapeutic agent and its target have been chosen, the next step is to ensure that the drug can reach its pharmacological target from the site of accumulation of the construct. As will be discussed in the next section on linkages, many drug targeting constructs will find their way into the lysosomal compartment, in which the complex will be degraded and the parent drug liberated from the carrier. Since most pharmacological targets are outside the lysosomal compartment, the liberated drug will have to traverse the lysosomal membrane in order to exert its pharmacological activity. Little is known about this final step in the targeting process. Most likely, the passage of the drug across the membrane is mediated by passive diffusion, but carrier-mediated transport via one of the many transporters in the lysosomal membrane cannot be excluded [52]. Several characteristics of the drug can influence its passive diffusion rate across the membrane, such as its lipophilicity and the presence of charged functional groups. The acidic pH in the lysosomes favours the diffusion of weak acids into the cytosol, but can inhibit the transport of basic compounds such as amines. Particularly when the lysosomal conditions are hostile towards the drug (e.g. when the drug has a peptide-like structure), this might result in a rapid degradation of the drug before it can exert its therapeutic activity. For such compounds, an alternative method of targeting should be considered, or a membrane-translocating modality should be included in the targeting construct (see Section 11.9.2).

A final important factor that contributes to the therapeutic success of a drug targeting preparation is the kinetics of processing of the delivered drug by the target cell. A rapid elimination of the free drug, either by metabolism at the target site or by (carrier-mediated) transport to other sites in the body, might prevent the drug from reaching effective therapeutic concentrations. Although rapid metabolism of the drug at the target site seems disadvantageous, this process can contribute to a high selectivity of targeted versus non-targeted pharmacological actions. This principle has been applied in the so-called soft-drug approach, which uses pro-drugs that are activated into rapidly-metabolized compounds. Since the latter compounds display very short elimination half-lives, their actions remain predominantly restricted to the site at which they are activated [53].

When the drug is cleared from the target site by redistribution to other tissues, the route of elimination can be relevant in relation to possible side-effects of the drug. When a drug is delivered into cells of the excretory organs, the liver and kidney, the elimination of the free compound may take place directly via the bile and urine respectively, thus preventing systemic redistribution. In most other tissues, the delivered drug can be eliminated only via the systemic circulation. Redistribution of the drug might eventually lead to effects on tissues other than the target tissue, especially if the clearance of the drug from the body is low. Furthermore, relatively high concentrations of the drug will accumulate in the organs that are responsible for the clearance of the drug.

As mentioned in Section 11.2, a special class of proteinaceous targeting constructs are those in which a therapeutic protein is used as the active drug substance. In such a preparation, the protein is redirected to the target tissue by the attachment of site-directing ligands such as those discussed in Section 11.3. For instance, interferon beta (IFN-β) can be redirected to the liver by enzymatic desialylation in a procedure similar to that described earlier for fetuin (Section 11.3.1). The resultant asialo-IFN-β was found to have an *in vivo* anti-viral effect when tested in a hepatitis B model in athymic nude mice [54].

A well-studied example of a proteinaceous drug is SOD. This enzyme is capable of deactivating reactive oxygen species such as superoxide radical O_2^-, and has been proposed as a potential therapeutic for liver fibrosis. The attachment of liver-directing ligands such as galactose (Gal-SOD) or mannose (Man-SOD) resulted in an increased distribution of the protein to hepatocytes and Kupffer cells respectively [47]. Disappointingly, Man-SOD showed no anti-inflammatory effect when tested in a rat model of liver disease, despite significant intrahepatic accumulation [46]. A possible explanation is that Man-SOD is rapidly endocytosed and degraded, which limits its therapeutic effects.

Several other modifications have been explored for the targeted delivery of SOD. Chemical modification with hydrophilic monomethoxypolyethyleneglycol polymers (MPEG) resulted in a derivative with an increased molecular weight of 130 kDa, and hence a reduced renal elimination rate. The MPEG-SOD preparations reduced arthritic lesions in a complete adjuvant arthritis model in the rat, while native SOD did not show an anti-inflammatory effect [55].

Another antioxidative enzyme that has been targeted to the liver is catalase (CAT). Succinylation and mannosylation resulted in an increased accumulation of the protein in the non-parenchymal cells of the liver. Furthermore, the CAT derivatives reduced hepatic injury in an ischaemia/reperfusion injury model [56].

The advances in recombinant DNA techniques can have a great impact on the preparation of (modified) therapeutic proteins. Technically, most recombinant preparations can be regarded as proteins in which the pharmacological activity is encoded in the protein backbone. An illustrative example of a recombinant targeting construct is the fusion protein of SOD and the non-toxic fragment of tetanus toxin which will be discussed in more detail in Section 11.8.2.2 [57]. By means of the targeting properties of the toxin fragment, this hybrid protein is selectively endocytosed by neuronal cells, and as such might be used for the treatment of cerebral ischaemia/reperfusion injury [58].

11.5 The Linkage Between Drug and Carrier

Proteins contain many different functional groups that can be used for conjugation reactions. The more hydrophobic residues are normally situated in the core of the protein and only become available after disruption of the tertiary structure of the protein. Hydrophilic amino acid residues are exposed at the protein surface and, consequently, these groups are normally the target for the conjugation reaction. In addition, some carrier proteins, for instance IgG molecules, contain surface-exposed glycosyl residues, which may also be used for the conjugation of drug molecules.

Many conjugation procedures are based on nucleophilic substitution reactions, in which an activated electrophilic group of the drug reacts with a nucleophilic group of the protein [59]. This reaction is preferred over the activation of side-chain residues of the protein, since such activated groups might react with nucleophilic residues in the protein, resulting in internal cross-linking and polymerization of the carrier.

Of the hydrophilic side chain residues, the cysteinyl thiol group is the most reactive. However, cysteine residues are involved in the formation of disulfide bridges between loops of the protein backbone, and therefore only few proteins contain free thiol groups. For instance, serum albumin and SOD, both proteins with an uneven number of cysteine amino acids, have a single free thiol group [47,60]. Such a residue can be used for the site-specific conjugation of drug molecules, as was demonstrated for HSA. A thiol-derivative of the cytostatic drug mitomycin was conjugated via the cys34 residue of HSA in a 1 : 1 molar ratio [61].

Following cysteine residues, the second most reactive groups in a protein are the primary amino groups, available in both lysine residues and at the N-terminus of the protein backbone. Since primary amino groups are present in abundance in proteins, most conjugation re-

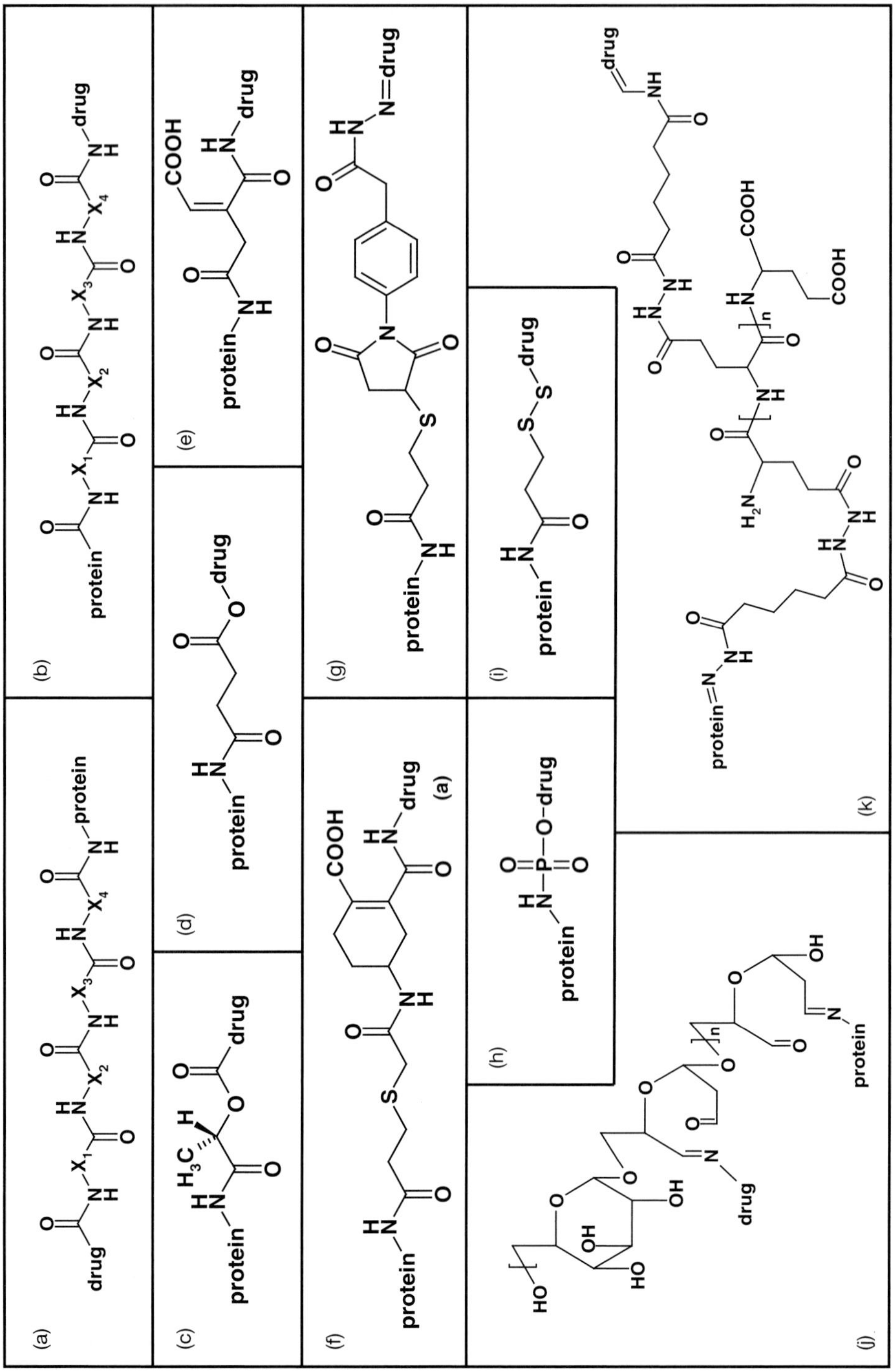
(a)
(b)
(c)
(d)
(e)
(f)
(g)
(h)
(i)
(j)
(k)

actions are directed towards these groups [62]. Although the direct coupling of a drug substance to an amino group might seem the most straightforward approach from a synthesis point of view, most constructs developed so far contain an indirect linkage which is achieved via a spacer molecule between the drug molecule and the protein residue. Spacers are used for the conjugation of drug molecules for two reasons, namely to facilitate the conjugation reaction and to introduce a biodegradable linkage between the drug and the carrier. Exceptionally, small drug molecules linked in a non-degradable manner, show biological activity, for instance as active metabolites containing a fragment of the protein backbone. This is illustrated by the drug targeting preparations naproxen-LZM and naproxen-HSA, which are degraded to naproxen-lysine [63]. The naproxen-lysine derivative showed pharmacological activity similar to that of naproxen. However, most small drug molecules in the current drug targeting preparations require the regeneration of the parent drug in order to be therapeutically effective. The following sections will discuss general approaches to the insertion of bioreversible linkages into a drug–protein conjugate.

How can the properties of the target site be utilized to facilitate the regeneration of the parent drug? Both enzymatic and chemical degradation of the linkage between the drug and carrier may result in the release of the drug substances. Several of the more universally applicable approaches will be presented in the following sections (Figure 11.3). Figure 11.4 shows representative examples of the drug molecules that have been coupled via these linkages.

11.5.1 Intracellular Degradation

Many drug–protein conjugates accumulate in the target organ by receptor-mediated or adsorptive endocytosis and subsequent transport to the lysosomes [64]. This cellular compartment contains a variety of enzymes, e.g. peptidases and esterases, which are capable of degrading drug targeting constructs [65]. Table 11.3 lists some peptide spacers which have been

◄

Figure 11.3. Different types of biodegradable linkages that have been used for the conjugation of small drug molecules. (a and b) Amide linkages. Peptide spacers can be used in the conjugation of drug molecules that contain a carboxylic acid group (a) or a primary amino group (b). X_{1-4} denotes the consecutive amino acids that are present in the spacer. (c and d) Ester linkages. (c) L-lactic acid spacer used for linking drug molecules containing a carboxylic acid group. (d) Linkage of drug molecules via their hydroxyl groups using a succinic acid spacer. (e–h) Acid-sensitive linkages. (e and f) Linkages formed with the *cis*-aconityl linker and a second-generation dicarboxylic acid spacer. The latter spacer forms a similar type of linkage with the amino group of the drug molecule as the *cis*-Aco spacer. (g) Schiff base acid-sensitive linkage. This linkage is formed with a carbonyl group of the drug molecule, resulting in an imino bond with the spacer. (h) Acid-sensitive phosphamide linkage formed between a phosphate group of the drug molecule and an amino group of the carrier protein. (i) Disulfide linkages. This type of linkage can be formed with drug molecules that contain free thiol groups or, alternatively, with drug derivatives in which a free thiol group has been introduced (see Figure 11.4, colchicine). (j and k) Polymeric linkages. Multiple drug molecules can be covalently attached to a single functional group of the carrier when polymeric bridges are used. (j) Dextran polymer which has been derivatized with drug molecules by oxidative amination, and subsequently reacted to an amino group of the protein. (k) The polymeric PGA bridge has been reacted with adipic hydrazide, after which these groups were used for the formation of a Schiff base linkage with carbonyl groups in the drug. Conjugation to the protein was achieved via the same adipic hydrazide groups at oxidized carbohydrate residues in the protein.

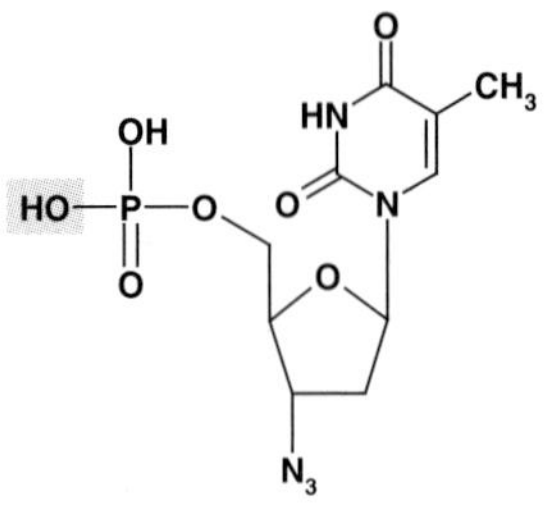

AZTMP

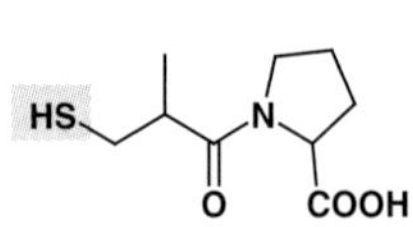

captopril

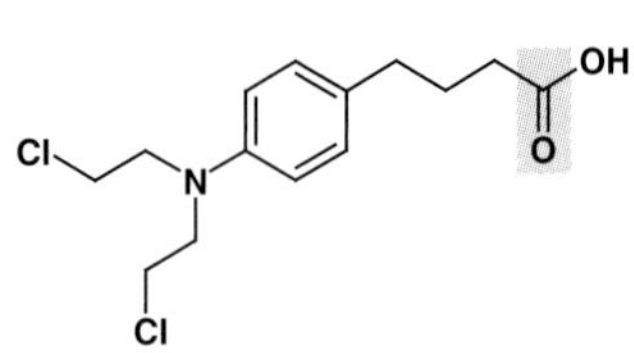

chlorambucil

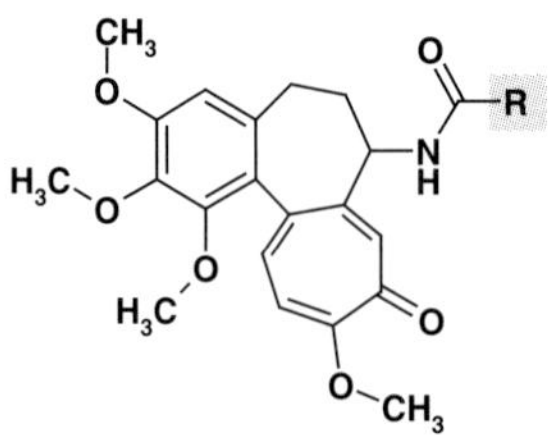

colchicine: R = CH_3

coupled derivative: R = C_2H_4-SH

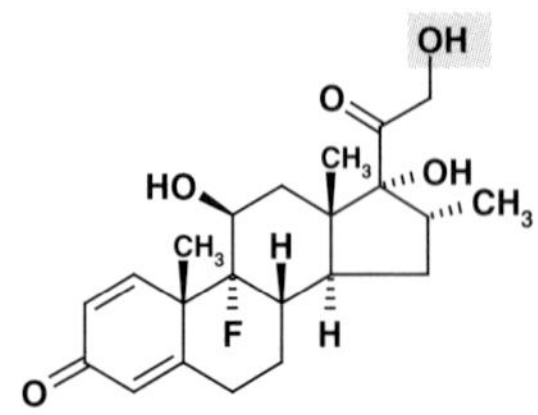

dexamethasone

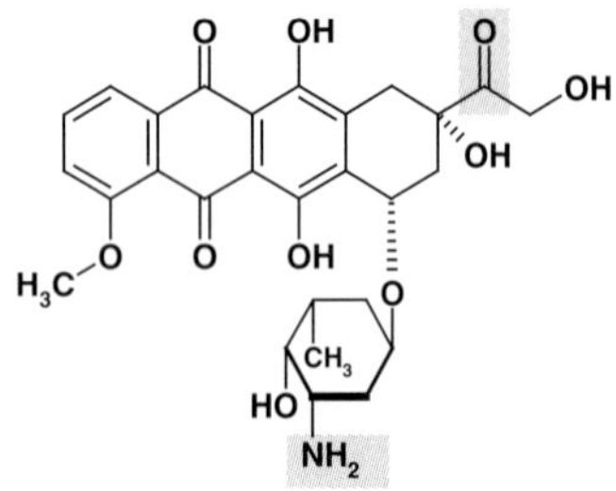

doxorubicin

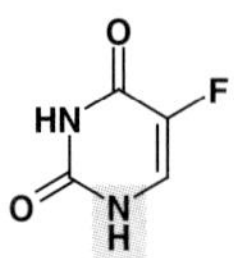

5-fluorouracil

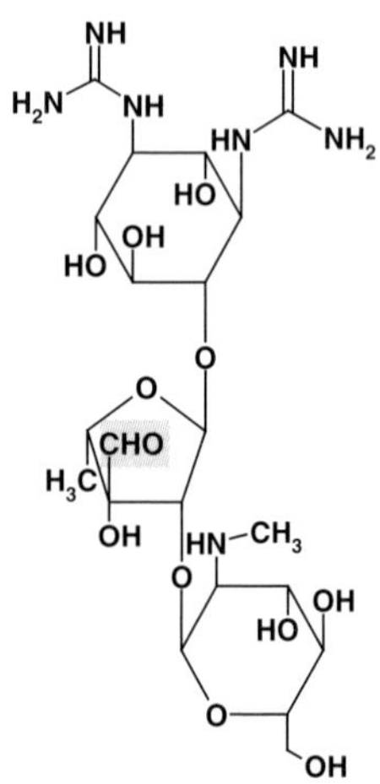

streptomycin

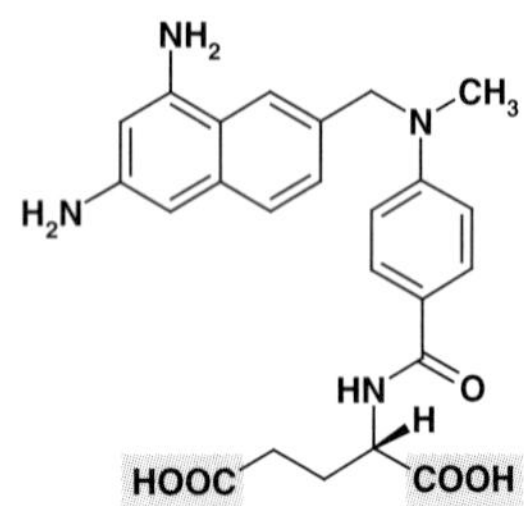

methotrexate

Figure 11.4. Overview of drug molecules that have been incorporated into drug targeting that are discussed in Section 11.5.1. The functional group of the drug molecule that is used for conjugation to the carrier has been high lighted.

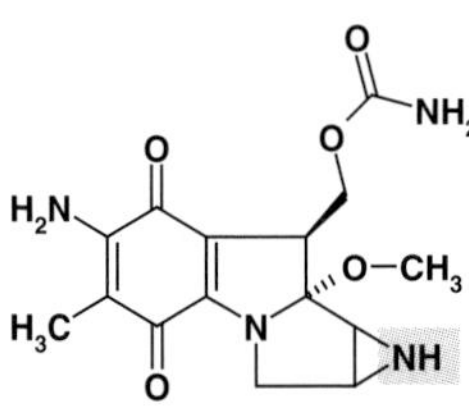

mitomycin C

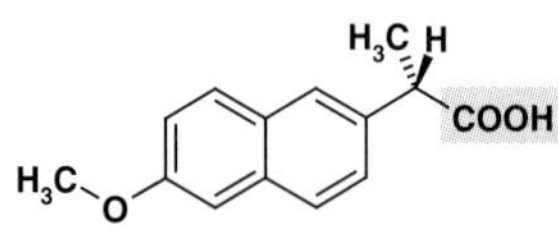

naproxen

Table 11.3. Peptide sequences that have been used as spacers in lysosomotropic drug delivery.

Drug	Spacer	References
Conjugation at carboxylic acid group of the drug		[66]
Methotrexate	ala-leu-ala-leu gly-gly-gly gly-gly-phe gly-phe-ala	
Conjugation at amino group of the drug		[67]
Doxorubicin, daunomycin	ala-leu-ala-leu gly-gly-gly-leu gly-gly-phe-gly gly-phe-leu-gly	
5-Fluorouracil	gly-phe-leu-gly-leu-gly	
Mitomycin C	ala-leu-ala-leu gly-phe-ala-leu gly-phe-leu-gly	
Primaquine	ala-leu-ala-leu	

developed for lysosomotropic drug delivery (i.e. delivery with constructs whose final destination is the lysosomes). The enzymes involved in the degradation of the spacers shown in Table 11.3 are normally not present in extracellular areas. For this reason, and since peptide bonds are very resistant towards chemical degradation, the linkages that are formed with these spacers are generally very stable in the circulation.

The peptide spacers shown in Figure 11.3a–b are susceptible to degradation by endopeptidases, which either disrupt the bond between the spacer and drug molecule or, alternatively, one of the internal peptide bonds of the spacer. In a subsequent degradation step the remaining part of the spacer is either attacked by other peptidases or is degraded via non-enzymatic processes [68,69]. Peptide spacers have been used for the delivery of drug molecules that contain peptide carboxylic acid groups, an example of which is the cytostatic drug methotrexate. It should be noted however that peptide spacers are ill-suited for the linkage of drug molecules which contain non-peptide carboxylic acid groups. Probably, the peptidases involved have high substrate specificities, resulting in the inability to degrade the amide bond that is formed with the drug molecule [63]. Peptide spacers have also been used for the conjugation of drug molecules to a primary amino group (Figure 11.3b). As can be concluded from the listed drugs in Table 11.3, the peptidases involved in the release of these drug molecules tolerate different types of non-peptide primary amino groups.

For drug molecules with carboxylic acid groups, ester bonds seem suitable linkages for lysosomal delivery. The anti-inflammatory drug naproxen has been conjugated to a proteinaceous carrier via an ester linkage by means of an α-hydroxy acid spacer (Figure 11.3c) [70]. Such a spacer, of which L-lactic acid is a typical example, contains in addition to the hydroxyl group a carboxyl group that can be used to link the spacer to the protein. Similar types of linkages can also be formed with a drug molecule containing an hydroxyl group and a spacer with two carboxylic acid groups, such as the succinate spacer (Figure 11.3d). Examples include the conjugates prepared with the corticosteroid dexamethasone and the taxoid paclitaxel [11,71,71,72].

The relatively acidic pH of the lysosomes (pH 5) has led to the development of several linkage types which are susceptible to acid-catalysed degradation. These acid-sensitive spacers are relatively stable at the neutral pH of the bloodstream, but become hydrolytically labile at lower pH values. Depending on the type of linkage and the functional group of the drug molecule, three subtypes can be distinguished: *cis*-aconityl linkers (*cis*-Aco) (Figure 11.3e–f), Schiff base hydrazones or imino linkages (Figure 11.3g), and phosphamides linkages (Figure 11.3h), respectively. The *cis*-Aco linker can be used for the conjugation of drug molecules with a primary amino group [73,74]. The amide linkage that is formed between the linker and the drug is chemically destabilized at lower pH values due to the flanking carboxylic acid group (anchiomeric assistance). When the *cis*-Aco spacer is conjugated via its flanking carboxylic acid group to the protein, the acid-sensitivity of the spacer is lost. In order to prevent such loss of pharmacological activity, several other *cis*-Aco spacers have been developed that are conjugated via a different chemistry to the protein (exemplified in Figure 11.3f) [75,76]. Often, these spacers contain maleimide and haloalkyl groups for the purpose of the final coupling to the protein, since these groups can be reacted under mild conditions with thiol groups which have been previously inserted into the protein.

The Schiff base hydrazone linkers form an imino-linkage between a carbonyl group of the drug and a hydrazine functionality of the spacer. Such linkages have been formed with drug molecules containing either a keto group (doxorubicin) [77,78], an aldehyde group (streptomycin) [79] or a carboxylic acid group (chlorambucil) [80] (Figure 11.4). Recently, the use of a branched spacer has been reported, which enables the attachment of multiple drug molecules to the spacer [81]. For the conjugation of the hydrazone spacer to a protein, strategies similar to those used in the case of the second generation *cis*-Aco spacers have been followed.

The third class of acid-catalysed linkages, the phosphamide linkages, are formed between a phosphate group and a primary amino group (Figure 11.3h). This type of linkage has been used for the lysosomotropic drug delivery of nucleotide analogues to the liver and macrophages [22,82,83].

Another type of linkage that is degraded intracellularly is the disulfide bond (Figure 11.3i). Disulfide linkages have been extensively used in the preparation of immunotoxins [84]. For instance, Ricin A immunotoxins are pharmacologically active only when they contain a biodegradable disulfide linkage [85]. With respect to the targeting of small drug molecules, disulfide linkages have been used in conjugates with the angiotensin converting enzyme inhibitor captopril, a drug molecule that contains a free thiol group, and in conjugates with derivatives of colchicine and methotrexate [14, 86–88].

A disadvantage of the disulfide linkage is its relative instability in the bloodstream. Disulfide bonds can be degraded by reducing enzymes or disrupted chemically by thiol-disulfide exchange with free thiol compounds such as gluthathione [89]. Although the latter process will proceed preferably intracellularly, due to the higher intracellular concentration of free thiol compounds, thiol-disulfide exchange can result in premature drug release. Disulfide spacers that contain sterically hindered disulfide bonds showed improved stability towards this non-enzymatic degradation [90,91].

One of the limitations of the use of a proteinaceous carrier is the relative small number of drug molecules that can be conjugated without gross alterations in the structure of the protein. For example, extensive derivatization of an antibody carrier may lead to the loss of its homing potential if the antigen recognition domain is affected. To circumvent this problem, dextran and poly-glutamic acid (PGA) polymers have been used as bridging molecules for the conjugation of cytostatic drugs (Figure 11.3j–k) [92,93]. These polymers were loaded with the drug and subsequently reacted with amino groups or carbohydrate residues of the carrier. This technique enabled the conjugation of about 80 doxorubicin molecules per protein in the case of the dextran bridge, and up to 100 doxorubicin molecules as a result of the PGA linkage. Efficacy studies with tumour cell lines and *in vivo* tumour xenograft models in mice demonstrated the potential of the above-described conjugates [93].

11.5.2 Extracellular Degradation

In addition to targeting constructs which are endocytosed and which release the active drug substance intracellularly, other constructs are activated outside the target cell. The latter approach is not appropriate for target tissues in which the extracellular fluid is rapidly removed by perfusion, since this would result in a reduction of the time that the drug remained at the target site and consequently systemic redistribution of the drug would occur. Thus, extracellular drug release is preferred for use in compartments or tissues where the rate of perfusion is low. Conditions of slow perfusion are associated with most solid tumours due to their poor lymphatic drainage. In addition, many tumour cells secrete proteolytic enzymes that are capable of degrading extracellular matrix in the process of tumour growth and metastasis. If such enzymes are present in the tumour and in minimal amounts in the extracellular fluid of other tissues, their presence can be exploited for the selective release of the drug in the tumour tissue. Examples of such enzymes include cathepsins, that are normally only present in the lysosomes, and matrix-degrading enzymes such as collagenase or plasminogen activators [67]. These enzymes are all peptidases, and therefore peptide linkers are feasible spacers for use with this approach. In the case of a secreted lysosomal enzyme, the same spacer sequences shown in Table 11.3 can be used for the linkers.

An attractive approach to drug targeting is the delivery of the drug-regenerating enzyme instead of the actual drug substance to the target site. This approach, also referred to as ADEPT (antibody directed enzyme pro-drug therapy), is based on a two-step targeting principle. In the first step, an enzyme is selectively delivered to the target site by means of an antibody–enzyme conjugate. In the second step, small-molecule pro-drugs are administered, which will subsequently be activated by the targeted enzyme [94,95].

A point to be noted regarding ADEPT relates to the plasma half-life of the enzyme–antibody construct. Generally, antibody–enzyme conjugates, are slowly cleared from the central circulation. Their sustained presence in the bloodstream will lead to non-target site activation of the pro-drug. Thus, the enzyme and pro-drug should be consecutively administered after a well-chosen time interval when the concentration of the antibody-enzyme conjugate is still high in the target tissue while being low in the central circulation and non-target tissues.

11.6 Recombinant DNA Approaches

The importance of recombinant DNA techniques for the synthesis of drug targeting constructs is rapidly increasing. This approach offers, in theory, the possibility of generating all three components of a drug targeting preparation as outlined in Figure 11.1. A carefully chosen cloning strategy results in a uniform end-product with optimum positioning of the different components. To obtain such a fully genetically engineered drug targeting construct, all three components must be peptides or proteins. With respect to the active drug substance this is likely to be an exception rather than the rule. Some constructs, such as immunotoxins and immunocytokines, that do fulfil these requirements have been studied extensively in drug targeting and will be described in detail.

By using recombinant DNA techniques, modifications in the protein backbone, such as additions, deletions and alterations of amino acids, are easily achieved. These modifications can contribute to improved pharmacokinetic properties of the construct. Additions may consist of the introduction of residues that allow covalent conjugation of drug molecules. Deletions of amino acids can employed to remove membrane-bound regions of a protein, thereby increasing its solubility. Single amino acid modifications can be used to minimize antibody responses and alter the binding specificity and/or the three-dimensional structure of a certain protein.

The final requirement of a recombinant DNA approach to the preparation of a drug delivery construct, is the ability to produce large amounts of the protein. This can by achieved by bacterial, fungal, insect and mammalian expression systems. The choice of system depends on how the expression of the protein is regulated, the required purity and yield of the protein, and whether the protein is toxic to certain types of producer cells. Furthermore, individual scientists may prefer a particular type of expressing system, depending on laboratory facilities, safety considerations and production costs. The possibilities and limitations of different expression systems will be discussed and general guidelines which need to be taken into account when choosing an appropriate strategy, will be mentioned briefly. Thereafter, with the aid of several examples, the development and applications of drug delivery constructs obtained using recombinant DNA technology will be described.

11.7 Recombinant DNA Expression Systems

11.7.1 Heterologous Gene Expression in *Escherichia coli*

The Gram-negative bacterium *E. coli* is probably the most widely used host for heterologous protein production. An obvious advantage of this system is its simplicity. The genetics are well characterized, the cells grow fast allowing rapid production and analysis of the expressed protein, and transformation is simple and requires minimal amounts of DNA.

In *E. coli* foreign genes are normally cloned using inducible promoters such as the *lac* promoter that is regulated by the *lac* repressor and induced by isopropyl β-D-thiogalactopyranoside (IPTG). This controls gene expression and prevents loss of the gene in situations where production of the protein might be toxic to the cells. Stronger synthetic promoters, derived from the *lac* system, *tac* and *trc* promoters, are commercially available. Other commonly used promoters include T7 RNA polymerase promoter and promoters that are regulated by temperature shift, such as the temperature sensitive λP_L promoter and the cold shock promoter *cspA* [96]. The latter promoter is especially beneficial for proteolytically-sensitive proteins since proteolysis is reduced at low temperature. Additionally, promoters that are activated by a decrease in temperature may provide a partial solution to another frequently encountered problem, namely misfolding and denaturation of proteins. Since cultivation under low temperature favours correct protein folding, this problem is less likely to occur.

The (over)expression of proteins in the cytoplasm of *E. coli* often leads to the formation of insoluble aggregates known as inclusion bodies. In fact this can simplify the purification protocol but at the same time often requires *in vitro* refolding of the protein into its active form, which can sometimes be difficult to achieve. Besides lowering the culture temperature (see above), the solubility of the expressed protein can be improved by constructing fusion proteins. Commercial systems suitable for fusion to maltose-binding protein (MBP), thioredoxin and glutathione S-transferase are available. Not only are fusion partners used to increase the solubility of the protein of interest, but they can also facilitate its purification. Additionally, poly-histidine tags are commonly used for efficient purification via immobilized metal affinity chromatography. Although the presence of the poly-His (and other) tags is acceptable in many cases because they rarely alter protein structure or function, their removal may be required in some therapeutic applications. The liberation of the heterologous protein from its fusion part (or affinity tag) is theoretically possible but needs expensive proteases and is very seldom complete. This generally results in reduced yields of the active product.

A new insight into the problem of insolubility of heterologous proteins has evolved from further information regarding the *in vivo* function of molecular chaperones in protein folding. As reviewed by Baneyx, co-expression of chaperones in the bacterial system can improve the folding and hence the yield of heterologous proteins [96].

An alternative approach is to direct secretion of proteins into the periplasmic space by using a signal peptide sequence that is removed during the translocation process [97,98]. Various signal sequences, derived from naturally occurring secretory proteins, including PelB, β-lactamase and alkaline phosphatase can be used for secretion of heterologous proteins. The periplasm is an oxidizing environment, containing enzymes necessary for the formation and rearrangement of disulfide bonds. This is especially relevant for the recombinant production of antibodies which require disulfide bonds for activity [99].

An important point which should be taken into account when expressing eukaryotic genes in *E. coli*, is the difference in codon usage between prokaryotes and eukaryotes. For instance, the arginine codons AGA and AGG are common in eukaryotic genes but rarely found in *E. coli.* This problem can be solved, either by site-directed mutagenesis or by co-overexpression of the gene encoding $tRNA^{Arg(AGG/AGA)}$. Another and more important limitation of *E. coli* as an expression system for eukaryotic proteins, is its inability to glycosylate proteins. Therefore, if glycosylation is required, other expression systems should be used.

11.7.2 Fungal Expression Systems

Fungi, both filamentous fungi and yeast, are often the expression system of choice when a high yield of eukaryotic protein is desired. Fungi grow rapidly on cheap medium and gene manipulation is not difficult. In contrast to *E. coli*, fungi are able to carry out post-translational modifications, such as glycosylation, proteolytic processing, folding and disulfide bridge formation. By applying fermentation technology, clinically and industrially important proteins have been successfully expressed in fungi [100–102].

The most commonly used filamentous fungi for heterologous gene expression belong to *Aspergillus* and *Trichoderma* species. The transformation system is based on complementation of auxotrophic mutants or on dominant selection marker genes and results in the integration of the foreign gene into the host genome [103]. Filamentous fungi have effective secretory machinery, allowing for accumulation of proteins in the culture medium. In *Aspergillus*, expression cassettes consisting of the foreign gene fused to an endogenous glucoamylase and separated by a KEX-2 proteolytic site, have resulted in elevated expression levels [102,104]. The KEX-2 site is effectively cleaved by an endopeptidase in the endoplasmic reticulum during secretion resulting in the correctly processed protein accumulating in the medium.

Several yeast species have been engineered for heterologous protein production but the most commonly used for these purposes are the baker's yeast *Saccharomyces cerevisae* and the methylotropic yeast *Pichia pastoris.* Yeast systems utilize both integrated and extrachromosomal (non-integrated) vectors.

Strong inducible yeast promoters used for protein production in *S. cerevisae*, include *GAL*1, *GAL*5, and *GAL*7 promoters which are induced by galactose, and repressed by glucose. The wealth of information on its genetics and molecular biology has made *S. cerevisae* an excellent model organism for protein–ligand and protein–protein interactions. However, for an abundant expression of heterologous proteins, other yeast systems such as *P. pastoris* with its strong and highly regulated alcohol oxidase (AOX1) promoter, have been more successful. In the last 16 years expression of more than 300 foreign proteins have been reported using this system [105]. Examples of more than 100-fold higher protein yield of recombinant single chain antibodies in *P. pastoris* compared to *E. coli* have been reported [106,107].

In summary, fungal expression systems are often an excellent choice for eukaryotic protein expression. However, like any other system, fungi have their own limitations, including the inability to carry out the same types of glycosylation as higher eukaryotic organisms and, in some cases, problems related to accurate protein folding resulting in degradation of the protein.

11.7.3 Baculovirus Expression Systems

Baculoviruses are members of a large group of double-stranded DNA viruses which only infect invertebrates, including insects. The restricted host range makes baculoviruses safer than mammalian expression systems. The most widely used baculoviruses are *Autographa californica* nuclear polyhedrosis virus and the *Bombyx mori* nuclear polyhedrosis virus. The host cell most commonly used is Sf9, derived from the fall armyworm *Spodoptera frugiperda.*

Typically, the foreign gene is placed under the control of the extremely strong polyhedrin promoter, allowing for a highly efficient secretion of the heterologous protein into the insect cell culture medium. Glycosylation and other post-translational modifications occur in the insect cells. Up to 1998, more than 500 different heterologous proteins had been produced by the baculovirus expression vector, of which more than 95% had the correct post-translational modifications [108]. No doubt, this number has rapidly increased since. However, a limitation of the baculovirus system is that optimal expression levels require high-quality growth media, careful culturing and the expression of the foreign protein during the phase in which the producing cells are dying.

11.7.4 Stable Transformations of Insect Cells

Although not as popular as the baculovirus-system, stable transformations of insect cells can be used to circumvent the problems mentioned in the last paragraph. Common hosts include the fruitfly and mosquito. The expressed genes are often under the control of the *Drosophila* metallothionein promoter. Genes coding for resistance to antibiotics such as hygromycin and neomycin are used as selection markers.

11.7.5 Expression Using Mammalian Cells

The mammalian cell expression system contains all the necessary regulatory machinery for accurate and efficient processing and secretion of eukaryotic proteins, although there may be species differences. Foreign DNA is introduced into the cells either via virus infection or directly, employing chemical (for instance lipocomplexes or calcium phosphate) and physical (electroporation or microinjection) methods. The transcriptional control elements (enhancers and promoters) are complex and vary between mammalian cell types. However, simian virus 40 (SV40) and human cytomegalovirus (CMV) promoters are active in many cell types and are therefore commonly used. Obtaining stable transfected cell lines can be time consuming and therefore a transient expression system is often used for initial analysis. Typically COS (African green monkey kidney) and CHO (Chinese hamster ovary) cells are used for this purpose. An obvious advantage of mammalian cell expression is the possibility of advanced glycosylation. Generally, the yield of heterologous proteins produced in a mammalian cell system is much less than in other expression systems. However, for some proteins, the use of mammalian cells may solve the problems observed in prokaryotic and lower eukaryotic organisms with regard to accurate folding and modifications.

11.7.6 Expression Systems: Concluding Remarks

The choice of an expression system for the production of a drug delivery construct is of vital importance but at the same time a difficult task. Several general considerations when choosing an appropriate expression system are outlined in Table 11.4. The use of microorganisms (bacteria and fungi) results in high yield of the product and they are therefore often preferred by researchers. However, for highly specific therapeutic applications, the use of microorganisms is less favourable since they are unable to carry out the post-translational modifications necessary for activity of the protein. For instance, glycosylation is not possible in *E. coli* and although possible in fungi, it differs from that in mammalian systems. Use of insect

Table 11.4. Advantages and limitations of various expression systems.

Expression system	Advantages	Disadvantages
E. coli	Economical, fast, easy, high yield, well characterized genetics, large number of cloning vectors	Insolubility and misfolding of proteins, no glycosylation possible, difference in codon usage between prokaryotes and eukaryotes
Fungi	Economical, fast, easy, high yield, well characterized genetics (yeast), glycosylation possible, able to secrete correctly folded and processed proteins	Glycosylation and other post-translational modifications are often different to mammalian systems
Insect cells		
baculovirus	high yield, safe due to restricted host range, able to perform most of the post-translational modifications carried out by mammalian cells	Controlled culture conditions are required, expression peaks when cells are dying
stable transformants	Stable	Time consuming, relatively low yield
Mammalian cells	Advanced post-translational modifications, signals for synthesis, processing and excretion are correctly recognized	Time consuming, relatively low yield Complex regulatory system

and mammalian systems can, at least in part, overcome these limitations, but these systems are more expensive and difficult to manipulate due to complex regulatory systems. The choice of appropriate expression organism depends on the individual protein and its applications. In the past few years, the fundamental insights into the mechanisms of production, stability and cellular locations of proteins have increased greatly. This knowledge will help researchers working in the field of drug targeting to rationalize their choice of expression systems.

11.8 Recombinant DNA Constructs

11.8.1 Antibody-based Constructs

As stated above, a coding sequence for carrier, homing device and active drug can be designed together in one fusion construct. However, even if this construct consisted of a small carrier, a very short recognition sequence as the homing device and a small proteinaceous drug substance, the final design would encode for a relatively large protein. Due to the size of the construct, one could expect problems regarding the stable maintenance of the encoding gene in a certain expression system, in addition to problems with respect to accurate synthesis, export and folding of the recombinant protein. The smaller the total size of the recombinant protein, and the smaller the changes made to the construct as compared to a natural protein, the less likely it will be for these problems to occur. A significant reduction in re-

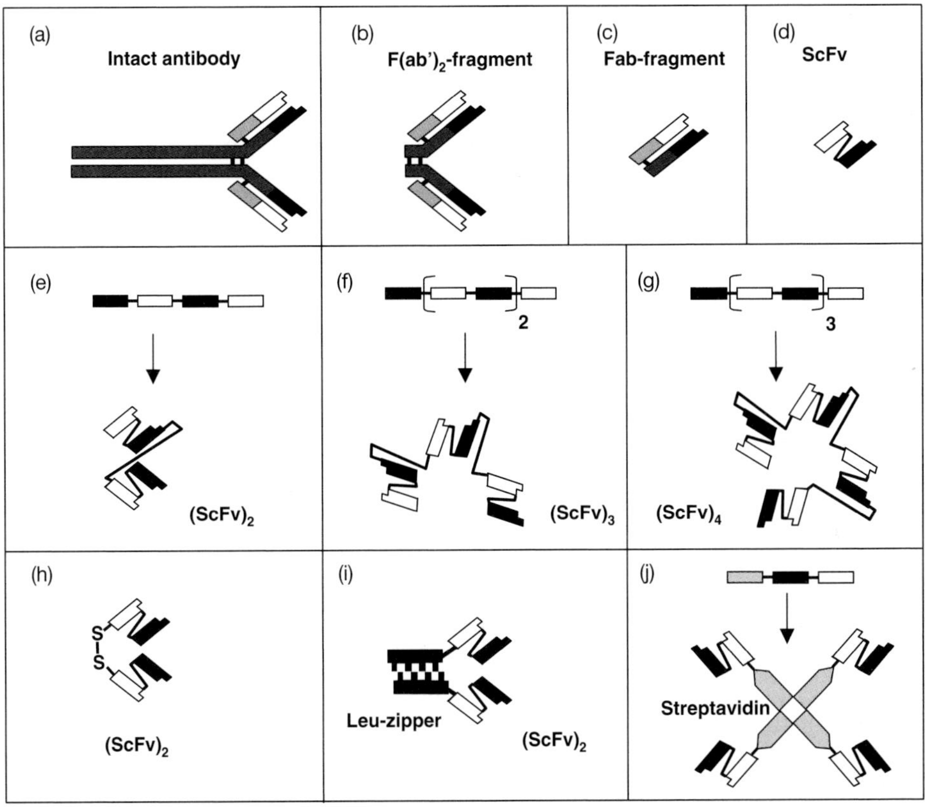

Figure 11.5. Schematic representation of genetically engineered antibody constructs for drug targeting. Intact antibodies consisting of two heavy and two light chains (a) can be converted into divalent $F(ab')_2$ fragments (b) or to monovalent Fab fragments (c). These fragments are stabilized via disulfide bridges. Alternatively, the variable heavy and light chain fragments are linked via a flexible linker resulting in a monovalent ScFv (d). Di-, tri- and tetravalent scFv fragments can be constructed by connecting two, three or four scFv fragments with peptide linkers (e–g) or by introducing a S–S bridge between the individual scFv fragments (h). Non-covalent interactions between scFv fragments are created by introducing leucine zipper sequences into the construct (i) or via streptavidin–biotin interactions (j).

combinant protein size, and thereby in complexity of the construct can be achieved when both the carrier and homing device functions are intrinsic properties of one protein.

Antibodies make up a group of proteins which can be considered to have the properties of both a carrier and a homing device and, as a result, have been used in many drug targeting studies [18,109]. However, the relatively large size (150 kDa) of whole IgG molecules hampers tissue penetration of these molecules. Several modifications of the original antibody structure can be carried out to reduce the size of the IgG molecule. For instance, in natural antibodies (Figure 11.5a) the Fc-region is necessary to activate T-cells of the immune system. Since this function of the antibody is not required in most drug targeting constructs, these domains have been removed by recombinant cloning techniques [110]. The resulting $F(ab')_2$, and Fab fragments, with molecular weight of around 100–110 and 50–55 kDa respectively, are linked together by one or more disulfide bridges (Figure 11.5b, c). Even smaller antibody-derived fragments, the so-called single chain Fv (scFv) molecules (Mw around 30 kDa), built up of a V_H and a V_L region linked by a flexible peptide linker, have been constructed (Figure 11.5d).

By fusion of a therapeutically active protein to a site other than the antigen recognition domain, these antibody fragments are able to function as carriers with intrinsic homing devices. However, a major drawback of these smaller fragments is the loss of the bivalent character normally present in the antibody, which ensures high avidity (functional activity). In many cases the multivalent character has been restored by genetic engineering [17,18]. Both covalent and non-covalent interactions have been used to combine two or more ScFv molecules to so-called di-, tri- and tetravalent constructs (Figure 11.5e–j). Covalent interactions can be achieved with totally genetically-engineered constructs consisting of two (or more) scFv moieties connected by a peptide linker, for instance repetitive sequences of the Gly_4-Ser motif [111]. Alternatively, covalent interaction between the subunits is established by the introduction of a cysteine residue at the C-terminus of the monovalent molecules or via chemical cross-linking reagents [17]. Non-covalent interactions have been created with specific sequences such as the leucine zipper domains (Fos- and Jun-fragments) which can interact with each other [112]. Another approach is the assembly of multiple scFv molecules using streptavidin-scFv fusion proteins [113]. This strategy will result in the production of tetrameric complexes due to the non-covalent assembly of four streptavidin moieties. In addition, bi-

Table 11.5. Immunotoxins, immunocytokines, cytotoxins and toxin-targeted constructs.

Type of construct	Targeting moiety	Effector moiety	Examples	References
Immunotoxins	Antibody or antibody fragment (scFv, Fab, F(ab')$_2$)	Bacterial or plant toxin	3B3-PE	[114]
			αTac-DT	[115]
			BerH2-SAP	[116, 117]
			OM124-PAP	[117, 118]
Immunocytokines	Antibody or antibody fragment	Cytokine, interleukin or growth factor	αEpCAM-IL-12	[119]
			αHer2/*neu*-IL-12	[120]
			RM4-TNFα	[121]
			RM4-IFNγ	[122]
Cytotoxins	Cytokine, interleukin, growth factor or the receptor binding domains of these proteins	Toxin, toxic protein or apoptosis-inducing protein	IL-2-DAB	[123]
			TGFα-PE40	[124, 125]
			IL-2-BAX	[126]
			bFGF-SAP	[127]
			bFGF-RNase	[128]
Toxin-targeted constructs	Receptor binding domain of toxin	ROS-scavenging protein, CTL-epitopes (e.g. gp120 HIV)	TT-SOD	[57]
			LFn-Ova$_{257\text{-}264}$	[129]
			LFn-LLO$_{91\text{-}99}$	[130]
			LF254-gp120	[131]

Antibodies have been used as the targeting moiety for the delivery of active drug substances like toxins (immunotoxins) or cytokines and other immunomodulatory proteins (immunocytokines). For reviews on immunotoxins see references [132–135]. Cytokines or their receptor binding domains have also been used as targeting moieties for toxins and other cell-killing proteins (cytotoxins). For reviews on cytotoxins see references [132, 134]. The receptor binding domain of toxins has been used to target other effector molecules, such as enzymes or CTL epitopes, to the cells expressing receptors for the toxin.

otin-labelled molecules can be attached to the recombinant protein via the biotin binding site of the construct.

Recombinant drug targeting constructs utilizing the targeting moiety of antibodies have predominantly involved toxins and cytokines as the active drug substance (Table 11.5). The bacterial toxins most commonly used for immunotoxin constructs are diphtheria toxin (DT), pseudomonas exotoxin (PE) and to a lesser extent shiga(-like) toxin (ST/SLT). These toxins all have the same overall composition: a receptor binding moiety, a fragment involved in membrane translocation and a toxic or catalytic domain. In order to avoid interactions with cells that bear the 'normal' toxin receptor, the receptor-binding domain is removed, resulting in so-called truncated toxins. Genetic coupling of such chemical 'bombs' to an antibody or scFv, results in a highly selective and potent drug targeting construct.

A variant of the original immunotoxin approach is the so-called immunocytokines. In these constructs the antibody targeting moiety is maintained, but the toxin as the effector molecule is replaced by a cytokine. In contrast to toxins, cytokines are often proteins endogenously produced in man. If both the antibody and cytokine are of human origin, then no foreign proteins are introduced which could provoke an antibody response from the host immune system when the drug targeting preparation is clinically applied.

Selective targeting of very potent cytokines may be an attractive approach to overcome the many side-effects seen after general systemic administration of such compounds [122]. Most cytokines are LMWPs and as such rapidly eliminated by renal glomerular filtration. Consequently, high doses are necessary to obtain locally effective concentrations.

The potential of immunocytokines was elegantly demonstrated in two separate studies using IL-12 for anti-tumour therapy. IL-12, a potent stimulator of natural killer cells and cytotoxic T-lymphocytes (CTL), activates the immune system to eradicate the cancer cells. IL-12

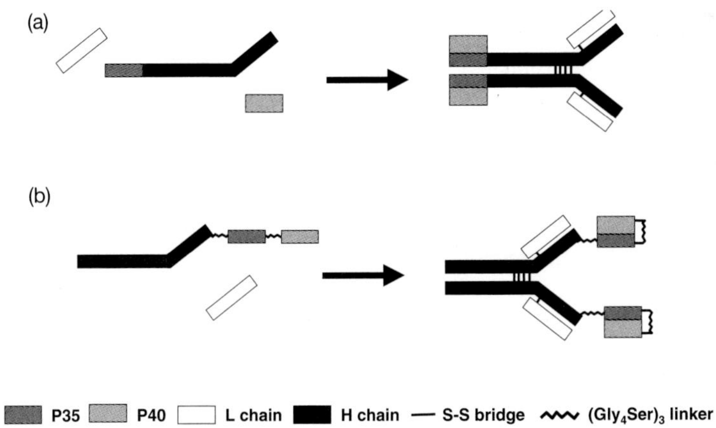

Figure 11.6. Schematic diagram showing the assembly of IL-12 protein for antibody-based drug delivery. (a) The mature sequences of the p35 subunit of IL-12 are fused to the C-terminus of the heavy chain of a tumour-specific antibody and co-expressed with the antibody light chain and the p40 subunit of IL-12. Formation of the final immunocytokine requires the creation of disulfide bridges between the antibody chains and interactions of p35 and p40 subunits of IL-12 [119]. (b) Alternatively the IgG heavy chain and both subunits of IL-12 can be linked via flexible linkers allowing for equimolar assembly of IL-12 [120].

is a heterodimeric protein composed of a p35 and a p40 subunit. A different approach for the final assembly of the IL-12 protein was followed in the two studies (Figure 11.6). Gillies *et al.* prepared a recombinant fusion construct of the p35 unit of IL-12 and a humanized anti-tumour antibody [119]. Co-expression of this construct with the p40 subunit yielded the final immunocytokine. In the second study, Peng *et al.* used recombinant single chain IL-12 (scIL-12), in which both subunits of the cytokine are linked via a flexible peptide linker, for the preparation of an IL-12–antibody construct [120]. This latter approach ensures the correct equimolar assembly of IL-12, and may confer stability to the fusion protein. Both approaches proved successful when tested in tumour xenograft models.

11.8.2 Receptor-targeted Constructs

Instead of utilizing the specific interaction between an antibody and its antigen epitope as the homing mechanism of a drug targeting construct, one could choose from a whole variety of specific interactions between structures on the target cell surface (receptors) and other molecules (ligands). Here some examples of constructs or potential construct moieties for drug targeting based on receptor–ligand interactions will be discussed.

11.8.2.1 Cytotoxins

The immunotoxins and immunocytokines have already been discussed. Another type of construct which makes use of both cytokines and toxins is the cytotoxin (Table 11.5). In these constructs the cytokine moiety is responsible for the targeting function, in contrast to the immunocytokines in which the cytokine moiety is the active drug compound. An example of a cytotoxin is the DT-IL-2 construct. The over-expression of the high affinity receptor for IL-2 on activated T-cells, B-cells and macrophages was utilized to selectively kill these cells. The receptor-binding domain of DT was replaced by the N-terminal IL-2 fragment to form the DAB486-IL-2 construct [123,134]. Upon binding to the IL-2 receptor, the construct was internalized by receptor-mediated endocytosis, and produced its toxic activity in the cytosol of the target cell.
Moreover, given its safe and well-tolerated behaviour in phase I/II clinical studies, and based on further information regarding the minimal structural requirements for the membrane translocation moiety and proteolytic activation, a second generation construct DAB389-IL-2 is currently being evaluated in a phase III clinical study [136]. These results have shown that immunotoxins and targeted cytotoxins can be used safely. However, the use of toxins is not always without risks and side-effects. For example, DT390-anti-CD3sFv and DT390-IL-3 showed, despite their selectivity, toxic side-effects in mouse models [137,138].

11.8.2.2 Toxin-targeted Constructs

As discussed in Section 11.8.1, many toxins of bacterial or plant origin are built up of different moieties or subunits which mediate binding, membrane translocation and catalytic or toxic activity. Anthrax and tetanus toxin are examples of toxins whose binding and membrane translocation moieties, but not their toxic moiety, have been used in drug targeting constructs.

Anthrax toxin is a bacterial toxin from *Bacillus anthracis* consisting of three parts: protective antigen (PA), lethal factor (LF) and edema factor (EF). Both LF and EF compete for binding sites on the PA protein. The PA protein binds with high affinity to an as yet unknown receptor on macrophages and related cell types. When PA is internalized by the target cells, it functions as a shuttle protein for either EF or LF. Intracellularly, in the acidic environment of the endosome, EF and LF are capable of entering the cytosol by pH-dependent pore formation [139].

The potential use of anthrax toxin as a delivery system aimed at antigen-presenting cells is most clearly demonstrated in the delivery of HIV gp120-derived peptides [131]. The N-terminal domain of LF was genetically fused to the gp120 portion of the HIV envelope protein. When administered in combination with recombinant PA, this construct elicited a specific cytotoxic T-lymphocyte immune response towards the HIV gp120 protein. This study and others, in which other peptide epitopes were delivered to antigen-presenting cells, imply a general application for the anthrax toxin as a peptide vaccine delivery vehicle [129,139].

A second example of a toxin that has been used as targeting device is tetanus toxin. Tetanus toxin is a potent neurotoxin, which can undergo uptake in the nerve endings of motor neurones and subsequent retrograde transport into the central nervous system. The non-toxic C-fragment of tetanus toxin (TTC, 451 amino acids), has been used to increase the neuronal uptake of the therapeutic protein SOD [57]. Following intravenous infusion, the recombinant hybrid protein reduced the occurrence of ischaemia-induced cerebral infarction in rats [58].

11.8.2.3 TfR-directed Constructs

The most frequently reported target for translocation of proteinaceous drugs across the blood–brain barrier is the transferrin receptor [140] (see also Chapter 2). Recently, a systematic study has reported on the genetic fusion of human transferrin and NGF [140]. This work clearly demonstrated the importance of controlling the positions of the different components within a drug targeting construct, and how this could be obtained by recombinant DNA technology. Only when NGF was cloned at the N-terminus of transferrin did the fusion protein retain the activities of both component molecules. In addition, a relatively long flexible linker between the two moieties, designed to promote dimerization, was required for functional activity. Direct attachment of NGF to transferrin probably prevented dimerization by steric hindrance.

The use of the TfR for drug targeting can be extended beyond the blood–brain barrier since all cells that have a high requirement for iron, such as actively proliferating tissues and tumour cells, express large numbers of transferrin receptors on their surface [19]. Cell activation, induced for instance by HIV-1 replication, also upregulates the levels of TfR expression. The delivery of anti-viral agents via the transferrin uptake pathway into HIV-infected cells has been reported [141]. DNA sequences, encoding nine amino acid residues, cleavable by HIV-1 protease, were cloned into the human transferrin gene. After uptake of these constructs, the recombinant protein could function as a competitive substrate for HIV-1 protease. This approach made use of the fact that surface-exposed loops of globular proteins can often tolerate insertions of additional amino acids without altering the function of the protein [142]. Molecular modelling was used to select candidate insertion sites in surface-exposed loops of transferrin that were distant from the biologically active domains.

Evidently, resolution of the three-dimensional structures of proteins will aid in the design of rational approaches for constructing drug conjugates, as demonstrated by the above-mentioned examples. Evaluation of molecular structure at this level may prove to be one of the more successful approaches used in the design of recombinant drug conjugates.

11.9 Recombinant Domains as Building Blocks for Drug Targeting Constructs

With the growing knowledge about protein structure–function relationships and the availability of new techniques like phage display, we are now able to select small proteinaceous sequences that could function as building blocks for recombinant targeting constructs. Such building blocks can function as a targeting moiety (homing device), a membrane translocation moiety and/or an active drug substance. In order to facilitate the construction of recombinant preparations, it seems reasonable to assemble the final construct from smaller subunits.

11.9.1 Targeting Domain

Probably the smallest sequence known to be responsible for receptor recognition is the RGD-tripeptide, initially discovered in fibronectin [143]. However, the specificity of the interaction with different integrins, the counter receptors of RGD sequences on the cell surface, is established by the flanking sequences of the RGD motif and the conformation of the tripeptide. In other words, the presentation of the RGD sequence is important for specific recognition by individual integrins.

Insertion of RGD sequences as targeting domains into protein carriers, is an attractive approach for integrin-directed targeting. Studies with RGD sequences of viral origin cloned into solvent-exposed regions of β-galactosidase demonstrated binding and internalization of the active chimeric enzyme into mammalian cells [144]. Likewise, introduction of an RGD motif into the capsid protein of adenoviruses was shown to increase the cell-specific delivery of adenoviral vectors as gene delivery vehicles [145]. Fusion proteins containing RGD sequences are likely to be effective delivery systems but the clinical relevance thereof awaits further analysis. The RGD motif can form the targeting domain, but at the same time can also function as the active drug, since its binding to the receptor may result in prevention or disruption of the natural ligand–integrin interaction, and consequently in a therapeutic response.

The use of phage-display techniques has identified peptide ligands with specific affinity to cell surface receptors or specific tissues (see Table 11.2). Such peptide homing devices can be genetically inserted into a recombinant protein backbone. For some of these peptides, the surface receptor has not yet been elucidated. Further investigation into the cellular processing of the receptor and its bound ligand is essential to ensure a rational design for the targeting constructs that are being developed.

11.9.2 Membrane Translocation Domain

Most of the targeting domains mentioned above are aimed at cell surface molecules for the obvious reason of accessibility. In many cases these target receptors are able to internalize together with the bound ligand. However, this process will deliver the construct to the lysosomes, a compartment in which enzymes and low pH will result in degradation of proteins. In order to escape this aggressive environment, a membrane translocation domain might be introduced so that the drug delivery preparation can cross the cell membrane in a receptor-independent manner.

Membrane translocation domains have been identified in toxins and viruses and derived from signal sequences of secreted proteins. When derived from a signal sequence the translocation domain contains hydrophobic sequences [146–148] while the toxin and viral translocation domains contain mostly basic residues [149,150].

In terms of targeting, membrane translocation domains lack specificity for particular cells or tissues. Therefore, these domains should be combined with targeting domains such as those discussed in the previous section. In such a construct, the targeting domain will ensure a rapid accumulation at the surface of a specific cell type and the translocation domain will facilitate entry into the cytosol of the target cells.

11.9.3 Assembly Domain

Although feasible and resulting in highly uniform end-products, the construction and synthesis of a complete drug targeting preparation as one genetic construct has one major disadvantage: lack of flexibility. If a construct does not show the expected results, the whole process of designing and production has to be repeated. Therefore, the use of assembly domains in the individual components of the drug targeting construct can be advantageous. Recently several studies have reported on the flexibility of such constructs containing an avidin/streptavidin moiety for non-covalent binding to biotinylated proteins. This approach was followed for the delivery of biotinylated compounds across the blood–brain barrier using a genetic fusion protein of avidin and an anti-TfR antibody [151]. A similar approach was used to engineer the RGD cell adhesion sequence into accessible surface regions of streptavidin without disrupting the biotin binding properties [152].

Even greater flexibility was achieved by genetic fusion of streptavidin with protein A [153,154]. Protein A specifically binds the Fc domain of IgG immunoglobulins of almost all mammals without inhibiting the antigen binding activity of the antibody. The streptavidin–protein A fusion construct was used for the assembly of complexes of biotinylated β-galactosidase and different monoclonal antibodies specific for tumour cell receptors. As a result these complexes were efficiently delivered into several cancer cell lines [154].

11.10 Concluding Remarks

This chapter has presented many approaches to the preparation of drug targeting constructs. In these constructs, a protein or part of it may function as a carrier for attached drug molecules, as a specific targeting moiety, or as a therapeutically active substance. Two entirely different approaches have been followed in the preparation of proteinaceous drug targeting constructs. First, chemical derivatization of existing proteins with site-directing ligands and/or drug molecules, and second, the engineering and expression of recombinant DNA constructs. Depending on the type of construct required, each approach has its own unique advantages. For instance, the engineering of protein backbone structures is most accurately performed by recombinant techniques. On the other hand, chemical approaches can be used for the attachment of small organic drug molecules, offering numerous opportunities for therapeutic intervention which cannot be matched by proteinaceous drug substances. Therefore, with reference to the title of this chapter, proteinaceous drug targeting constructs should preferably be prepared by chemical *and* recombinant DNA techniques, rather than by the exclusive use of either one.

Acknowledgements

R. J. Kok and S. A. Ásgeirsdóttir are members of UNYPHAR, a network collaboration between the universities of Groningen, Leiden and Utrecht and the pharmaceutical company Yamanouchi.

References

[1] Forssen E, Willis M, *Adv. Drug Deliv. Rev* **1998,** *29*, 249–271.
[2] Seymour LW, *Adv. Drug Deliv. Rev.* **1994,** *14*, 89–111.
[3] Brocchini S, Duncan R, In: *Encyclopedia of Controlled Drug Delivery* (Ed. Mathiowitz E), pp. 786–816. John Wiley and Sons, New York, **1999.**
[4] Meijer DKF, Molema G, Moolenaar F, De Zeeuw D, Swart PJ, *J. Control. Rel.* **1996,** *39*, 163–172.
[5] Maack T, Johnson V, Kau ST, Figuereido J, Sigulem D, *Kidney Int.* **1979,** *16*, 251–270.
[6] Stehle G, Wunder A, Schrenk HH, Hartung G, Heene DL, Sinn H, *Anti-Cancer Drugs* **1999,** *10*, 785–790.
[7] Maeda H, Wu J, Sawa T, Matsumura Y, Hori K, *J. Control. Rel.* **2000,** *65*, 271–284
[8] Stehle G, Sinn H, Wunder A, Schrenk HH, Schutt S, Maierborst W, Heene DL, *Anti-Cancer Drugs* **1997,** *8*, 677–685.
[9] Franssen EJF, Jansen RW, Vaalburg M, Meijer DKF, *Biochem. Pharmacol.* **1993,** *45*, 1215–1226.
[10] Melgert BN, Wartna E, Lebbe C, Albrecht C, Molema G, Poelstra K, Reichen J, Meijer DKF, *J. Drug Target.* **1998,** *5*, 329–342.
[11] Melgert BN, Olinga P, Jack VK, Molema G, Meijer DKF, Poelstra K, *J. Hepatol.* **2000,** *32*, 603–611.
[12] Haas M, Meijer DKF, Moolenaar F, De Jong PE, De Zeeuw D, In: *International Yearbook of Nephrology 1996* (Eds. Andreucci VE, Fine LG), pp. 3–11. Oxford University Press, Oxford, UK, **1996.**
[13] Franssen EJF, Moolenaar F, De Zeeuw D, Meijer DKF, *Adv. Drug Deliv. Rev.* **1994,** *14*, 67–88.
[14] Kok RJ, Grijpstra F, Walthuis RB, Moolenaar F, De Zeeuw D, Meijer DKF, *J. Pharmacol. Exp. Ther.* **1999,** *288*, 281–285.
[15] Haas M, Kluppel ACA, Wartna ES, Moolenaar F, Meijer DKF, De Jong PE, De Zeeuw D, *Kidney Int.* **1997,** *52*, 1693–1699.

[16] Milstein C, Waldmann H, *Curr. Opin. Immunol.* **1999,** *11*, 589–591.
[17] Hudson PJ, Curr. Opin. Immunol. **1999**, 11, 548–557.
[18] Colcher D, Pavlinkova G, Beresford G, Booth BJ, Choudhury A, Batra SK, *Q. J. Nucl. Med.* **1998,** *42*, 225–241.
[19] Wagner E, Curiel D, Cotten M, *Adv. Drug Deliv. Rev.* **1994,** *14*, 113–135.
[20] Pardridge WM, *Adv. Drug Deliv. Rev.* **1999,** *36*, 299–321.
[21] Feener EP, King GL, *Adv. Drug Deliv. Rev.* **1998,** *29*, 197–213.
[22] Fiume L, Busi C, Di Stefano G, Mattioli A, *Adv. Drug Deliv. Rev.* **1994,** *14*, 51–65.
[23] Schwartz AL, *CRC. Crit. Rev. Biochem.* **1984,** *16*, 207–223.
[24] Van der Sluijs P, Meijer DKF, In: *Liver Diseases. Targeted Diagnosis and Therapy Using Specific Receptors and Ligands* (Eds Wu GY, Wu CH), pp. 235–264. Marcel Dekker, Inc., New York, **1991.**
[25] Wadhwa MS, Rice KG, *J. Drug Target.* **1995,** *3*, 111–127.
[26] Stowell CP, Lee YC, *Adv. Carbohydr. Chem. Biochem.* **1980,** *37*, 225–281.
[27] Monsigny M, Roche A-C, Midoux P, Mayer R, *Adv. Drug Deliv. Rev.* **1994,** *14*, 1–24.
[28] Gabius S, Kayser K, Bovin NV, Yamazaki N, Kojima S, Kaltner H, Gabius H-J, *Eur. J. Pharm. Biopharm.* **1996,** *42*, 250–261
[29] Lerchen HG, Baumgarten J, Piel N, Kolb-Bachofen V, *Angew. Chem. Int. Ed. Engl.* **1999,** *38*, 3680–3683.
[30] Weikert S, Papac D, Briggs J, Cowfer D, Tom S, Gawlitzek M, Lofgren J, Mehta S, Chisholm V, Modi N, Eppler S, Carroll K, Chamow S, Peers D, Berman P, Krummen L, *Nature Biotechnol.* **1999,** *17*, 1116–1121.
[31] Mulligan MS, Warner RL, Rittershaus CW, Thomas LJ, Ryan US, Foreman KE, Crouch LD, Till GO, Ward PA, *J. Immunol.* **1999,** *162*, 4952–4959.
[32] Sudimack J, Lee RJ, *Adv. Drug Deliv. Rev.* **2000,** *41*, 147–162.
[33] Beljaars L, Molema G, Schuppan D, Geerts A, De Bleser PJ, Weert B, Meijer DKF, *J. Biol. Chem.* **2000,** *275*, 12743–12751.
[34] Koivunen E, Wang B, Ruoslahti E, *Biotechnology (NY)* **1995,** *13*, 265–270.
[35] Samoylova TI, Smith BF, *Muscle Nerve* **1999,** *22*, 460–466.
[36] Burg MA, Pasqualini R, Arap W, Ruoslahti E, Stallcup WB, *Cancer Res.* **1999,** *59*, 2869–2874.
[37] Rajotte D, Ruoslahti E, J. Biol. Chem. **1999**, *274*, 11593–11598.
[38] Martens CL, Cwirla SE, Lee RY, Whitehorn E, Chen EY, Bakker A, Martin EL, Wagstrom C, Gopalan P, Smith CW, *et al., J. Biol. Chem.* **1995,** *270*, 21129–21136.
[39] Arap W, Pasqualini R, Ruoslahti E, *Science* **1998,** *279*, 377–380.
[40] Ellerby HM, Arap W, Ellerby LM, Kain R, Andrusiak R, Del Rio G, Krajewski S, Lombardo CR, Rao R, Ruoslahti E, Bredesen DE, Pasqualini R, *Nature Med.* **1999,** *5*, 1032–1038.
[41] Pasqualini R, Koivunen E, Kain R, Lahdenranta J, Sakamoto M, Stryhn A, Ashmun RA, Shapiro LH, Arap W, Ruoslahti E, *Cancer Res.* **2000,** *60*, 722–727.
[42] Mao S, Gao C, Lo CH, Wirsching P, Wong CH, Janda KD, *Proc. Natl Acad. Sci. USA* **1999,** *96*, 6953–6958.
[43] Ruoslahti E, *Annu. Rev. Cell Dev. Biol.* **1996,** *12*, 697–715.
[44] Furitsu H, Ogawara K, Fujita T, Yamashita F, Takakura Y, Sezaki H, Hashida M, *Int. J. Pharm.* **1997,** *151*, 15–26.
[45] Kok RJ, Grijpstra F, Nederhoed KH, Moolenaar F, De Zeeuw D, Meijer DKF, *Drug Delivery* **1999,** *6*, 1–8.
[46] Swart PJ, Hirano T, Kuipers ME, Ito Y, Smit C, Hashida M, Nishikawa M, Beljaars L, Meijer DKF, Poelstra K, *J. Hepatol.* **1999,** *30*, 1034–1043.
[47] Inoue M, *Methods Enzymol.* **1994,** *233,* 212–221.
[48] Jansen RW, Molema G, Pauwels R, Schols D, De Clercq E, Meijer DKF, *Mol. Pharmacol.* **1991,** *39*, 818–823.
[49] Kuipers ME, Swart PJ, Witvrouw M, Esté JA, Reymen D, De Clercq E, Meijer DKF, *J. Drug Target.* **1999,** *6*, 323–335.
[50] Swart PJ, Beljaars E, Smit C, Pasma A, Schuitemaker H, Meijer DKF, *J. Drug Target.* **1996,** *4*, 109–116.
[51] Swart PJ, Beljaars L, Kuipers ME, Smit C, Nieuwenhuis P, Meijer DKF, *Biochem. Pharmacol.* **1999,** *58*, 1425–1435.
[52] Lloyd JB, *Adv. Drug Deliv. Rev.* **2000,** *41*, 189–200.
[53] Bodor N, *J. Control. Rel.* **1999,** *62*, 209–222.
[54] Eto T, Takahashi H, *Nature Med.* **1999,** *5*, 577–581.

[55] Conforti A, Caliceti P, Sartore L, Schiavon O, Veronese F, Velo GP, *Pharmacol Res.* **1991,** *23*, 51–56.
[56] Yabe Y, Nishikawa M, Tamada A, Takakura Y, Hashida M, *J. Pharmacol. Exp. Ther.* **1999,** *289*, 1176–1184.
[57] Francis JW, Hosler BA, Brown Jr RH, Fishman PS, *J. Biol. Chem* **1995,** *270*, 15434–15442.
[58] Francis JW, Ren J, Warren L, Brown Jr RH, Finklestein SP, *Exp. Neurol.* **1997,** *146*, 435–443.
[59] Brinkley M, *Bioconjug. Chem.* **1992,** *3*, 2–13.
[60] Carter DC, Ho JX, *Adv. Protein Chem* **1994,** *45*, 153–203.
[61] Yasuzawa T, Tomer KB, *Bioconjug. Chem.* **1997,** *8*, 391–399.
[62] Means GE, Feeney RE, *Bioconjug. Chem.* **1990,** *1*, 2–12.
[63] Franssen EJF, Koiter J, Kuipers CAM, Bruins AP, Moolenaar F, De Zeeuw D, Kruizinga WH, Kellogg RM, Meijer DKF, *J. Med. Chem.* **1992,** *35*, 1246–1259.
[64] McGraw TE, Maxfield FR, In: *Targeted Drug Delivery* (Ed. Juliano RL), pp. 11–41. Springer-Verlag, Berlin, **1991.**
[65] Barrett AJ, Heath MF, *Lysosomes: A Laboratory Handbook*. Elsevier, Amsterdam, **1977.**
[66] Fitzpatrick JJ, Garnett MC, *Anticancer Drug Des.* **1995,** *10,* 1–9.
[67] Soyez H, Schacht E, Van Der Kerken S, *Adv. Drug Deliv. Rev.* **1996,** *21*, 81–106.
[68] Rejmanova P, Kopacek J, Pohl J, Baudys M, Kostka V, *Makromol. Chem.* **1983,** *184*, 2009–2020.
[69] Duncan R, Cable HC, Lloyd JB, Rejmanová P, Kopacek J, *Makromol. Chem.* **1983,** *184*, 1997–2008.
[70] Franssen EJF, Moolenaar F, De Zeeuw D, Meijer DKF, *Pharm. Res.* **1993,** *10*, 963–969.
[71] Dosio F, Brusa P, Crosasso P, Arpicco S, Cattel L, *J. Control. Rel.* **1997,** *47*, 293–304.
[72] Fiume L, Chiricolo M, Busi C, Mattioli A, Spinelli C, Spinosa G, Bartolini G, Minghetti L, Orlandi M, Tomasi V, *Pharm. Acta Helv.* **1989,** *64*, 351–352.
[73] Shen W-C, Ryser J-P, *Biochem. Biophys. Res. Commun.* **1981,** *102*, 1048–1054.
[74] Franssen EJF, Moolenaar F, De Zeeuw D, Meijer DKF, *Int. J. Pharm.* **1992,** *80*, R15–R19.
[75] McIntyre GD, Scott Jr CF, Ritz J, Blattler WA, Lambert JM, *Bioconjug. Chem.* **1994,** *5*, 88–97.
[76] Blattler WA, Kuenzi BS, Lambert JM, Senter PD, *Biochemistry* **1985,** *24*, 1517–1524.
[77] Mueller BM, Wrasidlo WA, Reisfeld RA, *Bioconjug. Chem.* **1990,** *1*, 325–330.
[78] Kruger M, Beyer U, Zahn H, Unger C, Kratz F, *Chem. Pharm. Bull.* **1997,** *45*, 399–401.
[79] Coessens V, Schacht E, Domurado D, *J. Control. Rel.* **1996,** *38*, 141–150.
[80] Beyer U, Roth T, Schumacher P, Maier G, Unold A, Frahm AW, Fiebig HH, Unger C, Kratz F, *J. Med. Chem.* **1998,** *41*, 2701–2708.
[81] King HD, Yurgaitis D, Willner D, Firestone RA, Yang MB, Lasch SJ, Hellstrom KE, Trail PA, *Bioconjug. Chem.* **1999,** *10*, 279–288.
[82] Midoux P, Negre E, Roche A-C, Mayer R, Monsigny M, Balzarini J, De Clercq E, Mayer E, Ghaffar A, Gangemi JD, *Biochem. Biophys. Res. Commun.* **1990,** *167*, 1044–1049.
[83] Kuipers ME, Swart PJ, Hendriks MMWB, Meijer DKF, *J. Med. Chem.* **1995,** *38*, 883–889
[84] Wong SS, In: *Chemistry of Protein Conjugation and Cross-Linking*, (Ed. Wong SS), pp. 147–194. CRC Press, Boca Raton USA, **1993.**
[85] Cumber AJ, Westwood JH, Henry RV, Parnell GD, Coles BF, Wawrzynczak EJ, *Bioconjug. Chem.* **1992,** *3*, 397–401.
[86] Plourde R, Phillips AT, Wu CH, Hays RM, Chowdhury J, Chowdhury N, Wu GY, *Bioconjug. Chem.* **1996,** *7*, 131–137.
[87] Shen WC, Ryser HJP, La Manna L, *J. Biol. Chem.* **1985,** *260*, 10905–10908.
[88] Umemoto N, Kato Y, Hara T, *Cancer Immunol. Immunother.* **1989,** *28*, 9–16.
[89] Arunachalam B, Phan UT, Geuze HJ, Cresswell P, *Proc. Natl Acad. Sci. USA* **2000,** *97*, 745–750.
[90] Thorpe PE, Wallace PM, Knowles PP, Relf MG, Brown AN, Watson GJ, Knyba RE, Wawrzynczak EJ, Blakey DC, *Cancer Res.* **1987,** *47*, 5924–5931.
[91] Arpicco S, Dosio F, Brusa P, Crosasso P, Cattel L, *Bioconjug. Chem.* **1997,** *8,* 327–337.
[92] Galun E, Shouval D, Adler R, Shahaar M, Wilchek M, Hurwitz E, Sela M, *Hepatology* **1990,** *11*, 578–584.
[93] Adler R, Hurwitz E, Wands JR, Sela M, Shouval D, *Hepatology* **1995,** *22*, 1482–1487.
[94] Sherwood RF, *Adv. Drug Deliv. Rev.* **1996,** *22*, 269–288.
[95] Niculescu-Duvaz I, Springer CJ, *Adv. Drug Deliv. Rev.* **1997,** *26*, 151–172.
[96] Baneyx F, *Curr. Opin. Biotechnol.* **1999,** *10*, 411–421.
[97] Ferenci T, Silhavy TJ, *J. Bacteriol.* **1987,** *169*, 5339–5342.
[98] von Heijne G, *Curr. Opin. Cell Biol.* **1990,** *2*, 604–608.

[99] Verma R, Boleti E, George AJ, *J. Immunol. Methods* **1998,** *216*, 165–181.
[100] Verdoes JC, Punt PJ, Van den Hondel CAMJJ, *Appl. Microbiol. Biotechnol.* **1995,** *43*, 195–205.
[101] Archer DB, Peberdy JF, Crit. Rev. Biotechnol. **1997**, *17*, 273–306.
[102] Ward PP, Piddington CS, Cunningham GA, Zhou X, Wyatt RD, Conneely OM, *Biotechnology (NY)* **1995,** *13*, 498–503.
[103] Turner G, *Prog. Ind. Microbiol.* **1994,** *29*, 641–665.
[104] Mikosch T, Klemm P, Gassen HG, van den Hondel CA, Kemme M, *J. Biotechnol.* **1996,** *52*, 97–106.
[105] Cereghino GP, Cregg JM, *Curr. Opin. Biotechnol.* **1999,** *10*, 422–427.
[106] Ridder R, Schmitz R, Legay F, Gram H, *Biotechnology (NY)* **1995,** *13*, 255–260.
[107] Eldin P, Pauza ME, Hieda Y, Lin G, Murtaugh MP, Pentel PR, Pennell CA, *J. Immunol. Methods* **1997,** *201*, 67–75.
[108] Glick BR, Pasternak JJ, In: *Molecular Biotechnology: Principles And Applications of Recombinant DNA* (Eds Glick BR, Pasternak JJ), pp. 145–169. American Society for Microbiology, Washington DC, **1998.**
[109] Kosmas C, Linardou H, Epenetos AA, *J. Drug Target.* **1993,** *1*, 81–91.
[110] Segal DM, Weiner GJ, Weiner LM, *Curr. Opin. Immunol.* **1999,** *11*, 558–562.
[111] Atwell JL, Breheney KA, Lawrence LJ, McCoy AJ, Kortt AA, Hudson PJ, *Protein Eng.* **1999,** *12*, 597–604.
[112] Kostelny SA, Cole MS, Tso JY, *J. Immunol.* **1992,** *148*, 1547–1553.
[113] Dubel S, Breitling F, Kontermann R, Schmidt T, Skerra A, Little M, *J. Immunol. Methods* **1995,** *178*, 201–209.
[114] Bera TK, Kennedy PE, Berger EA, Barbas CF, Pastan I, *Mol. Med.* **1998,** *4*, 384–391.
[115] Chaudhary VK, Gallo MG, Fitzgerald DJ, Pastan I, *Proc. Natl Acad. Sci. USA* **1990,** *87*, 9491–9494.
[116] Grossbard ML, Fidias P, Kinsella J, O'Toole J, Lambert JM, Blattler WA, Esseltine D, Braman G, Nadler LM, Anderson KC, *Br. J. Haematol.* **1998,** *102*, 509–515.
[117] Terenzi A, Bolognesi A, Pasqualucci L, Flenghi L, Pileri S, Stein H, Kadin M, Bigerna B, Polito L, Tazzari PL, Martelli MF, Stirpe F, Falini B, *Br. J. Haematol.* **1996,** *92*, 872–879.
[118] Bolognesi A, Tazzari PL, Olivieri F, Polito L, Lemoli R, Terenzi A, Pasqualucci L, Falini B, Stirpe F, *Br. J. Haematol.* **1998,** *101*, 179–188.
[119] Gillies SD, Lan Y, Wesolowski JS, Qian X, Reisfeld RA, Holden S, Super M, Lo KM, *J. Immunol.* **1998,** *160*, 6195–6203.
[120] Peng LS, Penichet ML, Morrison SL, *J. Immunol.* **1999,** *163*, 250–258.
[121] Xiang J, Moyana T, Qi Y, *J. Biotechnol.* **1997,** *53*, 3–12.
[122] Xiang J, *Hum. Antibodies* **1999,** *9*, 23–36.
[123] Murphy JR, vanderSpek JC, *Semin. Cancer Biol.* **1995,** *6*, 259–267.
[124] Chaudhary VK, Fitzgerald DJ, Adhya S, Pastan I, *Proc. Natl Acad. Sci. USA* **1987,** *84*, 4538–4542.
[125] Siegall CB, Xu YH, Chaudhary VK, Adhya S, Fitzgerald D, Pastan I, FASEB J. **1989**, *3*, 2647–2652.
[126] Aqeilan R, Yarkoni S, Lorberboum-Galski H, *FEBS Lett.* **1999,** *457*, 271–276.
[127] Yu C, Cunningham M, Rogers C, Dinbergs ID, Edelman ER, *J. Pharm. Sci.* **1998,** *87*, 1300–1304.
[128] Futami J, Seno M, Ueda M, Tada H, Yamada H, *Protein Eng.* **1999,** *12*, 1013-1019.
[129] Ballard JD, Doling AM, Beauregard K, Collier RJ, Starnbach MN, *Infect. Immun.* **1998,** *66*, 615–619.
[130] Doling AM, Ballard JD, Shen H, Krishna KM, Ahmed R, Collier RJ, Starnbach MN, *Infect. Immun.* **1999,** *67*, 3290–3296.
[131] Goletz TJ, Klimpel KR, Arora N, Leppla SH, Keith JM, Berzofsky JA, *Proc. Natl Acad. Sci. USA* **1997,** *94*, 12059–12064.
[132] Strom TB, Anderson PL, Rubin-Kelley VE, Williams DP, Kiyokawa T, Murphy JR, *Semin. Immunol.* **1990,** *2*, 467–479.
[133] Kreitman RJ, Pastan I, *Adv. Drug Deliv. Rev.* **1998,** *31*, 53–88.
[134] Murphy JR, *Curr. Opin. Struct. Biol.* **1996,** *6,* 541–545.
[135] Kreitman RJ, *Curr. Opin. Immunol.* **1999,** *11*, 570–578.
[136] Frankel AE, Kreitman RJ, Sausville EA, *Clin. Cancer Res.* **2000,** *6,* 326–334.
[137] Vallera DA, Panoskaltsis-Mortari A, Blazar BR, *Protein Eng.* **1997,** *10*, 1071–1076.
[138] Vallera DA, Seo SY, Panoskaltsis-Mortari A, Griffin JD, Blazar BR, *Protein Eng.* **1999,** *12*, 779–785.
[139] Ballard JD, Collier RJ, Starnbach MN, *Proc. Natl Acad. Sci. USA* **1996,** *93*, 12531–12534.

[140] Park E, Starzyk RM, McGrath JP, Lee T, George J, Schutz AJ, Lynch P, Putney SD, *J. Drug Target.* **1998,** *6*, 53–64.
[141] Ali SA, Joao HC, Hammerschmid F, Eder J, Steinkasserer A, *J. Biol. Chem.* **1999,** *274*, 24066–24073.
[142] Finkelstein AV, *Curr. Opin. Struct. Biol.* **1997,** *7,* 60–71.
[143] Pierschbacher MD, Ruoslahti E, *Nature* **1984,** *309*, 30–33.
[144] Villaverde A, Feliu JX, Aris A, Harbottle RP, Benito A, Coutelle C, *Biotechnol. Bioeng.* **1998,** *59*, 294–301.
[145] Reynolds P, Dmitriev I, Curiel D, *Gene Ther.* **1999,** *6*, 1336–1339.
[146] Lin YZ, Yao SY, Hawiger J, *J. Biol. Chem.* **1996,** *271*, 5305–5308.
[147] Zhang L, Torgerson TR, Liu XY, Timmons S, Colosia AD, Hawiger J, Tam JP, *Proc. Natl Acad. Sci. USA* **1998,** *95*, 9184–9189.
[148] Rojas M, Donahue JP, Tan Z, Lin Y-Z, *Nature Biotechnol.* **1998,** *16*, 370–375.
[149] Blanke SR, Milne JC, Benson EL, Collier RJ, *Proc. Natl Acad. Sci. USA* **1996,** *93*, 8437–8442.
[150] Schwarze SR, Dowdy SF, *Trends Pharmacol. Sci.* **2000,** *21*, 45–48.
[151] Penichet ML, Kang YS, Pardridge WM, Morrison SL, Shin SU, *J. Immunol* **1999,** *163*, 4421–4426.
[152] McDevitt TC, Nelson KE, Stayton PS, *Biotechnol. Prog.* **1999,** *15*, 391–396.
[153] Sano T, Cantor CR, *Biotechnology (NY)* **1991,** *9*, 1378–1381.
[154] Ohno K, Levin B, Meruelo D, *Biochem .Mol. Med.* **1996,** *58*, 227–233.

12 Use of Human Tissue Slices in Drug Targeting Research

Peter Olinga, Geny M. M. Groothuis

12.1 Introduction

It has long been recognized that *in vitro* research can provide valuable information on basic mechanisms with respect to kinetics and efficacy of drug targeting concepts. Such *in vitro* research includes the use of isolated cells, cell lines and perfused organs. In this chapter the introduction of tissue slices into drug targeting research will be discussed. A brief history of the slice technique will first be given. Until now most research on the slice technique has been focused on the metabolism and transport of drugs and this topic will therefore be summarized before embarking on a discussion of the contribution of the tissue slice technique to the area of drug targeting research.

In vitro research began with organ culture of embryonic organ rudiments [1]. The slice technique, using slices of tumour and liver tissue, was performed as early as 1923 by Otto Warburg [2] and in the following years by H. A. Krebs [3], who investigated the metabolism of amino acids in liver slices of cats, dogs and rats. These liver slices were prepared manually with limited reproducibility and viability [4]. After a decline in the application of slices in liver research in favour of the use of isolated hepatocytes as well as isolated perfused liver preparation, the development of the Krumdieck slicer in the 1980s led to a 'comeback' of the technique enabling the production of reproducible and viable liver slices [5]. This technology induced a renaissance of the slice technology. The development of these *in vitro* preparations has been of paramount importance for research on *human* liver function. As most of the research with tissue slices concerned the liver, this chapter will focus on the use of liver as the target tissue and only briefly mention the use of slices from other tissues in the concluding section of the chapter.

The most abundant cell type in the liver is the hepatocyte, other cells in the liver are the non-parenchymal cells: Kupffer cells, the resident macrophages of the liver, endothelial cells and stellate cells. These cells have been discussed in more detail in Chapter 4.

Since a high yield isolation procedure of rat hepatocytes was described in 1969 [6], hepatocytes have become the model of choice for drug transport studies in the liver *in vitro* [7]. With this procedure, isolated hepatocytes from many species have been prepared, including hepatocytes from rat, mouse, chicken, dog, fish, hamster, pig, cow, sheep and monkey liver (for an extensive review see reference [8]). Before 1976, only relatively small numbers of human hepatocytes could be isolated, due to the use of non-perfusion techniques [9]. Bojar *et al.* [10] were the first to use a perfusion technique on human livers which greatly enhanced the yield of hepatocytes. In principle the procedure that is now commonly used, is based on the one described by Seglen [6] for rat hepatocytes. Either a biopsy wedge with intact capsula on three sides [11,12], or single lobes of liver [13] and even entire human livers [14] are used. One or more cannulas are inserted into (branches of) the portal vein(s) [12,13] and the tissue is perfused with collagenase. In general, the yield of human hepatocytes (5–20 x 10^6 hepatocytes g^{-1} liver [12, 15–18]) is low compared to rat hepatocytes (60–70 x 10^6 hepatocytes g^{-1} liver, average yield in our laboratory).

The first studies with isolated human hepatocytes concentrated on the characterization of these hepatocytes, as well as on the improvement of the isolation procedure, and the possibilities of culturing these cells [16,19–24]. Thereafter studies were performed to investigate the metabolism of drugs [25–28], in which emphasis was often put on the activity and concentration of cytochrome P450 isoforms. Nowadays, hepatocytes are more generally used in metabolic studies of specific compounds, in order to unravel potential species differences and

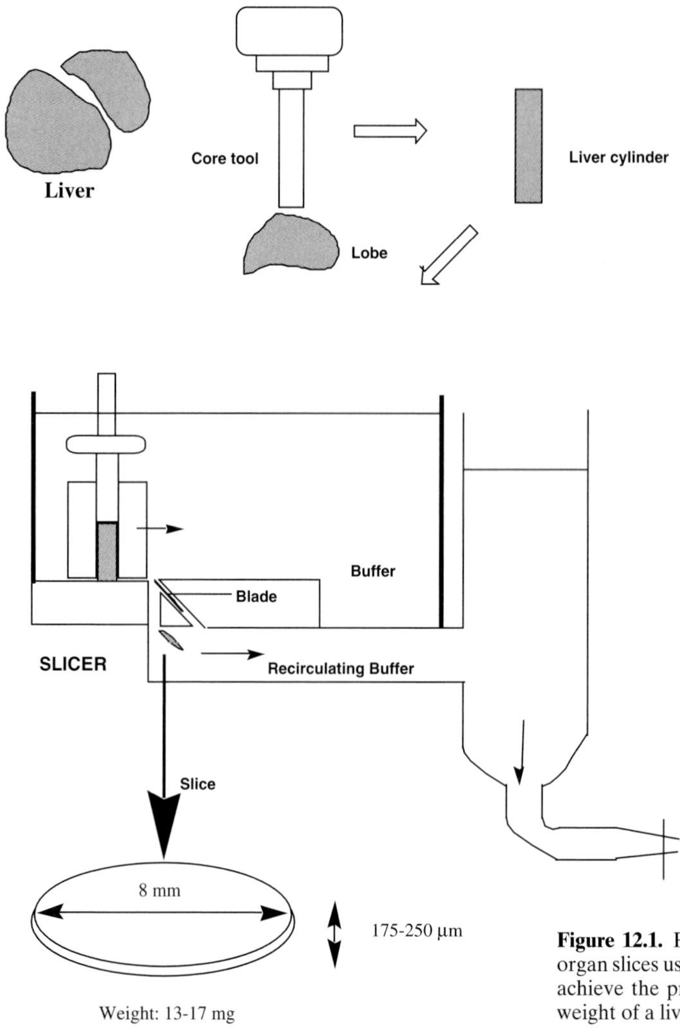

Figure 12.1. Production of precision-cut organ slices using the Krumdieck slicer to achieve the preferred thickness and wet weight of a liver slice.

are also extensively used in a variety of other research fields including pharmacology and toxicology.

Isolation from human liver of other cell types, such as Kupffer, endothelial and stellate cells has also been developed and extensively reviewed [29–31].

Although the isolated (human) liver cells have been shown to be valuable in the study of mechanisms of drug transport in drug targeting research, the isolation procedures involves digestion to disrupt the cell-to-cell contacts. Clearly, enzymatic digestion may also damage plasma membranes and transport systems therein. In addition, for hepatocytes the normal polarity is lost after isolation. For instance it has been reported that by using collagenase digestion to isolate hepatocytes, the amount of asialoglycoprotein receptor present on the membrane of the hepatocyte is reduced [32]. Recently, Ikejima *et al.* [33] showed that Kupffer cells isolated by the standard isolation procedure with collagenase and pronase had lost their CD14 receptor, presumably an important receptor in the uptake of lipopolysaccharide.

In contrast to isolated liver cells, no enzymatic digestion is necessary for the isolated perfused liver preparation. The isolated perfused liver preparation, extensively used to study rat liver functions [34], has also been employed with human liver tissue, but the application is limited by the fact that only pieces of tissue that are encapsulated with liver capsula can be used. The use of the isolated perfused human liver preparation has been discussed further in Chapter 4. Another *in vitro* liver preparation that can be used without the need for enzymatic digestion is the liver slice (Figure 12.1). One of the main features of slices is that the original architecture of the organ is retained in the slice. Therefore, in liver slices, the different cell types of the (human) liver; i.e. hepatocytes, Kupffer-, endothelial and stellate cells are still present in contact with their original matrix environment, which enables the function of all cell types present and their normal intercellular communication to be studied. In addition, studies on cell selective distribution of carriers and drug–carrier conjugates can be performed in liver slices, which is of major importance in drug targeting research. The slice technique itself can also be used on other solid organs such as kidney, lung, intestine and even brain; the use of slices from these organs will be described in the concluding section of this chapter, whilst the greater part of the chapter will describe the use of liver slices in drug targeting research.

12.2 Preparation of Liver Slices

Liver slices were initially prepared manually using razor blades or mechanical instruments such as the Stadie Riggs tissue slicer [4]. The reproducibility of the thickness of the slices at that time was largely dependent on the skills of the operator. Of note is the fact that the minimal slice thickness that could be produced was about 0.5 mm. This dimension appeared to limit the penetration of nutrients and oxygen into the inner cell layers: central necrosis in the slice occurred during incubation [35].

The introduction of the Krumdieck slicer enabled a more optimal and reproducible preparation of liver slices (Figure 12.1). With this technique the thickness of the slices is adjustable to a value as low as 100 μm. The slicing procedure itself is performed in a buffer assuring minimal trauma of the tissue. In addition, the Krumdieck slicer provides a rapid and automated production of slices with reproducible thickness. Recently, the so-called 'Brendel slicer' was introduced, which has largely the same characteristics as the Krumdieck slicer, but offers the advantage of more constant oxygenation. However, with this technique the slices have to be prepared manually [36]. Liver slices from both tissue slicers have been evaluated. No significant differences were observed in levels of protein, potassium, total glutathione (i.e. GSH and GSSG), reduced glutathione (GSH) and cytochrome P450 and activities of 7-ethoxyresorufin O-deethylase and 7-benzoxyresorufin O-debenzylase in freshly cut rat liver slices produced by either of the two tissue slicers [37].

To prepare liver slices with the Krumdieck slicer, cylindrical cores of tissue are first isolated from the liver specimens (Figure 12.1). These tissue cores are prepared preferably by advancing a sharp rotating metal tube into the liver tissue using a drilling press, thus assuring the preparation of accurately cylindrical cores. If a biopsy punch is used to prepare the cores, it is difficult to obtain a uniform cylindrical shape.

The cores are subsequently placed in the slicer, and the slicing procedure is performed by advancing the core over an oscillating knife in a controlled environment (Figure 12.1). Cold (4°C) Krebs–Henseleit buffer (pH = 7.4, saturated with 95% O_2 and 5% CO_2) supplemented with 25 mM glucose is commonly used in preparing the slices [35,38–40], but Williams' medium E [41], Earle's balanced salt solutions [37], Sacks preservation medium [42] and V-7 preservation buffer [43,44] are also used.

All these buffers have a glucose concentration of 25 mM, which seems to be essential for the viability of the slices.

The optimal thickness for liver slices, in order to retain their viability during culture, is approximately 175–250 μm. Price *et al.* [45] reported that the optimal thickness of liver slices for drug metabolism studies should be 175 μm. In slices thicker than 250 μm the inner cell layers suffer from a lack of oxygen and substrates, and in slices thinner than 175 μm the ratio of damaged cells in the outer cell layers to the living cell mass becomes unfavourable [36,43,44,46–48]. For cryopreservation slightly thicker slices were reported to give better results [40], although recent developments show that slices of approximately 200–250 μm can also be successfully used for cryopreservation (unpublished observation).

12.3 Incubation and Culture of Liver Slices

12.3.1 Incubation Systems

Previously, liver slices were incubated in static organ cultures [1]. Hart *et al.* [49] cultured rat liver slices for 24 h spread out on wet filter paper, floating on top of the incubation medium. Several slice-containing vessels were placed in a box with saturated 95% O_2 and 5% CO_2 at 37°C. However, the slices employed were rather thick (approximately 0.3 mm) and only the upper cell layers (0.2 mm) in the slice contained viable cells. Together with the introduction of the Krumdieck slicer [5,46], a new incubation technique for slices, the dynamic organ culture system (DOC), was introduced [35]. The main characteristic of this system is the intermittent exposure of the slice to incubation medium and the gas phase. The DOC is in fact a modified version of the Trowell incubation system [1].

Meanwhile many incubation systems have been developed, mostly based on either DOC or culturing the slices in multi-well incubation systems [38,40–43,50–52] and all have been used in pharmacological and toxicological research [36]. The most remarkable phenomenon emerging from these studies is the observation that the liver slices can be cultured for up to 72 h with the maintenance of their biotransformation activities [43]. In contrast, the use of primary suspensions of hepatocytes for metabolic and transport studies is restricted to a few hours [25]. Culturing of the hepatocytes allows experiments to last for a longer period of time (up to approximately 5–7 days). The hepatocytes form monolayers and develop bile canalicular-like spaces in between the cells [20]. However, specific liver functions such as albumin secretion, transport activity and cytochrome P450 activity decrease considerably during incubation [53,54]. After 24 h of culturing, drug metabolism activity will already have decreased by about 50%. This is very likely due to de-differentiation on the level of gene transcription [55]. In recent years much effort has been put into the improvement of the culture conditions of hepatocytes by adding extracellular matrix components or by co-culturing with other cell types in order to maintain their differentiation status [54,56–63]. Although survival and functioning of these cells has been greatly improved, complete maintenance of differentiated isoenzymes patterns has not been achieved yet. In fact, the liver slices can be seen as the most natural co-culture system within the original matrix.

12.3.2 Evaluation of Incubation Systems

There are only a few studies published in which the various incubation systems for liver slices were evaluated. Smith *et al.* [35] showed that slices in dynamic organ culture maintain their viability, as measured by ATP and potassium concentration, up to 20 h. Connors *et al.* [64] used a 24-well incubation system, in which the medium was stirred with a magnetic stirrer. In this incubation system rat liver slices were cultured for 8 h and human liver slices for 9 h during which time a high potassium concentration was maintained in the slices. Connors *et al.* [65] reported that the 24-well incubation system and the dynamic organ culture gave similar metabolite patterns after 24 h of incubation with a somatostatin analogue. Vickers *et al.* [66] also used the 24-well incubation system for 24 h, but no viability parameters were described. Dogterom *et al.* [51] showed that in a 12-well culture plate, which is put on a gyratory shaker, rat liver slices maintain their viability up to 11 h as determined by potassium concentration and ATP content. However, an impairment of the rat liver slices in a 24-well incubation system on the gyratory shaker was seen after 11 h. This was explained by the insufficient agitation of the medium in the 24-well incubation system. Leeman *et al.* [41] described a modification of the dynamic organ culture: a netwell insert (200-µm polyester mesh carrier) placed in the wells of a six-well culture plate on a rocker platform. In this system, as with the DOC, the slices are intermittently exposed to the gas phase, which in this system is 40% O_2/5% CO_2/55% N_2 and to the medium. Using this incubation system, the 3[4,5-dimethyl-thiazole-2-yl]-2,5-diphenyltetrazolium bromide (MTT) reduction, a test for the cellular reduction capacity both in mitochondria and extramitochondrially involving NADH and NADPH [67], was maintained in the slices for up to 72 h [41]. Simple incubation of slices in a 25-ml Erlenmeyer flask in a shaking water bath was reported by de Kanter *et al.* to be successful over a 24-h period [40].

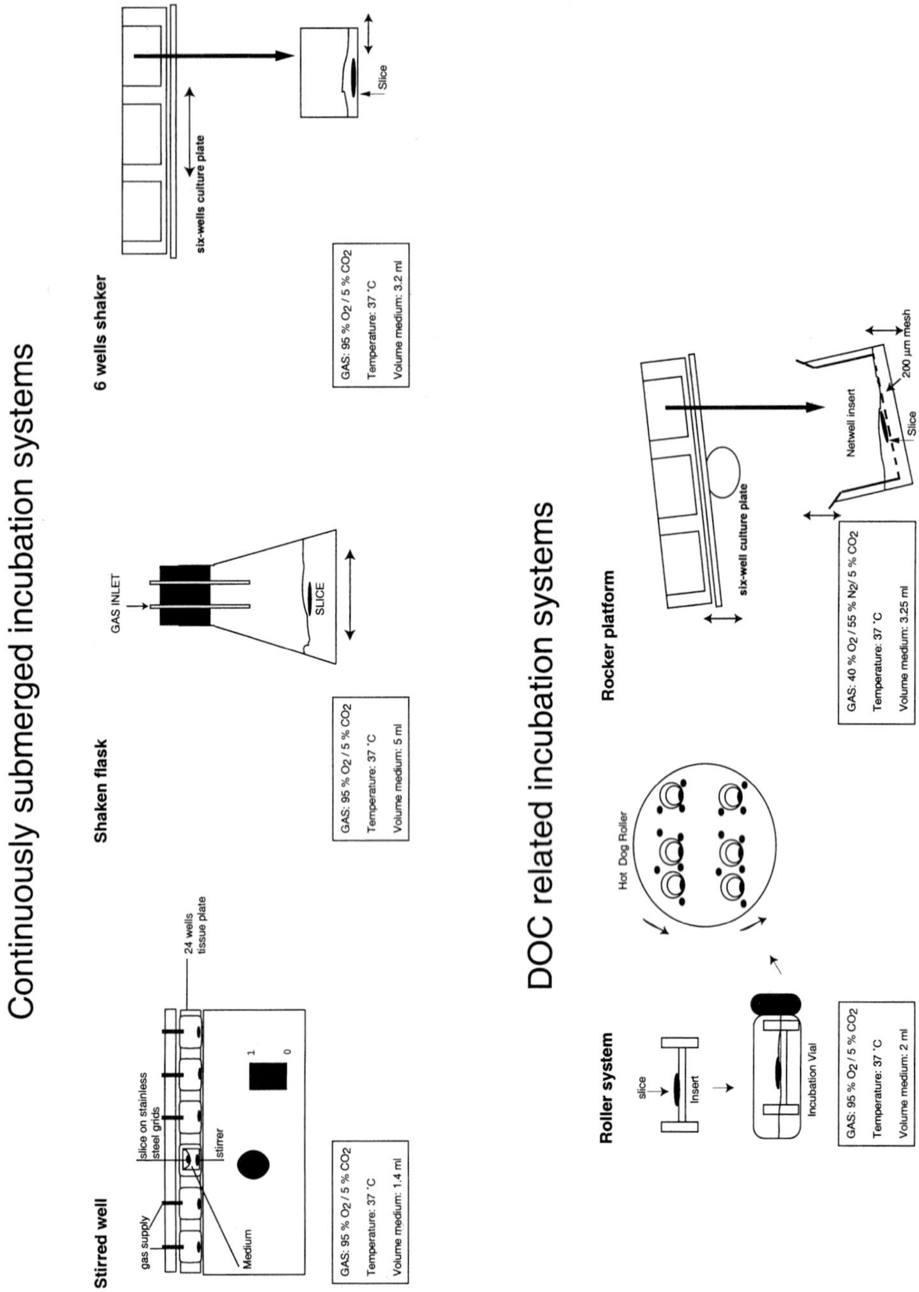

Figure 12.2. The five incubation systems for liver slices, divided into two groups: incubation systems continuously submerged in culture medium and dynamic organ culture-related incubation systems, where the liver slices are intermittently exposed to the medium and to the air.

Based on these various findings under a variety of conditions a thorough comparison of five incubation systems (Figure 12.2) was made by us in a collaborative study of four laboratories [52]. The five systems that were evaluated included: the shaken flask (a 25-ml Erlenmeyer flask in a shaking water bath [40]), the stirred well (24-well culture plate equipped with stainless steel grids and magnetic stirrers [38,64]), the rocker platform (a DOC system using six-well culture plates with Netwell inserts, rocked on a platform [41]), the roller system (dynamic organ culture rolled on an insert in a glass vial [35]) and the six-well shaker (six-well culture plates in a shaking water bath). In the rocker platform 40% O_2/5% CO_2/55% N_2 was used whereas in the other four systems 95% O_2/5% CO_2 was used to oxygenate the tissue. The liver slices were incubated in these incubation systems for 0.5, 1.5 and 24.5 h and subsequently subjected to viability and metabolic function tests. The viability of the incubated liver slices was evaluated by potassium content, MTT assay, energy charge, histomorphology and lactate dehydrogenase (LDH) leakage. Their metabolic functions were studied by determination of the metabolism of lidocaine (Figure 12.3), testosterone and antipyrine. Up to 1.5 h of incubation, all five incubation systems gave similar results with respect to viability and metabolic function of the slices. However, after 24 h, the shaken flask, the rocker platform and the six-well shaker incubation systems, appeared to be superior to the stirred well and the roller incubator. It is notable that the cytochrome P450-dependent metabolism of testosterone and lidocaine was retained at the same levels as found after 0.5 and 1.5 h of incubation in the shaken flask, rocker platform and six-well incubation systems. This suggests that the de-differentiation seen after 24 h in pure hepatocyte culture does not occur in slices for at least 24 h.

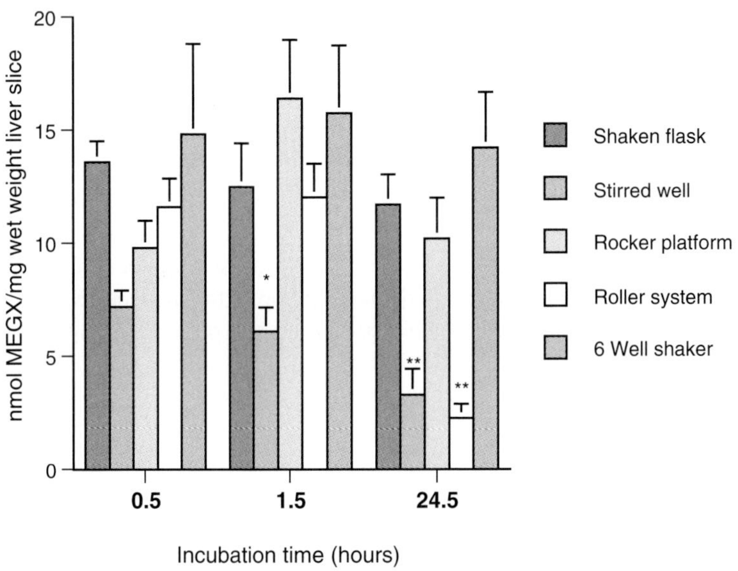

Figure 12.3. Metabolism of lidocaine to MEGX (in nmol MEGX mg^{-1} wet weight liver slice) in liver slices after different incubation times (h). *$p < 0.05$ versus shaken flask, rocker platform, roller system and six-well shaker. **$p < 0.05$ versus shaken flask, rocker platform and six-well shaker. Data are the mean of three separate experiments ± SEM.

Brendel's group compared two incubation systems, the roller system (dynamic organ culture (Figure 12.2)) and the 12-well plate culture (the plates were put on a gyratory shaker), with respect to their ability to maintain the functionality of rat liver slices over 72 h of culturing. The slices were evaluated with respect to ATP concentration, potassium retention, MTT reduction and protein synthesis, in addition to alanine transaminase (ALT) and LDH leakage. Metabolic function was investigated by oxidative *O*-deethylation of 7-ethoxycoumarin (7-EC) [43]. It was concluded that dynamic organ culture was superior to multi-well plate culture [43]. Recently, another comparison has been made between the DOC-system and 12-well system, showing that the 12-well system was superior to the DOC-system with regard to the metabolism of xenobiotics following long-term incubations (> 24 h) [68]. However, this study was performed in 95% air and 5% CO_2, which may have influenced the results obtained. A high oxygen percentage of at least 40% is essential for optimal incubation of liver slices, as will be described in more detail below. In addition, both sets of experiments were carried out using different incubation media, which also may have influenced the results obtained.

12.3.3 Incubation Systems for Human Liver Slices

Various incubation systems have also been tested using human liver slices, these include the 24-well plates with magnetic stirrers [38,66,69], the six-well plates in a shaker [52] and the DOC roller system [70,71], but no direct comparison has been made as yet. Human liver slices can be cultured for 72 h in DOC and maintain their ability to respond to specific inducers of cytochrome P450 such asmethylclofenapate and Aroclor 1254 [71].

12.3.4 Oxygenation and Culture Media for Liver Slice Incubation

In addition to the incubation system itself, the oxygen and nutrient concentration of the medium are also important for the viability of liver slices [46,47]. It appeared that a nutrient-enriched medium containing bicarbonate maintained the slice viability better than a simpler medium such as Krebs–HEPES buffer [44]. It was shown that K^+-retention, protein synthesis and LDH leakage was maintained in rat liver slices for 5 days in a Waymouth's/bicarbonate medium in a dynamic organ culture system [44]. Oxygenation of the hepatocytes, especially those in the centre of the slice, has been a major concern. In this respect it is important to note that oxygen at too high a concentration may be toxic, due to tissue damage by oxygen radicals, whereas excessively low levels of oxygen may result in ischaemia. Both 95% air/5% CO_2 and 95% O_2/5% CO_2 are commonly used. In long-term culture up to 5 days [43] the DOC system or DOC-related incubation systems are recommended, because of the intermittent exposure of the slice to culture medium and to the gas phase. This feature was claimed to be important for optimal gas exchange. In our experiments however, slices that were continuously submerged in medium performed equally well or even better than those in the DOC system or DOC-related system (six-well culture plate on the rocker platform [72]). It can be calculated that liver slices consume only 0.3–1% of the dissolved O_2 per minute. This implies that, provided that the medium is continuously oxygenated, the availability of O_2 is unlikely to be a limiting factor. In our laboratory we have carried out a study to investigate the effect of different oxygen percentages on ATP-levels and the rate of lidocaine metabolism in liver slices in the six-well incubation system. When rat liver slices were incubated for up to 24 h with oxygen percentages between 50 and 95%, no differences were observed in either ATP levels or the rate of lidocaine biotransformation. However, if liver slices were incubated for up to 24 h with 20% O_2 both ATP levels and the rate of lidocaine biotransformation were significantly decreased. Moreover, this also explains why the lower oxygen percentage (40%) used in the rocker platform system [72] did not seem to influence the functionality or viability of rat liver slices [52]. Thus it seems that the agitation of the medium and a sufficient O_2 supply is more influential with regard to the viability of the slices than their intermittent exposure to the medium and the gas phase.

12.3.5 Pre-incubation of Liver Slices

Another important issue in the incubation of liver slices is the potential benefits of pre-incubation. We showed that at least 1.5 h of incubation is necessary to restore K^+ and ATP levels [52]. It has been suggested that a change of medium after pre-incubation is useful in removing cell debris created by cells which were damaged during the slicing process. This is of special importance when leakage of cell components, such as LDH and ALT, is used as a marker of cell damage in toxicity studies. However, no conclusive evidence regarding the necessity and duration of pre-incubation has been published as yet.

In conclusion, for short incubations the choice of incubation system is not critical for slice viability which may be determined by other features, such as the volume of the medium, the duration of the sampling procedure and the costs. For studies where rapid sampling of the slices is necessary, for instance in studies on drug uptake, incubation in the shaken six-well system is recommended. In the 24-well system the agitation of the medium appeared insufficient, whereas in the DOC systems the uptake rate of the drug may be influenced by the limited supply of substrate from the medium during exposure to the gas phase. For longer term incubations, the choice of incubation system and medium seems to be more critical and further basic studies on slice technology need to be carried out to assure optimal long-term culture of liver slices. Among other conditions, incubation experiments should be undertaken using different oxygen concentrations and concomitant measurement of oxygen consumption, in order to establish the optimal oxygen concentration. The agitation of the medium should also be studied in more detail. Care should be taken when extrapolating results obtained with rat liver slices to human liver slices, since considerable species differences with respect to the influence of incubation systems on slice viability have been reported [43].

Finally, inter-laboratory standardization of incubation systems and culture media would increase the validity of comparisons made between results from different laboratories.

12.4 Viability and Functionality of Liver Slices

For pharmacological, toxicological and transport studies it is of utmost importance to assess not only the viability but also the functionality of the liver slices. This is essential both for end-point determination of toxic cell damage, and to assess the quality of the tissue during incubation. Several viability tests have been developed for liver slices, in line with those for isolated hepatocytes: K^+ retention, ATP content, energy charge, enzyme (LDH, ALT, AST) leakage, protein synthesis and MTT reduction [36,43,44,51,52,72,73]. Specific liver function tests include urea synthesis, albumin synthesis, gluconeogenesis, biotransformation of test substrates (such as testosterone and 7-ethoxycoumarin) and GSH concentration [40,74–77] Potassium retention is generally used to assess the viability of liver slices [36,43]. However, in our studies on the comparison of incubation systems and on cold storage of slices, the potassium concentration in the slices was retained while their metabolic capacity had clearly decreased [78]. This illustrates that in drug metabolism studies the rate of metabolism of a standard drug should be included as a viability test.

The determination of the energy charge (EC) is of limited value in assessing the viability of liver slices, because changes in ATP, ADP and AMP have to be quite large before a significant variation in EC is observed. Fisher *et al.* [43] proposed the following ranking of sensitivity for tests aimed at the detection of cellular viability: ATP content > K^+ retention > protein synthesis > enzyme leakage > MTT reduction.

Because these different viability tests all reflect different aspects of cell viability, the choice of test depends on the aim of the study. For toxicity studies where biotransformation is an important bioactivation or detoxification step, metabolic function tests should be included to judge the validity of the method, whereas viability tests are needed to assess toxic effects. Both positive and negative controls should be included in such studies. When human liver is used, the characterization of metabolic activity is especially important because of the large inter-individual variability associated with this property [75].

The viability and function tests described above are used to evaluate the hepatocytes within the slice. Up to now, tests to measure the viability of the non-parenchymal cells have not been reported. The presence of the latter cell types is one of the conceptual advantages of slices as compared to isolated hepatocytes. As some drug targeting devices are designed to target non-parenchymal cells in the liver, the development of tests for the sinusoidal cell types deserves more attention. For example, the uptake of substrates such as succinylated human serum albumin (Suc-HSA, which is specifically endocytosed by endothelial cells [79]), or hyaluronic acid [80], can be used to assess the functionality of endocytotic pathways in the endothelial cells in the liver [81]. Other modified proteins that are specifically taken up by Kupffer cells such as mannosylated HSA, may be used to assess the functionality of the endocytotic pathway in Kupffer cells [79]. Another parameter which can be used to assess the functionality of these non-parenchymal liver cells, is the excretion of cytokines in response to pro-inflammatory stimuli. Non-parenchymal cell function in liver slices will be described in more detail in the Section 12.7.

12.5 *In Vitro* Transport Studies

12.5.1 Transport in Hepatocytes

In the liver drugs are predominantly taken up by the hepatocytes, e.g. by carrier-mediated uptake, metabolized in the hepatocyte and excreted either via the bile canaliculus into the bile or back into the bloodstream, e.g. by carrier-mediated excretion.

The mechanisms of uptake and excretion of drugs by the liver has been widely studied using isolated hepatocytes and isolated perfused livers of rodents. The subject was extensively discussed by Oude Elferink *et al.* [82], summarized in a comprehensive special issue of the *Journal of Hepatology* in 1996 [83] and reviewed in several papers and chapters by our group [84,85]. In general, the rate and mechanism of drug uptake in isolated rat hepatocytes are very similar to those found *in vivo*. Only a few studies have investigated transport of drugs in human hepatocytes, and even fewer have used liver slices. However, studies in human and rat hepatocytes are hampered by the fact that the isolation procedures involve collagenase digestion for the disruption of cell-to-cell contacts. Clearly, this proteolytic enzyme may also damage plasma membranes and transport systems therein.

Jansen *et al.* [86] published an example of transport of a drug targeting preparation in human and rat hepatocytes. They found that the anti-viral drug ara-AMP coupled to lactosaminated human serum albumin, was taken up to the same extent by human and rat hepatocytes. This is one of the few examples where an equal rate of transport was found in both species. In general the uptake of drugs in human hepatocytes is slower than that in rat hepatocytes. *In vitro–in vivo* scaling calculations [87,88] showed that these differences in the uptake rate of drugs between rat and human cells, actually reflect inter-species differences rather than being due to differences in viability. Moreover, the differences described in rate and mechanism of drug transport in rat and man emphasize again that extrapolation to man of pharmacokinetic data obtained in rat, is hazardous. Most of the current knowledge on drug transport carriers is derived from experiments with rats. Therefore, more studies need to be carried out in human liver preparations to further elucidate the mechanisms of drug transport in the human liver and hence the relevance of animal data.

These results also indicate that human hepatocytes are an appropriate model in which to study inter-species differences and the mechanisms of hepatic transport in man.

In contrast to isolated hepatocytes, liver slices retain the cellular architecture of the liver without prior digestion with collagenase. This makes a systematic comparison of the data relating to transport of free drugs as well as drug targeting moieties from isolated hepatocytes and liver slices, an attractive model for studying the potential and limitations of the liver slice model in this area of research.

12.5.2 Transport in Liver Slices

Mechanisms of drug uptake in liver slices were studied *in vitro* as early as in 1963 by Schanker and Solomon [89]. The results obtained in these experiments are still valuable and show that the influence of temperature, anoxia, metabolic inhibitors and substrate inhibition can be successfully studied in this preparation. However, as mentioned before, at that time the preparation of reproducible precision-cut slices was not feasible. Therefore, the slice incubation technique was virtually abandoned in transport studies after the introduction of the successful isolation of rat hepatocytes.

In order to investigate the possibilities and limitations of the use of precision-cut liver slices prepared with a mechanical slicer in drug transport studies, different aspects of the mechanism of uptake of several classes of drugs in human and rat liver slices were investigated in our laboratory. Four model compounds which enter hepatocytes via entirely differ-

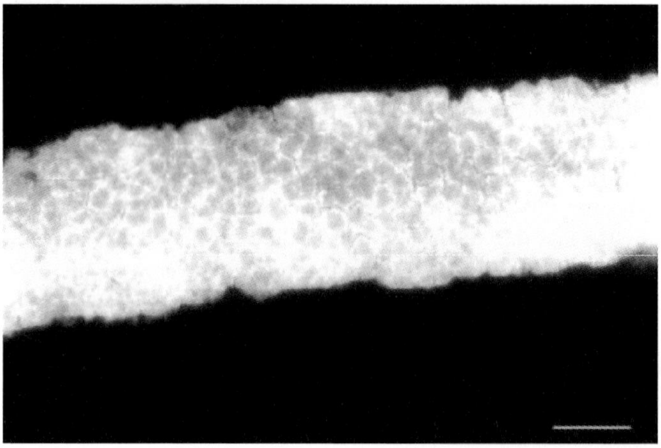

Figure 12.4. Rhodamine B (25 μM) distribution in the cross-section of a rat liver slice (± 250 μM) after 5 min incubation. Fluorescence microscopy, bar = 100 μM.

ent membrane transport mechanisms were investigated: the fluorescent dyes rhodamine B and lucigenin, the cardiac glycoside digoxin and the neo-glycoprotein lactosylated albumin. Receptor-mediated endocytosis into endothelial cells was studied with succinylated and aconylated albumin. The rate of penetration into the rat liver slice was studied with the lipophilic cationic compound rhodamine B (RB) and the hydrophilic organic cation lucigenin (LU). RB, which enters hepatocytes by passive diffusion was homogeneously distributed throughout the rat liver slice (250 μm thickness) within 5 min [81] (Figure 12.4). These results indicate that for very lipophilic components both the penetration rate into the slice and the diffusion rate into all the cells are rapid processes. In contrast, after incubation with LU, which is taken up by hepatocytes through adsorptive endocytosis, fluorescence in the inner cell layers could only be detected after 15 min. If the rates of uptake of drugs in liver slices are to be compared with parameters obtained *in vivo*, some difficulties may arise. The uptake rate of compounds into the slice may not only reflect the uptake rate of the cells involved, but may also be influenced by the rate of penetration of the substrates (i.e. diffusion through sinusoidal spaces) into the slice. From our results with the lipophilic agent rhodamine B it is clear that the penetration process into the slice takes at least 5 min. Therefore, for substrates that are taken up into hepatocytes relatively fast, penetration into the slice may limit the uptake of the drug by the inner cell layers in the slice.

Digoxin uptake into rat liver slices showed a temperature-dependent component, compatible with the involvement of carrier-mediated uptake mechanisms. Quinine markedly inhibited the uptake of digoxin, in contrast to its diastereomer quinidine, which only slightly inhibited the digoxin uptake in rat liver slices. This stereoselective inhibition is in line with results obtained in isolated rat hepatocytes and isolated perfused rat livers [90,91]. These results were also found after cryopreservation of the slices, indicating that carrier-specific phenomena can be studied after cryopreservation [92].

From these results it can be concluded that liver slices are a powerful tool for studying the mechanisms and specificity of carrier-mediated uptake of drugs and drug interactions which occur at the transport level.

Many drugs that are taken up and metabolized by hepatocytes are excreted via the bile canaliculi into the bile. One of the remaining topics in liver slice research is the question of whether liver slices are capable of bile excretion via the bile canaliculus. Thompson *et al.*[93] showed that slices are capable of excreting bile acids, however, there is a need for more experiments to determine whether this excretion takes place across the bile canalicular membrane.

12.6 The Use of Liver Slices in Drug Targeting Research

12.6.1 Distribution and Transport of Drug Targeting Devices

In our Institute, drug targeting devices are predominantly developed for the treatment off inflammatory diseases of the liver, such as fibrosis and cirrhosis. In order to increase the therapeutic efficacy of drugs, human serum albumin is used as a protein backbone; modifying this protein by the attachment of different sugar groups or peptide molecules targets these modified proteins to specific cell types in the liver as described in more detail in Chapter 4. Anti-inflammatory agents are subsequently coupled to the protein backbone to serve as effector moieties.

Further studies are needed to determine whether these liver-directed drug targeting preparations are actually delivered to the specific liver cells. In addition, experiments should be carried out to ascertain whether the drug coupled to a drug targeting device is released in the target cell in the liver, and to ensure that the drug is still active in these cells. Up until now the distribution of these modified proteins has been tested *in vivo* mainly in rats, and *in vitro* in the perfused rat liver or isolated liver cells (both parenchymal and non-parenchymal cells). Almost no studies have been performed in the target species, man. As the main aim of drug targeting research is the development of preparations for clinical use, we investigated whether precision-cut liver slices could be used to study the uptake of these modified proteins into target cells of the human liver.

The temperature-dependent uptake and immunohistochemical localisation of modified proteins, $Lactose_{27}$-Human Serum Albumin ($Lact_{27}$-HSA), Succinylated-Human Serum Albumin (Suc-HSA) and Aconylated Human Serum Albumin (Aco-HSA), in rat and human liver slices showed that large molecules can enter the slice and are probably taken up by receptor-mediated endocytosis (Figure 12.5). These large modified proteins were found distributed all over the liver slice, as was determined by immuno-histochemistry [81] (Figure 12.6). Recently, Beljaars *et al.* [94] showed that mannose-6-$phosphate_{21}$-Bovine Serum Albumin ($M6P_{21}$-BSA) is taken up by the non-parenchymal cells of human liver using slices from cirrhotic human livers. The non-parenchymal cells were identified as hepatic stellate cells (HSC) or endothelial cells. The involvement of hepatocytes, Kupffer cells, and bile duct epithelial cells was excluded. Until now, there has been little information concerning the presence of the mannose-6-phosphate receptor in human adult livers. The accumulation of M6P-modified albumin in human cirrhotic livers demonstrates, however, that M6P receptors are present in the human liver and that M6P-modified albumin may be useful as a drug carrier for the targeting of anti-fibrotic drugs in patients. In another study by Beljaars *et al.* [95],

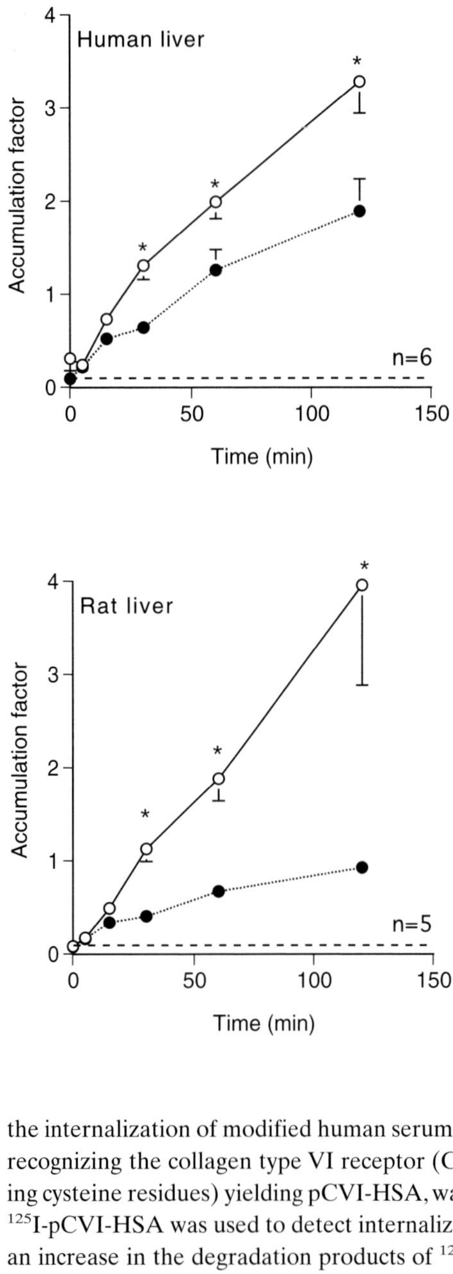

Figure 12.5. Uptake of ^{125}I-Suc-HSA in liver slices from humans and rats at 37°(○) and 4°C (•). The accumulation factor is defined as the concentration of the compound in the slices divided by the concentration in the medium. Each point is the mean of 5-6 separate experiments ± SEM. *n* = number of livers. *$p < 0.05$ versus 4°C. The dotted line represents the accumulation factor if ^{125}I-Suc-HSA is exclusively distributed within the sinusoids.

the internalization of modified human serum albumin (HSA) with 10 cyclic peptide moieties recognizing the collagen type VI receptor (C*GRGDSPC*, in which C* denotes the cyclizing cysteine residues) yielding pCVI-HSA, was studied in normal and cirrhotic rat liver slices. ^{125}I-pCVI-HSA was used to detect internalization in the liver slices. In contrast to ^{125}I-HSA, an increase in the degradation products of ^{125}I-pCVI-HSA was found over time, during the incubation of liver slices. These data show that pCVI-HSA is taken up and degraded in the cells of the liver slice. By immunohistochemistry it was shown that pCVI-HSA was specifically bound to rat HSC, in particular to the activated cells.

These distribution studies show that liver slices can be used to assess the level of uptake of a drug into the target cells in both healthy and diseased human liver. Isolation of non-

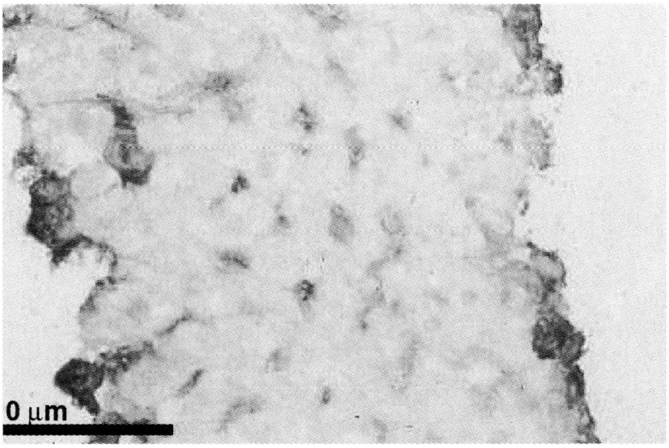

Figure 12.6. Fluorescein-labelled aconylated human serum albumin distribution in the cross-section of a rat liver slice (± 250 μM) after 120 min incubation. Bar = 100 μM.

parenchymal cells from diseased livers is experimentally very difficult, therefore the liver slice technique offers a unique opportunity to study drug targeting preparations in diseased liver.

12.7 Efficacy Testing of the Drug Targeting Device in the Liver

It is of paramount importance to ensure that the drug targeting device with the drug attached or incorporated, is not only taken up by the target cells in the liver, but is also released from the device and remains active within the target cell. In the case of the development of drug targeting strategies for inflammatory liver diseases, an *in vitro* system was needed that could be used to test the anti-inflammatory effect of the drug-targeting preparations in the human liver.

To set up and validate the *in vitro* systems we initiated a study with rat liver slices. Stimulation by lipopolysaccharide (LPS) in liver slices was used to evoke a pro-inflammatory response in the liver. Lipopolysaccharide (LPS), a component of Gram-negative bacterial cell walls (also called endotoxin), has been associated with tissue injury and sepsis. In the liver LPS activates the resident macrophages, the Kupffer cells, which results in cytokine release [96]. Furthermore, LPS is cleared by the liver, mainly by Kupffer cells [97]. One of the major features of endotoxic shock is the induction of nitric oxide synthase in the liver [98]. Inducible nitric oxide synthase (iNOS), the expression of which is induced by LPS and cytokines, produces nitric oxide (NO) in large quantities [99].

The induction of iNOS by LPS, as observed in the hepatocytes *in vivo* [99], cannot be achieved in pure cultures of isolated hepatocytes. In fact, the induction of hepatocyte-associated iNOS is found in pure cultures only after incubation with a mix of LPS and cytokines [100]. Induction of iNOS by LPS alone can be accomplished in co-cultures of hepatocytes and Kupffer cells with LPS [98]. These data indicate that induction of iNOS by LPS is mediated by cytokines released by the Kupffer cells in the liver and is thus a result of intercellular communications. However, the establishment of such co-cultures is technically complex while the cellular organization as present in the intact liver cannot as yet be achieved in culture. Therefore, to study the effects of LPS stimulation and the subsequent pharmacological intervention by targeted anti-inflammatory drugs in the whole liver *in vitro*, a system analogous to liver slices in which all cell types are present as in the original liver, would be the ideal.

Until recently only a few studies have focused on the activity or the viability of cell types other than hepatocytes within the liver slice [101,102]. Therefore, we attempted to establish whether non-parenchymal cells are still viable in the rat liver slice and whether they respond to LPS stimulation. Cytokine levels (tumour necrosis factor α (TNFα), interleukin 1β (IL-1β) and interleukin-10 (IL-10)) were measured in the incubation medium as a marker of non-parenchymal cell function. We found that in liver slices stimulated by LPS, IL-1β, TNFα and IL-10 are formed. In addition, the study was designed to elucidate the interaction between non-parenchymal and parenchymal cells in the liver after LPS induction. After LPS activation of rat liver slices iNOS was upregulated in the hepatocytes as determined by immunohistochemistry (Figure 12.7). This resulted in the production of NO, as measured by nitrate and nitrite (NO_x) in the incubation medium (Figure 12.8). In addition to studying the pro-inflammatory response on a protein level, enzyme induction was studied at the transcription level. By RT-PCR, changes in specific mRNA were studied during the LPS-induced pro-inflammatory response. In rat liver slices the mRNA of iNOS was already upregulated only 2–3 h after induction with LPS, whereas the enzyme iNOS was found after 5 h.

Anti-inflammatory drugs such as pentoxyfilline and dexamethasone inhibited the release of cytokines and thereby the induction of iNOS and the release of NO in the LPS-stimulated liver slices.

Melgert *et al.* [103] used this *in vitro* system to determine whether the drug targeting preparation containing the anti-inflammatory drug dexamethasone coupled to HSA, was still able to manifest its anti-inflammatory properties in liver slices. $Dexamethasone_{10}$–HSA and uncoupled dexamethasone showed effective inhibition of LPS-induced NO and TNFα pro-

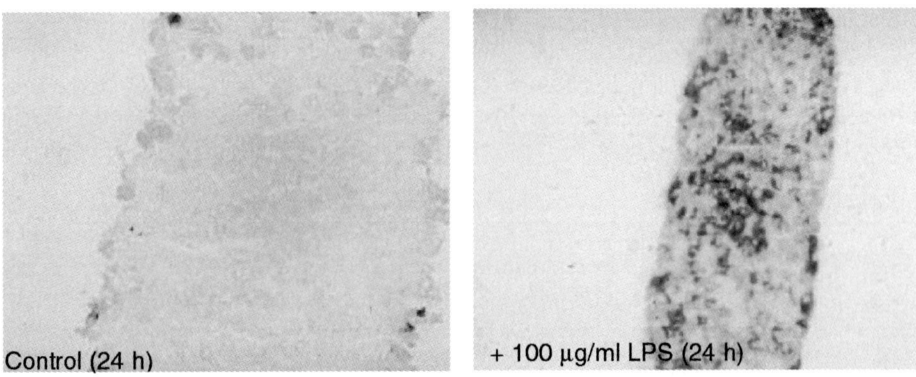

Figure 12.7. Cross-section of rat liver stained for iNOS. Left panel: control incubation after 24 h. Right panel: after 24 h incubation with 100 µg ml^{-1} LPS.

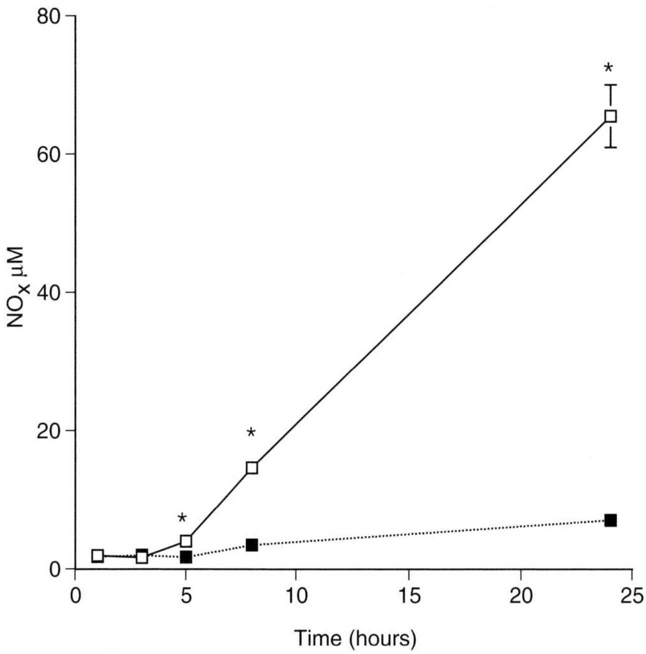

Figure 12.8. NO production in rat liver slices after incubation in the absence or presence of 100 μg ml^{-1} LPS for different time periods. The NO production is measured as nitrate/nitrite (NO_x) concentrations in the medium (μM). Control (■) and + 100 μg ml^{-1} LPS (□). Data are expressed as mean ± SEM of four experiments. *$p < 0.005$ represents a significant increase in NO production by liver slices due to stimulation by LPS.

duction (Figure 12.9) in the liver slice model. These results show that the conjugate dexamethasone$_{10}$–HSA is taken up intracellularly and that active dexamethasone is released.

Studies similar to those described above are now being carried out in human liver slices. LPS induction in human liver slices also increased TNFα production to the same extent as was found in rat liver slices [104] (Figure 12.10). Human liver slices also produced IL-6, IL-8 and IL-1β, although the latter to a lesser extent than that observed in the liver slices of rat origin. However, human liver slices produced less NO after LPS stimulation than those of the rat. More experiments will be undertaken to elucidate this species difference.

Taken together these results indicate that non-parenchymal cells are still active in the slice preparations and that intercellular communication is still intact. Furthermore, pharmacological intervention by anti-inflammatory drugs can be successfully studied in liver slices. Together with the results obtained in regard to drug transport, liver slices seem to be a very promising *in vitro* system for studying intercellular distribution, cellular processing and effectiveness of anti-inflammatory drugs coupled to a targeting device.

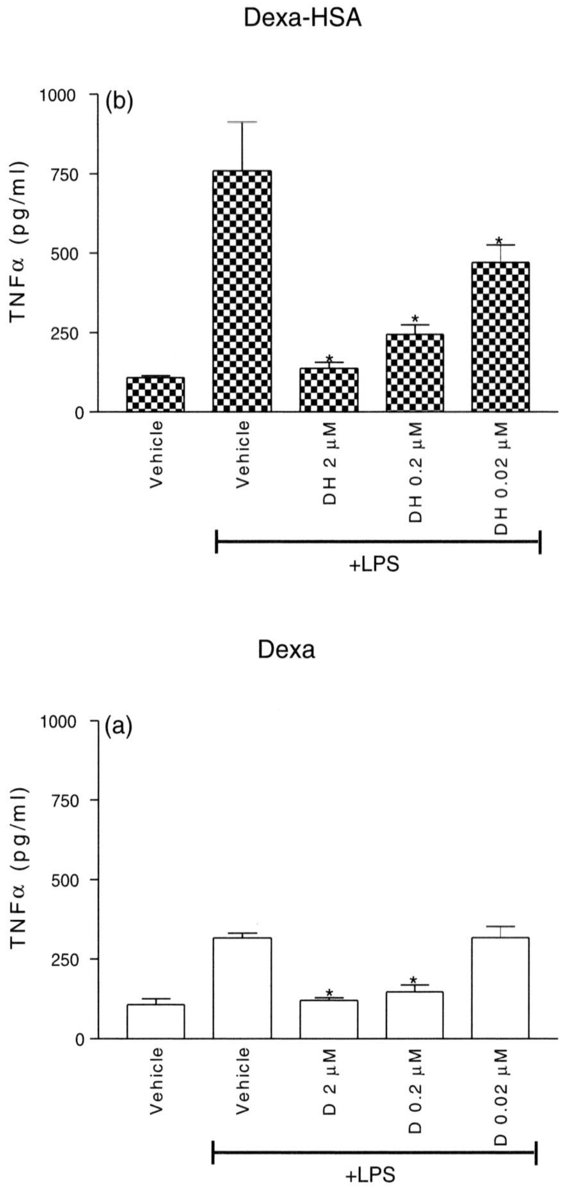

Figure 12.9. (a) TNFα production by rat liver slices ($n = 9$) after stimulation with 100 μg ml^{-1} LPS for 24 h in either the presence or absence of dexamethasone (D). Vehicle consists of PBS. *$p < 0.05$ versus vehicle + LPS. (b) TNFα production by rat liver slices ($n = 9$) after LPS stimulation for 24 h with or without $Dexa_{10}$-HSA (DH). Vehicle contains PBS and an equimolar amount of HSA. *$p < 0.05$ versus vehicle + LPS.

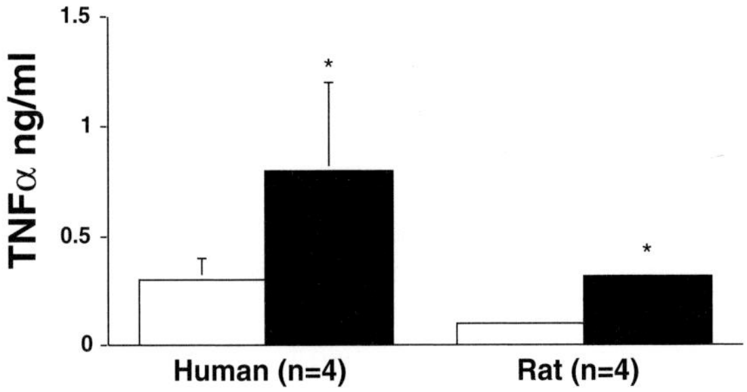

Figure 12.10. Production of TNFα by human and rat liver slices in culture medium after 24 h stimulation with or without 100 μg ml^{-1} LPS. White bar: control; black bar: 100 μg ml^{-1} LPS present. Data are the mean of four experiments ± SEM. *$p < 0.05$ versus control.

12.8 Tissue Slices from Other Organs

Precision-cut tissue slices have also been prepared from other organs apart from the liver. Kidney slices are prepared by the same method as liver slices [73]. Kidney slices from different species, including man are used in the study of the toxicology and metabolism of drugs [65,66,73,105–108], organic anion and cation transport [109,110], release of prostaglandin and noradrenalin [36,111], and also in the study of organ preservation [70,112]. Since region-selective slices (cortex or medulla slices) can be prepared from the kidney, toxicity and metabolism in different regions of the kidney can therefore be studied [36,108,113]. Lungs cannot be sliced directly but need to be filled with 1.5% (w/v) low melting agarose solution containing 0.9% (w/v) NaCl at 37°C and allowed to gel on ice [73]. Lung slices have been used for drug transport and toxicity studies [114–118]. Up until now slices from other organs have not been used in the (transport) study of drug targeting devices, but like liver slices, these *in vitro* preparations have the potential and advantages to be useful in the study of transport, cellular processing and efficacy of drug targeting devices. In addition, slices of tumours could be used to study drug targeting in cancer research.

12.9 Summary and Future Possibilities

Drug targeting preparations are designed to be used in man, however most research with these preparations is carried out in animals. Due to known species differences, the study of these preparations in man in an early stage of development is therefore of paramount importance. *In vitro* studies exploiting human tissue can be used to ensure that these drug targeting devices reach the desired target cells and once there, are effective. When cells in the liver are the main target, *in vitro* research should be undertaken using preparations of both healthy and diseased human liver. As was discussed earlier in this chapter, liver slices seem like the ideal *in vitro* preparation for this purpose. The original architecture of the liver is still intact in the slice, which enables normal intercellular communication and cell-selective distribution of drugs. Slices can also be used to study drug interactions and the mechanisms and specificity of carrier-mediated uptake of drugs. In addition, the distribution of the drug into different cell types in the tissue can easily be studied in preparations of organ slices, as can the efficacy of the drug which is coupled to the targeting device. Furthermore, metabolism and toxicity of the drug targeting device or the released drug can be determined in the human liver. And finally, an important aspect of this type of *in vitro* research in man is, that it will ultimately lead to a reduction in the use of animal experiments.

In future, drug targeting devices aimed at other human organs may also be studied using precision-cut tissue slices. The latest data/literature on precision-cut tissue slices can be found at http://www.farm.rug.nl/slice/open.html.

References

[1] Trowell OA, *Expert Cell Res.* **1959,** *16*, 118–147.
[2] Warburg O, *Biochem. Z*, **1923,** *142*, 317–333.
[3] Krebs HA, *Hoppe-Seyl. Z*, **1933,** *217*, 190–227
[4] Stadie WC, Riggs BC, *J. Biol. Chem.* **1944,** *154*, 687.
[5] Krumdieck CL, *Anal. Biochem.* **1980,** *104*, 118–123.
[6] Seglen PO, *Methods Cell Biol.* **1976,** *13*, 29–83.
[7] Skett P, Toxicol. *In Vitro* **1994,** *3*, 491–504.
[8] Berry MN, Edwards AM, Barritt GJ, *Isolated Hepatocytes Preparation, Properties and Applications,* Vol. 21. Elsevier, Amsterdam, New York, Oxford, **1991.**
[9] Guillouzo A, Oudea P, Le-Guilly Y, Oudea MC, Lenoir P, Bourel M, *Exp. Mol. Pathol.* **1972,** *16*, 1–15.
[10] Bojar H, Basler M, Fuchs F, Dreyfürst R, Staib W, Broelsch C, *J. Clin. Chem. Clin. Biochem.* **1976,** *14*, 527–532.
[11] Reese JA, Byard JL, *In Vitro* **1981,** *17*, 935–940.
[12] Groothuis GMM, Sandker GW, Pruim J, Weert B, Slooff MJH, Meijer DKF, *Toxicol. In Vitro* **1995,** *9*, 951–958.
[13] Dorko K, Freeswick PD, Bartoli F, Cicalese L, Bardsley BA, Tzakis A, Nussler AK, *Cell Transplant.* **1994,** *3*, 387–395.
[14] Moshage HJ, Rijntjes PJ, Hafkenscheid JC, Roelofs HM, Yap SH, *J. Hepatol.* **1988,** *7*, 34–44.
[15] Sandker GW, *Isolated human hepatocytes: a research tool for investigations on drug transport, drug metabolism and liver transplantation.* PhD thesis, University of Groningen, **1993.**
[16] Ryan CM, Carter EA, Jenkins RL, Sterling LM, Yarmush ML, Malt RA, Tompkins RG, *Surgery* **1993,** *113*, 48–54.
[17] Schröder AJ, Blaheta RA, Scholz M, Encke A, Markus BH, *Zentralbl. Chir.* **1994,** *119*, 127–138.
[18] Takahashi M, Matsue H, Matsushita M, Nakajima Y, Uchino J, *Artif. Organs* **1993,** *17*, 653–659.
[19] Trevisan A, Gudat F, Guggenheim R, Krey G, Durmuller U, Luond G, Duggelin M, Landmann J, Tondelli P, Bianchi L, *Hepatology* **1982,** *2*, 832–835.
[20] Ballet F, Bouma ME, Wang SR, Amit N, Marais J, Infante R, *Hepatology* **1984,** *4*, 849–854.
[21] Houssin D, Capron M, Celier C, Cresteil T, Demaugre F, Beaune P, *Life Sci.* **1983,** *33*, 1805–1809.
[22] Hsu IC, Lipsky MM, Cole KE, Su CH, Trump BF, *In Vitro Cell Dev. Biol.* **1985,** *21*, 154–160.
[23] Miyazaki K, Takaki R, Nakayama F, Yamauchi S, Koga A, Todo S, *Cell Tissue Res.* **1981,** *218*, 13–21.
[24] Strom SC, Jirtle RL, Jones RS, Novicki DL, Rosenberg MR, Novotny A, Irons G, McLain JR, Michalopoulos G, *J. Natl Cancer Inst.* **1982,** *68*, 771–778.
[25] Tee LB, Seddon T, Boobis AR, Davies DS, *Br. J. Clin. Pharmacol.* **1985,** *19*, 279—294.
[26] Guillouzo A, Beaune P, Gascoin MN, Begue JM, Campion JP, Guengerich FP, Guguen-Guillouzo C, *Biochem. Pharmacol.* **1985,** *34*, 2991–2995.
[27] Wiebkin P, Dees JH, Mathis JM, Prough RA, *Drug Metab. Dispos.* **1985,** *13*, 163–168.
[28] Blaauboer BJ, van-Holsteijn I, van-Graft M, Paine AJ, *Biochem. Pharmacol.* **1985,** *34*, 2405–2408.
[29] Friedman SL, Roll FJ, *Anal. Biochem.* **1987,** *161*, 207–218.
[30] Alpini G, Phillips JO, Vroman B, LaRusso NF, *Hepatology* **1994,** *20*, 494–514.
[31] Heuff G, Meyer S, Beelen RH, *J. Immunol. Methods* **1994,** *174*, 61–65.
[32] Zeitlin PL, Hubbard AL, *J. Cell. Biol.* **1982,** *92*, 634–647.

[33] Ikejima K, Enomoto N, Seabra V, Ikejima A, Brenner DA, Thurman RG, *Am. J. Physiol.* **1999,** *276,* G591–G598.
[34] Meijer DKF, Keulemans K, Mulder GJ, *Methods Enzymol.* **1981,** *77,* 81–94.
[35] Smith P, Krack G, McKee R, Johnson D, Gandolfi A, Hruby V, Krumdieck C, Brendel K, *In Vitro Cell Dev. Biol.* **1986,** *22,* 706–712.
[36] Parrish AR, Gandolfi AJ, Brendel K, *Life Sci.* **1995,** *57,* 1887–1901.
[37] Price RJ, Ball SE, Renwick AB, Barton PT, Beamand JA, Lake BG, *Xenobiotica* **1998,** *28,* 361–371.
[38] Olinga P, Merema MT, Meijer DKF, Slooff MJH, Groothuis GMM, *ATLA* **1993,** *21,* 466–468.
[39] Ekins S, Murray GI, Burke MD, Williams JA, Marchant NC, Hawksworth GM, *Drug Metab. Dispos.* **1995,** *23,* 1274–1279.
[40] de Kanter R, Koster HJ, *ATLA* **1995,** *23,* 653–665.
[41] Leeman W, van de Gevel I, Rutten A, *Toxicol. In Vitro* **1995,** *9,* 291–298.
[42] Fisher RL, Putnam CW, Koep LJ, Sipes IG, Gandolfi AJ, Brendel K, *Cryobiology* **1991,** *28,* 131–142.
[43] Fisher RL, Shaughnessy RP, Jenkins PM, Austin ML, Roth GL, Gandolfi AJ, Brendel K, *Toxicol. Methods* **1995,** *5,* 99–113.
[44] Fisher RL, Hasal SJ, Sanuik JT, Gandolfi AJ, Brendel K, *Toxicol. Methods* **1995,** *5,* 115–130.
[45] Price RJ, Renwick AB, Barton PT, Houston JB, Lake BG, *ATLA* **1998,** *26,* 541–548.
[46] Brendel K, Gandolfi A, Krumdieck C, Smith P, *TIPS* **1987,** *8,* 11–15.
[47] Smith PF, McKee R, Gandolfi AJ, Krumdieck CL, Fisher R, Brendel K, *Precision Cut Liver Slices: A New In Vitro Tool In Toxicology.* Telford Press, Caldwell, NJ, **1989.**
[48] Bach PH, Vickers AEM, Fisher R, Baumann A, Brittebo E, Carlile DJ, Koster HJ, Lake BG, Salmon F, Sawyer TW, Skibinski G, *ATLA* **1996,** *24,* 893–923.
[49] Hart A, Mattheyse J, Balinsky JB, *In Vitro* **1983,** *19,* 841–852.
[50] Goethals F, Deboyser D, Lefebvre V, de Coster I, Roberfroid M, *Toxicol. In Vitro* **1990,** *4,* 435–438.
[51] Dogterom P, *Drug Metab. Dispos.* **1993,** *21,* 699–704.
[52] Olinga P, Groen K, Hof IH, de Kanter R, Koster HJ, Leeman WR, Rutten AAJJL, van Twillert K, Groothuis GMM, *J. Pharm. Toxicol. Methods* **1997,** *38,* 59–69.
[53] Kwekkeboom J, Princen H, Van Voorthuizen EM, Meijer P, Kempen H, *Biochem. Biophys. Res. Commun.* **1989,** *162,* 619–625.
[54] Paine AJ, *Chem. Biol. Interact.* **1991,** *74,* 1–31.
[55] Padgham CR, Boyle CC, Wang XJ, Raleigh SM, Wright MC, Paine AJ, *Biochem. Biophys. Res. Commun.* **1993,** *197,* 599–605.
[56] Rogiers V, In: *Human Cells In In Vitro Pharmaco-Toxicology* (Eds Rogiers V, Sonck W, Shephard E, Vercruysse A), pp. 77–115. VUB Press, Brussels, **1993.**
[57] Miyazaki M, Handa Y, Oda M, Yabe T, Miyano K, Sato J, *Exp. Cell Res.* **1985,** *159,* 176–190.
[58] Dich J, Vind C, Grunnet N, *Hepatology* **1988,** *8,* 39–45.
[59] Jefferson DM, Clayton DF, Darnell Jr J, Reid LM, *Mol. Cell Biol.* **1984,** *4,* 1929–1934.
[60] Dunn JC, Tompkins RG, Yarmush ML, *J. Cell Biol.* **1992,** *116,* 1043–1053.
[61] Koebe HG, Pahernik S, Eyer P, Schildberg FW, *Xenobiotica* **1994,** *24,* 95–107.
[62] Saad B, Scholl FA, Thomas H, Schawalder H, Streit V, Waechter F, Maier P, *Eur. J. Biochem.* **1993,** *213,* 805–814.
[63] Guillouzo A, Morel F, Fardel O, Meunier B, *Toxicology* **1993,** *82,* 209–219.
[64] Connors S, Rankin DR, Gandolfi AJ, Krumdieck CL, Koep LJ, Brendel K, *Toxicology* **1990,** *61,* 171–184.
[65] Connors MS, Larrauri A, Dannecker R, Nufer R, Brendel K, Vickers AEM, *Xenobiotica* **1996,** *26,* 133–141.
[66] Vickers AEM, Fischer V, Connors S, Fisher RL, Baldeck JP, Maurer G, Brendel K, *Drug Metab. Dispos.* **1992,** *20,* 802–809.
[67] Berridge MV, Tan AS, *Arch. Biochem. Biophys.* **1993,** *303,* 474–482.
[68] Hashemi E, Dobrota M, Till C, Ioannides C, *Xenobiotica* **1999,** *29,* 11–25.
[69] Olinga P, Merema MT, Sandker GW, Slooff MJH, Meijer DKF, Groothuis GMM, In: *The World Congress on Alternatives and Animal Use in The Life Sciences: Education, Research, Testing* (Eds Goldberg AM, Van Zutphen LFM), pp. 197–203. Liebert, New York, **1995.**
[70] Fisher RL, Hasal SJ, Sanuik JT, Scott KS, Gandolfi AJ, Brendel K, *Cryobiology* **1993,** *30,* 250–261.
[71] Lake BG, Charzat C, Tredger JM, Renwick AB, Beamand JA, Price RJ, *Xenobiotica* **1996,** *26,* 297–306.

[72] Leeman WR, van de Gevel IA, Rutten AAJJL, *Toxicol. In Vitro* **1995,** *9*, 291–298.
[73] de Kanter R, Olinga P, de Jager MH, Merema MT, Meijer DKF, Groothuis GMM, *Toxicol. In Vitro* **1999,** *13*, 737–744.
[74] de Kanter R, Olinga P, Hof I, de Jager M, Verwillegen WA, Slooff MJ, Koster HJ, Meijer DK, Groothuis GM, *Xenobiotica* **1998,** *28*, 225–234.
[75] Olinga P, Merema M, Hof IH, de Jong KP, Slooff MJH, Meijer DKF, Groothuis GMM, *Drug Metab. Dispos.* **1998,** *26*, 5–11.
[76] Valentovic MA, Ball JG, Anestis DK, Rankin GO, *Toxicol. In Vitro* **1995,** *9*, 75–81.
[77] Miller MG, Beyer J, Hall GL, Degraffenried LA, Adams PE, *Toxicol. Appl. Pharmacol.* **1993,** *122*, 108–116.
[78] Olinga P, Merema MT, Hof IH, De Jager MH, De Jong KP, Slooff MJH, Meijer DKF, Groothuis GMM, *Xenobiotica* **1998,** *28*, 349–360.
[79] Jansen RW, Olinga P, Harms G, Meijer DKF, *Pharm. Res.* **1993,** *10*, 1611–1614.
[80] Sutto F, Brault A, Lepage R, Huet P-M, *J. Hepatol.* **1994,** *20*, 611–616.
[81] Olinga P, Hof IH, Smit M, de Jager MH, Nijenhuis E, Slooff MJH, Meijer DKF, Groothuis GMM, *J. Pharm. Toxicol. Methods*, **2001** (in press).
[82] Oude Elferink RPJ, Meijer DKF, Kuipers F, Jansen PLM, Groen AK, Groothuis GMM, *Biochim. Biophys. Acta* **1995,** *1241*, 215–268.
[83] Tiribelli C, Reichen J, Meijer DKF, *J. Hepatol.* **1996,** *24*, 11–34.
[84] Meijer DKF, In: *Handbook of Physiology. The Gastrointestinal System, Vol. III* (Eds Schulz J, Forte G, Rauner BB), pp. 717–758. Oxford University Press, New York, **1989.**
[85] Meijer DKF, Ziegler K, In: *Biological Barriers to Protein Delivery* (Eds Audus KL, Raub TJ), pp. 339–407. Plenum Press, New York and London, **1993.**
[86] Jansen RW, Kruijt JK, van Berkel TJ, Meijer DK, *Hepatology* **1993,** *18*, 146–152.
[87] Olinga P, Merema MT, Hof IH, Slooff MJH, Proost JH, Meijer DKF, Groothuis GMM, *J. Pharmacol. Exp. Ther.* **1998,** *285*, 506–510.
[88] Sandker GW, Weert B, Olinga P, Wolters H, Slooff MJH, Meijer DKF, Groothuis GMM, *Biochem. Pharmacol.* **1994,** *47*, 2193–2200.
[89] Schanker LS, Solomon HM, *Am. J. Physiol.* **1963,** *204*, 829–832.
[90] Hedman A, Meijer DKF, *J. Pharm. Sci.* **1998,** *87*, 457–461.
[91] Hedman A, Meijer DK, *J. Hepatol.* **1998,** *28*, 240–249.
[92] Olinga P, de Kanter R, Hof IH, de Jager MH, Slooff MJH, Koster HJ, Meijer DKF, Groothuis GMM, *Hepatology* **1996,** *24*, 468A.
[93] Thompson MB, Davis DG, Morris RW, *J. Lipid Res.* **1993,** *34*, 553–561.
[94] Beljaars L, Molema G, Weert B, Bonnema H, Olinga P, Groothuis GMM, Meijer DKF, Poelstra K, *Hepatology* **1999,** *29*, 1486–1493.
[95] Beljaars L, Molema G, Schuppan D, Geerts A, De Bleser PJ, Weert B, Meijer DKF, Poelstra K, *J. Biol. Chem.* **2000,** *275*, 12743–12751.
[96] Aono K, Isobe K, Kiuchi K, Fan ZH, Ito M, Takeuchi A, Miyachi M, Nakashima I, Nimura Y, *J. Cell Biochem.* **1997,** *65*, 349–358.
[97] Van Bossuyt H, De Zanger RB, Wisse E, *J. Hepatol.* **1988,** *7*, 325–337.
[98] Billiar TR, Curran RD, Ferrari FK, Williams DL, Simmons RL, *J. Surg. Res.* **1990,** *48*, 349–353.
[99] Vos TA, Gouw AS, Klok PA, Havinga R, van Goor H, Huitema S, Roelofsen H, Kuipers F, Jansen PL, Moshage H, *Gastroenterology* **1997,** *113*, 1323–1333.
[100] Nussler AK, Di Silvio M, Liu ZZ, Geller DA, Freeswick P, Dorko K, Bartoli F, Billiar TR, *Hepatology* **1995,** *21*, 1552–1560.
[101] Luster MI, Germolec DR, Yoshida T, Kayama F, Thompson M, *Hepatology* **1994,** *19*, 480–488.
[102] Sultan K, Hartung J, Bade EG, *FEBS Lett.* **1996,** *394*, 51–54.
[103] Melgert BN, Olinga P, Jack VK, Molema G, Meijer DKF, Poelstra K, *J. Hepatol.* **2000,** *32*, 603–611.
[104] Olinga P, Merema MT, de Jager MH, Slooff MJH, Limburg PC, Bijzet J, Poelstra K, Meijer DKF, Groothuis GMM, *Hepatology* **1998,** *28*, 184A.
[105] Stijntjes GJ, Commandeur JN, te Koppele JM, McGuinness S, Gandolfi AJ, Vermeulen NP, *Toxicology* **1993,** *79*, 67–79.
[106] Vickers AEM, Fisher RL, Brendel K, Guertler J, Dannecker R, Keller B, Fischer V, *Drug Metab. Dispos.* **1995,** *23*, 327–333.
[107] Foulkes EC, *Proc. Soc. Exp. Biol. Med.* **1996,** *211*, 155–162.
[108] Ruegg CE, Gandolfi AJ, Nagle RB, Krumdieck CL, Brendel K, *J. Pharmacol. Methods* **1987,** *17*, 111–123.

[109] Dutt A, Priebe TS, Teeter LD, Kuo MT, Nelson JA, *J. Pharmacol. Exp. Ther.* **1992,** *261*, 1222–1230.
[110] Drieman JC, Thijssen HH, Struyker-Boudier HA, *Br. J. Pharmacol.* **1993,** *108*, 204-208.
[111] Rump LC, Schwertfeger E, Schaible U, Schuster MJ, Frankenschmidt A, Schollmeyer P, *Eur. J. Clin. Pharmacol.* **1993,** *44 (Suppl. 1)*, S47–S49.
[112] Stubenitsky BM, Ametani M, Danielewicz R, Southard JH, Belzer FO, *Transplant. Int.* **1995,** *8,* 293–297.
[113] Gonzalez-Flecha B, Evelson P, Sterin-Speziale N, Boveris A, *Biochim. Biophys. Acta* **1993,** *1157*, 155–161.
[114] Hoet PH, Lewis CP, Demedts M, Nemery B, *Biochem. Pharmacol.* **1994,** *48*, 517–524.
[115] Placke ME, Fisher GL, *Toxicol. Appl. Pharmacol.* **1987,** *90*, 284–298.
[116] Price RJ, Renwick AB, Wield PT, Beamand JA, Lake BG, *Arch. Toxicol.* **1995,** *69*, 405–409.
[117] Price RJ, Renwick AB, Beamand JA, Esclangon F, Wield PT, Walters DG , Lake BG, *Food Chem. Toxicol.* **1995,** *33*, 233–237.
[118] Fisher RL, Smith MS, Hasal SJ, Hasal KS, Gandolfi AJ, Brendel K, *Hum. Exp. Toxicol.* **1994,** *13*, 466–471.

13 Pharmacokinetic/Pharmacodynamic Modelling in Drug Targeting

Johannes H. Proost

13.1 Introduction

13.1.1 Drug Targeting and Effectiveness: The Role of Pharmacokinetics

The key issue in drug targeting is the improvement of the effectiveness of the intended drug therapy in comparison to conventional drug administration. In the present context, effectiveness is defined as the net benefit of drug administration, that is, the balance of the therapeutic drug effect and any harmful effect, including minor and major side-effects and toxicity. For the sake of simplicity, any harmful effect of the drug will be referred to as toxicity throughout this chapter. Also, effectiveness may be defined in terms of the increased apparent potency and/or therapeutic effect of the administered drug. A drug targeting system produces a larger and/or more prolonged pharmacologic effect than an equimolar dose of the free drug, and a lower single dose and/or dosing rate is needed to reach the same effect.

To demonstrate an improvement in effectiveness, relevant and reliable measures of the effect of drug administration should be available. Generally speaking, the best measure of the effectiveness of drug therapy should be a measure of the ultimate goal: the benefit to the patient. Although this approach is conceptually sound and logical, in practice the leap between the experimental development of drug targeting preparations and the ultimate benefit to the patient is extremely large. First, the experiments in the early development phase of a drug are usually carried out in small laboratory animals. Second, the pathogenesis in these animals is, in general, different from that in the patient for whom the therapy is intended. Third, the measures of effectiveness and toxicity may differ between laboratory animals and man. In laboratory animals the range of potential parameters or end-points is practically not limited, since the complete animal can be analysed after sacrifice. On the other hand, the eventual goal of the therapy, efficacy in the patient, cannot be readily assessed in objective terms.

Despite limitations, the measurement of the effectiveness of drug therapy in laboratory animals, for example, by the reduction in size of a solid tumour or decreased levels of surrogate tumour markers, is indispensable for the development of drug targeting preparations. However, during rational drug development it is not sufficient to ascertain that one drug preparation is more effective than another, it is important to find out the reasons why this is the case. This will enable the introduction of further improvements in the process of optimizing the design of the drug preparation. In the understanding of the effectiveness of drug therapy, pharmacokinetics (relating drug administration to drug concentration at the site of action) and pharmacodynamics (relating drug concentration to drug effect) are essential elements [1,2]. It should be stressed that PK and PK/PD modelling (Section 13.2) are essential tools in the evaluation of the effectiveness of any dosing strategy, including drug targeting. In some cases, it is sufficient to measure the drug concentration in the target tissue, if the relationship between concentration and drug effect is relatively simple (Section 13.2.5). In other case, however, the complex relationship between concentration and effect, for example due to dependence on time, dose, rate of drug administration, drug concentration, or rate of drug concentration change, requires the application of PK/PD modelling [3]. A detailed review of the pharmacodynamic aspects of drug delivery has been published recently [4].

A second, and equally important, application of pharmacokinetics in the field of drug targeting is the evaluation of the potentials and limitations in the drug targeting approach in relation to the properties of the drug and the drug–carrier conjugate. The theoretical framework designed by Stella and Himmelstein [5], and explored further by Hunt *et al.* [6], Boddy *et al.* [7], and others, is a useful tool to investigate the desirable properties of the drug and the drug–carrier conjugate, including the selection of therapeutic agents to be targeted by the chosen drug carrier.

Finally, pharmacokinetic and pharmacokinetic/pharmacodynamic modelling can be used for the purpose of prediction of the concentration–time profile of the drug and drug–carrier conjugate after repeated administration from single dose data, as well as for the prediction of the dose needed to maintain the concentration at the target site within a therapeutic window.

13.1.2 Pro-drugs and Drug–Carrier Conjugates

From the point of view of pharmacokinetics, there is no principal difference between pro-drugs and drug–carrier conjugates [5,6,8]. In both cases the active drug is administered as a part of a molecule that has pharmacokinetic and pharmacodynamic properties that are usually largely different from that of the active moiety ('drug'). The kinetics of pro-drugs which are converted in the body to the active form by conjugation (for example, the formation of the phosphorylated active forms of nucleoside analogues) can be modelled using the same approach. In general, however, the pharmacokinetic properties of pro-drugs and drug–carrier conjugates are quite different due to their different physicochemical properties, for example, with respect to their molecular weights, hydrophilic/hydrophobic character or exposed functional groups, among other structural features. The pharmacokinetic properties of pro-drugs and drug–carrier conjugates can be further optimized during the development phase of a product, the objective of which is to improve the pharmacokinetic properties by conjugation of the compounds to targeting devices which have a wide variety of physicochemical properties, without affecting the intrinsic potency of the coupled agent. In contrast, the pharmacokinetic properties of the active drug itself cannot usually be modified without affecting its pharmacologic profile. However, there are examples in which novel compounds were designed with improved pharmacokinetic properties without loss of potency or changes in the spectrum of activity.

Drug targeting technology may increase the drug concentration at the target site, may decrease the drug concentration at the sites where toxicity may occur, may prolong the retention time at the target site, and thus may improve the efficacy of drug administration.

For the sake of simplicity, in this chapter it is assumed that both the drug–carrier conjugate and the drug carrier itself, or the pro-drug, do not exert any pharmacologic or toxicologic effect, and that any therapeutic or toxic effect is due to the released or activated drug. This does not take into account the possibility of using intrinsically active carriers, which not only deliver the coupled drug to the appropriate site, but also contribute to the overall therapeutic effect, an approach known as 'dual targeting' [9].

13.1.3 Scope of this Chapter

The aim of this chapter is to provide an overview of the application of PK and PK/PD modelling and analysis to the field of drug targeting research. For those readers not familiar with the general principles of pharmacokinetics and pharmacodynamics, modelling, simulation, and data analysis, these topics are described in some detail in Section 13.2. These methods can be used for advanced PK and PK/PD modelling and analysis, as well as for conventional analysis of plasma concentration–time profiles of drugs, drug carriers, and drug–carrier conjugates. Conventional pharmacokinetic approaches, including descriptive methods for the evaluation of the concentration or concentration ratio profiles in different tissues, are not dealt with in this chapter.

The particular models used in drug targeting research are dealt with in Section 13.3, and quantitative measures of the effectiveness of drug targeting are described in Section 13.4, followed by a discussion relating to their application in Section 13.5.

Drug targeting by direct regional drug administration, controlled drug release, and pharmacokinetic modelling and analysis of *in vitro* experimental data, are outside the scope of this chapter. For the sake of completeness, some references to relevant papers in these areas are given in Section 13.6. After a short section (13.7) on software for pharmacokinetic modelling and data analysis, the perspectives of the application of PK and PK/PD modelling are discussed (Section 13.8).

13.2 Pharmacokinetics and Pharmacodynamics, Modelling, Simulation, and Data Analysis

13.2.1 Pharmacokinetics

13.2.1.1 Pharmacokinetic Processes

'Pharmacokinetics' (PK) can be defined as the study of the mechanisms and kinetics of drug disposition in the body (acronym 'LADME'), and includes the following:

- Liberation of drug from the dosage form. For example, the dissolution of drug from a tablet;
- Absorption, the transport from the site of administration to the general circulation. For example the transport of a drug from the gastrointestinal lumen, via the portal vein and the liver to the central venous blood pool;

- Distribution of the drug throughout the body, characterized by the volume of distribution (V) which is defined as the amount of drug in the body divided by the drug concentration in plasma;
- Metabolism, the biotransformation of the drug into metabolites, which may be inert, active, or toxic;
- Excretion of the intact drug, and its metabolites, into urine and faeces.

The term 'elimination' is used as a common term for the disappearance of the drug from the body by either metabolism or excretion. The term 'clearance' (CL) is used as a measure of the collective capacity of the eliminating organs to remove a certain drug, and is defined as the rate of drug elimination (amount/time) divided by the drug concentration in plasma, and indicates the volume of plasma that is cleared from the drug per unit of time (dimension volume/time). The elimination rate constant (k) is defined as the rate of drug elimination (amount/time) by the amount of drug in the body, and is equal to the clearance divided by volume of distribution (CL/V). The (elimination) half-life ($t_{1/2}$) is the time taken for the plasma concentration, as well as for the amount of drug in the body, to fall by 50%, and is approximately equal to $0.7/k$ [10].

In the field of drug targeting, the LADME processes refer to both the drug–carrier conjugate and the active drug. Liberation would refer to the release of the drug from a drug–carrier conjugate or the conversion of a pro-drug to the active moiety.

13.2.1.2 Transport Mechanisms

The transport mechanisms that operate in distribution and elimination processes of drugs, drug–carrier conjugates and pro-drugs include convective transport (for example, by blood flow), passive diffusion, facilitated diffusion and active transport by carrier proteins, and, in the case of macromolecules, endocytosis. The kinetics of the particular transport processes depend on the mechanism involved. For example, convective transport is governed by fluid flow and passive diffusion is governed by the concentration gradient, whereas facilitated diffusion, active transport and endocytosis obey saturable Michaelis–Menten kinetics.

13.2.1.3 Perfusion and Permeability

Both distribution of the drug within the body and elimination from the body require two sequential steps: the transport of the drug by blood flow to the organ or tissue (perfusion), and transport from the capillary to the tissue, and then to receptors on or in the cells of the tissue. The latter processes are governed by the permeability of the barriers between the capillary lumen and the receptor site, and may imply passive or carrier-mediated membrane passage. If there are hardly any barriers for the transport to the tissue, that is, if permeability is high, the supply of drug by the blood flow, that is, the perfusion of the organ or tissue may become the rate-limiting step of transport. In this case a large fraction of the drug present in blood is transported to the tissue, so the extraction ratio is high. On the other hand, if the perfusion is high, and the barriers for the transport within the tissue are considerable, permeability, may become the rate-limiting step of transport. In this case a small fraction of the drug present in blood is transported to the tissue, and the extraction ratio is low. So, depending on the rate-limiting step, the net transport may be perfusion-limited or permeability-limited. In any case, the upper limit to the rate of delivery is provided by the product of blood flow and the blood concentration of the drug.

13.2.1.4 Plasma Protein Binding and Tissue Binding

Many drugs are partly bound to plasma proteins, primarily albumin for acidic drugs and α_1-acid-glycoprotein for basic drugs, and to various macromolecular structures in the tissues [10]. An extensive discussion of the influence of plasma protein binding and tissue binding on pharmacokinetics is beyond the scope of this chapter. However, it should be noted that the binding of drugs has a major influence on pharmacokinetic and PK/PD modelling [11]. The unbound drug concentration is the driving force for transport within the body, including distribution, metabolism, and excretion, and for interaction with receptors, and thus for the pharmacologic effect. Unfortunately, the majority of PK and PK/PD models do not take into account plasma protein binding and tissue binding, and describe only the total drug concentration. It should be noted that this approach may lead to erroneous interpretations of PK and PK/PD.

13.2.2 Pharmacodynamics

'Pharmacodynamics' (PD) can be defined as the study of the mechanisms of drug action, including the relationship between drug concentration at the site of action and the drug effect. In many cases drug action is the result of the interaction of the drug and a receptor. However, many PD models (Section 13.2.5) do not take into account the precise mechanism of action, and are applicable to both receptor-mediated drug effects and effects initiated by other mechanisms.

The effectiveness of drug targeting should be evaluated by taking into account not only pharmacokinetic aspects, but also the pharmacodynamic aspects. The latter include the concentration–effect relationship in the target tissue and at the sites where toxicity may occur [7,12]. The therapeutic effect of the drug and its toxic effect may be different with regard to their mechanisms, and hence their concentration–effect relationship may also be different, both qualitatively (different PD models) and quantitatively (different model parameters).

13.2.3 Model and Modelling

The relationship between drug administration and the drug concentration at the site of action (PK) and the relationship between drug concentration at the site of action and the drug effect (PD), may be quantified by mathematical models describing the PK and PD processes involved in the drug activity profile. Combining PK and PD models allows the quantification of the relationship between drug administration and drug action (PK/PD models).

PK models (Section 13.2.4), PD models (Section 13.2.5), and PK/PD models (Section 13.2.6) can be used in two different ways, that is, in simulations (Section 13.2.7) and in data analysis (Section 13.2.8). Simulations can be performed if the model structure and its underlying parameter values are known. In fact, for any arbitrary dose or dosing schedule the drug concentration profile in each part of the model can be calculated. The quantitative measures of the effectiveness of drug targeting (Section 13.4) can also be evaluated. If actual measurements have been performed in *in-vivo* experiments in laboratory animals or man, the relevant model structure and its parameter values can be assessed by analysis of plasma disappearance curves, excretion rate profiles, tissue concentration data, and so forth (Section 13.2.8).

13.2.4 Pharmacokinetic Models

There are many types of PK models, which can be divided in two classes.

13.2.4.1 Compartmental Models

These are relatively simple models describing drug transport between compartments which are not necessarily specified in a physiological or anatomical context. The quantity of drug in each compartment is assumed to be evenly distributed throughout the volume of the compartment, and the rates of drug elimination and transport to other compartments are assumed to be proportional to the drug concentration in the original compartment. In pharmacokinetic literature these compartments are called well-stirred compartments [10,13,14].

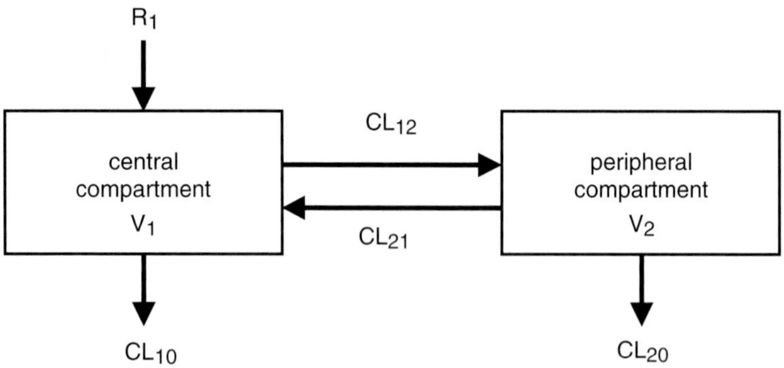

Figure 13.1. Compartmental model based on clearance and volume (Section 13.2.4.1). The drug is administered at a rate R_1 into the central compartment, which is characterized by a volume of distribution V_1. The drug is transported to and from the peripheral compartment with inter-compartmental clearance CL_{12} and CL_{21}, respectively (usually it is assumed that there is no net transport between the two compartments if the concentrations in both compartments are equal; in this case $CL_{21} = CL_{12}$). The peripheral compartment is characterized by a volume of distribution V_2. Elimination may take place from both compartments and is characterized by clearance CL_{10} and CL_{20}, respectively.

These models have relatively few parameters, and the parameters have a limited physiological or anatomical meaning. For example, a compartmental volume relates the quantity of the drug to its concentration in a compartment, and does not refer to an anatomically- or physiologically-defined area of the body.

The differential equations defining a compartmental model are derived from logical and simple principles. As an example, consider a model with two compartments as depicted in Figure 13.1. The change in the quantity of a drug in a compartment is the net result of the rate of entry of the drug, that is, the sum of the amount of drug administered to the compartment (for example, an intravenous infusion) or formed within the compartment (for example, release from a drug–carrier conjugate) and the rate of transport from other compartments, reduced by the rate of exit, that is, the sum of the rates of removal from the compartment by elimination or by transport to other compartments.

The rate of transport from a certain compartment is governed by the concentration in that compartment and a proportionality constant, denoted (elimination or distribution) clearance (dimension: volume time^{-1}) as formulated below.

$$V_1 \cdot \frac{dC_1}{dt} = R_1 + CL_{21} \cdot C_2 - CL_{12} \cdot C_1 - CL_{10} \cdot C_1 \quad (13.1)$$

where V_1 is the apparent volume of compartment 1, C_1 is the drug concentration in compartment 1, R_1 is the rate of drug administration or drug release in compartment 1, CL_{12} is the distribution clearance from compartment 1 to compartment 2, and CL_{10} is the elimination clearance from compartment 1.

Usually, it is assumed that there is no net transport between two compartments if the concentrations in both compartments are equal; in this specific case $CL_{21} = CL_{12}$.

Similar equations can be written for compartment 2. The same principle can be applied to any compartmental model, irrespective of its complexity.

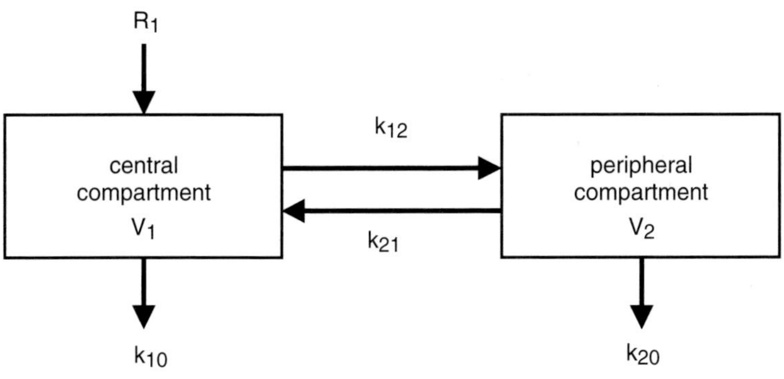

Figure 13.2. Compartmental model based on rate constants (Section 13.2.4.1). The drug is administered at a rate R_1 into the central compartment, which is characterized by a volume of distribution V_1. The drug is transported to and from the peripheral compartment with rate constants k_{12} and k_{21}, respectively. The peripheral compartment is characterized by a volume of distribution V_2 (usually it is assumed that there is no net transport between the two compartments if the concentrations in both compartments are equal; in this case $k_{21} \cdot V_2 = k_{12} \cdot V_1$). Elimination may take place from both compartments and is characterized by rate constants k_{10} and k_{20}, respectively.

Usually, the differential equations are written in a different form, by relating the rate of transport from a compartment to the quantity of the drug in that compartment and a rate constant (dimension: time^{-1}), as depicted in Figure 13.2. and formulated as follows.

$$\frac{dA_1}{dt} = R_1 + k_{21} \cdot A_2 - k_{12} \cdot A_1 - k_{10} \cdot A_1 \tag{13.2}$$

where A_1 is the quantity of the drug in compartment 1, k_{12} and k_{21} are distribution rate constants and k_{10} is the elimination rate constant.

Comparing Eq. 13.1 and 13.2, it follows that a rate constant k_{xy} is equal to CL_{xy} / V_x.

From the assumption that there is no net transport between two compartments if the concentrations in both compartments are equal, it follows that
$k_{21} \cdot V_2 = k_{12} \cdot V_1$ $(= CL_{21} = CL_{12})$.

Eq. 13.1 and 13.2 are mathematically equivalent, and thus may be used arbitrarily without affecting the modelling results. However, Eq. 13.1 (and Figure 13.1) is preferred since it reflects better the mechanistic basis, as drug transport is governed by drug concentration, both for passive diffusion according to Fick's Law, and for carrier-mediated transport. In the case of the latter, the terms referring to the rate of transport from compartment x to compartment y,

$$CL_{xy} \cdot C_x \tag{13.3}$$

should be replaced by their Michaelis–Menten equivalent

$$\frac{Vmax_{xy}}{Km_{xy} + C_x} \cdot C_x \tag{13.4}$$

where $Vmax_{xy}$ is the maximum transport rate between compartments x and y, and Km_{xy} is the Michaelis–Menten constant of the transport between x and y.

An example of the use of Michaelis–Menten kinetics in a compartmental model is given in the model of Stella and Himmelstein [5], depicted in Figure 13.3.

13.2.4.2 Physiologically-based Pharmacokinetic (PB-PK) Models

These are relatively complex models describing drug transport between blood and a series of physiological and/or anatomical entities, for example, organs, tissues, or cells [15–20]. PB-PK models are characterized by a relative large number of parameters. In many cases, several of these can be estimated from physiology or anatomy (for example, blood flow and volumes), others may be obtained from *in vitro* experiments (for example, partition coefficients between water and tissue), or by experiments in isolated tissues (for example, binding and metabolism in isolated liver cells or slices; see Chapter 12). In principle, PB-PK models are well adapted to take into account the extracellular and/or intracellular events in the disposition of the targeting device.

The number of compartments in a physiologically-based pharmacokinetic model may vary between two (in drug targeting: a target compartment and a non-target compartment) and 10 or more, depending on the desired degree of differentiation. The more compartments, the greater the ability of the model to define the true behaviour of the drug. However, the increased number of parameters increases the problem of assigning reliable values to these parameters, both in simulation (Section 13.2.7) and in data analysis (Section 13.2.8). As a general rule, the number of compartments should be chosen carefully, according to the parsimony principle: start the modelling with the simplest model that can discriminate the processes of interest. If the chosen model does not provide satisfactory results (in terms of credible predictions or satisfactory goodness-of-fit), the model can be explored further by adding compartments or connections in a step-by-step procedure.

On the other hand, PB-PK models are frequently used in toxicokinetics for a different purpose, that is, the model should be able to explain the drug distribution over a large number of tissues as measured from *in vivo* animal studies, with the eventual goal of data extrapolation to man. In this case, the starting point is a model including each organ and tissue from which measurements are available. If necessary, the number of compartments can be reduced by combining compartments with similar properties. A detailed description of the process of explicit (or formal) combining has been given by Nestorov *et al.* [19] and Weiss [20].

The principles of PB-PK modelling will be explained using the model of Hunt [6], depicted in Figure 13.4, a PB-PK model suited for evaluation of drug targeting strategies (Sections 13.3.2 and 13.4). For this model, the following set of differential equations describing the drug transport (mass per unit of time) can be written according to mass balance:

$$V_C \cdot \frac{dC_C}{dt} = R_C + Q_R \cdot \frac{C_R}{K_R} + Q_T \cdot \frac{C_T}{K_T} + Q_E \cdot \frac{C_E}{K_E} - (Q_R + Q_T + Q_E) \cdot C_C \tag{13.5}$$

$$V_R \cdot \frac{dC_R}{dt} = R_R + Q_R \cdot C_C - Q_R \cdot \frac{C_R}{K_R} - CL_R \cdot C_R \tag{13.6}$$

$$V_T \cdot \frac{dC_T}{dt} = R_T + Q_T \cdot C_C - Q_T \cdot \frac{C_T}{K_T} - CL_T \cdot C_T \tag{13.7}$$

$$V_E \cdot \frac{dC_E}{dt} = R_E + Q_E \cdot C_C - Q_E \cdot \frac{C_E}{K_E} - CL_E \cdot C_E \tag{13.8}$$

where V is the volume of the compartment, C is the drug concentration, Q is the blood (or plasma, whichever is the reference fluid) flow, K is the tissue/blood partition coefficient, CL is the (elimination) clearance, and R is the rate of drug input; subscripts C, R, T and E refer to the central compartment, response (or target) compartment, toxicity compartment, and elimination compartment, respectively (Note: Hunt *et al.* [6] did not include the partition coefficient K as such, in their equations. Rather, their tissue concentrations refer to a blood or plasma concentration which is in equilibrium with the tissue concentration, equal to the ratio C/K; consequently, their tissue volumes refer to apparent volumes, equal to the product $K \cdot V$).

13.2.4.3 Compartmental Models Versus Physiologically-based Models

Although compartmental models and physiologically-based models may at first, seem quite different, and are usually treated as two different classes of models, both approaches are actually similar [17]. When appropriately defined, probably any PB-PK model can be written as a compartmental model and vice versa. This can be seen by comparing the models in Figures 13.1 and 13.3, and their mathematical descriptions in Eq. 13.1 and 13.5.

The major difference between both approaches is not in the mathematical or pictorial description, but in the interpretation of the parameters. In compartmental modelling, the starting point is the parameters that do not necessarily have a particular anatomical or physiological meaning. This meaning, however, may become clear after a careful analysis of the data, including measurements in different organs and tissues. On the other hand, PB-PK starts with a model with physiologically meaningful parameters. It should be stated that, when applying a PB-PK model to real data, the identification of the parameters may become a major problem in the interpretation (see Section 13.2.8.4).

13.2.4.4 Principles of Modelling

In both types of models, the quantity or concentration of the drug in various sites of the body is described by mathematical equations quantifying drug administration, drug transport and drug elimination. These mathematical representations are usually in the form of differential equations, which can be solved numerically. In some simple cases an explicit analytical solution of the differential equations can be obtained, thus facilitating the calculations. The numerical procedure of solving the differential equations is more generally applicable, but is complicated by the necessity to find a compromise between accuracy and speed of execution. However, using modern, user-friendly software and fast-performing hardware, this is much less of an issue today (see Section 13.7).

13.2.5 Pharmacodynamic Models

Pharmacodynamic (PD) models are used to describe the relationship between drug concentration and drug effect. An overview of various PD models can be found in the literature [21]. The essential elements will be treated in the following sections.

13.2.5.1 Sigmoid E_{max} Model

For simplicity, a linear relationship between concentration and effect is often assumed, reducing the problem of PK/PD to the pharmacokinetics. However, the concentration–effect relationship of any drug tends towards a plateau, and a sigmoidal model (sigmoid E_{max} model or Hill equation) is more appropriate [21–24]:

$$E = E_0 + E_{max} \cdot \frac{C_e^\gamma}{C_e^\gamma + \mathrm{EC}_{50}^\gamma} \tag{13.9}$$

where E is the drug effect (arbitrary unit; same unit as E_0 and E_{max}), E_0 is the drug effect in the absence of drug (typically zero, or baseline effect), E_{max} is the maximum achievable drug effect, C_e is the drug concentration at the effector site, γ is a dimensionless value, indicating the gradient of the concentration–effect relationship, and EC_{50} is the drug concentration at which the drug effect is 50% of the maximum effect E_{max}.

In some cases more complex relationships between drug concentration and effect may occur, for example, when indirect drug effects, threshold concentration, all-or-none effect, time effects, or development of tolerance have to be taken into account.

13.2.5.2 Growth/Kill Models

For antibiotic and anti-tumour drugs, more complex models should be applied, taking into account the growth of microorganisms and tumour cells in the absence and presence of the drug.

The following PD model has been proposed for antibiotic drugs [25,26]:

$$\frac{dN}{dt} = \left\{ \lambda \cdot \left(1 - \frac{N}{N_{max}}\right) - E_{max} \cdot \frac{C_e^\gamma}{C_e^\gamma + EC_{50}^\gamma} \right\} \cdot N \tag{13.10}$$

where N is the number of microorganisms, λ is the microbial growth rate in the absence of drug, N_{max} is the maximum number of microorganisms that can be reached, E_{max} is the maximum achievable killing rate, C_e is the drug concentration at the effector site, γ is a dimensionless value, indicating the gradient of the concentration–effect relationship, and EC_{50} is the drug concentration at which the killing rate is 50% of its maximum value E_{max}. If N_0 is the initial number of microorganisms at time zero, N reflects the number of microorganisms at time t.

For anti-tumour drugs, Ozawa *et al.* [27] proposed the following models. For cell cycle phase non-specific drugs (type I drug), the cytotoxic activity depends on the drug exposure, as reflected in the area under the intracellular concentration–time profile (AUC), and can be modelled using the following formula [2,28]:

$$\frac{dN}{dt} = \{k_s - k_r - k \cdot fu_T \cdot (C_T - C_{min})\} \cdot N \tag{13.11}$$

where N is the number of tumour cells, k_s is the cell proliferation rate constant of the tumour cells in the absence of drug, k_r is the rate constant of physiological cell degradation, k is the drug-induced cell killing rate constant, fu_T is the unbound fraction (unbound concentration divided by total concentration) within the cells, C_T is the drug concentration within the cells, and C_{min} is the minimum concentration required for the cell killing effect [28].

The cytotoxic activity of cell cycle phase specific drugs (type II drugs) is time-dependent, and is different for cells in the sensitive phase (N_S) and in the resistant phase (N_R), as described as follows [2,27,28]:

$$\frac{dN_s}{dt} = 2\,k_{RS} \cdot N_R - \{k_{SR} + k_r + k_1 \cdot fu_T \cdot (C_T - C_{min})\} \cdot N_S \tag{13.12}$$

$$\frac{dN_R}{dt} = k_{RS} \cdot N_S - \{k_{RS} + k_r + k_2 \cdot fu_T \cdot (C_T - C_{min})\} \cdot N_R \tag{13.13}$$

where k_1 and k_2 are the drug-induced cell killing rate constants for sensitive and resistant cells, respectively, and k_{SR} and k_{RS} are the cell-cycle traverse rate constants from S-phase to R-phase, and from R-phase to S-phase, respectively.

(Note the typing errors in Nakai *et al.* [28]; in their equations 16 and 19, the second term k_{SR} should be replaced by k_{RS}; similarly, in equations 17 and 20, the second term k_{RS} should be replaced by k_{SR}).

13.2.5.3 Empirical PK/PD Relationships

Many empirical relationships between PK and PD have been described in the literature. Several of these empirical relationships have been reviewed by Kobayashi *et al.* [29].

13.2.6 Pharmacokinetic/Pharmacodynamic (PK/PD) Models

PK/PD models are obtained by combining a PK model (Section 13.2.4) and a PD model (Section 13.2.5), allowing the quantification of the relationship between drug administration and drug action. The principles of PK/PD modelling will be dealt with briefly. For a more detailed treatise, some excellent reviews can be found in the literature [21].

Usually, the target compartment of the PK model is the site where the active drug is released. If there is a negligible diffusion barrier between the site of drug release and the site of drug action (effector site), the drug action (Eq. 13.9) is governed by the concentration in the target compartment. In other cases the site of action may be more remote from the site of drug release, and the concentration at both sites may be different due to a diffusion barrier. In such cases, an extra compartment (effect compartment, effector site) can be added to the model as a link between the 'driving' compartment (here, the site of drug release) and the drug effect [23,24]:

$$\frac{dC_e}{dt} = k_{e0} \cdot (C - C_e) \tag{13.14}$$

where C_e is the drug concentration in the effect compartment, C is the concentration in the driving compartment, and k_{e0} is the transfer rate constant between the effect compartment and the driving compartment.

13.2.7 Simulations

If an appropriate model is selected and the model parameters are known, the time course of the drug concentration in each compartment (PK models) and the drug effect (PK/PD models) can be calculated for any dosing regimen. In addition, the relevant measures of the effectiveness of drug targeting can be calculated (see Section 13.4).

Usually, the most appropriate model and values of model parameters are not known, in which case, the relevant information is obtained from simulations with various models and parameter values, based on reasonable estimates and on previously obtained experimental data. Despite their limitations, such simulations can be helpful in drug design and development, including the prediction and evaluation of drug targeting strategies. Using PK or PK/PD models, the effect of drug targeting can be quantified, taking into account not only the process modified in order to target the drug, but also the kinetics of the carrier–drug conjugate and the active drug after liberation. Such a modelling process might precede the experimentation process, in order to gain insight into the potential benefit of a drug targeting concept in comparison to traditional administration of the drug.

13.2.8 Data Analysis by Modelling

Data analysis by modelling may be applied for various reasons, for example:

- To condense the data, thus obtaining a model with relatively few model parameters instead of one with a large number of measurements.
- To explore mechanisms involved in the process under investigation (for example, carrier-mediated transport).
- To make predictions (for example, to predict the dose needed to maintain the concentration of the drug at the target site within a therapeutic window).

13.2.8.1 Model Building

The process of data analysis by modelling implies building a model from measurements, which involves two steps: (a) building the model structure, and (b) assessment of model parameters.

First, the simplest model which includes the minimum number of compartments and model parameters must be defined. For this model, the parameters are estimated from a set of measurements obtained by non-linear regression or curve-fitting (Section 13.2.8.3). The purpose of this process is to find a set of model parameters which best fits the measurements (Section 13.2.8.2). If the goodness-of-fit is acceptable (Section 13.2.8.5), the model can be evaluated by comparison with other models (Section 13.2.8.6).

13.2.8.2 Defining the Objective Function

The first step in 'curve fitting' is to define the 'best fit'. Usually, the criterion for 'best fit' is a weighted least-squares criterion, based on statistical grounds [30–34]. Assuming that the errors in the measured concentrations are normally distributed, the best fitting set of parameters is obtaining by minimization of the following objective function, *OBJ:*

$$OBJ = \sum_{i=1}^{n} \left\{ \frac{(C_{\text{meas},\,i} - C_{\text{calc},\,i})^2}{\sigma_i^2} + \ln(\sigma_i^2) \right\} \qquad (13.15)$$

where

n = total number of concentration measurements
$C_{\text{meas},i}$ = measured concentration at time point i ($i = 1,2,...,n$)
$C_{\text{calc},i}$ = calculated concentration at time point i
σ_i = standard deviation of the measurement at time point i

In Eq. 13.15, the squared standard deviations (variances) act as 'weights' of the squared residuals. The standard deviations of the measurements are usually not known, and therefore an arbitrary choice is necessary. It should be stressed that this choice may have a large influence of the final 'best' set of parameters. The scheme for appropriate weighting and, if appropriate, transformation of data (for example logarithmic transformation to fulfil the requirement of homoscedastic variance) should be based on reasonable assumptions with respect to the error distribution in the data, for example as obtained during validation of the plasma concentration assay. The choice should be checked afterwards, according to the procedures for the evaluation of goodness-of-fit (Section 13.2.8.5).

Usually, the standard deviation of the measurement is dependent on the magnitude of the concentration. The most commonly applied assumption is that the standard deviation is proportional to the concentration, which is either the measured concentration (also referred to as 'data-based weighting'), or the calculated concentration (model-based weighting).

Many software packages provide only a limited selection of weighting procedures. The most commonly applied weighting procedure is based on the assumption that the standard deviation of the concentration is proportional to the measured concentration. In that case, the objective function may be simplified to:

$$OBJ = \sum_{i=1}^{n} \frac{(C_{\text{meas},\,i} - C_{\text{calc},\,i})^2}{(C_{\text{meas},\,i})^2} \tag{13.16}$$

Alternatively, the following objective function may be used, assuming that the errors in the measured concentrations are log-normally distributed:

$$OBJ = \sum_{i=1}^{n} [\ln(C_{\text{meas},\,i}) - \ln(C_{\text{calc},\,i})]^2 \tag{13.17}$$

Since both Eq. 13.16 and 13.17 assume a constant coefficient of variation of the measurement error, these equations provide similar (but not identical) results.

13.2.8.3 Searching the Best-fitting Set of Parameters

The best-fitting set of parameters can be found by minimization of the objective function (Section 13.2.8.2). This can be performed only by iterative procedures. For this purpose several minimization algorithms can be applied, for example, Simplex, Gauss–Newton, and the Marquardt methods. It is not the aim of this chapter to deal with non-linear curve-fitting extensively. For further reference, excellent papers and books are available [18].

The fitting procedure may be performed by any suitable minimization algorithm. In theory, the final parameter set depends only on the objective criterion (Section 13.2.8.2) and is not dependent on the minimization algorithm, nor on the initial set of parameter values (except for rounding-off errors). However, in practice, the minimization algorithm may fail to reach the minimum of the objective function, and may end in a local minimum. In this respect, minimization algorithms may vary widely. An algorithm which is insensitive to the choice of the initial estimates, is said to be robust, which is a highly desirable property. To lower the risk of convergence to a local minimum of the objective function, the convergence criterion (or stop criterion, for example the relative improvement of the objective function) chosen should be sufficiently small, and the fitting procedure may be repeated with a different set of initial estimates, or a different minimizing algorithm.

13.2.8.4 Identification of Model Parameters

The procedure to obtain the best fitting set of model parameters (Section 13.2.8.3) can be performed only if each model parameter is uniquely identifiable from the measurements [35–38]. This implies that the same set of model parameters is obtained, irrespective of the initial set (Section 13.2.8.3). In some cases one or more model parameters cannot be identified uniquely, because the measurement data do not contain enough 'information' on that particular parameter, for example:

- In the model depicted in Figure 13.1, the parameters CL_{10} and CL_{20} cannot be obtained uniquely if only measurement data from compartment 1 are available. There is an infinite number of parameter sets yielding exactly the same concentration profile in compartment 1. Only if certain constraints are imposed (for example, $CL_{20} = 0$ or $CL_{20} = CL_{10}$), can the model parameters be identified uniquely.
- In the case of Michaelis–Menten kinetics (Eq. 13.4), $V\mathrm{max}_{xy}$ and $K\mathrm{m}_{xy}$ cannot be assessed uniquely if the concentration C_x is far below the value of $K\mathrm{m}_{xy}$; in that case, Eq. 13.4 reduces to Eq. 13.3, and only one parameter CL_{xy}, or the ratio $V\mathrm{max}_{xy}/\mathrm{Km}_{xy}$ can be calculated uniquely.
- In the model shown in Figure 13.1, if CL_{12} is very large compared to CL_{10}, and if no data shortly after administration of a bolus dose are available, the model would behave as a single compartment model, and CL_{12} will not identifiable; also, only the sum of the volumes rather than both volumes separately can be assessed.

The problem of identification grows rapidly with increasing complexity of the model. In some cases this problem can be solved by an appropriate experimental design. As a general rule, the problem is reduced if the concentrations in compartments other than the central plasma compartment can be measured. Also, simultaneous measurement of drug and drug–carrier conjugate or pro-drug is a condition for identification of the models described in Section 13.3.

Jacquez and Perry [37] developed the program IDENT to investigate the identification of model parameters. In most cases, problems of identification can be detected by inspection of the standard errors of the model parameters (the standard error of a model parameter is a measure of the credibility of the parameter value, which is provided by the most fitting programs). A high standard error (for example, more than 50% of the parameter value) indicates that the parameter value cannot be assessed from the data, most likely due to an identification problem. In that case, the parameter value itself is meaningless, and thus the parameter set should be discarded (see Section 13.2.8.5).

13.2.8.5 Goodness-of-Fit

After fitting the parameters of a model to a set of measurement data, criteria for the goodness-of-fit are required. There will always be some differences between the measured data and the values calculated from the model. These differences may be due to the following causes:

- Measurement errors in the data, for example, inevitable analytical errors implicit in the chosen analytical method. In general, measurement errors are random errors, and their order of magnitude may be known from the precision of the assay, as assessed during the validation of the assay. If the magnitude of the measurement errors is comparable to the precision of the assay, the goodness-of-fit is acceptable. The possibility of problems in the case of measurements close to the detection limit of the assay should be taken into account. In this case, the relative errors in the analysis may be significantly larger than over the usual range.
- Model mis-specification. If an inappropriate model is chosen (for example, a model with too small a number of compartments, or an incorrect structure), it will not be able to describe the measurements adequately, resulting in systematic deviations between the measurements (for example, plasma or tissue concentrations) and the values calculated from the model. Such systematic deviation can be detected by the visual methods described below.
- Other errors in the procedure, such as failure to distinguish between carrier-bound and unbound drug, as well as errors in dosing, deviations in the time of measurement, incorrect sampling procedure, exchange of samples, mistakes in dilution during sample treatment, and so forth. These types of error are the most problematic, and no general solution can be given.

There are several methods available for the assessment of goodness-of-fit, however, there are no exact and objective criteria for its evaluation. This is due to the following: (1) goodness-of-fit is not a single property, and cannot be expressed in a single value, and (2) numerical measures of goodness-of-fit do no have an absolute meaning. Therefore there is a dependence on somewhat subjective criteria. To ensure maximal objectivity, the criteria for accepting a set of model parameters obtained by the fitting procedure as a valid result should be defined explicitly before the analysis is initiated. In practice, however, this condition is hardly applicable during the development of new drug targeting preparations, taking into account the complexity of the modelling procedure.

The following criteria could be used to ensure an acceptable goodness-of-fit:

- Visual inspection of the observed and calculated data should not reveal any significant lack of fit.
- Residuals (difference between observed and calculated data) or normalized residuals (residuals divided by the corresponding standard deviation) should be scattered randomly around zero, by visual inspection.
- Normalized residuals should be neither diverging nor converging when plotted against time or plotted against (logarithm of) concentration, by visual inspection.
- Residuals should not be serially correlated, as identified by visual inspection or by an appropriate statistical test (for example a Run's test).

- The standard error of each relevant parameter should be lower than a predefined value (for example 50% of the parameter value). High standard errors may reflect problems in the identification (see Section 13.2.8.4).
- In the case of any of the calculated pharmacokinetic parameters being seen as physiologically unfeasible, the analysis should be interpreted with care, and should not be presented without comments.
- Outlying data points should be dealt with explicitly, and should not be discarded unless felt to be physiologically impossible. The impact of elimination of the outlying points on the parameter estimates should be investigated.

Non-compliance with one or more of these criteria may indicate that an inappropriate model or an inappropriate weighting scheme was chosen.

13.2.8.6 Model Selection

There may be more than one plausible model structure that can be used to describe the data. In that case, any plausible model is analysed in a similar way. If the goodness-of-fit of more than one model is acceptable (see Section 13.2.8.5), a procedure for selecting the 'best' model is required.

It is common practice to compare the results of different models, each yielding an acceptable goodness-of-fit, according to the following procedure. First, the models are classified hierarchically in a tree structure. The more complex models are considered as extensions of the simpler models, by adding extra parameters, for example, an extra compartment, a extra degradation step, or a time lag. It can be said that the simpler model is a special case of the more complex model, for example because one or more parameters have a fixed value (in general a zero value). Then, starting with the simplest model (see Section 13.2.8.1), the models are compared in pairs according to their hierarchical relationship. Such a comparison can be based on statistical criteria, for example [39]:

(a) An *F*-test by which the following value is calculated:

$$F = \frac{WRSS_S - WRSS_C}{WRSS_C} \cdot \frac{N - P_C}{P_C - P_S} \qquad (13.18)$$

where N is the number of measurements, P is the number of model parameters, and $WRSS$ is the weighted residual sum of squares (see Section 13.2.8.2); the subscript s refers to the simpler model and the subscript c to the more complex model.

If the value F exceeds the tabulated value of Fischer's F-distribution for $(P_c - P_s)$ and $(N - P_c)$ degrees of freedom, and a confidence level of (usually) 95% ($\alpha = 0.05$), then the complex model fits significantly better to the data than the simpler model. If not, the 'parsimony' principle dictates that the simpler model should be accepted as the 'best' model.
(b) Akaike Information Criterion (AIC). For each model the AIC is calculated according to the following equation:

$$AIC = N \cdot \ln (WRSS) + 2P \qquad (13.19)$$

The model with the lowest AIC value is accepted as the 'best' model.

13.3 Pharmacokinetic Models for Drug Targeting

In 1980, Stella and Himmelstein [5] introduced the principles of pharmacokinetic modelling into the field of pro-drugs and site-specific delivery. In 1986, Hunt *et al.* [6] extended the model of Stella and Himmelstein by taking into account a specific area where toxicity occurs. Their work may be considered as the frame of reference for later work in this area.

13.3.1 Model of Stella and Himmelstein

The model of Stella and Himmelstein [5], depicted in Figure 13.3, was originally derived for pro-drugs, and may be considered as a minimal model for evaluating the pharmacokinetics of pro-drugs and drug targeting systems. Since the drug–carrier conjugate (DC) and the active drug (D) are different entities, and since the model should be able to discriminate between the target site and non-target sites, a pharmacokinetic model for drug–carrier conjugates should include, at least, the following compartments (Figure 13.3):

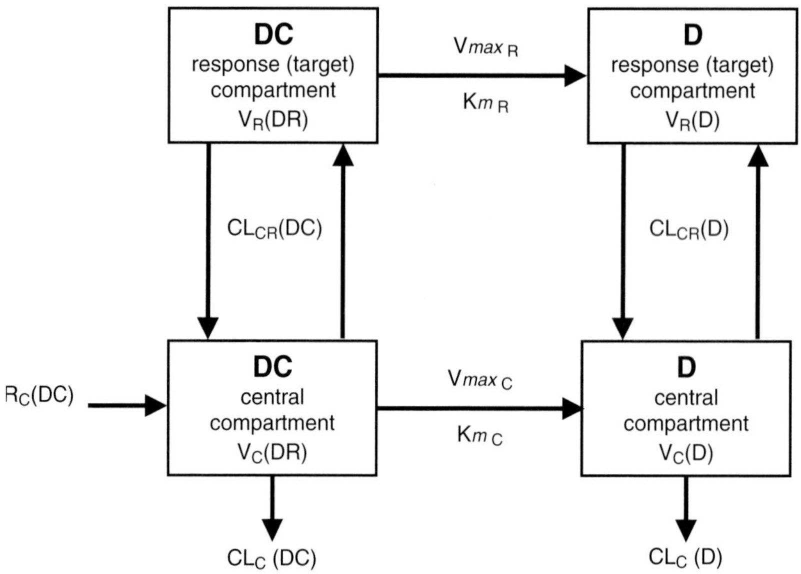

Figure 13.3. Model of Stella and Himmelstein, adapted from reference [5] (Section 13.3.1). The drug–carrier conjugate (DC) is administered at a rate R_C(DC) into the central compartment of DC, which is characterized by a volume of distribution V_C(DC). DC is transported with an inter-compartmental clearance CL_{CR}(DC) to and from the response (target) compartment with volume V_R(DC), and is eliminated from the central compartment with a clearance CL_C(DC). The active drug (D) is released from DC in the central and response compartments via saturable processes obeying Michaelis–Menten kinetics defined by *V*max and *K*m values. D is distributed over the volumes V_C(D) and V_R(D) of the central and response compartment, respectively. D is transported with an inter-compartmental clearance CL_{CR}(D) between the central compartment and response compartment, and is eliminated from the central compartment with a clearance CL_C(D).

- The DC in non-target sites, including the plasma compartment (central compartment). The drug–carrier conjugate in the non-target sites may be modelled as a single compartment, if DC distributes rapidly over its distribution volume. That is, the plasma compartment in the case of large macromolecular carriers which cannot cross the endothelial lining of blood vessels, or the extracellular volume in the case of smaller compounds which do have the opportunity to extravasate. Both administration (except for delivery to the target site) and elimination occur in this DC central compartment. If necessary, one or two peripheral compartments may be added to improve the accuracy of predicting the distribution of the DC concentrations in the body.
- DC in target sites (response compartment). The drug–carrier conjugate is targeted by some specific mechanism to the target site. The volume of this compartment depends on the specificity of this mechanism, and can also be influenced by the disease process which is being targeted. This value may range from a few milliliters (for a focal inflammation or infection site) to several liters (for a widely spread disease) [40].
- D in target sites (response compartment). Ideally, the target site of the drug is located very near to the DC target site, where the drug is released or activated. In that case, the target sites of DC and D are identical, and any amount of drug released or activated may exert its effect immediately. However, if the drug is released or activated at sites distant from the target site of the drug, an additional compartment is required. This will have a detrimental effect on the efficacy of the targeting strategy [40].
- D in non-target sites, including the plasma compartment (central compartment). After release or activation of the drug in or near the target site, the drug will be transported to the plasma compartment by diffusion and/or convection. Also, some drug may be released or activated at non-target sites. The central compartment in which the drug distributes may vary, depending on its physicochemical properties: from a minimum value equal to the plasma compartment to high values, exceeding the physical volume due to excessive binding of the drug to tissue components [10]. Elimination of the drug occurs from this central compartment. If necessary, one or two peripheral compartments may be added to enable a more accurate prediction of the drug concentrations in the body to be made. If drug D is converted to active metabolites, additional compartments to account for the fate of these metabolites may be required.

The central compartment may also include the sites where toxicity occurs. Alternatively, these sites may be modelled as compartments connected to the central compartment, analogous to the D target-site compartment. If the drug is released or activated within these toxicity sites, and/or the drug carrier is targeted to some degree to these sites, an additional DC compartment should be modelled, analogous to the DC target site.

Using the model shown in Figure 13.3, the main processes involved are the following.

13.3.1.1 Disposition of DC

This refers to the distribution and elimination of the DC, excluding transport to the target site (and toxicity site, as shown in Figure 13.4). In general, it is desirable that the DC is not rapidly eliminated from the circulation. This is necessary both to minimize the exposure of eliminating organs to the targeted drug (in the case of release or activation of the drug in the eliminating organs) and to maximize the therapeutic availability, that is, to maximize the fraction of the dose that reaches the target site. Ideally, the elimination of the DC should take place exclusively in the target organ (see Sections 13.3.1.2 and 13.3.1.3). Elimination at other sites reduces the efficiency of targeting. For example, elimination of the DC without release of the active drug implies a waste of a sophisticated and expensive product, and elimination of the DC with release of active drug in non-target tissue increases the concentration of D in non-target tissue, thus lowering the Drug Targeting Index (DTI, see Section 13.4.2) [12].

In general, optimization of the disposition of DC is the simplest part of the design of a drug targeting system. There are many methods by which the retention time of the drug carrier within the target tissue can be prolonged (described in other chapters of this book), and the assessment of the disposition of the DC by measuring its plasma or blood concentration–time profile is relatively simple.

13.3.1.2 Delivery of the DC to the Target Site

The focus of research in the field of drug targeting is mainly on the selectivity of the DC for the target site (see other chapters in this book), undoubtedly because it is the essential process in drug targeting.

Although the delivery of the DC to the target site is a critical step in the efficacy of drug targeting, it is not necessarily the most critical step. In fact, each of the processes involved may be critical, and it is possible that in practice, given the sophisticated principles which are applied to increase the selectivity of DC for the target site, one of the other processes involved (see Sections 13.3.1.1. and 13.3.1.3.6) is responsible for the eventual failures of drug targeting systems.

The general principles of drug transport have been described in Sections 13.2.1.2 and 13.2.1.3. The rate of delivery of the DC to the target site is not critical in absolute terms. Even if the delivery rate is low, as long as the selectivity of the DC for the target site is high efficient targeting is assured. In this case, it will take more time to reach effective steady-state drug levels in the target organ and the duration of the drug effect will be prolonged. However, if the selectivity of DC for the target site is less than 100%, the slow delivery of the DC to the target site may decrease the efficacy of drug targeting [12].

If the rate of delivery of the DC to the target site is high compared to the release of the activated drug (see Section 13.3.1.3), accumulation of DC at the target site might result in a loss of DC from the target site either back into the blood, or via lymphatic drainage, and thus in a decrease of the efficiency of targeting. An example of this is the occurrence of retro-endocytosis in which the endocytosed DC is refluxed back into the systemic circulation before it has released the drug to be targeted [41]. The possibility that receptor-mediated endocytosis is a bidirectional rather than a unidirectional process is often not taken into account in drug targeting models. Yet, a secondary release of endocytosed material through reactivation of endocytotic vesicles in the plasma membrane has been demonstrated in various cell types. This process may be of quantitative importance, especially if trafficking to and association with lysosomes is slow or if proteolytic degradation is rate-limiting.

13.3.1.3 Release or Activation of D at the Target Site

The release of the drug from the drug–carrier conjugate or the formation of the active moiety from the pro-drug, is also reported to be of major concern in many research papers on drug targeting (see other chapters in this book). Although this process cannot be measured *in vivo* in most cases, valuable information can be obtained from *in vitro* measurements, for example, using cultured cell lines. The mechanism of release or activation may contribute to the selectivity of the target site, for example, in case of the formation of the phosphorylated active forms of nucleoside analogues by viral enzymes. In the case of release or activation by enzymatic processes, the rate of release or activation may be limited by saturation of the enzyme capacity.

In general, rapid release contributes to the efficiency of drug targeting, although in many cases the rate of release may not be critical as a result of rate limitation in the delivery of the DC to the target site. With respect to the selectivity of the target site for the release or activation in comparison to non-target sites, it can be stated that the more selective the delivery of the DC to the target site, the less important is the selectivity of the release or activation at the target site.

13.3.1.4 Removal of D from the Target Site

Little attention has been paid to this aspect in most research papers on drug targeting, despite several reports on the importance of removal of the released drug from the target site [6,12,42]. Most likely, this lack of attention is related to at least three factors: removal of drug from the target site (1) cannot be manipulated by the design of the drug targeting system, (2) is difficult to measure, and (3) may be the bottleneck of the efficiency of drug targeting strategies.

A striking example of the importance of the secondary removal of a drug from the target site are the multi-drug-resistance (MDR) processes of tumour cells. For instance antineoplastic drugs can be extruded efficiently by P-glycoprotein [43]. Although this example is only relevant for a limited category of drugs, it clearly demonstrates that the delivery of a drug into a cell does not guarantee an optimal retention time in the cell nor an optimal drug effect. Levy [42] assumed that in many cases drug elimination from the target site will be much more rapid that drug elimination from the body, as exemplified for intracerebroventricular injections of barbiturates.

Removal of active drug from the target site can occur by two different routes: direct removal by a local elimination process (typically a metabolic process, causing inactivation of the drug), or removal by the blood flow following diffusion out of the target cell. In both cases removal reduces the efficiency of drug targeting by lowering the concentration of the active drug at the target site. However, in case of local removal, the active drug is not transported to non-target tissues, and does not contribute to toxicity. In contrast, removal by the bloodstream causes an increase of concentration in the non-target tissue, thus lowering the targeting efficiency further. It may be noted that a high blood flow to the target tissue favours the delivery of the DC to the target site, but also increases the rate of removal of the active drug. Ideally, the active drug should be eliminated slowly from the target site, so that effective drug levels are maintained at the site after administration of relatively low maintenance doses of the drug targeting system. Obviously, a low, but substantial, rate of active drug removal is a prerequisite for controlling the intensity and duration of the drug effect.

13.3.1.5 Release of D at Non-target Sites

The release or activation of a drug at non-target sites is the undesirable counterpart of release at the target site (see Section 13.3.1.3), determining the selectivity of the release or activation of the drug, and thus of the drug targeting.

In principle, the release at non-target sites can be measured by the same methods as used for the target site (Section 13.3.1.3). In practice, however, measurement must be limited to one or more selected cell lines, which do not necessarily include all the non-target sites in which the drug may be released outside the intended target tissue.

13.3.1.6 Disposition of D

The distribution and elimination characteristics of the active drug have a profound influence on the efficiency of drug targeting. This topic is dealt with in Section 13.5.

13.3.2 Model of Hunt

Hunt *et al.* [6] extended the model of Stella and Himmelstein by adding two compartments, that is, a specific area where toxicity occurs, and an elimination compartment consisting of the liver and kidney (Figure 13.4). Their model may be regarded as a simplified physiologically-based pharmacokinetic model (Section 13.2.4.2). It is assumed that the drug concentration in the blood or plasma (whichever is the reference fluid) exiting the compartments is a function of the concentrations within the corresponding compartments. This implies that tissue perfusion rather than permeability of the drug between blood and tissue is assumed to be the rate-limiting step of the transport between the central compartment and each of the other compartments (Section 13.2.1.3).

The model does not use a pharmacokinetic model for the drug–carrier conjugate. Instead, the release of the active drug is modelled as an input function in each of the four compartments: 'blood' (central compartment), 'response' (target site), 'toxicity', and 'elimination'. The aim of their model was the derivation of the Therapeutic Availability (TA, Section 13.4.1) and the Drug Targeting Index (Section 13.4.2). For the particular derivations, several simplifications were made; for example, the aforementioned drug input replaces the model for DC. As a result, their simplified approach cannot be used for the prediction of the time course of the drug concentrations in the various compartments. This limitation does not affect the ability of the model to evaluate steady-state drug concentrations, which can also be used as an appropriate measure of the average concentration over a single dose interval after repeated administration. However, their analysis does not give insight into the time course of drug action, that is, the time needed to reach steady state, and the duration of the

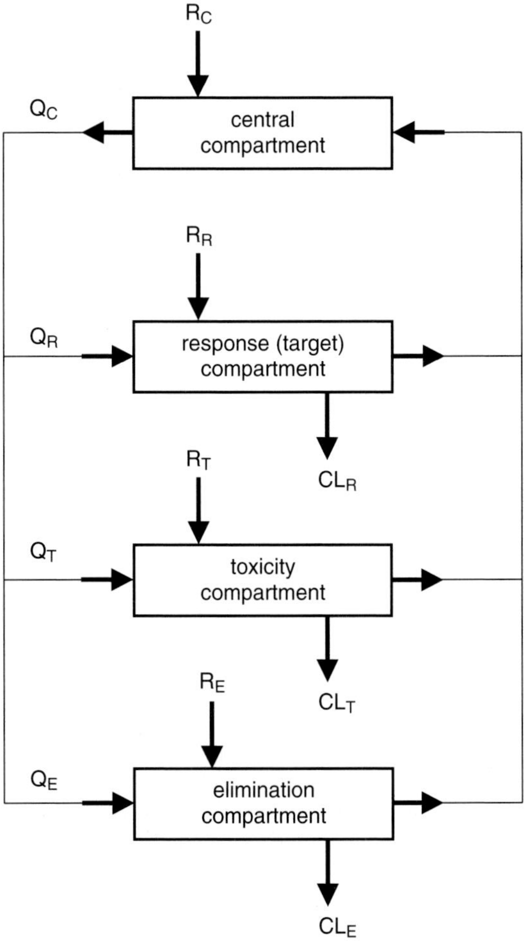

Figure 13.4. Model of Hunt, adapted from reference [6] (Section 13.3.2). The model describes the fate of the active drug (D) only. It consists of four compartments: A central compartment (V_C) representing blood and all other tissues not accounted for by the other three compartments, a response (target) compartment (V_R) representing all tissues containing target sites for the desired response, a toxicity compartment (V_T) representing tissues where the cascade of events leading to a toxic response is initiated, and an elimination compartment (V_E) representing the eliminating organs, excluding the elimination sites in the response and toxicity compartments. Each compartment is characterized by a volume of distribution V, a blood flow Q (where $Q_C = Q_R + Q_T + Q_E$), and, except for the central compartment, an elimination clearance CL. Drug delivery is designated by input functions R in the four compartments. Drug targeting and conventional drug administration are modelled by changing the relative contributions of R_C, R_R, R_T, and R_E. When $R_C = R_T = R_E = 0$, the drug carrier is an ideal target-specific carrier.

drug effect after the conclusion of drug administration. To evaluate the time course of action, a full model, including the pharmacokinetic behaviour of the drug–carrier conjugate, should be used, for example the model of Stella and Himmelstein (Section 13.3.1) or the model of Boddy *et al.* (Section 13.3.3).

Boddy and Aarons [44] used a simplified model, in which the toxicity sites are included in the systemic (non-target) tissues to allow derivation of the Drug Targeting Index which was less restricted by model assumptions. Their approach was criticized by Siegel *et al.* [45]. When some drug release from the carrier occurs at either the central, toxicity or elimination regions, the DTI is affected by the specific fraction of the drug delivered to each region. Since these regions are combined by Boddy and Aarons into a single systemic region, their model cannot account for this effect. Siegel *et al.* came to the conclusion that the model of Hunt *et al.* and the model of Boddy and Aarons are only equivalent in the case of an ideal carrier.

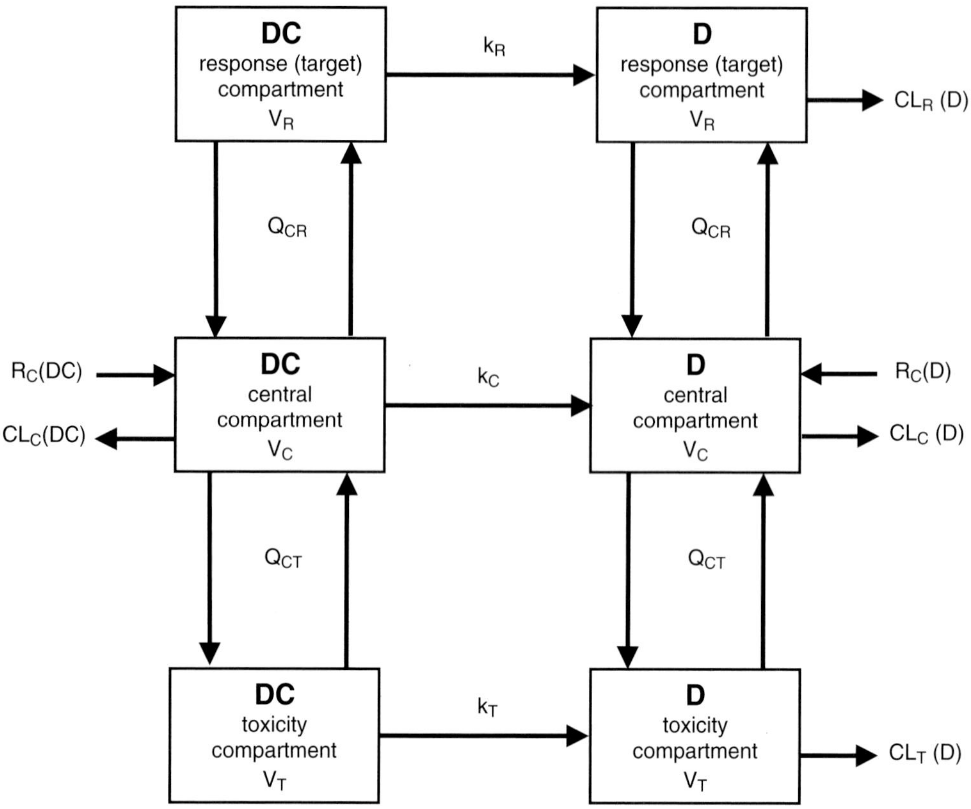

Figure 13.5. Model of Boddy, adapted from reference [7] (Section 13.3.3). The drug–carrier conjugate (DC) is administered at a rate R_C(DC) into the central compartment, which is characterized by a volume of distribution V_C. DC is transported by blood flow Q_{CR} to and from the response (target) compartment, characterized by a volume of distribution V_R, and by blood flow Q_{CT} to and from the toxicity compartment, characterized by a volume of distribution V_T. DC is eliminated from only the central compartment with a clearance CL_C(DC). The active drug (D) is released from DC in the central, response and toxicity compartments with first-order rate constants k_C, k_R and k_T, respectively. The D is distributed over these compartments in a manner similar to the DC. The D is eliminated from these compartments with a clearance of CL_C(D), CL_R(D) and CL_T(D), respectively. Conventional drug administration can be simulated by the input of D at a rate R_C(D) into the central compartment.

13.3.3 Model of Boddy

Boddy *et al.* [7] extended the model of Hunt by incorporating the pharmacokinetic behaviour of the drug–carrier conjugate, analogous to the model of Stella and Himmelstein. Therefore this model, depicted in Figure 13.5, is suited to the evaluation of the time course profile of drug concentrations in each compartment, in contrast to the model of Hunt. Also, they reduced the model by incorporating the elimination compartment within the central compartment. Furthermore, the model was simplified by assuming that the volumes of distribution are the same for the drug–carrier conjugate and the active drug. It is assumed that the blood flowing from the compartments carries the DC and D at concentrations equal to those within the compartments. In this specific case, the inter-compartmental clearances for both DC and D can be equated to the blood flow to the response and toxicity compartments. The authors presented their model as a three-compartment model. However, since each compartment may contain both the drug and the drug–carrier conjugate (which are treated as different entities), it might be more appropriate to refer to it as a six-compartment model. Finally, Boddy *et al.* [7] extended the model to a PK/PD model by incorporation of the pharmacodynamic equations analogous to Eq. 13.9 describing the therapeutic and toxic effect. Therefore this model is not only suitable for the evaluation of the effectiveness of drug targeting with respect to drug concentrations, but also with regard to the balance between therapeutic and toxic drug effects.

13.3.4 Model of Rowland and McLachlan

Rowland and McLachlan [46] also extended the model of Hunt to allow the evaluation of the effect of a permeability barrier and plasma protein binding on drug uptake. The model of Rowland and McLachlan is based on the work of Aubrée-Lecat *et al.* [47], who investigated the influence of various parameters on the amount of a macromolecular drug taken up by the target tissue. In the model of Rowland and McLachlan the target site (target tissue) consists of three distinct sites: blood, interstitium, and cells. Although the paper refers mainly to regional drug delivery (for example, intra-arterial injection), their model allows the evaluation of the effect of a permeability barrier and plasma protein binding on the removal of the drug from the target site, which may be a critical step in the effectiveness of drug targeting (see Section 13.4). The influence of permeability and plasma protein binding is evaluated using the DTI (Section 13.4.2) as a measure of effectiveness of drug targeting.

13.4 Measures of Effectiveness of Drug Targeting

For practical purposes, measures of the effectiveness of drug targeting are required. Such measures have been derived based on the pharmacokinetic profiles of the drug targeting system and that of the active drug.

13.4.1 Therapeutic Availability (TA)

Hunt *et al.* [6] introduced the term Therapeutic Availability as the ratio of the fraction of the dose reaching the target sites, if the dose is administered as the drug–carrier conjugate, to the fraction of the dose which reaches the same sites if an equal dose of the active drug is administered intravenously, as formulated below.

$$TA = \frac{AUC_{target(DC)}}{AUC_{target(D)}} \tag{13.20}$$

which is equivalent to

$$TA = \frac{Css_{target(DC)}}{Css_{target(D)}} \tag{13.21}$$

where AUC denotes the area under the curve (normalized for dose) after a single dose or over one dosing interval in the case of steady state; *C*ss denotes the average steady-state concentration; the subscript 'target' refers to the target site; DC refers to administration of the drug–carrier conjugate, and D to the intravenous administration of the active drug.

In contrast to the parameter Bioavailability (the fraction of the administered dose reaching the systemic circulation), the Therapeutic Availability may exceed a value of 1. A value exceeding 1 implies that the drug has been targeted successfully, the concentration at the target site being higher than that following a conventional intravenous administration of the same dose, and (see gueries) that a lower dose can be administered to reach the same concentration. Thus, an increase in TA is effectively equivalent to an increase in potency.

Boddy *et al.* [7] defined the Therapeutic Availability as the ratio of the rate of input of free drug divided by that of the drug carrier for the same degree of maximal therapeutic effect. When considering steady-state conditions, this definition is equivalent to that shown in Eq. 13.20 and 13.21.

13.4.2 Drug Targeting Index (DTI)

Hunt *et al.* [6] also introduced the Drug Targeting Index, which was defined as the ratio of drug delivered to the target and toxicity sites when the drug–carrier conjugate is administered, divided by the same ratio when the active drug is administered intravenously and is formulated as follows.

$$DTI = \frac{AUC_{target(DC)}/AUC_{tox(DC)}}{AUC_{target(D)}/AUC_{tox(D)}} \tag{13.22}$$

which is equivalent to

$$DTI = \frac{Css_{target(DC)}/Css_{tox(DC)}}{Css_{target(D)}/Css_{tox(D)}} \tag{13.23}$$

where AUC denotes the area under the curve (normalized for dose) after a single dose or over one dosing interval in the case of steady state; *C*ss denotes the average steady-state concentration; subscripts 'target' and 'tox' refer to the target and toxicity sites, respectively; DC refers to administration of the drug–carrier conjugate, and D to the intravenous administration of the active drug.

The concept of DTI is based on a pharmacokinetic model analogous to the model shown in Figure 13.4. Boddy and Aarons [44] used a simplified model, corresponding to Figure 13.3, in which the toxicity sites are included in the systemic (non-target) tissues, and the formula for DTI should be modified accordingly.

Comparing Eq. 13.20 and Eq. 13.23, it follows that the DTI is equal to the ratio of the Therapeutic Availability at the target site (Eq. 13.20 or 13.21) and the corresponding equation for the toxicity site. In the case of an ideal carrier, that is, if the drug is released only, and completely, at the target site, and if the target site does not contribute to the elimination of the drug, the value of the DTI is identical to that of TA [6]. This could be inferred from the recognition that the AUC at the toxicity sites, after administration of the drug–carrier conjugate, is the same as that after intravenous administration of the same dose of the free drug.

Hunt *et al.* [6] demonstrated that the DTI is also equivalent to the ratio of the therapeutic index (abbreviated to TI in Hunt *et al.*'s paper; in this chapter TI is defined differently, see Section 13.4.3) of the drug–carrier conjugate and that of the free drug. The therapeutic index (also called the therapeutic ratio) is a statistical measure defined as the ratio of the median toxic dose to the median effective dose [22].

Hunt *et al.* [6] considered the DTI the best measure of the effectiveness of the carrier.

13.4.3 Targeting Index (TI)

Boddy *et al.* [7] introduced the term Targeting Index as the ratio of the toxic effect when a drug–carrier conjugate is administered divided by the toxic effect when the free drug is given intravenously at a rate producing the same drug concentration (and thus the same drug effect) in the target site.

The Targeting Index is the 'effect' analogue of the Drug Targeting Index, which can be defined as the ratio of the drug concentration in the toxicity compartment when a drug–carrier conjugate is administered divided by the drug concentration in the toxicity compartment when the free drug is given intravenously at a rate producing the same drug concentration (and thus the same drug effect) in the target site; this definition is equivalent to Eq. 13.23.

Note that Hunt *et al.* [6] used the abbreviation TI for the Therapeutic Index (see Section 13.4.2).

13.5 Evaluation of Effectiveness of Drug Targeting Using PK and PK/PD Modelling

13.5.1 Effectiveness of an Ideal Carrier

An example of the application of the model of Boddy is depicted in Figure 13.6, showing the concentrations of active drug in the response and toxicity compartments after repeated administration of an hypothetical drug–carrier conjugate. After the first dose the concentration

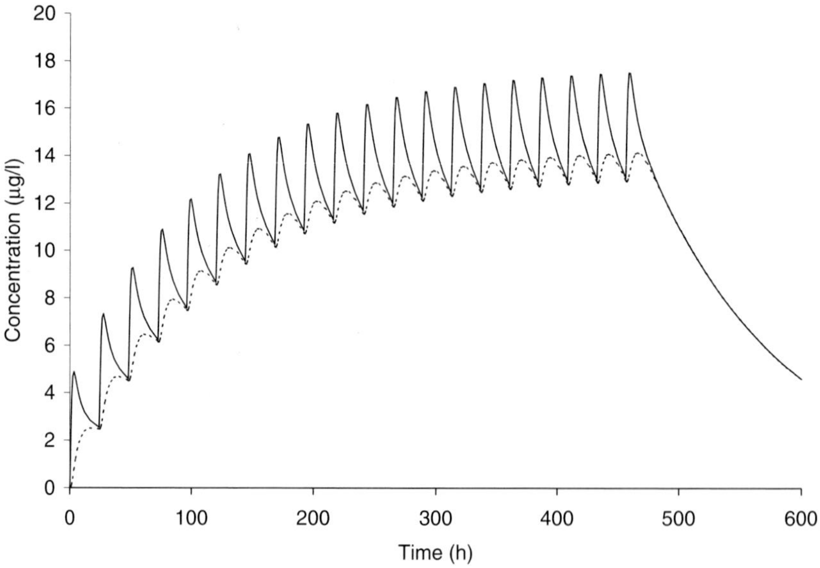

Figure 13.6. Simulation example using the Model of Boddy. The solid line represents the concentration of active drug in the response compartment, the dashed line the concentration in the toxicity compartment. A hypothetical drug–carrier conjugate is administered every 24 h in a dose of 100 µg active drug. After the first dose the concentration in the response compartment is markedly higher than that in the toxicity compartment. After repeated administration the average concentration in the response compartment is only slightly higher than that in the toxicity compartment. The simulation was performed using the program Ph\EdSim (MediWare, Groningen, The Netherlands).

in the response compartment is markedly higher than that in the toxicity compartment. After repeated administration, however, the average concentration in the response compartment is only slightly higher than that in the response compartment. This implies that the Drug Targeting Index is low (close to 1).

The effectiveness of drug targeting can be demonstrated using the model of Hunt [6]. It is assumed that the drug carrier is ideal, that is, it delivers the active drug only to the target site and the active drug can reach the other sites only by transport from the target site. Furthermore, it is also assumed that drug elimination takes place both in the elimination compartment and in the target site. In this case, the equation for the Drug Targeting Index reduces to (equation 23 in the paper by Hunt *et al.* [6]):

$$\mathrm{DTI} = 1 + \frac{CL}{Q_{\mathrm{R}} \cdot (1\text{-}E_{\mathrm{R}})} \tag{13.24}$$

where CL is the (total body) clearance of the drug, Q_{R} is the flow through the target site (subscript 'R' refers to the response site), and E_{R} is the extraction ratio of the drug in the target site; CL and Q_{R} refer to blood or plasma, whichever is the reference fluid.

Similar equations have been derived by other authors, using various types of compartmental models [40,48] (Section 13.2.4.1) or simplified physiological models [44,45,48] (Section 13.2.4.2), demonstrating that the choice of the model is not critical in the derivation of the DTI. Therefore the criticism of Siegel *et al.* [45] does not seem relevant for the appreciation of the DTI. It should be noted that the equations for the DTI in several references [7,44,48] are seemingly different from Eq. 31.24, by using the clearance from the non-target sites instead of the total body clearance. In fact, these equations are identical after appropriate rearrangements.

Apart from drugs directed at hepatocytes and renal tubular cells, in the case of the majority of conventional drugs, it is likely that the drug is not eliminated directly from the target site, and therefore $E_R = 0$, which further simplifies Eq. 13.24 to:

$$\text{DTI} = 1 + \frac{CL}{Q_R} \tag{13.25}$$

Eq. 13.25 is a more convenient form of Hunt's equation (16) [6].

13.5.2 Implications of the DTI Concept

The simplicity of Eq. 13.25 is striking: the DTI is determined by only two parameters. This can be understood qualitatively from the following reasoning. The total dose of the drug is delivered at the target site. If the drug did not leave the target compartment (that is, $Q_R = 0$), the drug concentration at the target site would remain constant, and the DTI would be infinitely high. The faster the removal of the drug from the target site (increase of Q_R), the lower the AUC at the target site, and the higher the AUC at non-target sites, including the toxicity site, and thus the DTI is lowered. If there is no clearance of the active drug (that is, $CL = 0$), it would eventually distribute evenly over the body, and there would be no net drug targeting effect (DTI = 1). The faster the removal of the drug which is present outside the target site (increase of CL), the lower the concentration and the AUC in the non-target compartments, including the toxicity compartment, and thus the higher the DTI.

The impact of Eq. 13.25 can be rather impressive, as was demonstrated in the paper of Hunt *et al.* [6]. These authors evaluated Eq. 13.25 by expressing both CL and Q_R as percentages of the cardiac output (normal value for an adult man 5 l min^{-1}). The total blood flow through the liver (1.5 l min^{-1}) and kidneys (1.2 l min^{-1}) is 54% of the cardiac output. Since a high extraction ratio in both the liver and the kidneys is unlikely, a clearance corresponding to 40% of the cardiac output may be considered as a maximum value for drugs eliminated by liver and kidney (for drugs eliminated by other mechanisms, a higher value might be possible, however). If we postulate that a DTI value of 5 is a minimum for a successful targeting strategy, it follows that for a drug with a clearance of 40% of the cardiac output, the blood flow through the target organ must be lower than 10% of the cardiac output in order to obtain sufficient targeting efficiency. The upper limit for Q_R is even lower for drugs with a lower clearance, and decreases further rapidly in the more realistic cases where the drug carrier is not ideal. For a detailed analysis see Hunt *et al.* [6].

In conclusion, drug targeting to tissues that receive a relatively large fraction of the cardiac output is unlikely to create effective targeting if transport of the free drug to and from the target cell can occur. Particularly after multiple dosing, a steady state will be reached in which the increase in free drug concentration in the target tissue compared to that in the plasma and that in the toxicity related tissue, will be moderate.

It follows from Eq. 13.24 that the DTI may be higher than in the aforementioned analysis if $E_R > 0$. However, for conventional drugs it is unlikely that there is a significant elimination of the drug at the target site. Therefore the extensive evaluations of cases where $E_R > 0$ [6, 7, 46] do not seem applicable to targeting conventional drugs. On the other hand, for newer types of drugs, including protein drugs, antisense oligonucleotides and plasmid DNA, a major fraction of the drug may be eliminated by lysosomal degradation at the target site [8]. Therefore these types of drug are ideal candidates for drug targeting, even with the limitation of having to cross the cell membrane or other barriers (Section 13.5.3).

It should be noted that the conclusions inferred from the aforementioned pharmacokinetic models are valid only under the condition that the rather strict assumptions on which the particular model is based, are justified. In general, however, it is not likely that deviation from these conditions will alter the general conclusions seriously. Despite the lack of knowledge concerning many of the parameters of the complete model (Figure 13.4), the concept of DTI allows an evaluation of the potential benefit of drug targeting, by providing insight into the critical values of parameters which are of real concern in the design of drug targeting systems.

The extended model of Boddy (Section 13.3.3) may be helpful for the investigation of the dependence of DTI, TI and TA on various model parameters. Boddy *et al.* [7] demonstrated that the DTI is independent of the rate of drug release from the carrier (provided that is takes place only in the target site) or the rate of elimination of the drug–carrier conjugate by other mechanisms which do not lead to the release of the drug. Neither of these parameters influences the pharmacokinetics of the free drug in the steady state. However, the TA is affected by both parameters. TA increases either as the rate of drug release increases, or as the rate of elimination of the DC decreases.

These authors also showed that an increase in the amount of drug eliminated directly from the target site (that is, an increase of E_R) does not always increase the benefits of drug targeting, as in the case of the release of the free drug in the central or toxicity compartment. Levy [42] commented on some pharmacokinetic considerations of targeted drug delivery. He assumed that drug elimination from the target site (see Section 13.3.1) will, in many case, be much more rapid that drug elimination from the body. Furthermore, he assumed that, following targeted delivery, a drug will be eliminated from the site of action to the rest of the body, which will act, at least initially, as an infinite sink. Consequently, it may be expected that, after a single dose, (1) the duration of action of a targeted bolus dose will be much shorter than the duration of action of a conventionally-administered, equipotent bolus dose, and (2) the elimination of a targeted drug from the site of action is not affected by changes in the disposition of the active drug from the non-target compartments. During a steady state, however, the situation is completely different, since the assumption of sink conditions in the systemic circulation is no longer valid, and thus systemic clearance becomes a major determinant of the effectiveness of drug targeting, as reflected by Eq. 13.24 or 13.25. Levy [42] correctly stated that the selectivity of the drug targeting process will be lost as drug concentrations in the systemic circulation rise. Such a loss of selectivity can be minimized by increasing the selectivity and specificity of targeting (that is, by using an ideal carrier) or by reducing the rate of removal of drug from the target site. The latter is reflected in the increase of DTI by reducing Q_R in Eq. 13.24 or 13.25.

13.5.3 Drug Candidates for Effective Targeting

From a pharmacokinetic point of view, drug targeting strategies may be applied to increase the Drug Targeting Index and/or Therapeutic Availability. Although drug targeting systems may increase DTI and TA for any drug, there are large differences between drugs with respect to the efficacy of drug targeting in comparison to conventional drug administration. Generally speaking, effective targeting can be achieved only if there is an explicit need for targeting. If the active drug cannot reach the response sites, for example due to diffusion barriers, a targeting strategy is necessary, and is more likely to be effective than the active drug, even in the absence of barriers. Also, the efficacy of drug targeting is dependent on the pharmacodynamic properties of the drug (Section 13.2.2). For example, if the peak concentration at the response site is more important for the therapeutic effect than the average concentration, drug targeting may be more effective than predicted from Eq. 13.24 or 13.25, since these equations refer to the AUC or average steady-state concentration (Eq. 13.22 and 13.23). This situation may occur for antibiotic and antineoplastic drugs if the drug effect is dependent on the cell cycle phase (Section 13.2.5.2). The same applies to drugs which are used therapeutically in a single dose or over a short period of time.

The theoretical analysis in Section 13.5.2 demonstrates that suitable candidates for combination with a targeted drug carrier should exhibit a rapid elimination from the non-target compartments, or in other words, they should have a high plasma clearance rate. However, drugs exhibiting rapid elimination, for example due to a high extraction ratio in the liver, are poor candidates for conventional therapy, and it is unlikely that such drugs will pass the conventional screening systems used during drug development [6]. Therefore, currently available drugs used in conventional therapy, are, in general, poor candidates for effective drug targeting with drug carrier systems.

From a pharmacokinetic point of view, the newer generation of therapeutic peptides may be much better suited to drug targeting methods. This class of compounds tends to have a high clearance rate which probably occurs to a large extent at the target site ($E_R > 0$). The rapid clearance rate implies the necessity for frequent dosing to maintain a therapeutic effect, and the use of large doses to compensate for the high loss of drug. Also, transport of the active drug to the response site may be hampered by diffusion barriers. These problems may be solved by applying these peptides in combination with an appropriate drug carrier system.

13.5.4 Limitations of PK and PK/PD Modelling

Despite the promising applications of PK and PK/PD modelling in many fields of drug research [3,49] and drug utilization [29], awareness of their limitations is necessary. Among others, the following aspects may interfere with successful PK and PK/PD modelling and analysis:

- Inter-individual variability in pharmacokinetics, in particular with respect to clearance [10]. Little is known about the variability of transport to the target site and removal from the target site. Population analysis approaches in PK and PK/PD [50–52] may be helpful in identifying models and model parameters by analysing the data from a group of animals or subjects simultaneously rather than individually.

- Inter-individual variability in pharmacodynamics. Although over the last 30 years it has been the common view that the opposite holds true, Levy demonstrated that PD variability in humans is extensive, reproducible and usually more pronounced than PK variability [53,54].
- Intra-individual variability, for example due to up- or downregulation of various proteins after multiple dosing, including receptors, carriers and enzymes, resulting in, amongst other effects, the development of tolerance.
- Complexity of the models, making selection of the most appropriate model difficult (Section 13.2.8.6).
- Difficulty in identifying model parameters (see Section 13.2.8.4).
- Binding of drug or drug–carrier conjugate to plasma proteins and various other non-target tissues [46,55], which potentially act as a slow release compartments (see Section 13.2.1.4).

13.6 Examples of PK Modelling in Drug Targeting

Until now, applications of drug targeting models to real data have been scarce. With the exception of the aforementioned papers dealing with models and theoretical simulations, only a few examples of PK modelling in the area of drug targeting were found in the literature. A brief description of the type of investigations reported in these papers will be given below.

13.6.1 *In Vivo* Studies

Nishikawa *et al.* [56] investigated the pharmacokinetic behaviour of dextran in mice. Dextran is a drug carrier with several suitable properties, including high solubility, low immunogenicity, and a long history as a plasma expander in clinical use. They applied it in a physiological model including Michaelis–Menten saturable processes for hepatic uptake and extra-hepatic elimination, and were able to demonstrate that dextran was taken up by the same receptor as galactosylated albumin.

Zhu *et al.* [57,58] applied PB-PK modelling to radiolabelled tumour-targeted monoclonal antibodies for radioimmunodetection and radioimmunotherapy.

Sakaeda *et al.* [59] described an example of physiologically-based modelling for a lung-specific pro-drug of the antibiotic drug ceftazidime.

Mishina and Jusko [60] studied the pharmacokinetics and pharmacodynamics of methylprednisolone encapsulated in liposomes in the rat. The data were analysed using PK and PK/PD models, taking into account the plasma concentration time profile, an effect compartment analogous to that described by Sheiner [23,24], interaction of the drug with receptors, and a PD model analogous to that defined by Eq. 13.9, using the concentration of the drug–receptor complex as the determinant of the drug effect, that is, the suppression of lymphocyte proliferation. Although the authors succeeded in modelling the drug effect, their approach cannot be considered as a true drug targeting model, since it does not take into account a target site as described in the models described in Section 13.3 (Figures 13.3–13.5). The authors state that their liposomes were taken up by macrophages of the RES, but their model did not include a compartment to which the liposomes are targeted. Instead, their effect compartment was directly linked to the plasma compartment, for both the liposome-encapsulated drug and the free drug. Therefore the specific drug targeting effects were not taken into account. The major difference between the PK/PD properties of liposome-encapsulated drugs and the free drug is the duration of their presence in plasma. Moreover, the authors did not evaluate the effectiveness of their targeting approach by using the measures DTI or TA. From their results it can be concluded that the approach was successful in reducing the dosing rate needed to maintain a desired effect, implying that the TA was increased considerably. However, the authors did not provide any information regarding a possible improvement in the DTI, and it cannot be excluded that the DTI was actually close to unity.

Swart *et al.* [61] investigated the pharmacokinetics of negatively-charged serum albumins with a potent anti-HIV-1 activity *in vitro*, in relation to their potential applicability for dual targeting [9], in monkeys. Using the program MW\Pharm [62], they evaluated the implications of the pharmacokinetic behaviour for dosage regimens which would maintain the concentration within a therapeutic range calculated from *in vitro* data.

13.6.2 *In Vitro* Studies

Sugiyama and Kato [63] described a model for the receptor-mediated endocytosis of the polypeptide hormones epidermal growth factor (EGF) and hepatocyte growth factor (HGF) in isolated perfused rat liver and in isolated rat hepatocytes, to estimate the efficiency of drug targeting using these polypeptide hormones as potential drug carriers.

Chan and Murphy [64] applied a mathematical model describing the kinetics of cellular trafficking of monoclonal antibodies against melanoma cells, and of immunotoxins targeted by the antibodies. The model allowed the assessment of equilibrium and kinetic constants by fitting it to the data obtained from *in vitro* cultured cell experiments.

13.6.3 Regional Drug Administration

Gallo *et al.* [65] applied a physiologically-based pharmacokinetic model to the targeting of anti-cancer drugs to the brain following intra-arterial administration in glioma-2 bearing rats.

Malhotra *et al.* [66] modelled the route of administration-based enhancement of delivery of EAB 515, a hydrophilic N-methyl-D-aspartate antagonist, to the brain.

Stevens *et al.* [55] determined the relationship between the DTI and pharmacokinetic parameters of three non-steroidal anti-inflammatory drugs using the rat air pouch inflammation model. The derivation of the DTI was modified to take into account the simultaneous influx of plasma albumin, to which the drugs were extensively bound.

PK/PD evaluation of pulmonary administration has been described in several papers by Hochhaus *et al.* [67,68].

13.6.4 Controlled Drug Delivery

The application of PK and PD to controlled drug delivery has been promoted by, among others, Breimer and Danhof [3,49], and Hoffman [4].

Grass *et al.* [69] evaluated the performance of controlled release dosage forms of the antithrombic drug ticlopidine using computer simulations based on data from *in vitro* intestinal permeability studies in various sections of the intestine of rabbit and monkey.

13.6.5 Pharmacokinetic Properties of Macromolecular Carrier Systems

The pharmacokinetic properties of macromolecular carrier systems for targeted drug delivery were reviewed by Takakura and Hashida [8].

13.7 Software for PK and PK/PD Modelling

Where scientists working in drug development have not encountered PK and PK/PD modelling, this lack of familiarity with these useful techniques may have limited their application. Also, the complexity of the models needed for predicting drug concentrations after administration of a drug targeting preparation, cannot be handled by many commercially-available computer programs for 'routine' PK and PK/PD modelling. As a result, this field was investigated mainly by researchers with a broad experience in modelling and who wrote their own computer programs.

However, the recent availability of powerful, flexible, and user-friendly computer programs have brought these techniques within the reach of every scientist. A major breakthrough was the introduction of graphics packages which facilitated structural model building on screen, without the necessity for the user to write the mathematical model equation. Such graphics programs are included, among others, in the following packages:

- ACSL Tox (www.pharsight.com)
- Ph\EdSim (MediWare, Groningen, The Netherlands; e-mail: vanessene@cs.com)
- SAAM II [70] (Department of Bioengineering, University of Washington, Seattle, WA; http://courses.washington.edu/rfka)
- ModelMaker (www.modelkinetix.com)

In many of the currently available computer programs, however, the user still has to write the model equations in the case of more complex models such as that required for application to drug targeting. These programs include:

- Adapt II (Biomedical Simulations Resources, University of Southern California, Los Angeles, CA; www.usc.edu/dept/biomed/BMSR/index.html)
- Boomer (www.boomer.org)
- Kinetica (www.innaphase.com)

- Simusolv (Dow Chemical Co., Midland MI; http://software-guide.com/cdprod1/swhrec/016/481.shtml)
- WinNonlin (www.pharsight.com)

A regularly updated list of pharmacokinetic software can be found at www.boomer.org/pkin/soft.html.

13.8 Perspectives and Conclusions

A promising concept for the targeting of a particular drug to a target tissue is a prerequisite, but not sufficient on its own, for the development of an effective and successful drug targeting system. During the early days of drug targeting, and perhaps even now, many researchers seemed to be focused primarily on the mechanism of targeting, for understandable reasons. To date however, the increase in the therapeutic repertoire is still disappointing, certainly in view of the enormous efforts which have been made in this area. It is tempting to ascribe this, at least in part, to the dearth of information in the area of pharmacokinetic and pharmacodynamic principles, despite the publication of several early papers [5,6,42].

On the other hand, it should be noted that many of the scientists who are experienced in the field of PK and PK/PD modelling and analysis, did not enter the area of drug targeting which is a challenging new field, but also a treacherous quagmire of unwarranted simplifications and over-dimensioned models. A literature survey shows that the contribution of PK and PK/PD to the area of drug targeting is mainly found in theoretical frameworks which have been developed to evaluate the potential benefit and limitations of drug targeting (Section 13.5). Until now, applications of drug targeting models to real data have been scarce (Section 13.6). It is not obvious whether this lack of successful PK and PK/PD analyses in the literature reflects the unpredictable interactions of drug targeting systems and the living organism or the lack of scientific sophistication. It may thus be argued that many challenges still remain before the effectiveness of drug targeting systems by appropriate PK and PK/PD analysis can be ultimately established. Population approaches in PK and PK/PD [50–52] may be helpful in identifying models and model parameters by analysing data from groups of animals or subjects simultaneously rather than individually.

Nowadays, there can be no doubt that the application of pharmacokinetic and pharmacodynamic principles in the design of effective drug targeting systems is essential [1,2]. This may be illustrated by the following examples.

(a) Pharmacokinetic simulations have taught us to estimate quantitatively the potential benefits, and limitations, of drug targeting (Section 13.5).

(b) Pharmacokinetic simulations may identify the critical steps in the process (Section 13.3). It has become clear that targeting of the drug to a specific tissue is not always the critical step, as rapid removal of the active drug from the target organ may limit the beneficial effects of the targeting.

(c) Pharmacokinetic simulations have made clear which drugs are suitable candidates for drug targeting and which are not (Section 13.5.3). Drugs which are rapidly cleared from the body when administered in their free form, are poor candidates for conventional drug therapy, but are probably the best candidates for drug targeting. Although the latter aspect had already been stated by Hunt *et al.* in 1986 [6], their message did not seem to be understood by many researchers, given the enormous volume of literature on the targeting of conventional drugs with a relatively low clearance rate. The fact that this basic premise does not seem to have been acknowledged by research workers in this area, may well be related to the limited number of successful drug targeting procedures which have been reported thus far.

Looking to the future, the newer generation of therapeutic peptides that have been identified using molecular biology technology rather than classical pharmacology, including protein drugs, antisense oligonucleotides and plasmid DNA, are excellent candidates for drug targeting, for at least three reasons. First, targeting may be necessary for this type of compound to reach the target sites, due to their physicochemical properties. Second, the clearance rate of these compounds is generally high, and therefore the DTI will be high. Third, the high clearance rate requires frequent dosing to maintain a therapeutic effect, and large doses to compensate for the high loss of the drug. Drug targeting may significantly increase the duration of action and the apparent potency.

In addition, the successful application of drug targeting to this new generation of drugs may provide the impetus for further research and development, since well-designed drug targeting strategies may be used for the delivery of potentially therapeutic compounds which cannot be utilized in the currently available conventional drug delivery systems. The results of future developments may challenge researchers in every discipline involved in drug development.

In the field of PK and PK/PD modelling, the newer generation of drugs provide opportunities to extend the area of research. These newer drugs may also raise many problems, for example in the analysis of these compounds in tissues. This is particularly true in relation to the selectivity of assays towards inactive or active metabolites, and with respect to the levels of sensitivity required to detect very small amounts of highly active compounds. Also, PK modelling may become more detailed with respect to the target and toxicity sites. Physiologically-based PK/PD modelling may be necessary to evaluate the effectiveness under changed (patho-)physiological conditions (for example, changes in blood perfusion or pH of the target site, changes in receptor density and the development of tolerance). PK and PK/PD modelling are valuable tools for quantifying the beneficial effects of changes in the construction of drug–carrier conjugates, including the optimization of dosing schedules for such sophisticated drug targeting systems.

In conclusion, it may be expected that further development of PK and PK/PD will contribute to the successful development of new drug targeting products.

References

[1] Sugiyama Y, *Adv. Drug Deliv. Rev.* **1996,** *19*, 333–334.
[2] Suzuki H, Nakai D, Seita T, Sugiyama Y, *Adv. Drug Deliv. Rev.* **1996,** *19*, 335–357.
[3] Breimer DD, *J. Drug Target* **1996,** *3*, 411–415.
[4] Hoffman A, *Adv. Drug Deliv. Rev.* **1998,** *33*, 185–199.
[5] Stella VJ, Himmelstein KJ, *J. Med. Chem.* **1980,** *23*, 1275–1282.
[6] Hunt CA, MacGregor RD, *Pharm. Res.* **1986,** *3*, 333–344.
[7] Boddy A, Aarons L, Petrak K, *Pharm. Res.* **1989,** *6*, 367–372.
[8] Takakura Y, Hashida M, *Pharm. Res.* **1996,** *13,* 820–831.
[9] Meijer DKF, Jansen RW, Molema G, *Antiviral Res.* **1992,** *18*, 215–258.

[10] Rowland M, Tozer TN, *Clinical Pharmacokinetics – Concepts and Applications*. Williams and Wilkins, Media PA, **1995.**
[11] Proost JH, Wierda JMKH, Meijer DKF, *J. Pharmacokinet. Biopharm.* **1996,** *24*, 45–77.
[12] Boddy A, Aarons L, *Adv. Drug Deliv. Rev.* **1989,** *3*, 155–163.
[13] Rowland M, Benet LZ, Graham GG, *J. Pharmacokinet. Biopharm.* **1973,** *1*, 123–135.
[14] Rowland M, Leitch D, Fleming G, Smith B, *J. Pharmacokinet. Biopharm.* **1984,** *12*, 129–147.
[15] Rowland M, *Drug Metab. Rev.* **1984,** *15*, 55–74.
[16] Bischoff KB, *Bull. Math. Biol.* **1986,** *48*, 309–322.
[17] Balant LP, Gex-Fabry M, *Xenobiot.* **1990,** *20*, 1241–1257.
[18] Bourne DWA, *Mathematical Modeling of Pharmacokinetic Data*. Technomic Publishing, Lancaster PA, **1995.**
[19] Nestorov IA, Aarons LJ, Arundel PA, Rowland M, *J. Pharmacokinet. Biopharm.* **1998,** *26*, 21–46.
[20] Weiss M, *Eur. J. Pharm. Sci.* **1999,** *7*, 119–127.
[21] Meibohm B, Derendorf H, *Int. J. Clin. Pharmacol. Ther.* **1997,** *35*, 401–413.
[22] Goldstein A, Aronov L, Kalman SM, *Principles of Drug Action: The Basis of Pharmacology*. John Wiley & Sons, New York, **1974.**
[23] Sheiner LB, Stanski DR, Vozeh S, Miller RD, Ham J, *Clin. Pharmacol. Ther.* **1979,** *25*, 358–371.
[24] Holford NHG, Sheiner LB, *Clin. Pharmacokinet.* **1981,** *6*, 429–453.
[25] Zhi J, Nightingale CH, Quintilliani R, *J. Pharm. Sci.* **1986,** *75*, 1063–1067.
[26] Mouton JW, Vinks AATMM, Punt NC, *Antimicrob. Agents Chemother.* **1997,** *41*, 733–738.
[27] Ozawa S, Sugiyama Y, Mitsuhashi J, Inaba M, *Cancer Res.* **1989,** *49*, 3823–3828.
[28] Nakai D, Fuse E, Suzuki H, Inaba M, Sugiyama Y, *J. Drug Target.* **1996,** *3*, 443–453.
[29] Kobayashi K, Jodrell DI, Ratain MJ, *Cancer Surv.* **1993,** *17*, 51–78.
[30] Sheiner LB, *J. Pharmacokinet. Biopharm.* **1984,** *12*, 93–118.
[31] Sheiner LB, *J. Pharmacokinet. Biopharm.* **1985,** *13*, 514–540.
[32] Sheiner LB, *J. Pharmacokinet. Biopharm.* **1986,** *14*, 539–555.
[33] Peck CC, Beal SL, Sheiner LB, Nichols AI, *J. Pharmacokinet. Biopharm.* **1984,** *12*, 545–558.
[34] Sheiner LB, Beal SL, *J. Pharmacokinet. Biopharm.* **1985,** *13*, 185–201.
[35] Jacquez JA, *FASEB Fed. Proc.* **1987,** *46*, 2477–2480.
[36] Godfrey KR, Chapman MJ, *J. Pharmacokinet. Biopharm.* **1989,** *17*, 229–267.
[37] Jacquez JA, Perry T, *Am. J. Physiol.* **1990,** *258*, E727–E736.
[38] Godfrey KR, Chapman MJ, Vajda S, *J. Pharmacokinet. Biopharm.* **1994,** *22*, 229–251.
[39] Ludden TM, Beal SL, Sheiner LB, *J. Pharmacokinet. Biopharm.* **1994,** *22*, 431–445.
[40] McMartin C, *Biotherapy* **1991,** *3*, 9–23.
[41] Magnusson S, Faerevik I, Berg T, In: *Hepatic Endocytosis of Lipids and Proteins* (Eds Windler E,Greten H), pp. 189–195. W. Zuckschwerdt Verlag, München, **1992.**
[42] Levy G, *Pharm. Res.* **1987,** *4*, 3–4.
[43] Zhu BT, *Mol. Carcinog.* **1999,** *25*, 1–13.
[44] Boddy AV, Aarons LJ, *J. Pharmacokinet. Biopharm.* **1991,** *19*, 355–362.
[45] Siegel RA, MacGregor RD, Hunt CA, *J. Pharmacokinet. Biopharm.* **1991,** *19*, 363–373.
[46] Rowland M, McLachlan A, *J. Pharmacokinet. Biopharm.* **1996,** *24*, 369–387.
[47] Aubrée-Lecat A, Duban MC, Demignot S, Domurado M, Fournié P, Domurado D, *J. Pharmacokinet. Biopharm.* **1993,** *21*, 75–98.
[48] Stella VJ, Kearney AS, In: *Targeted Drug Delivery* (Ed. Juliano RL), pp. 71–103. Springer-Verlag, Berlin, **1991.**
[49] Breimer DD, Danhof M, *Clin. Pharmacokinet.* **1997,** *32*, 259–267.
[50] Sheiner LB, Ludden TM, *Annu. Rev. Pharmacol. Toxicol.* **1992,** *32*, 185–209.
[51] Aarons L, Balant LP, Mentré F, Morselli PL, Rowland M, Steimer J-L, Vozeh S, *Eur. J. Clin. Pharmacol.* **1996,** *49*, 251–254.
[52] Vozeh S, Steimer J-L, Rowland M, Morselli P, Mentré F, Balant LP, Aarons L, *Clin. Pharmacokinet.* **1996,** *30*, 81–93.
[53] Levy G, *Clin. Pharmacokinet.* **1998,** *34*, 323–333.
[54] Levy G, *Adv. Drug Deliv. Rev.* **1998,** *33*, 201–206.
[55] Stevens AJ, Martin SW, Brennan BS, McLachlan A, Gifford LA, Rowland M, Houston JB, *Pharm. Res.* **1995,** *12*, 1987–1996.
[56] Nishikawa M, Yamashita F, Takakura Y, Hashida M, Sezaki H, *J. Pharm. Pharmacol.* **1992,** *44*, 396–401.
[57] Zhu H, Baxter LT, Jain RK, *J. Nucl. Med.* **1997,** *38*, 731–741.
[58] Zhu H, Jain RK, Baxter LT, *J. Nucl. Med.* **1998,** *39*, 65–76.

[59] Sakaeda T, Fukumura K, Takahashi K, Matsumura S, Matsuura E, Hirano K, *J. Drug Target.* **1998,** *6*, 261–272.
[60] Mishina EV, Jusko WJ, *Pharm. Res.* **1996,** *13*, 141–145.
[61] Swart PJ, Schutten M, Smit C, Van Amerongen G, Proost JH, Osterhaus ADME, Meijer DKF, *Drug Del.* **1996,** *3*, 165–171.
[62] Proost JH, Meijer DKF, *Comput. Biol. Med.* **1992,** *22*, 155–163.
[63] Sugiyama Y, Kato Y, *Drug Dev. Ind. Pharm.* **1994,** *20*, 591–614.
[64] Chan MC, Murphy RM, *Cancer Immunol. Immunother.* **1999,** *47*, 312–319.
[65] Gallo JM, Varkonyi P, Hassan EE, Groothuis DR, *J. Pharmacokinet. Biopharm* **1993,** *21*, 575–592.
[66] Malhotra BK, Brundage RC, Lemaire M, Sawchuk RJ, *J. Drug Target.* **1997,** *4*, 277–288.
[67] Hochhaus G, Schmidt E-W, Rominger KL, Möllmann H, *Pharm. Res.* **1992,** *9*, 291–297.
[68] Hochhaus G, Mollmann H, Derendorf H, Gonzalez-Rothi RJ, *J. Clin. Pharmacol.* **1997,** *37*, 881–892.
[69] Grass GM, Bozarth CA, Vallner JJ, *J. Drug Target.* **1994,** *2*, 23–33.
[70] Barrett PHR, Bell BM, Cobelli C, Golde H, Schumitzky A, Vicini P, Foster DM, *Metabolism* **1998,** *47*, 484–492

14 Drug Targeting Strategy: Scrutinize the Concepts Before Screening the Constructs

Dirk K. F. Meijer

14.1 Introduction

The current problems in controlling cancer, severe infections and chronic degenerative diseases, as well as the lack of effective and safe pharmacotherapeutic measures for such disorders, have renewed interest in the options of targeting drugs, peptides, genes and anti-sense material to sites of disease.

Structure–activity relationship studies and rational drug design procedures have led to the synthesis of many novel drugs that are highly potent. Yet, at the same time, they can exhibit severe toxicity since they are also accessible to non-target cells. In fact, the physicochemical features of drugs that dictate their pharmacologic activity also determine their distribution patterns in the body. To overcome the undesirable side-effects and to 'uncouple' the pharmacokinetic behaviour of the drug from its 'pharmacodynamic profile', the drug can be directed to its site of action and/or be diverted from sites where it will be potentially toxic by coupling to macromolecular carriers. The chosen carrier is then supposed to determine the fate of the coupled drug in the body.

The design and development of potential carriers for cell-specific delivery of therapeutics should be based on a detailed knowledge of recognition sites on the surface of target cells as well as on insight into the internalization and further cellular disposition of such macromolecules.

In the drug targeting approach, two different strategies can be distinguished: *passive targeting* and *active targeting*. In the case of *passive targeting*, the carrier-associated drug is, for instance, delivered to macrophages, resulting in gradual degradation of the carrier and slow release of the liberated drug from the cells. Through size-restricted extravasation of the carrier, the carrier–drug complex tends to stay in the systemic circulation and is, at least partly, prevented from distribution to sites where it may have a toxic effect. *Active targeting*, on the other hand, should lead to higher therapeutic concentrations at the site of action through cell-specific delivery via the macromolecular carrier. In principle, in active targeting, the dose of the drug can be reduced and the side-effects will thus be decreased.

Drug delivery research in practice requires professional planning and careful avoidance of intrinsic pitfalls. Some essential guidelines, derived from our own experience, are listed in Table 14.1. They should not only be taken into account before embarking on a drug delivery project, but should also be integrated during the developmental phases of the drug innovation process.

Table 14.1. General guidelines in drug targeting research.

- It is preferable to test the effects of drug-targeting preparations on the whole body as early as possible
- It is advisable to test drug targeting preparations with regard to possible immunogenicity at an early stage of development
- Cell-specific distribution of the drug-targeting preparations as well as the rate of drug release from the carrier should be studied both in the healthy and pathological situation
- Drug loading of the carrier should be carefully balanced: sufficient drug molecules should be internalized to obtain therapeutic levels. However, excessive loading may corrupt the cell specificity of the carrier
- The chosen carrier should be non-toxic, as should its degradation products
- The chosen carrier should be capable of traversing anatomical barriers in the body en route to the particular target tissues in the diseased state
- Since parenteral administration is required, drug targeting formulations should provide distinct advantages in efficacy and safety compared with the non-targeted drug
- Special attention should be paid to the patenting of targeting constructs: the unique combination of drug, linker and carrier may provide options for product protection
- The large gap between pre-clinical and clinical research can be (at least partly) bridged by screening of the disposition properties as well as the efficacy of drug delivery preparations in (diseased) human tissues *in vitro*
- Aspect of therapy costs of drug targeting preparations should be carefully weighed in relation to the present state of the art in the therapy and cost containment aspects of health care

14.2 Receptor-based drug targeting

The success of drug targeting with macromolecular carriers is intimately dependent on the selectivity of the cellular targets in the body. Other crucial factors are the anatomical and/or pathological barriers that have to be passed en route to these recognition sites and the events following receptor recognition and internalization of the drug conjugate: intracellular routing encompassing carrier degradation and drug release.

Table 14.2 lists a number of receptors for macromolecules that have been identified so far and that are more or less specific for the organ/tissue or even the cell type indicated. Some of these receptors are lectins which recognize oligosaccharide chains in a specific geometric arrangement with a specific type of terminal sugar or otherwise clustered sugars and/or randomly exposed sugars with sufficient density. Others are receptors for cytokines, growth factors and adhesion molecules which bind specific peptides that can be used as a homing devices. Such receptors can select their substrates on the basis of the specific conformation presented by the functional groups and the charge density of such macromolecules. Selectivity in binding can also be based on multivalency in sugar or peptide recognition.

Although such receptors often provide mechanisms for internalization followed by intracellular transport to compartments where degradation takes place, the rates of these processes can be markedly different in various cell types. In some cases only external binding occurs and consequently, the microclimate of the cell membrane at which local release of the drug from the carrier takes place, should provide sufficient driving force to ensure uptake of the drug into the target cell.

Table 14.2. Organ and tissue selective distribution of potential drug carriers based on receptor recognizing principles: a few examples.

Organ/Tissue		**Carriers**	**Species**	**Diseases aimed at**
Liver	Hepatocytes	Lactosaminated (H)SA Arabinogalactan Asialoglycoproteins	Rat, man Man Rat, man	Hepatitis B and C Liver cancer
	Kupffer cells	Mannosylated (H)SA	Rat	Inflammation, sepsis
	Endothelial cells	Negatively charged (H)SA	Rat, man	Organ rejection, I/R damage
	Stellate cells	Man - 6 P (H)SA RGD oligopeptides - (H)SA	Rat, man Rat	Liver fibrosis Liver fibrosis
	Cholangiocytes	pol.IgA, alkaline phosphatase	Man, rat	Peribiliary cirrhosis
Kidney	Tubular cells	Low MW proteins (Lysozyme)	Rat, man	Nephrotic syndrome Renal cancer
	Mesangial cells	IgA (asialo) Anti- Thy1-Ab	Rat Rat, man	Renal fibrosis
Brain	BBB endothelia	Transferrin Anti-Transferrin-R Ab Anti-Insulin-R Ab	Rat, man Rat Rat	CNS infections Parkinson's disease Alzheimer disease Brain tumours
Lung	Alveolar macrophages	Glucosylated proteins	Rat	Lung cancer Lung infections
	Endothelial cells	Anti-CD31 Ab	Rat	Lung inflammation/ cancer
Blood cells	Monocytes/Macrophages	ß-Glucans, Mannosylated proteins	Rat Rat	HIV infections Ovarian cancer
	T-lymphocytes	HIV-gp120, IGF-I, sCD4 Anti-CD3 Ab	Rat, man	HIV infections Rejection transpl. organs
	B-lymphocytes	Anti-CD20 Ab	man	B-cell cancer
Blood vessels	Endothelia Tumour vasculature	Lactoferrin, OxLDL Anti-VEGF-R Ab, VEGF	Rat	Atherosclerosis Solid tumours
Intestines	Enterocytes	Dimeric IgG1	Rat	Colitis, Crohn's disease
		Lactoferrin (enteral)	Rat, man	

Ab, Antibody; BBB, Blood Brain Barrier; CNS, Central Nervous System; HIV, Human Immunodeficiency Virus; HSA, Human Serum Albumin; IGF, Insulin Growth Factor; I/R, Ischaemia/Reperfusion; MW, Molecular Weight; OxLDL, Oxidized Low Density Lipoprotein; -R, -receptor; sCD4, soluble CD4; VEGF, Vascular Endothelial Growth Factor.

With regard to the specificity of sugar–lectin interactions, it should be noted that interactions of sugar-based compounds with lectins on different cell types seem to be determined by the recognition of either a randomly presented sugar with sufficient density on the protein or a particular sugar arranged in an antennary structure. Alternatively, high affinity binding may involve recognition of a combination of different sugars. The use of one sugar type in inhibition experiments may therefore give an false picture of the true recognition sites.

A general warning should be given with regard to the design of drug targeting preparations for anti-infective drugs. If endocytosis is required for cellular delivery, infected cells may be less active in this respect, e.g. due to depletion of energy-rich metabolites or decreased expression of cell surface receptors. The efficiency of the delivery process may thus be decreased during infection, in particular after extensive transformation of cells leading to gross changes in the surface molecules and/or the ability of the cell to degrade the drug–carrier complex.

Both receptor density and affinity for a given substrate as well as the presence of competing endogenous ligands, determine the extent of carrier–receptor occupation and thus the extraction of the carrier–drug complex by the target tissue. Endogenous ligands can include tumour antigens and soluble receptor molecules that are shed during the disease and its treatment, and may partly inactivate or neutralize the chosen drug carrier delivery system.

Finally, continuous exposure of certain receptors to their macromolecular ligands can lead to rapid downregulation of cell surface receptors, especially if receptor recycling within the cells is incomplete. Fortunately, expression of many receptors, for example for certain cytokines, growth hormones and adhesion factors, can be extensively upregulated in the disease process and this can result in disease-induced drug-targeting.

Down- and upregulation of receptors should therefore be taken into account in predicting the pharmacokinetics of macromolecular carriers upon chronic administration. For instance, when the particular receptors to be targeted are present on more than one cell type in the body, and up- or downregulation in these cells occurs at different rates, tissue specificity for drug–carrier complexes in the body may change with time during chronic dosing. Also the therapeutic effects attained may influence selective distribution through changes in receptor expression and/or carrier degradation.

Some of the drug carriers which are currently being developed, provide intrinsic therapeutic activity that may add to the effect of the coupled drug, a principle called dual targeting. Such multi-active drug targeting preparations may offer the advantages of synergistic effects and for instance, counteraction of drug resistance, in addition to improving the specificity of distribution within the body.

The rate-limiting steps in the distribution of drug targeting preparations throughout the body, can be elegantly simulated using appropriate (patho)physiology-based predictive models. Computer-assisted modelling can give further insight and a more accurate prediction of drug levels at the target and non-target sites. Such simulations should certainly include multiple dose regimens for obvious practical reasons. In general: drug delivery scientists should be less attracted by superficial (*in vitro*) concepts but rather should look for a realistic prediction of the particular value of the chosen targeting procedures in the *in vivo* setting.

14.3 Concluding remarks

Although, so far, some promising results have been achieved *in vitro* in the 'targeting' of various categories of drugs, it often remains unclear which fraction of the chosen carrier really enters the target cells *in vivo*. This was extensively studied for lactosaminated HSA in liver (hepatocyte) targeting, and for some monoclonal antibodies. However, much work remains to be done with regard to carriers of the particle type, antibody preparations and (neo-)glycoproteins as well as derivatized polyaminoacids and polymers. Rapid screening and structure optimization for such potential carriers should be performed in various species, both in the healthy and diseased state.

Coupling of drugs to macromolecular carrier systems *a priori* implies that parenteral formulations have to be used. Although parenteral dosing is quite acceptable for short-term and even long-term clinical use (e.g. insulin and other peptide drugs), it is clear that drug targeting preparations should have obvious advantages compared with the parent drug in order to justify their development. These advantages could include much higher potency, shorter treatment periods, therapy of intracellular infections in the case of poorly penetrating drugs, and/or a major reduction in the dosing frequency and toxicity.

It should be emphasized that site-specific drug delivery does not prevent the build up of steady-state plasma concentrations of the parent drug: even if the release rate for the drug in the target cells is slow, some of the targeted drug will tend to enter the general circulation. However, the plasma levels will be generally lower and the local concentration in the target tissue higher. It should also be taken into account that delivery procedures can lead to a shift in toxicity patterns. For example, inclusion of daunomycin in pegylated liposomes may reduce cardiac toxicity but, at the same time, may induce macrophage toxicity, since after multiple dosing liposomes (even surface modified) will finally end up in the monocyte-phagocyte system.

There is a current tendency to develop carriers on the basis of polypeptides and other polymeric carriers with rather simple structures. For instance, polylysines, polyhydroxymethyl-acrylamide and polylactic acid material with variations in charge and molecular weight can be tailor-made and equipped with clustered recognition sites. The biocompatibility of such carrier systems with chronic dosing should, however, be more clearly established.

In conclusion, it can be stated that the opportunities for targeting drugs seem to be abundant. Nevertheless, the manipulation of drug distribution in the diseased state in humans will require a multidisciplinary effort on the part of cell biologists, biochemists, molecular biologists, pharmacologists, pharmaceutical technologists and clinicians, before the many innovative technologies can be put into practice. In this respect, it is crucial that drug delivery technology is more structurally integrated in the overall activity of industrial drug innovation.

It is unlikely that novel compounds with promising pharmacodynamic profiles will at the same time possess completely adequate pharmacokinetic properties. Of note, attractive and potent drugs which exhibit unfavourable kinetic properties or show severe side-effects and toxicity, may be prematurely eliminated from the test bench, in spite of all the R & D money spent.

In general one should, in an early developmental phase, combine the available pharmacological and drug delivery know-how to aim for novel therapeutic modalities that display high efficacy and selectively. In other words, drug targeting options should be considered more as a high-tech extension of the process of drug design and development and less as an art of trouble-shooting in retrospect.

Index